CAMBRIDGE LIBRARY COLLECTION

Books of enduring scholarly value

Physical Sciences

From ancient times, humans have tried to understand the workings of the world around them. The roots of modern physical science go back to the very earliest mechanical devices such as levers and rollers, the mixing of paints and dyes, and the importance of the heavenly bodies in early religious observance and navigation. The physical sciences as we know them today began to emerge as independent academic subjects during the early modern period, in the work of Newton and other 'natural philosophers', and numerous sub-disciplines developed during the centuries that followed. This part of the Cambridge Library Collection is devoted to landmark publications in this area which will be of interest to historians of science concerned with individual scientists, particular discoveries, and advances in scientific method, or with the establishment and development of scientific institutions around the world.

Outlines of Astronomy

Sir John Frederick William Herschel (1792–1871) – astronomer, mathematician, chemist – was one of the most important English scientists of the nineteenth century. Son of the famous astronomer William Herschel and nephew of Caroline, he was persuaded by his father to pursue the astronomical investigations William could no longer undertake; John's subsequent career resulted in a knighthood and a lifetime of accolades. Outlines of Astronomy (1849), an updated and expanded version of his 1833 Treatise on Astronomy (also reissued in this series), went through eleven editions in two decades and was translated into several languages. Outlines examines terrestrial and celestial phenomena, providing the reader with a wide range of knowledge about the physical world as a whole. The work is an important textbook, the object of which 'is not to convince or refute opponents, nor to inquire ... for principles of which we are all the time in full possession – but simply to teach what is known'.

Cambridge University Press has long been a pioneer in the reissuing of out-of-print titles from its own backlist, producing digital reprints of books that are still sought after by scholars and students but could not be reprinted economically using traditional technology. The Cambridge Library Collection extends this activity to a wider range of books which are still of importance to researchers and professionals, either for the source material they contain, or as landmarks in the history of their academic discipline.

Drawing from the world-renowned collections in the Cambridge University Library, and guided by the advice of experts in each subject area, Cambridge University Press is using state-of-the-art scanning machines in its own Printing House to capture the content of each book selected for inclusion. The files are processed to give a consistently clear, crisp image, and the books finished to the high quality standard for which the Press is recognised around the world. The latest print-on-demand technology ensures that the books will remain available indefinitely, and that orders for single or multiple copies can quickly be supplied.

The Cambridge Library Collection will bring back to life books of enduring scholarly value (including out-of-copyright works originally issued by other publishers) across a wide range of disciplines in the humanities and social sciences and in science and technology.

Outlines of Astronomy

JOHN FREDERICK WILLIAM HERSCHEL

CAMBRIDGE UNIVERSITY PRESS

Cambridge, New York, Melbourne, Madrid, Cape Town, Singapore,
São Paolo, Delhi, Dubai, Tokyo

Published in the United States of America by Cambridge University Press, New York

www.cambridge.org
Information on this title: www.cambridge.org/9781108013772

© in this compilation Cambridge University Press 2010

This edition first published 1864
This digitally printed version 2010

ISBN 978-1-108-01377-2 Paperback

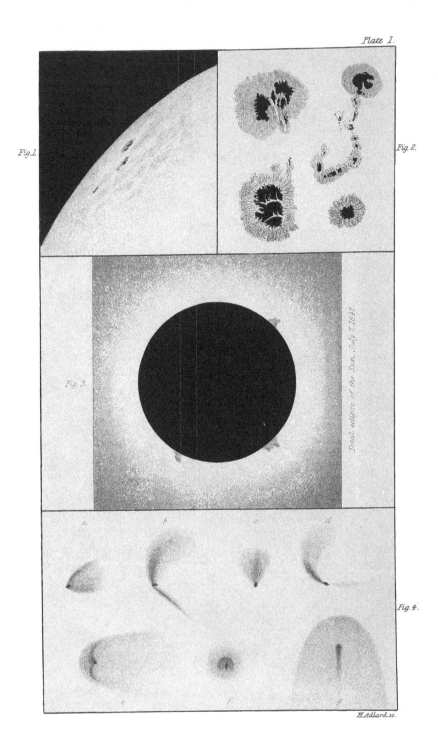

Plate I.

Fig.1

Fig.2

Fig.3.

Total eclipse of the Sun, July 7, 1842.

Fig.4.

H. Adlard, sc.

London; Longman, Brown, Green & Longmans.

OUTLINES

OF

ASTRONOMY:

BY

SIR JOHN F. W. HERSCHEL, BART. K.H.

M.A. D.C.L. F.R.S.L. & E. Hon. M.R.I.A. F.R.A.S. F.G.S. M.C.U.P.S.

Correspondent or Honorary Member of the Imperial, Royal, and National, Academies of Sciences
of Berlin, Brussels, Copenhagen, Göttingen, Haarlem, Massachussets (U.S), Modena, .
Naples, Paris, Petersburg, Stockholm, Turin, and Washington (U.S.);
the Italian and Helvetic Societies;
the Academies, Institutes, &c., of Albany (U.S.), Bologna, Catania, Dijon, Lausanne,
Nantes, Padua, Palermo, Rome, Venice, Utrecht, and Wilna;
the Philomathic Society of Paris; Asiatic Society of Bengal; South African Lit. and Phil. Society;
Literary and Historical Society of Quebec; Historical Society of New York;
Royal Medico-Chirurgical Soc., and Inst. of Civil Engineers, London;
Geographical Soc. of Berlin; Astronomical and
Meteorological Soc. of British Guiana;
&c. &c. &c.

LONDON:

PRINTED FOR

LONGMAN, BROWN, GREEN, AND LONGMANS,

PATERNOSTER-ROW;

AND JOHN TAYLOR, UPPER GOWER STREET.

1849.

PREFACE.

THE work here offered to the Public is based upon and may be considered as an extension, and, it is hoped, an improvement of a treatise on the same subject, forming Part 43. of the Cabinet Cyclopædia, published in the year 1833. Its object and general character are sufficiently stated in the introductory chapter of that volume, here reprinted with little alteration; but an opportunity having been afforded me by the Proprietors, preparatory to its re-appearance in a form of more pretension, I have gladly availed myself of it, not only to correct some errors which, to my regret, subsisted in the former volume, but to remodel it altogether (though in complete accordance with its original design as *a work of explanation*); to introduce much new matter in the earlier portions of it; to re-write, upon a far more matured and comprehensive plan, the part relating to the lunar and planetary perturbations, and to bring the subjects of sidereal and nebular astronomy to the level of the present state of our knowledge in those departments.

The chief novelty in the volume, as it now stands, will be found in the manner in which the subject of Perturbations is treated. It is not — it cannot be made *elementary*, in the sense in which that word is understood in these days of light reading. The chap-

ters devoted to it must, therefore, be considered as
addressed to a class of readers in possession of somewhat more mathematical knowledge than those who
will find the rest of the work readily and easily accessible; to readers desirous of preparing themselves,
by the possession of a sort of *carte du pays*, for a
campaign in the most difficult, but at the same time
the most attractive and the most remunerative of all
the applications of modern geometry. More especially they may be considered as addressed to students
in that university, where the " Principia" of Newton
is not, nor ever will be, put aside as an obsolete
book, behind the age; and where the grand though
rude outlines of the lunar theory, as delivered in the
eleventh section of that immortal work, are studied
less for the sake of the theory itself than for the
spirit of far-reaching thought, superior to and dis-
encumbered of technical aids, which distinguishes
that beyond any other production of the human in-
tellect.

In delivering a rational as distinguished from a
technical exposition of this subject, however, the
course pursued by Newton in the section of the
Principia alluded to, has by no means been servilely
followed. As regards the perturbations of the nodes
and inclinations, indeed, nothing equally luminous
can ever be substituted for his explanation. But as
respects the other disturbances, the point of view
chosen by Newton has been abandoned for another,
which it is somewhat difficult to perceive why he did
not, himself, select. By a different resolution of the
disturbing forces from that adopted by him, and by
the aid of a few obvious conclusions from the laws
of elliptic motion which would have found their

PREFACE.

place, naturally and consecutively, as corollaries of
the seventeenth proposition of his first book (a proposition which seems almost to have been prepared
with a special view to this application), the momentary change of place of the upper focus of the disturbed ellipse is brought distinctly under inspection;
and a clearness of conception introduced into the
perturbations of the excentricities, perihelia, and
epochs which the author does not think it presumption to believe can be obtained by no other method,
and which certainly is not obtained by that from
which it is a departure. It would be out of keeping
with the rest of the work to have introduced into
this part of it any algebraic investigations; else it
would have been easy to show that the mode of procedure here followed leads direct, and by steps (for
the subject) of the most elementary character, to the
general formulæ for these perturbations, delivered
by Laplace in the Mécanique Céleste.*

The reader will find one class of the lunar and
planetary inequalities handled in a very different
manner from that in which their explanation is usually presented. It comprehends those which are
characterized as incident on the epoch, the principal
among them being the annual and secular equations
of the moon, and that very delicate and obscure part
of the perturbational theory (so little satisfactory in
the manner in which it emerges from the analytical
treatment of the subject), the constant or permanent
effect of the disturbing force in altering the disturbed
orbit. I will venture to hope that what is here
stated will tend to remove some rather generally

* Livre ii. chap. viii. art. 67.

A 3

diffused misapprehensions as to the true bearings of
Newton's explanation of the annual equation.*

If proof were wanted of the inexhaustible fertility
of astronomical science in points of novelty and interest, it would suffice to adduce the addition to the
list of members of our system of no less than eight
new planets and satellites during the preparation of
these sheets for the press. Among them is one
whose discovery must ever be regarded as one of the
noblest triumphs of theory. In the account here
given of this discovery, I trust to have expressed
myself with complete impartiality; and in the exposition of the perturbative action on Uranus, by
which the existence and situation of the disturbing
planet became revealed to us, I have endeavoured,
in pursuance of the general plan of this work, rather
to exhibit a rational view of the dynamical action,
than to convey the slightest idea of the conduct of
those masterpieces of analytical skill which the researches of Messrs. Leverrier and Adams exhibit.

To the latter of these eminent geometers, as well
as to my excellent and esteemed friend the Astronomer Royal, I have to return my best thanks for
communications which would have effectually relieved some doubts I at one period entertained had
I not succeeded in the interim in getting clear of
them as to the *compatibility* of my views on the
subject of the annual equation already alluded to,
with the tenor of Newton's account of it. To my
valued friend, Professor De Morgan, I am indebted
for some most ingenious suggestions on the subject
of the mistakes committed in the early working of

* Principia, lib. i. prop. 66. cor. 6.

the Julian reformation of the calendar, of which I should have availed myself, had it not appeared preferable, on mature consideration, to present the subject in its simplest form, avoiding altogether entering into minutiæ of chronological discussion.

J. F. W. HERSCHEL.

Collingwood, April 12. 1849.

CONTENTS.

PART I.

CHAPTER I.

CHAPTER II.

CHAPTER III.

CHAPTER VIII.

CHAPTER IX.

OF THE SOLAR SYSTEM.

CHAPTER X.

OF THE SATELLITES.

CHAPTER XI.

OF COMETS.

PART II.

OF THE PLANETARY PERTURBATIONS.

CHAPTER XII.

CHAPTER XIII.

THEORY OF THE AXES, PERIHELIA, AND EXCENTRICITIES.

CHAPTER XIV.

PART III.

OF SIDEREAL ASTRONOMY.

CHAPTER XV.

CHAPTER XVI.

in giving the same result. Principles on which the investigation of the solar motion depends. Absolute velocity of the Sun's motion. Supposed revolution of the whole sidereal system round a common centre. Systematic parallax and aberration. Effect of the motion of light in altering the apparent period of a binary star   - Page 554

CHAPTER XVII.

OF CLUSTERS OF STARS AND NEBULÆ.

Of clustering groups of stars. Globular clusters. Their stability dynamically possible. List of the most remarkable. Classification of nebulæ and clusters. Their distribution over the heavens. Irregular clusters. Resolvability of nebulæ. Theory of the formation of clusters by nebulous subsidence. Of elliptic nebulæ. That of Andromeda. Annular and planetary nebulæ. Double nebulæ. Nebulous stars. Connection of nebulæ with double stars. Insulated nebulæ of forms not wholly irregular. Of amorphous nebulæ. Their law of distribution marks them as outliers of the galaxy. Nebulæ and nebulous group of Orion — of Argo — of Sagittarius — of Cygnus. The Magellanic clouds. Singular nebula in the greater of them. The zodiacal light. Shooting stars   -   -   -   -   - 591

PART IV.

OF THE ACCOUNT OF TIME.

CHAPTER XVIII.

Natural units of time. Relation of the sidereal to the solar day affected by precession. Incommensurability of the day and year. Its inconvenience. How obviated. The Julian Calendar. Irregularities at its first introduction. Reformed by Augustus. Gregorian reformation. Solar and lunar cycles. Indiction. Julian period. Table of Chronological eras. Rules for calculating the days elapsed between given Dates. Equinoctial time.   -   -   -   -   - 622

APPENDIX.

ERRATA AND ADDENDA.

Page 143. line 14. for "imaginably," read "imaginable."

line 29. in denominator for "000," read "10,000."

144. line 2. in denominator for "000," read "10,000."

line 18. for "89,400," read "86,400."

187. line 17. for "Z P E," read "P Z E."

209. line 17. for "is the time," read "is to the time."

242. line 30. for "A F," read "E F."

253. line 2. for "F H *h f*," read "F B H*f*."

line 4. for "K M E," read "K M *k*."

line 7. for "D S," read "E S."

line 8. for "23,984 such," read "23,984, and the semidiameter of the Sun 111½ such."

261. End of art. 432. Insert as follows :—"M. Arago has shown, from a comparison of rain, registered as having fallen during a long period, that a slight preponderance in respect of quantity falls near the new Moon over that which falls near the full. This would be a natural and necessary consequence of a preponderance of a cloudless sky about the full, and forms, therefore, part and parcel of the same meteorological fact.

291. line 13. for "frequently those," read "frequently ; those."

319. line 1. for "172° 32′," read "173° 32′."

line 34. for "state unstable," read "state of unstable."

361. note, line 7 from bottom, for "semiaxis," read "log. semiaxis."

372. note ‡, for "1837," read "1537."

376. line 2. from bottom, for "move in one direction round the Sun," read "circulate in one direction."

OUTLINES

OF

ASTRONOMY.

INTRODUCTION.

(1.) EVERY student who enters upon a scientific pursuit,
especially if at a somewhat advanced period of life, will find
not only that he has much to learn, but much also to un-
learn. Familiar objects and events are far from presenting
themselves to our senses in that aspect and with those connections under which science requires them to be viewed, and
which constitute their rational explanation. There is, therefore, every reason to expect that those objects and relations
which, taken together, constitute the subject he is about to
enter upon will have been previously apprehended by him,
at least imperfectly, because much has hitherto escaped his
notice which is essential to its right understanding : and not
only so, but too often also erroneously, owing to mistaken
analogies, and the general prevalence of vulgar errors. As a
first preparation, therefore, for the course he is about to
commence, he must loosen his hold on all crude and hastily
adopted notions, and must strengthen himself, by something
of an effort and a resolve, for the unprejudiced admission of
any conclusion which shall appear to be supported by careful
observation and logical argument, even should it prove of a
nature adverse to notions he may have previously formed for
himself, or taken up, without examination, on the credit of

B

others. Such an effort is, in fact, a commencement of that intellectual discipline which forms one of the most important ends of all science. It is the first movement of approach towards that state of mental purity which alone can fit us for a full and steady perception of moral beauty as well as physical adaptation. It is the " euphrasy and rue " with which we must "purge our sight" before we can receive and contemplate as they are the lineaments of truth and nature.

(2.) There is no science which, more than astronomy, stands in need of such a preparation, or draws more largely on that intellectual liberality which is ready to adopt whatever is demonstrated, or concede whatever is rendered highly probable, however new and uncommon the points of view may be in which objects the most familiar may thereby become placed. Almost all its conclusions stand in open and striking contradiction with those of superficial and vulgar observation, and with what appears to every one, until he has understood and weighed the proofs to the contrary, the most positive evidence of his senses. Thus, the earth on which he stands, and which has served for ages as the unshaken foundation of the firmest structures, either of art or nature, is divested by the astronomer of its attribute of fixity, and conceived by him as turning swiftly on its centre, and at the same time moving onwards through space with great rapidity. The sun and the moon, which appear to untaught eyes round bodies of no very considerable size, become enlarged in his imagination into vast globes, — the one approaching in magnitude to the earth itself, the other immensely surpassing it. The planets, which appear only as stars somewhat brighter than the rest, are to him spacious, elaborate, and habitable worlds; several of them much greater and far more curiously furnished than the earth he inhabits, as there are also others less so ; and the stars themselves, properly so called, which to ordinary apprehension present only lucid sparks or brilliant atoms, are to him suns of various and transcendent glory — effulgent centres of life and light to myriads of unseen worlds. So that when, after dilating his thoughts to comprehend the grandeur of those

ideas his calculations have called up, and exhausting his imagination and the powers of his language to devise similes and
metaphors illustrative of the immensity of the scale on which
his universe is constructed, he shrinks back to his native
sphere; he finds it, in comparison, a mere point; so lost—
even in the minute system to which it belongs — as to be invisible and unsuspected from some of its principal and remoter
members.

(3.) There is hardly any thing which sets in a stronger
light the inherent power of truth over the mind of man,
when opposed by no motives of interest or passion, than the
perfect readiness with which all these conclusions are assented
to as soon as their evidence is clearly apprehended, and the
tenacious hold they acquire over our belief when once admitted. In the conduct, therefore, of this volume, I shall
take it for granted that the reader is more desirous to learn
the system which it is its object to teach as it now stands,
than to raise or revive objections against it; and that, in
short, he comes to the task with a willing mind; an assumption which will not only save the trouble of piling argument
on argument to convince the sceptical, but will greatly
facilitate his actual progress; inasmuch as he will find it at
once easier and more satisfactory to pursue from the outset
a straight and definite path, than to be constantly stepping
aside, involving himself in perplexities and circuits, which,
after all, can only terminate in finding himself compelled to
adopt the same road.

(4.) The method, therefore, we propose to follow in this
work is neither strictly the analytic nor the synthetic, but
rather such a combination of both, with a leaning to the
latter, as may best suit with a *didactic* composition. Its
object is not to convince or refute opponents, nor to inquire,
under the semblance of an assumed ignorance, for principles
of which we are all the time in full possession — but simply
to *teach* what is *known*. The moderate limit of a single
volume, to which it will be confined, and the necessity of
being on every point, within that limit, rather diffuse and
copious in explanation, as well as the eminently matured and

ascertained character of the science itself, render this course both practicable and eligible. Practicable, because there is now no danger of any revolution in astronomy, like those which are daily changing the features of the less advanced sciences, supervening, to destroy all our hypotheses, and throw our statements into confusion. Eligible, because the space to be bestowed, either in combating refuted systems, or in leading the reader forward by slow and measured steps from the known to the unknown, may be more advantageously devoted to such explanatory illustrations as will impress on him a familiar and, as it were, a practical sense of the sequence of phenomena, and the manner in which they are produced. We shall not, then, reject the analytic course where it leads more easily and directly to our objects, or in any way fetter ourselves by a rigid adherence to method. Writing only to be understood, and to communicate as much information in as little space as possible, consistently with its *distinct* and *effectual* communication, no sacrifice can be afforded to system, to form, or to affectation.

(5.) We shall take for granted, from the outset, the Copernican system of the world; relying on the easy, obvious, and natural explanation it affords of all the phenomena as they come to be described, to impress the student with a sense of its truth, without either the formality of demonstration or the superfluous tedium of eulogy, calling to mind that important remark of Bacon : — " Theoriarum vires, arcta et quasi se mutuo sustinente partium adaptatione, quâ quasi in orbem cohærent, firmantur * ; " not failing, however, to point out to the reader, as occasion offers, the contrast which its superior simplicity offers to the complication of other hypotheses.

(6.) The preliminary knowledge which it is desirable that the student should possess, in order for the more advantageous perusal of the following pages, consists in the familiar prac-

---

* " The confirmation of theories relies on the compact adaptation of their parts, by which, like those of an arch or dome, they mutually sustain each other, and form a coherent whole." This is what Dr. Whewell expressively terms the *consilience* of inductions.

tice of decimal and sexagesimal arithmetic; some moderate acquaintance with geometry and trigonometry, both plane and spherical; the elementary principles of mechanics; and enough of optics to understand the construction and use of the telescope, and some other of the simpler instruments. Of course, the more of such knowledge he brings to the perusal, the easier will be his progress, and the more complete the information gained; but we shall endeavour in every case, as far as it can be done without a sacrifice of clearness, and of that useful brevity which consists in the absence of prolixity and episode, to render what we have to say as independent of other books as possible.

(7.) After all, I must distinctly caution such of my readers as may commence and terminate their astronomical studies with the present work (though of such, — at least in the latter predicament, — I trust the number will be few), that its utmost pretension is to place them on the threshold of this particular wing of the temple of Science, or rather on an eminence exterior to it, whence they may obtain something like a general notion of its structure; or, at most, to give those who may wish to enter a ground-plan of its accesses, and put them in possession of the pass-word. Admission to its sanctuary, and to the privileges and feelings of a votary, is only to be gained by one means, — *sound and sufficient knowledge of mathematics, the great instrument of all exact inquiry, without which no man can ever make such advances in this or any other of the higher departments of science as can entitle him to form an independent opinion on any subject of discussion within their range.* It is not without an effort that those who possess this knowledge can communicate on such subjects with those who do not, and adapt their language and their illustrations to the necessities of such an intercourse. Propositions which to the one are almost identical, are theorems of import and difficulty to the other; nor is their evidence presented in the same way to the mind of each. In teaching such propositions, under such circumstances, the appeal has to be made, not to the pure and abstract reason, but to the sense of analogy — to practice and experience: principles and modes of action have to be established not by direct argument from acknowledged axioms, but by continually recurring to the sources from which the axioms themselves have been drawn; viz. examples; that is to say, by bringing forward and dwelling on simple and familiar instances in which the same principles and the same or similar modes of action take place: thus erecting, as it were, in each particular case, a separate induction, and constructing at each step a little body of science to meet its exigencies. The difference is that of pioneering a road through an untraversed country and advancing at ease along a broad and beaten highway; that is to say, if we are determined to make ourselves distinctly understood, and will appeal to reason at all. As for the method of *assertion*, or a direct demand on the *faith* of the student (though in some complex cases indispensable, where illustrative explanation would defeat its own end by becoming tedious and burdensome to both parties), it is one which I shall neither willingly adopt nor would recommend to others.

(8.) On the other hand, although it is something new to abandon the road of mathematical demonstration in the treatment of subjects susceptible of it, and to teach any considerable branch of science entirely or chiefly by the way of illustration and familiar parallels, it is yet not impossible that those who are already well acquainted with our subject, and whose knowledge has been acquired by that confessedly higher practice which is incompatible with the avowed objects of the present work, may yet find their account in its perusal, —for this reason, that it is always of advantage to present any given body of knowledge to the mind in as great a variety of different lights as possible. It is a property of illustrations of this kind to strike no two minds in the same manner, or with the same force; because no two minds are stored with the same images, or have acquired their notions of them by similar habits. Accordingly, it may very well happen, that a proposition, even to one best acquainted with it, may be placed not merely in a new and uncommon, but in a more impressive and satisfactory light by such a course -- some obscurity may be dissipated, some inward misgivings cleared up, or even some links supplied which may lead to the perception of connections and deductions altogether unknown before. And the probability of this is increased when, as in the present instance, the illustrations chosen have not been studiously selected from books, but are such as have presented themselves freely to the author's mind as being most in harmony with his own views; by which, of course, he means to lay no claim to originality in all or any of them beyond what they may really possess.

(9.) Besides, there are cases in the application of mechanical principles with which the mathematical student is but too familiar, where, when the data are before him, and the numerical and geometrical relations of his problems all clear to his conception, — when his forces are estimated and his lines measured, — nay, when even he has followed up the application of his technical processes, and fairly arrived at his conclusion, — there is still something wanting in his mind — not in the evidence, for he has examined each link, and finds the chain complete — not in the principles, for those he well knows are too firmly established to be shaken — but precisely in the *mode of action.* He has followed out a train of reasoning by logical and technical rules, but the signs he has employed are not pictures of nature, or have lost their original meaning as such to his mind: he has not seen, as it were, the process of nature passing under his eye in an instant of time, and presented as a consecutive whole to his imagination. A familiar parallel, or an illustration drawn from some artificial or natural process, of which he has that direct and individual impression which gives it a reality and associates it with a name, will, in almost every such case, supply in a moment this deficient feature, will convert all his symbols into real pictures, and infuse an animated meaning into what was before a lifeless succession of words and signs. I cannot, indeed, always promise myself to attain this degree of vividness of illustration, nor are the points to be elucidated themselves always capable of being so *paraphrased* (if I may use the expression) by any single in-

stance adducible in the ordinary course of experience; but
the object will at least be kept in view; and, as I am very
conscious of having, in making such attempts, gained for
myself much clearer views of several of the more concealed
effects of planetary perturbation than I had acquired by
their mathematical investigation in detail, it may reasonably
be hoped that the endeavour will not always be unattended
with a similar success in others.

(10.) From what has been said, it will be evident that our
aim is not to offer to the public a technical treatise, in which
the student of practical or theoretical astronomy shall find
consigned the minute description of methods of observation,
or the formulæ he requires prepared to his hand, or their de-
monstrations drawn out in detail. In all these the present
work will be found meagre, and quite inadequate to his
wants. Its aim is entirely different; being to present in each
case the mere ultimate *rationale* of facts, arguments, and
processes; and, in all cases of mathematical application,
avoiding whatever would tend to encumber its pages with
algebraic or geometrical symbols, to place under his inspection that central thread of common sense on which the pearls
of analytical research are invariably strung; but which, by
the attention the latter claim for themselves, is often concealed from the eye of the gazer, and not always disposed in
the straightest and most convenient form to follow by those
who string them. This is no fault of those who have conducted the inquiries to which we allude. The contention of
mind for which they call is enormous; and it may, perhaps,
be owing to their experience of *how little* can be accomplished
in carrying such processes on to their conclusion, by mere
ordinary *clearness of head;* and how necessary it often is to
pay more attention to the purely mathematical conditions
which ensure success, — the hooks-and-eyes of their equations and series, — than to those which enchain causes with
their effects, and both with the human reason, — that we
must attribute something of that indistinctness of view which
is often complained of as a grievance by the earnest student,
and still more commonly ascribed ironically to the native

cloudiness of an atmosphere too sublime for vulgar comprehension. We think we shall render good service to both
classes of readers, by dissipating, so far as lies in our power,
that accidental obscurity, and by showing ordinary untutored
comprehension clearly what it *can,* and what it *cannot,* hope
to attain.

CHAPTER I.

GENERAL NOTIONS. — APPARENT AND REAL MOTIONS. — SHAPE AND
SIZE OF THE EARTH. — THE HORIZON AND ITS DIP. — THE AT-
MOSPHERE. — REFRACTION. — TWILIGHT. — APPEARANCES RE-
SULTING FROM DIURNAL MOTION — FROM CHANGE OF STATION
IN GENERAL. — PARALLACTIC MOTIONS. — TERRESTRIAL PARAL-
LAX. — THAT OF THE STARS INSENSIBLE. — FIRST STEP TO-
WARDS FORMING AN IDEA OF THE DISTANCE OF THE STARS. —
COPERNICAN VIEW OF THE EARTH'S MOTION. — RELATIVE
MOTION. — MOTIONS PARTLY REAL, PARTLY APPARENT. — GEO-
CENTRIC ASTRONOMY, OR IDEAL REFERENCE OF PHÆNOMENA
TO THE EARTH'S CENTRE AS A COMMON CONVENTIONAL
STATION.

(11.) THE magnitudes, distances, arrangement, and motions
of the great bodies which make up the visible universe, their
constitution and physical condition, so far as they can be
known to us, with their mutual influences and actions on each
other, so far as they can be traced by the effects produced,
and established by legitimate reasoning, form the assemblage
of objects to which the attention of the astronomer is directed.
The term astronomy * itself, which denotes the *law* or rule of
the *astra* (by which the ancients understood not only the
stars properly so called, but the sun, the moon, and all the
visible constituents of the heavens), sufficiently indicates this;
and, although the term astrology, which denotes the *reason,
theory,* or *interpretation* of the stars†, has become degraded in
its application, and confined to superstitious and delusive attempts to divine future events by their dependence on pre-

* Αστηρ, *a star;* νομος, *a law;* or νεμειν, to tend, as a shepherd his flock ; so
that αστρονομος means "shepherd of the stars." The two etymologies are, however, coincident.
† Λογος, *reason,* or *a word,* the vehicle of reason ; the interpreter of thought.

tended planetary influences, the same meaning originally attached itself to that epithet.

(12.) But, besides the stars and other celestial bodies, the earth itself, regarded as an individual body, is one principal object of the astronomer's consideration, and, indeed, the chief of all. It derives its importance, in a practical as well as theoretical sense, not only from its proximity, and its relation to us as animated beings, who draw from it the supply of all our wants, but as the station from which we see all the rest, and as the only one among them to which we can, in the first instance, refer for any determinate marks and measures by which to recognize their changes of situation, or with which to compare their distances.

(13.) To the reader who now for the first time takes up a book on astronomy, it will no doubt seem strange to class the earth with the heavenly bodies, and to assume any community of nature among things apparently so different. For what, in fact, can be more apparently different than the vast and seemingly immeasurable extent of the earth, and the stars, which appear but as points, and seem to have no size at all? The earth is dark and opaque, while the celestial bodies are brilliant. We perceive in it no motion, while in them we observe a continual change of place, as we view them at different hours of the day or night, or at different seasons of the year. The ancients, accordingly, one or two of the more enlightened of them only excepted, admitted no such community of nature ; and, by thus placing the heavenly bodies and their movements without the pale of analogy and experience, effectually intercepted the progress of all reasoning from what passes here below, to what is going on in the regions where they exist and move. Under such conventions, astronomy, as a science of cause and effect, could not exist, but must be limited to a mere registry of appearances, unconnected with any attempt to account for them on reasonable principles, however successful to a certain extent might be the attempt to follow out their order of sequence, and to establish empirical laws expressive of this order. To get rid of this prejudice, therefore, is the first step towards acquiring a knowledge of what is really the case ; and the student has made his first effort towards the acquisition of sound knowledge, when he has learnt to familiarize himself with the idea that the earth, after all, *may* be nothing but a great star.   How correct such an idea may be, and with what limitations and modifications it is to be admitted, we shall see presently.

(14.) It is evident, that, to form any just notions of the arrangement, in space, of a number of objects which we cannot approach and examine, but of which all the information we can gain is by sitting still and watching their evolutions, it must be very important for us to know, in the first instance, whether what we call sitting still is *really* such : whether the station from which we view them, with ourselves, and all objects which immediately surround us, be not itself in motion, unperceived by us ; and if so, of what nature that motion is.   The apparent places of a number of objects, and their apparent arrangement with respect to each other, will of course be materially dependent on the situation of the spectator among them ; and if this situation be liable to change, unknown to the spectator himself, an appearance of change in the respective situations of the objects will arise, without the reality.   If, then, such be actually the case, it will follow that *all* the movements we think we perceive among the stars will not be real movements, but that some part, at least, of whatever changes of relative place we perceive among them must be merely apparent, the results of the shifting of our own point of view ; and that, if we would ever arrive at a knowledge of their real motions, it can only be by first investigating our own, and making due allowance for its effects.   Thus, the question whether the earth is in motion or at rest, and if in motion, what that motion is, is no idle inquiry, but one on which depends our only chance of arriving at true conclusions respecting the constitution of the universe.

(15.) Nor let it be thought strange that we should speak of a motion existing in the earth, unperceived by its inhabitants : we must remember that it is of the earth *as a whole*, with all that it holds within its substance, or sustains on its surface, that we are speaking; of a motion common to the solid mass beneath, to the ocean which flows around it, the air that rests upon it, and the clouds which float above it in the air. Such a motion, which should displace no terrestrial object from its relative situation among others, interfere with no natural processes, and produce no sensations of shocks or jerks, might, it is very evident, subsist undetected by us. There is no peculiar sensation which advertises us that we are *in motion*. We perceive *jerks*, or *shocks*, it is true, because these are sudden *changes* of motion, produced, as the laws of mechanics teach us, by sudden and powerful forces acting during short times ; and these forces, applied to our bodies, are what we *feel*. When, for example, we are carried along in a carriage with the blinds down, or with our eyes closed (to keep us from seeing external objects), we perceive a tremor arising from inequalities in the road, over which the carriage is successively lifted and let fall, but we have no sense of *progress*. As the road is smoother, our sense of motion is diminished, though our rate of travelling is accelerated. Railway travelling, especially by night or in a tunnel, has familiarized every one with this remark. Those who have made aeronautic voyages testify that with closed eyes, and under the influence of a steady breeze communicating no oscillatory or revolving motion to the car, the *sensation* is that of perfect rest, however rapid the transfer from place to place.

(16.) But it is on shipboard, where a great system is maintained in motion, and where we are surrounded with a multitude of objects which participate with ourselves and each other in the common progress of the whole mass, that we feel most satisfactorily the identity of sensation between a state of motion and one of rest. In the cabin of a large and heavy vessel, going smoothly before the wind in still water, or drawn along a canal, not the smallest indication acquaints us with the way it is making. We read, sit, walk, and perform every customary action as if we were on land. If we throw a ball into the air, it falls back into our hand ; or if we drop it, it alights at our feet. Insects buzz around us as in the

free air; and smoke ascends in the same manner as it would
do in an apartment on shore.   If, indeed, we come on deck,
the case is, in some respects, different; the air, not being
carried along with us, drifts away smoke and other light
bodies — such as feathers abandoned to it — apparently, in
the opposite direction to that of the ship's progress; but, in
reality, *they* remain at rest, and we leave them behind in the
air.   Still, the illusion, so far as massive objects and our own
movements are concerned, remains complete; and when we
look at the shore, we then perceive the effect of our own
motion transferred, in a contrary direction, to external objects
— *external, that is, to the system of which we form a part.*

" Provehimur portu, terræque urbesque recedunt."

(17.) In order, however, to conceive the earth as in motion, we must form to ourselves a conception of its shape and
size.   Now, an object cannot have shape and size unless it
is *limited* on all sides by some definite outline, so as to admit
of our imagining it, at least, disconnected from other bodies,
and existing insulated in space.   The first rude notion we
form of the earth is that of a flat surface, of indefinite extent
in all directions from the spot where we stand, *above* which
are *the air* and *sky ;* below, to an indefinite profundity, solid
matter.   This is a prejudice to be got rid of, like that of the
earth's immobility ; — but it is one much easier to rid ourselves of, inasmuch as it originates only in our own mental
inactivity, in not questioning ourselves *where* we will place
a limit to a thing we have been accustomed from infancy
to regard as immensely large ; and does not, like that, originate in the testimony of our senses unduly interpreted.
On the contrary, the direct testimony of our senses lies the
other way.   When we see the sun set in the evening in the
west, and rise again in the east, as we cannot doubt that it is
the *same* sun we see after a temporary absence, we must do
violence to all our notions of solid matter, to suppose it to
have made its way *through* the substance of the earth.   It
must, therefore, have gone *under* it, and that not by a mere
subterraneous *channel ;* for if we notice the points where it

sets and rises for many successive days, or for a whole year, we shall find them constantly shifting, round a very large extent of the horizon; and, besides, the moon and stars also set and rise again in *all* points of the visible horizon. The conclusion is plain: the earth cannot extend indefinitely in depth downwards, nor indefinitely in surface laterally; it must have not only bounds in a horizontal direction, but also an *under side* round which the sun, moon, and stars can pass; and that side must, at least, be so far like what we see, that it must have a sky and sunshine, and a day when it is night to us, and *vice versâ;* where, in short,

— "redit à nobis Aurora, diemque reducit.
Nosque ubi primus equis oriens afflavit anhelis,
Illic sera rubens accendit lumina Vesper."   *Georg.*

(18.) As soon as we have familiarized ourselves with the conception of an earth without *foundations* or fixed supports — existing insulated in space from contact of every thing external, it becomes easy to imagine it in motion — or, rather, difficult to imagine it otherwise; for, since there is nothing to *retain* it in one place, should any causes of motion exist, or any *forces* act upon it, it must obey their impulse. Let us next see what obvious circumstances there are to help us to a knowledge of the *shape* of the earth.

(19.) Let us first examine what we can actually *see* of its shape. Now, it is not on land (unless, indeed, on uncommonly level and extensive plains), that we can see any thing of the *general* figure of the earth; — the hills, trees, and other objects which roughen its surface, and break and elevate the line of the horizon, though obviously bearing a most minute proportion to the *whole* earth, are yet too considerable with respect to ourselves and to that small portion of it which we can see at a single view, to allow of our forming any judgment of the form of the whole, from that of a part so disfigured. But with the surface of the sea or any vastly extended level plain, the case is otherwise. If we sail out of sight of land, whether we stand on the deck of the ship or climb the mast, we see the surface of the sea — not losing itself in distance and mist, but terminated by a sharp, clear, well-defined line or *offing* as it is called, which runs all round us in a circle, having our station for its centre. That this line is really a circle, we conclude, first, from the perfect apparent similarity of all its parts; and, secondly, from the fact of all its parts appearing at the same distance from us, and that, evidently, a moderate one; and thirdly, from this, that its apparent *diameter*, measured with an instrument called the *dip sector*, is the same (except under some singular atmospheric circumstances, which produce a temporary distortion of the outline), in whatever direction the measure is taken,—properties which belong only to the circle among geometrical figures. If we ascend a high eminence on a plain (for instance, one of the Egyptian pyramids), the same holds good.

(20.) Masts of ships, however, and the edifices erected by man, are trifling eminences compared to what nature itself affords; Ætna, Teneriffe, Mowna Roa, are eminences from which no contemptible *aliquot* part of the whole earth's surface can be seen; but from these again—in those few and rare occasions when the transparency of the air will permit the real boundary of the horizon, the true sea-line, to be seen—the very same appearances are witnessed, but with this remarkable addition, viz. that the angular *diameter* of the visible area, as measured by the dip sector, is materially *less* than at a lower level; or, in other words, that the *apparent size* of the earth has sensibly diminished as we have receded from its surface, while yet the *absolute quantity* of it seen at once has been increased.

(21.) The same appearances are observed universally, in every part of the earth's surface visited by man. Now, the figure of a body which, however seen, appears always *circular*, can be no other than a sphere or globe.

(22.) A diagram will elucidate this. Suppose the earth to be represented by the sphere L H N Q, whose centre is C, and let A, G, M be stations at different elevations above various points of its surface, represented by *a*, *g*, *m* respectively. From each of them (as from M) let a line be drawn, as M N *n*,

a tangent to the surface at N, then will this line represent
the visual ray along which the spectator at M will see the
visible horizon; and as this tangent sweeps round M, and
comes successively into the positions M O o, M P p, M Q q,
the point of contact N will mark out on the surface the
circle N O P Q. The area of the spherical surface compre-

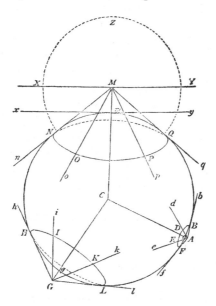

hended within this circle is the portion of the earth's surface
visible to a spectator at M, and the angle N M Q included
between the two extreme visual rays is the measure of its
apparent angular diameter. Leaving, at present, out of consideration the effect of refraction in the air below M, of which
more hereafter, and which always tends, in some degree, to
increase that angle, or render it more *obtuse*, this is the angle
measured by the dip sector. Now, it is evident, 1st, that as
the point M is more elevated above m, the point immediately
below it on the sphere, the visible area, *i. e.* the spherical
segment or slice N O P Q, increases; 2dly, that the distance
of the visible *horizon* * or boundary of our view from the eye,

* Ὁρίζω, to terminate.

viz. the line M N, increases; and, 3dly, that the angle N M Q becomes *less obtuse*, or, in other words, the apparent angular diameter of the earth diminishes, being nowhere so great as 180°, or two right angles, but falling short of it by some sensible quantity, and that more and more the higher we ascend. The figure exhibits three states or stages of elevation, with the horizon, &c. corresponding to each, a glance at which will explain our meaning; or, limiting ourselves to the larger and more distinct, M N O P Q, let the reader imagine *n* N M, M Q *q* to be the two legs of a ruler jointed at M, and kept extended by the globe N *m* Q between them. It is clear, that as the joint M is urged home towards the surface, the legs will open, and the ruler will become more nearly *straight*, but will not attain *perfect* straightness till M is brought fairly up to contact with the surface at *m*, in which case its whole length will become a *tangent* to the sphere at *m*, as is the line *x y*.

(23.) This explains what is meant by the *dip of the horizon*. M *m*, which is perpendicular to the general surface of the sphere at *m*, is also the direction in which a *plumb-line** would hang; for it is an observed fact, that in all situations, in every part of the earth, the direction of a plumb-line is exactly perpendicular to the surface of still water; and, moreover, that it is also exactly perpendicular to a line or surface truly adjusted by a *spirit-level*.* Suppose, then, that at our station M we were to adjust a line (a wooden ruler for instance) by a spirit-level, with perfect exactness; then, if we suppose the direction of this line indefinitely prolonged both ways, as X M Y, the line so drawn will be at right angles to M *m*, and therefore parallel to *x m y*, the tangent to the sphere at *m*. A spectator placed at M will therefore see not only all the vault of the sky *above* this line, as X Z Y, but also that portion or zone of it which lies between X N and Y Q; in other words, his sky will be more than a hemisphere by the zone Y Q X N. It is the angular breadth of this redundant zone — the angle Y M Q, by which the *visible* horizon appears depressed below the direction of a spirit-level — that is called

---

* See these instruments described in Chap. III.

the *dip of the horizon.* It is a correction of constant use in nautical astronomy.

(24.) From the foregoing explanations it appears, 1st, That the general figure of the earth (so far as it can be gathered from this kind of observation) is that of a sphere or globe. In this we also include that of the sea, which, wherever it extends, covers and fills in those inequalities and local irregularities which exist on land, but which can of course only be regarded as trifling deviations from the general outline of the whole mass, as we consider an orange not the less round for the roughness on its rind. 2dly, That the appearance of a *visible* horizon, or sea-offing, is a consequence of the curvature of the surface, and does not arise from the inability of the eye to follow objects to a greater distance, or from atmospheric indistinctness. It will be worth while to pursue the general notion thus acquired into some of its consequences, by which its consistency with observations of a different kind, and on a larger scale, will be put to the test, and a clear conception be formed of the manner in which the parts of the earth are related to each other, and held together as a whole.

(25.) In the first place, then, every one who has passed a little while at the sea side is aware that objects may be seen perfectly well beyond the *offing* or visible horizon — but not the *whole* of them. We only see their upper parts. Their bases where they rest on, or rise out of the water, are hid from view by the spherical surface of the sea, which protrudes between them and ourselves. Suppose a ship, for instance, to sail directly away from our station; — at first, when the distance of the ship is small, a spectator, S, situated at some certain height above the sea, sees the whole of the ship, even to the *water line* where it rests on the sea, as at A. As it recedes it diminishes, it is true, in apparent size, but still the *whole* is seen down to the water line, till it reaches the *visible* horizon at B. But as soon as it has passed this distance, not only does the visible portion still continue to diminish in apparent *size,* but the hull begins to disappear bodily, as if sunk below the surface. When it has reached a certain

distance, as at C, its hull has entirely vanished, but the masts
and sails remain, presenting the appearance c. But if, in
this state of things, the
spectator quickly as-
cends to a higher sta-
tion, T, whose visible
horizon is at D, the
hull comes again in
sight ; and, when he
descends again, he loses

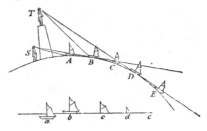

it. The ship still receding, the lower sails seem to sink below
the water, as at d, and at length the whole disappears : while yet
the distinctness with which the last portion of the sail d is
seen is such as to satisfy us that were it not for the interposed
segment of the sea, A B C D E, the distance T E is not so
great as to have prevented an equally perfect view of the
whole.

(26.) The history of aëronautic adventure affords a curious
illustration of the same principle. The late Mr. Sadler, the
celebrated aëronaut, ascended on one occasion in a balloon
from Dublin, and was wafted across the Irish Channel, when,
on his approach to the Welsh coast, the balloon descended
nearly to the surface of the sea. By this time the sun was
set, and the shades of evening began to close in. He threw
out nearly all his ballast, and suddenly sprang upwards to a
great height, and by so doing brought his horizon to *dip*
below the sun, producing the whole phenomenon of a
western sunrise. Subsequently descending in Wales, he of
course witnessed a second sunset on the same evening.

(27.) If we could measure the heights and exact distance
of two stations which could barely be discerned from each
other over the edge of the horizon, we could ascertain
the actual size of the earth itself : and, in fact, were it
not for the effect of refraction, by which we are enabled
to see in some small degree *round* the interposed segment (as
will be hereafter explained), this would be a tolerably good
method of ascertaining it. Suppose A and B to be two
eminences, whose perpendicular heights A a and B b (which

for simplicity, we will suppose to be exactly equal) are known, as well as their exact horizontal interval aDb, by measurement ; then it is clear that D, the visible horizon of both, will lie just half-way between them, and if we suppose aDb to be the sphere of the earth, and C its centre in the figure C D b B, we know D b, the length of the arch

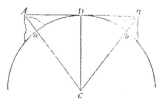

of the circle between D and b, — viz. half the measured interval, and b B, the excess of its secant above its radius — which is the height of B, — data which, by the solution of an easy geometrical problem, enable us to find the length of the radius D C. If, as is really the case, we suppose both the heights and distance of the stations inconsiderable in comparison with the size of the earth, the solution alluded to is contained in the following proposition : —

The earth's diameter bears the same proportion to the distance of the visible horizon from the eye as that distance does to the height of the eye above the sea level.

When the stations are unequal in height, the problem is a little more complicated.

(28.) Although, as we have observed, the effect of refraction prevents this from being an exact method of ascertaining the dimensions of the earth, yet it will suffice to afford such an approximation to it as shall be of use in the present stage of the reader's knowledge, and help him to many just conceptions, on which account we shall exemplify its application in numbers. Now, it appears by observation, that two points, each ten feet above the surface, cease to be visible from each other over still water, and in average atmospheric circumstances, at a distance of about 8 miles. But 10 feet is the 528th part of a mile, so that half their distance, or 4 miles, is to the height of each as 4×528 or $2112 : 1$, and therefore in the same proportion to 4 miles is the length of the earth's diameter. It must, therefore, be equal to

4 × 2112 = 8448, or, in round numbers, about 8000 miles, which is not very far from the truth.

(29.) Such is the first rough result of an attempt to ascertain the earth's magnitude; and it will not be amiss, if we take advantage of it to compare it with objects we have been accustomed to consider as of vast size, so as to interpose a few steps between it and our ordinary ideas of dimension. We have before likened the inequalities on the earth's surface, arising from mountains, valleys, buildings, &c. to the roughnesses on the rind of an orange, compared with its general mass. The comparison is quite free from exaggeration. The highest mountain known hardly exceeds five miles in perpendicular elevation: this is only one 1600th part of the earth's diameter; consequently, on a globe of sixteen inches in diameter, such a mountain would be represented by a protuberance of no more than one hundredth part of an inch, which is about the thickness of ordinary drawing-paper. Now, as there is no entire continent, or even any very extensive tract of land, known, whose general elevation above the sea is any thing like half this quantity, it follows, that if we would construct a correct model of our earth, with its seas, continents, and mountains, on a globe sixteen inches in diameter, the whole of the land, with the exception of a few prominent points and ridges, must be comprised on it within the thickness of thin writing-paper; and the highest hills would be represented by the smallest visible grains of sand.

(30.) The deepest mine existing does not penetrate half a mile below the surface: a scratch, or pin-hole, duly representing it, on the surface of such a globe as our model, would be imperceptible without a magnifier.

(31.) The greatest depth of sea, probably, does not very much exceed the greatest elevation of the continents; and would, of course, be represented by an excavation, in about the same proportion, into the substance of the globe : so that the ocean comes to be conceived as a mere film of liquid, such as, on our model, would be left by a brush dipped in colour, and drawn over those parts intended to represent the sea : only, in so conceiving it, we must bear in mind that the resemblance extends no farther than to proportion in point of quantity. The mechanical laws which would regulate the distribution and movements of such a film, and its adhesion to the surface, are altogether different from those which govern the phenomena of the sea.

(32.) Lastly, the greatest extent of the earth's surface which has ever been seen at once by man, was that exposed to the view of MM. Biot and Gay-Lussac, in their celebrated aëronautic expedition to the enormous height of 25,000 feet, or rather less than five miles. To estimate the proportion of the area visible from this elevation to the whole earth's surface, we must have recourse to the geometry of the sphere, which informs us that the convex surface of a spherical segment is to the whole surface of the sphere to which it belongs as the versed sine or thickness of the segment is to the diameter of the sphere ; and further, that this thickness, in the case we are considering, is almost exactly equal to the perpendicular elevation of the point of sight above the surface. The proportion, therefore, of the visible area, in this case, to the whole earth's surface, is that of five miles to 8000, or 1 to 1600. The portion visible from Ætna, the Peak of Teneriffe, or Mowna Roa, is about one 4000th.

(33.) When we ascend to any very considerable elevation above the surface of the earth, either in a balloon, or on mountains, we are made aware, by many uneasy sensations, of an insufficient supply of *air*. The barometer, an instrument which informs us of the weight of air incumbent on a given horizontal surface, confirms this impression, and affords a direct measure of the rate of diminution of the quantity of air which a given space includes as we recede from the surface. From its indications we learn, that when we have ascended to the height of 1000 feet, we have left below us about one-thirtieth of the whole mass of the atmosphere : — that at 10,600 feet of perpendicular elevation (which is rather less than that of the summit of Ætna*) we have ascended

* The height of Ætna above the Mediterranean (as it results from a barome-

through about one-third; and at 18,000 feet (which is nearly that of Cotopaxi) through one-half the material, or, at least, the ponderable body of air incumbent on the earth's surface. From the progression of these numbers, as well as, *à priori*, from the nature of the air itself, which is *compressible,* i. e. capable of being condensed or crowded into a smaller space in proportion to the incumbent pressure, it is easy to see that, although by rising still higher we should continually get above more and more of the air, and so relieve ourselves more and more from the pressure with which it weighs upon us, yet the amount of this additional relief, or the *ponderable quantity* of air surmounted, would be by no means in proportion to the additional height ascended, but in a constantly decreasing ratio. An easy calculation, however, founded on our experimental knowledge of the properties of air, and the mechanical laws which regulate its dilatation and compression, is sufficient to show that, at an altitude above the surface of the earth not exceeding the hundredth part of its diameter, the tenuity, or rarefaction, of the air must be so excessive, that not only animal life could not subsist, or combustion be maintained in it, but that the most delicate means we possess of ascertaining the existence of *any air at all* would fail to afford the slightest perceptible indications of its presence.

(34.) Laying out of consideration, therefore, at present, all nice questions as to the probable existence of a definite limit to the atmosphere, beyond which there is, absolutely and rigorously speaking, *no* air, it is clear, that, for all practical purposes, we may speak of those regions which are more distant above the earth's surface than the hundredth part of its diameter as void of air, and of course of clouds (which are nothing but visible vapours, diffused and *floating* in the air, sustained by it, and rendering it *turbid* as mud does water). It seems probable, from many indications, that the greatest height at which visible clouds *ever exist* does not exceed ten miles; at which height the density of the air is about an eighth part of what it is at the level of the sea.

---

trical measurement of my own, made in July, 1821, under very favourable circumstances) is 10,872 English feet. — *Author.*

(35.) We are thus led to regard the atmosphere of air, with the clouds it supports, as constituting a coating of equable or nearly equable thickness, enveloping our globe on all sides; or rather as an aërial ocean, of which the surface of the sea and land constitutes the bed, and whose inferior portions or strata, within a few miles of the earth, contain by far the greater part of the whole mass, the density diminishing with extreme rapidity as we recede upwards, till, within a very moderate distance (such as would be represented by the sixth of an inch on the model we have before spoken of, and which is not more in proportion to the globe on which it rests, than the downy skin of a peach in comparison with the fruit within it), all sensible trace of the existence of air disappears.

(36.) Arguments, however, are not wanting to render it, if not absolutely certain, at least in the highest degree probable, that the surface of the aërial, like that of the aqueous ocean, has a real and definite limit, as above hinted at; beyond which there is positively *no* air, and above which a fresh quantity of air, could it be added from without, or carried aloft from below, instead of dilating itself indefinitely upwards, would, after a certain very enormous but still finite enlargement of volume, sink and merge, as water poured into the sea, and distribute itself among the mass beneath. With the truth of this conclusion, however, astronomy has little concern; all the effects of the atmosphere in modifying astronomical phenomena being the same, whether it be supposed of definite extent or not.

(37.) Moreover, whichever idea we adopt, within those limits in which it possesses any appretiable density its constitution is the same over all points of the earth's surface; that is to say, on the great scale, and leaving out of consideration temporary and local causes of derangement, such as winds, and great fluctuations, of the nature of waves, which prevail in it to an immense extent. In other words, the law of diminution of the air's density as we recede upwards *from the level of the sea* is the same in every column into which we may conceive it divided, or from whatever point of

the surface we may set out.   It may therefore be considered
as consisting of successively superposed strata or layers, each
of the form of a spherical shell, concentric with the general
surface of the sea and land, and each of which is *rarer*, or specifically lighter, than that immediately beneath it; and *denser*,
or specifically heavier, than that immediately above it.   This,
at least, is the kind of distribution which alone would be consistent with the laws of the equilibrium of fluids.   Inasmuch,
however, as the atmosphere is not in perfect equilibrium,
being always kept in a state of circulation, owing to the excess of heat in its equatorial regions over that at the poles,
some slight deviation from the rigorous expression of this law
takes place, and in peculiar localities there is reason to believe
that even considerable permanent depressions of the contours
of these strata, below their general or spherical level, subsist.
But these are points of consideration rather for the meteorologist than the astronomer.   It must be observed, moreover,
that with this distribution of its strata the inequalities of
mountains and valleys have little concern.   These exercise
hardly more influence in modifying their general spherical
figure than the inequalities at the bottom of the sea interfere
with the general sphericity of its surface.   They would exercise absolutely none were it not for their effect in giving
another than horizontal direction to the currents of air constituting winds, as shoals in the ocean throw up the currents which sweep over them towards the surface, and so
in some small degree tend to disturb the perfect level of that
surface.

(38.) It is the power which air possesses, in common with
all transparent media, of *refracting* the rays of light, or bending them out of their straight course, which renders a knowledge of the constitution of the atmosphere important to the
astronomer.   Owing to this property, objects seen obliquely
through it appear otherwise situated than they would to the
same spectator, had the atmosphere no existence.   It thus
produces a false impression respecting their places, which
must be rectified by ascertaining the amount and direction of
the displacement so apparently produced on each, before we

can come at a knowledge of the true directions in which they are situated from us at any assigned moment.

(39.) Suppose a spectator placed at A, any point of the earth's surface K A *k*; and let L *l*, M *m*, N *n*, represent the successive strata or layers, of decreasing density, into which we may conceive the atmosphere to be divided, and which are spherical surfaces concentric with K *k*, the earth's surface. Let S represent a star, or other heavenly body, beyond the utmost limit of the atmosphere. Then, if the air were away, the spectator would see it in the direction of the straight line A S. But, in reality, when the ray of light S A reaches the atmosphere, suppose at *d*, it will, by the laws of optics, begin to bend *downwards*, and take a more inclined direction, as *d c*. This bending will at first be imperceptible,

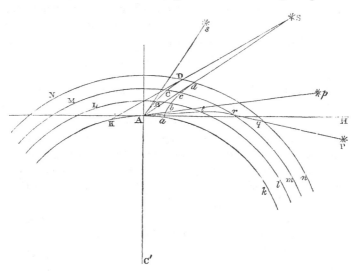

owing to the extreme tenuity of the uppermost strata; but as it advances downwards, the strata continually increasing in density, it will continually undergo greater and greater *refraction* in the same direction; and thus, instead of pursuing the straight line S *d* A, it will describe a curve S *d c b a*, continually more and more concave downwards, and will reach the earth, not at A, but at a certain point *a*, nearer to

S. *This* ray, consequently, will not reach the spectator's eye. The ray by which he will see the star is, therefore, not S *d* A, but another ray which, had there been no atmosphere, would have struck the earth at K, a point *behind* the spectator; but which, being bent by the air into the curve S D C B A, actually strikes on A. Now, it is a law of optics, that an object is seen in the direction which the visual ray has at the instant of *arriving at the eye*, without regard to what may have been otherwise its course between the object and the eye. Hence the star S will be seen, not in the direction A S, but in that of A *s*, a *tangent* to the curve S D C B A, at A. But because the curve described by the refracted ray is concave downwards, the tangent A *s* will lie *above* A S, the unrefracted ray: consequently the object S will appear more elevated above the horizon A H, when seen through the refracting atmosphere, than it would appear were there no such atmosphere. Since, however, the disposition of the strata is the same in all directions around A, the visual ray will not be made to deviate *laterally*, but will remain constantly in the same vertical plane, S A C', passing through the eye, the object, and the earth's centre.

(40.) The effect of the air's refraction, then, is to *raise* all the heavenly bodies higher above the horizon in appearance than they are in reality. Any such body, situated actually *in* the true horizon, will appear *above* it, or will have some certain apparent *altitude* (as it is called). Nay, even some of those actually below the horizon, and which would therefore be invisible but for the effect of refraction, are, by that effect, raised above it and brought into sight. Thus, the sun, when situated at P below the true horizon, A H, of the spectator, becomes visible to him, as if it stood at *p*, by the refracted ray P *q r t* A, to which A *p* is a tangent.

(41.) The exact estimation of the amount of atmospheric refraction, or the strict determination of the angle S A *s*, by which a celestial object at any assigned altitude, H A S, is raised in appearance above its true place, is, unfortunately, a very difficult subject of physical inquiry, and one on which geometers (from whom alone we can look for any information on the subject) are not yet entirely agreed. The difficulty arises from this, that the *density* of any stratum of air (on which its refracting power depends) is affected not *merely* by the superincumbent pressure, but also by its *temperature* or degree of heat. Now, although we know that as we recede from the earth's surface the temperature of the air is constantly diminishing, yet the *law*, or amount of this diminution at different heights, is not yet fully ascertained. Moreover, the refracting power of air is perceptibly affected by its *moisture ;* and this, too, is not the same in every part of an aërial column ; neither are we acquainted with the laws of its distribution. The consequence of our ignorance on these points is to introduce a corresponding degree of uncertainty into the determination of the amount of refraction, which affects, to a certain appretiable extent, our knowledge of several of the most important *data* of astronomy. The uncertainty thus induced is, however, confined within such very narrow limits as to be no cause of embarrassment, except in the most delicate inquiries, and to call for no further allusion in a treatise like the present.

(42.) A " Table of Refractions," as it is called, or a statement of the amount of apparent displacement arising from this cause, at all altitudes, or in every situation of a heavenly body, from the horizon to the *zenith**, or point of the sky vertically above the spectator, and, under all the circumstances in which astronomical observations are usually performed which may influence the result, is one of the most important and indispensable of all astronomical tables, since it is only by the use of such a table we are enabled to get rid of an illusion which must otherwise pervert all our notions respecting the celestial motions. Such have been, accordingly, constructed with great care, and are to be found in every collection of astronomical tables. Our design, in the present treatise, will not admit of the introduction of tables; and we must, therefore, content ourselves here, and in similar cases, with referring the reader to works especially destined to

* From an Arabic word of this signification. See this term technically defined in Chap. II.

furnish these useful aids to calculation. It is, however, desirable that he should bear in mind the following general notions of its amount, and law of variations.

(43.) 1st. In the *zenith* there is no refraction. A celestial object, situated vertically over head, is seen in its true direction, as if there were no atmosphere, at least if the air be tranquil.

2dly. In descending from the *zenith* to the horizon, the refraction continually increases. Objects near the horizon appear more elevated by it above their true directions than those at a high altitude.

3dly. The *rate* of its increase is nearly in proportion to the tangent of the apparent angular distance of the object from the zenith. But this rule, which is not far from the truth, at moderate *zenith distances*, ceases to give correct results in the vicinity of the horizon, where the law becomes much more complicated in its expression.

4thly. The average amount of refraction, for an object half-way between the zenith and horizon, or at an apparent altitude of 45°, is about 1' (more exactly 57″), a quantity hardly sensible to the naked eye; but at the visible horizon it amounts to no less a quantity than 33', which is rather more than the greatest apparent diameter of either the sun or the moon. Hence it follows, that when we see the lower edge of the sun or moon just *apparently* resting on the horizon, its whole disk is in reality below it, and would be entirely out of sight and concealed by the convexity of the earth, but for the bending round it, which the rays of light have undergone in their passage through the air, as alluded to in art. 40.

5thly. That when the barometer is higher than its average or mean state, the amount of refraction is greater than its mean amount; when lower, less: and,

6thly. That in one and the same state of the barometer the refraction is greater, the colder the air. The variation, owing to these two causes, from its mean amount (at temp. 55°, pressure 30 inches), are about one 420th part of that amount for each degree of the thermometer of Fahrenheit, and one 300th for each tenth of an inch in the height of the barometer.

(44.) It follows from this, that one obvious effect of refraction must be to shorten the duration of night and darkness, by actually prolonging the stay of the sun and moon
above the horizon. But even after they are set, the influence
of the atmosphere still continues to send us a portion of their
light; not, indeed, by direct transmission, but by *reflection*
upon the vapours, and minute solid particles which float in it,
and, perhaps, also on the actual material atoms of the air
itself. To understand how this takes place, we must recollect,
that it is not only by the direct light of a luminous object
that we see, but that whatever portion of its light which
would not otherwise reach our eyes is intercepted in its
course, and thrown back, or laterally, upon us, becomes to
us a means of illumination. Such reflective obstacles always
exist floating in the air. The whole course of a sun-beam
penetrating through the chink of a window-shutter into a
dark room is *visible* as a bright line in the air: and even if it
be stifled, or *let out* through an opposite crevice, the light
scattered through the apartment from this source is sufficient
to prevent entire darkness in the room. The luminous lines
occasionally seen in the air, in a sky full of partially broken
clouds, which the vulgar term "the sun drawing water," are
similarly caused. They are sunbeams, through apertures in
clouds, partially intercepted and reflected on the dust and
vapours of the air below. Thus it is with those solar rays
which, after the sun is itself concealed by the convexity of
the earth, continue to traverse the higher regions of the
atmosphere above our heads, and pass through and out of it,
without directly striking on the earth at all. Some portion
of them is intercepted and reflected by the floating particles
above mentioned, and thrown back, or laterally, so as to reach
us, and afford us that secondary illumination, which is twilight. The course of such rays will be immediately understood from the annexed figure, in which A B C D is the
earth; A a point on its surface, where the sun S is in the
act of setting; its last lower ray S A M just grazing the
surface at A, while its superior rays S N, S O, traverse the
atmosphere above A without striking the earth, leaving it

finally at the points P Q R, after being more or less bent in
passing through it, the lower most, the higher less, and that
which, like S R O, merely grazes the exterior limit of the
atmosphere, not at all.   Let us consider several points,
A, B, C, D, each more remote than the last from A, and each
more deeply involved in the *earth's shadow*, which occupies
the whole space from A beneath the line A M.   Now, A just
receives the sun's last direct ray, and, besides, is illuminated

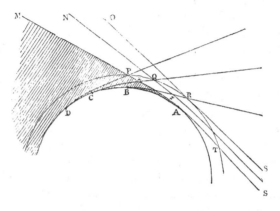

by the whole reflective atmosphere P Q R T.   It therefore
receives twilight from the whole sky.   The point B, to which
the sun has set, receives no direct solar light, nor any, direct
or reflected, from all that part of *its* visible atmosphere which
is below A P M; but from the lenticular portion P R *x*,
which is traversed by the sun's rays, and which lies above
the visible horizon B R of B, it receives a twilight, which is
strongest at R, the point immediately below which the sun
is, and fades away gradually towards P, as the luminous
part of the atmosphere thins off.   At C, only the last or
thinnest portion, P Q *z* of the lenticular segment, thus illuminated, lies above the horizon, C Q, of that place; here,
then, the twilight is feeble, and confined to a small space in
and near the horizon, which the sun has quitted, while at D
the twilight has ceased altogether.

(45.) When the sun is above the horizon, it illuminates the atmosphere and clouds, and these again disperse and scatter a portion of its light in all directions, so as to send some of its rays to every exposed point, from every point of the sky. The generally diffused light, therefore, which we enjoy in the daytime, is a phenomenon originating in the very same causes as the twilight. Were it not for the reflective and scattering power of the atmosphere, no objects would be visible to us out of direct sunshine; every shadow of a passing cloud would be pitchy darkness; the stars would be visible all day, and every apartment, into which the sun had not direct admission, would be involved in nocturnal obscurity. This scattering action of the atmosphere on the solar light, it should be observed, is increased by the irregularity of temperature caused by the same luminary in its different parts, which, during the daytime, throws it into a constant state of undulation, and, by thus bringing together masses of air of very unequal temperatures, produces partial reflections and refractions at their common boundaries, by which some portion of the light is turned aside from the direct course, and diverted to the purposes of general illumination.

(46.) From the explanation we have given, in arts. 39 and 40, of the nature of atmospheric refraction, and the mode in which it is produced in the progress of a ray of light through successive strata, or layers, of the atmosphere, it will be evident, that whenever a ray passes *obliquely* from a higher level to a lower one, or *vice versâ*, its course is not rectilinear, but concave downwards; and of course any object seen by means of such a ray, must appear deviated from its true place, whether that object be, like the celestial bodies, entirely beyond the atmosphere, or, like the summits of mountains seen from the plains, or other terrestrial stations at different levels seen from each other, immersed in it. Every difference of level, accompanied, as it must be, with a difference of density in the aërial strata, must also have, corresponding to it, a certain amount of refraction; less, indeed, than what would be produced by the *whole* atmosphere, but still often of very appretiable, and even considerable, amount. This refraction

D

between terrestrial stations is termed *terrestrial refraction*, to distinguish it from that total effect which is only produced on celestial objects, or such as are beyond the atmosphere, and which is called celestial or astronomical refraction.

(47.) Another effect of refraction is to distort the visible forms and proportions of objects seen near the horizon. The sun, for instance, which at a considerable altitude always appears round, assumes, as it approaches the horizon, a flattened or oval outline; its horizontal diameter being visibly greater than that in a vertical direction. When very near the horizon, this flattening is evidently more considerable on the lower side than on the upper; so that the apparent form is neither circular nor elliptic, but a species of oval, which deviates more from a circle below than above. This singular effect, which any one may notice in a fine sunset, arises from the rapid rate at which the refraction increases in approaching the horizon. Were every visible point in the sun's circumference equally raised by refraction, it would still appear circular, though displaced; but the lower portions being *more* raised than the upper, the vertical diameter is thereby shortened, while the two extremities of its horizontal diameter are equally raised, and in parallel directions, so that its apparent length remains the same. The dilated size (generally) of the sun or moon, when seen near the horizon, beyond what they appear to have when high up in the sky, has nothing to do with refraction. It is an illusion of the judgment, arising from the terrestrial objects interposed, or placed in close comparison with them. In that situation we view and judge of them as we do of terrestrial objects — in detail, and with an acquired habit of attention to parts. Aloft we have no associations to guide us, and their insulation in the expanse of sky leads us rather to undervalue than to over-rate their apparent magnitudes. Actual measurement with a proper instrument corrects our error, without, however, dispelling our illusion. By this we learn, that the sun, when just on the horizon, subtends at our eyes almost exactly the same, and the moon a materially *less* angle, than when seen at a great altitude in the sky, owing to its greater distance from us in the former situation as compared with the latter, as will be explained farther on.

(48.) After what has been said of the small extent of the atmosphere in comparison with the mass of the earth, we shall have little hesitation in admitting those luminaries which people and adorn the sky, and which, while they obviously form no part of the earth, and receive no support from it, are yet not borne along at random like clouds upon the air, nor drifted by the winds, to be external to our atmosphere. As such we have considered them while speaking of their refractions — as existing in the immensity of space beyond, and situated, perhaps, for any thing we can perceive to the contrary, at enormous distances from us and from each other.

(49.) Could a spectator exist unsustained by the earth, or any solid support, he would see around him at one view the whole contents of space — the visible constituents of the universe : and, in the absence of any means of judging of their distances from him, would refer them, in the directions in which they were seen from his station, to the concave surface of an imaginary sphere, having his eye for a centre, and its surface at some vast indeterminate distance. Perhaps he might judge those which appear to him large and bright, to be nearer to him than the smaller and less brilliant; but, independent of other means of judging, he would have no warrant for this opinion, any more than for the idea that all were equidistant from him, and *really* arranged on such a spherical surface. Nevertheless, there would be no impropriety in his referring their places, geometrically speaking, to those points of such a purely imaginary sphere, which their respective visual rays intersect; and there would be much advantage in so doing, as by that means their appearance and relative situation could be accurately measured, recorded, and mapped down. The objects in a landscape are at every variety of distance from the eye, yet we lay them all down in a picture on one plane, and at one distance, in their actual *apparent proportions,* and the likeness is not taxed with incorrectness, though a man in the foreground should be represented larger than a mountain in the distance. So it is to a spectator of the heavenly bodies pictured, *projected*, or mapped down on that imaginary sphere we call the *sky* or *heaven*. Thus, we may easily conceive that the moon, which appears to us as large as the sun, though less bright, *may* owe that apparent equality to its greater proximity, and *may* be really much less; while both the moon and sun may only appear larger and brighter than the stars, on account of the remoteness of the latter.

(50.) A spectator on the earth's surface is prevented, by the great mass on which he stands, from seeing into all that portion of space which is below him, or to see which he must look in any degree downwards. It is true that, if his place of observation be at a great elevation, the dip of the horizon will bring within the scope of vision a little more than a hemisphere, and refraction, wherever he may be situated, will enable him to look, as it were, a little round the corner; but the zone thus added to his visual range can hardly ever, unless in very extraordinary circumstances, exceed a couple of degrees in breadth, and is always ill seen on account of the vapours near the horizon. Unless, then, by a change of his geographical situation, he should shift his horizon (which is always a plane passing through his eye, and touching the spherical convexity of the earth); or unless, by some movements proper to the heavenly bodies, they should of themselves come above his horizon; or, lastly, unless, by some rotation of the earth itself on its centre, the point of its surface which he occupies should be carried round, and presented towards a different region of space; he would never obtain a sight of almost one half the objects external to our atmosphere. But if any of these cases be supposed, more, or all, may come into view according to the circumstances.

(51.) A traveller, for example, shifting his locality on our globe, will obtain a view of celestial objects invisible from his original station, in a way which may be not inaptly illustrated by comparing him to a person standing in a park close to a large tree. The massive obstacle presented by its trunk cuts off his view of all those parts of the landscape which it occupies as an object; but by walking round it a complete successive view of the whole panorama may be obtained. Just in the same way, if we set off from any station, as London, and travel southwards, we shall not fail to notice that many celestial objects which are never seen from London come successively into view, as if rising up above the horizon, night after night, from the south, although it is in reality our horizon, which, travelling with us south-wards round the sphere, sinks in succession beneath them. The novelty and splendour of fresh constellations thus gra-

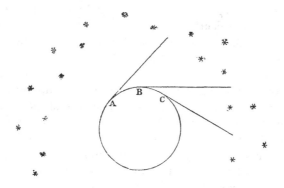

dually brought into view in the clear calm nights of tropical climates, in long voyages to the south, is dwelt upon by all who have enjoyed this spectacle, and never fails to impress itself on the recollection among the most delightful and in-teresting of the associations connected with extensive travel. A glance at the accompanying figure, exhibiting three suc-cessive stations of a traveller, A, B, C, with the horizon cor-responding to each, will place this process in clearer evidence than any description.

(52.) Again: suppose the earth itself to have a motion of rotation on its centre. It is evident that a spectator at rest (as it appears to him) on any part of it will, unperceived by himself, be carried round with it: unperceived, we say, because his horizon will constantly contain, and be limited by, the same terrestrial objects. He will have the same landscape constantly before his eyes, in which all the familiar objects in it, which serve him for landmarks and directions, retain, with respect to himself or to each other, the same invariable situations. The perfect smoothness and equality of the motion of so vast a mass, in which every object he sees around him participates alike, will (art. 15.) prevent his entertaining any suspicion of his actual change of place. Yet, with respect to external objects, — that is to say, all celestial ones which do not participate in the supposed rotation of the earth, — his horizon will have been all the while shifting in its relation to them, precisely as in the case of our traveller in the foregoing article. Recurring to the figure of that article, it is evidently the same thing, so far as their visibility is concerned, whether he has been carried by the earth's rotation successively into the situations A, B, C; or whether, the earth remaining at rest, he has transferred himself personally along its surface to those stations. Our spectator in the park will obtain precisely the same view of the landscape, whether he walk round the tree, or whether we suppose it sawed off, and made to turn on an upright pivot, while he stands on a projecting step attached to it, and allows himself to be carried round by its motion. The only difference will be in his view of the tree itself, of which, in the former case, he will see every part, but, in the latter, only that portion of it which remains constantly opposite to him, and immediately under his eye.

(53.) By such a rotation of the earth, then, as we have supposed, the horizon of a stationary spectator will be constantly depressing itself below those objects which lie in that region of space towards which the rotation is carrying him, and elevating itself above those in the opposite quarter, admitting into view the former, and successively hiding the latter. As the horizon of every such spectator, however, appears *to him* motionless, all such changes will be referred by him to a motion in the objects themselves so successively disclosed and concealed. In place of his horizon approaching the stars, therefore, he will judge the stars to approach his horizon; and when it passes over and hides any of them, he will consider them as having sunk below it, or *set;* while those it has just disclosed, and from which it is receding, will seem to be rising above it.

(54.) If we suppose this rotation of the earth to continue in one and the same direction, — that is to say, to be performed round one and the same *axis*, till it has completed an entire revolution, and come back to the position from which it set out when the spectator began his observations, — it is manifest that every thing will then be in precisely the same relative position as at the outset: all the heavenly bodies will appear to occupy the same places in the concave of the sky which they did at that instant, except such as may have actually moved in the interim; and if the rotation still continue, the same phenomena of their successive rising and setting, and return to the same places, will continue to be repeated in the same order, and (if the velocity of rotation be uniform) in equal intervals of time, *ad infinitum.*

(55.) Now, in this we have a lively picture of that grand phenomenon, the most important beyond all comparison which nature presents, the daily rising and setting of the sun and stars, their progress through the vault of the heavens, and their return to the same apparent places at the same hours of the day and night. The accomplishment of this restoration in the regular interval of twenty-four hours is the first instance we encounter of that great law of *periodicity* *, which, as we shall see, pervades all astronomy; by which expression we understand the continual reproduction of the same phenomena, in the same order, at equal intervals of time.

(56.) A free rotation of the earth round its centre, if it exist and be performed in consonance with the same mechanical laws which obtain in the motions of masses of matter under our immediate control, and within our ordinary experience, must be such as to satisfy two essential conditions. It must be invariable in its direction *with respect to the sphere itself*, and uniform in its velocity. The rotation must be performed *round an axis* or diameter of the sphere, whose *poles* or extremities, where it meets the surface, correspond

* Περίοδος, a *going round*, a circulation or revolution.

always to the same points on the sphere. Modes of rotation of a solid body under the influence of external agency are conceivable, in which the poles of the imaginary line or axis about which it is at any moment revolving shall hold no fixed places on the surface, but shift upon it every moment. Such changes, however, are inconsistent with the idea of a rotation of a body of regular figure about its axis of symmetry, performed in free space, and without resistance or obstruction from any surrounding medium, or disturbing influences. The complete absence of such obstructions draws with it, of necessity, the strict fulfilment of the two conditions above mentioned.

(57.) Now, these conditions are in perfect accordance with what we observe, and what recorded observation teaches us, in respect of the diurnal motions of the heavenly bodies. We have no reason to believe, from history, that any sensible change has taken place since the earliest ages in the interval of time elapsing between two successive returns of the same star to the same point of the sky; or, rather, it is demonstrable from astronomical records that no such change *has* taken place. And with respect to the other condition, — *the permanence of the axis of rotation,* — the appearances which any alteration in that respect must produce, would be marked, as we shall presently show, by a corresponding change of a very obvious kind in the apparent motions of the stars; which, again, history decidedly declares them *not* to have undergone.

(58.) But, before we proceed to examine more in detail how the hypothesis of the rotation of the earth about an axis accords with the phenomena which the diurnal motion of the heavenly bodies offers to our notice, it will be proper to describe, with precision, in what that diurnal motion consists, and how far it is participated in by them all; or whether any of them form exceptions, wholly or partially, to the common analogy of the rest. We will, therefore, suppose the reader to station himself, on a clear evening, just after sunset, when the first stars begin to appear, in some open situation whence a good general view of the heavens can be obtained. He will then perceive, above and around him, as it were, a vast concave hemispherical vault, beset with stars of various magnitudes, of which the brightest only will first catch his attention in the twilight; and more and more will appear as the darkness increases, till the whole sky is over-spangled with them. When he has awhile admired the calm magnificence of this glorious spectacle, the theme of so much song, and of so much thought, — a spectacle which no one can view without emotion, and without a longing desire to know something of its nature and purport, — let him fix his attention more particularly on a few of the most brilliant stars, such as he cannot fail to recognize again without mistake after looking away from them for some time, and let him refer their apparent situations to some surrounding objects, as buildings, trees, &c., selecting purposely such as are in different quarters of his horizon. On comparing them again with their respective points of reference, after a moderate interval, as the night advances, he will not fail to perceive that they have changed their places, and advanced, as by a general movement, in a westward direction; those towards the eastern quarter appearing to rise or recede from the horizon, while those which lie towards the west will be seen to approach it; and, if watched long enough, will, for the most part, finally sink beneath it, and disappear; while others, in the eastern quarter, will be seen to rise as if out of the earth, and, joining in the general procession, will take their course with the rest towards the opposite quarter.

(59.) If he persist for a considerable time in watching their motions, on the same or on several successive nights, he will perceive that each star appears to describe, as far as its course lies above the horizon, a circle in the sky; that the circles so described are not of the same magnitude for all the stars; and that those described by different stars differ greatly in respect of the parts of them which lie above the horizon. Some, which lie towards the quarter of the horizon which is denominated the SOUTH *, only remain for a short time above

* We suppose our observer to be stationed in some northern latitude; somewhere in Europe, for example.

it, and disappear, after describing in sight only the small upper segment of their diurnal circle; others, which rise between the south and east, describe larger segments of their circles above the horizon, remain proportionally longer in sight, and set precisely as far to the westward of south as they rose to the eastward; while such as rise exactly in the east remain just twelve hours visible, describe a semicircle, and set exactly in the west. With those, again, which rise between the east and north, the same law obtains; at least, as far as regards the time of their remaining above the horizon, and the proportion of the visible segment of their diurnal circles to their whole circumferences. Both go on increasing; they remain in view more than twelve hours, and their visible diurnal arcs are more than semicircles. But the magnitudes of the circles themselves diminish, as we go from the east, northward; the greatest of all the circles being described by those which rise exactly in the east point. Carrying his eye farther northwards, he will notice, at length, stars which, in their diurnal motion, just graze the horizon at its north point, or only dip below it for a moment; while others never reach it at all, but continue always above it, revolving in entire circles round ONE POINT called the POLE, which appears to be the common centre of all their motions, and which alone, in the whole heavens, may be considered immoveable. Not that this point is marked by any star. It is a purely imaginary centre; but there is near it one considerably bright star, called the Pole Star, which is easily recognized by the very small circle it describes; so small, indeed, that, without paying particular attention, and referring its position very nicely to some fixed mark, it may easily be supposed at rest, and be, itself, mistaken for the common centre about which all the others in that region describe their circles; or it may be known by its configuration with a very splendid and remarkable *constellation* or group of stars, called by astronomers the GREAT BEAR.

(60.) He will further observe, that the apparent relative situations of all the stars among one another, is not changed by their diurnal motion. In whatever parts of their circles

they are observed, or at whatever hour of the night, they
form with each other the same identical groups or configurations, to which the name of CONSTELLATIONS has been given.
It is true, that, in different parts of their course, these groups
stand differently with respect to the horizon; and those
towards the north, when in the course of their diurnal movement they pass alternately above and below that common
centre of motion described in the last article, become actually
inverted with respect to the horizon, while, on the other
hand, they always turn the same points towards the pole.   In
short, he will perceive that the whole assemblage of stars
visible at once, or in succession, in the heavens, may be
regarded as one great constellation, which seems to revolve
with a uniform motion, as if it formed one coherent mass; or
as if it were attached to the internal surface of a vast hollow
sphere, having the earth, or rather the spectator, in its centre,
and turning round an axis inclined to his horizon, so as to pass
through that fixed point or *pole* already mentioned.

(61.) Lastly, he will notice, if he have patience to out-
watch a long winter's night, commencing at the earliest
moment when the stars appear, and continuing till morning
twilight, that those stars which he observed setting in the
west have again risen in the east, while those which were
rising when he first began to notice them have completed
their course, and are now set; and that thus the hemisphere,
or a great part of it, which was then above, is now beneath
him, and its place supplied by that which was at first under
his feet, which he will thus discover to be no less copiously
furnished with stars than the other, and bespangled with
groups no less permanent and distinctly recognizable.   Thus
he will learn that the great constellation we have above
spoken of as revolving round the pole is co-extensive with the
whole surface of the sphere, being in reality nothing less than
a universe of luminaries surrounding the earth on all sides,
and brought in succession before his view, and referred
(each luminary according to its own visual ray or direction
from his eye) to the imaginary spherical surface, of which
he himself occupies the centre. (See art. 49.)   There is

always, therefore (he would justly argue), a star-bespangled
canopy over his head, by day as well as by night, only that
the glare of daylight (which he perceives gradually to efface
the stars as the morning twilight comes on) prevents them
from being seen. And such is really the case. The stars
actually continue visible through telescopes in the day-
time ; and, in proportion to the power of the instrument, not
only the largest and brightest of them, but even those of
inferior lustre, such as scarcely strike the eye at night as at
all conspicuous, are readily found and followed even at noonday,—unless in that part of the sky which is very near the
sun,—by those who possess the means of pointing a telescope
accurately to the proper places. Indeed, from the bottoms
of deep narrow pits, such as a well, or the shaft of a mine,
such bright stars as pass the zenith may even be discerned by
the naked eye; and we have ourselves heard it stated by a
celebrated optician, that the earliest circumstance which drew
his attention to astronomy was the regular appearance, at a
certain hour, for several successive days, of a considerable
star, through the shaft of a chimney. Venus in our climate,
and even Jupiter in the clearer skies of tropical countries,
are often visible, without any artificial aid, to the naked eye
of one who knows nearly where to look for them. During
total eclipses of the sun, the larger stars also appear in their
proper situations.

(62.) But to return to our incipient astronomer, whom we
left contemplating the sphere of the heavens, as completed in
imagination beneath his feet, and as rising up from thence in
its diurnal course. There is one portion or segment of this
sphere of which he will not thus obtain a view. As there is
a segment towards the north, adjacent to the pole above his
horizon, in which the stars *never set*, so there is a corresponding
segment, about which the smaller circles of the more southern
stars are described, in which they *never rise*. The stars which
border upon the extreme circumference of this segment just
graze the southern point of his horizon, and show themselves
for a few moments above it, precisely as those near the circumference of the northern segment graze his northern

horizon, and dip for a moment below it, to re-appear immediately. Every point in a spherical surface has, of course, another diametrically opposite to it; and as the spectator's horizon divides his sphere into two hemispheres — a superior and inferior — there must of necessity exist a depressed pole to the south, corresponding to the elevated one to the north, and a portion surrounding it, perpetually beneath, as there is another surrounding the north pole, perpetually above it.

> " Hic vertex nobis semper sublimis ; at illum
> Sub pedibus nox atra videt, manesque profundi."—VIRGIL.
>
> One pole rides high, one, plunged beneath the main,
> Seeks the deep night, and Pluto's dusky reign.

(63.) To get sight of this segment, he must travel southwards. In so doing, a new set of phenomena come forward. In proportion as he advances to the south, some of those constellations which, at his original station, barely grazed the northern horizon, will be observed to sink below it and set; at first remaining hid only for a very short time, but gradually for a longer part of the twenty-four hours. They will continue, however, to circulate about the same point — that is, holding the same invariable position *with respect to them* in the concave of the heavens among the stars; but this point itself will become gradually depressed with respect to the spectator's horizon. The axis, in short, about which the diurnal motion is performed, will appear to have become continually less and less inclined to the horizon; and by the same degrees as the northern pole is depressed the southern will rise, and constellations surrounding it will come into view; at first momentarily, but by degrees for longer and longer times in each diurnal revolution — realizing, in short, what we have already stated in art. 51.

(64.) If he travel continually southwards, he will at length reach a line on the earth's surface, called *the equator*, at any point of which, indifferently, if he take up his station and recommence his observations, he will find that he has both the centres of diurnal motion in his horizon, occupying opposite points, the northern Pole having been depressed, and the southern raised, so that, in this geographical position, the diurnal rotation of the heavens will appear to him to be performed about a horizontal axis, every star describing half its diurnal circle above and half beneath his horizon, remaining alternately visible for twelve hours, and concealed during the same interval. In this situation, *no* part of the heavens is concealed from his *successive* view. In a night of twelve hours (supposing such a continuance of darkness possible at the equator) the whole sphere will have passed in review over him — the whole hemisphere with which he began his night's observation will have been carried down beneath him, and the entire opposite one brought up from below.

(65.) If he pass the equator, and travel still farther southwards, the southern pole of the heavens will become elevated above his horizon, and the northern will sink below it; and the more, the farther he advances southwards; and when arrived at a station as far to the south of the equator as that from which he started was to the north, he will find the whole phenomena of the heavens reversed. The stars which at his original station described their whole diurnal circles above his horizon, and never set, now describe them entirely below it, and never *rise*, but remain constantly invisible to him; and *vice versâ*, those stars which at his former station he never saw, he will now never cease to see.

(66.) Finally, if, instead of advancing southwards from his first station, he travel northwards, he will observe the northern pole of the heavens to become more elevated above his horizon, and the southern more depressed below it. In consequence, his hemisphere will present a less variety of stars, because a greater proportion of the whole surface of the heavens remains constantly visible or constantly invisible: the circle described by each star, too, becomes more nearly parallel to the horizon ; and, in short, every appearance leads to suppose that could he travel far enough to the north, he would at length attain a point *vertically under* the northern pole of the heavens, at which none of the stars would either rise or set, but each would circulate round the horizon in circles parallel to it. Many endeavours have been made to reach this point, which is called the north pole of the

earth, but hitherto without success; a barrier of almost in-
surmountable difficulty being presented by the increasing
rigour of the climate: but a very near approach to it has
been made; and the phenomena of those regions, though not
precisely such as we have described as what must subsist *at*
the pole itself, have proved to be in exact correspondence
with its near proximity. A similar remark applies to the
south pole of the earth, which, however, is more unap-
proachable, or, at least, has been less nearly approached, than
the north.

(67.) The above is an account of the phenomena of the
diurnal motion of the stars, as modified by different geographical situations, not grounded on any speculation, but
actually observed and recorded by travellers and voyagers.
It is, however, in complete accordance with the hypothesis
of a rotation of the earth round a fixed axis. In order to
show this, however, it will be necessary to premise a few observations on *parallactic motion* in general, and on the appearances presented by an assemblage of remote objects, when
viewed from different parts of a small and circumscribed
station.

(68.) It has been shown (art. 16.) that a spectator in
smooth motion, and surrounded by, and forming part of, a
great system partaking of the same motion, is unconscious of
his own movement, and transfers it in idea to objects external
and unconnected, in a contrary direction; those which he
leaves behind appearing to recede from, and those which he
advances towards to approach, him. Not only, however,
do external objects at rest appear in motion generally, with
respect to ourselves when we are in motion among them,
but they appear to move one among the other — they shift
their *relative* apparent places. Let any one travelling
rapidly along a high road fix his eye steadily on any object, but at the same time not entirely withdraw his attention from the general landscape, — he will see, or think he
sees, the whole landscape thrown into *rotation*, and moving
round that object as a centre; all objects between it and
himself appearing to move *backwards*, or the contrary way

to his own motion; and all beyond it, forwards, or in the
direction in which he moves: but let him withdraw his eye
from that object, and fix it on another, — a nearer one, for
instance, — immediately the appearance of rotation shifts
also, and the apparent centre about which this illusive
circulation is performed is transferred to the new object,
which, for the moment, appears to rest. This apparent
change of situation of objects with respect to one another,
arising from a motion of the spectator, is called a *parallactic
motion*. To see the reason of it we must consider that the
position of every object is referred by us to the surface of an
imaginary sphere of an indefinite radius, having our eye for
its centre; and, as we advance in any direction, A B, carry-

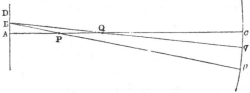

ing this imaginary sphere along with us, the visual rays A P,
A Q, by which objects are referred to its surface (at C, for
instance), shift their positions with respect to the line in whi h
we move, A B, which serves as an axis or line of reference,
and assume new positions, B P *p*, B Q *q*, revolving round
their respective objects as centres. Their intersections, therefore, *p*, *q*, with our visual sphere, will appear to recede on its
surface, but with different degrees of angular velocity in proportion to their proximity; the same distance of advance
A B subtending a greater angle, A P B = *c* P *p*, at the near
object P than at the remote one Q.

(69.) A consequence of the familiar appearance we have
adduced in illustration of these principles is worth noticing,
as we shall have occasion to refer to it hereafter. We observe that every object nearer to us than that on which our
eye is fixed appears to recede, and those farther from us to
advance in relation to one another. If then we did not know,
or could not judge by any other appearances, which of two
objects were nearer to us, this apparent advance or recess of

one of them, when the eye is kept steadily fixed on the other, would furnish a criterion. In a dark night, for instance, when all intermediate objects are unseen, the apparent relative movement of two lights which we are assured are themselves fixed, will decide as to their relative proximities. That which seems to advance with us and gain upon the other, or leave it behind it, is the farthest from us.

(70.) The apparent angular motion of an object, arising from a change of our point of view, is called in general *parallax*, and it is always expressed by *the angle* A P B *subtended at the object* P (see fig. of art. 68.) by a line joining the two points of view A B under consideration. For it is evident that the difference of angular position of P, with respect to the invariable direction A B D, when viewed from A and from B, is the difference of the two angles D B P and D A P ; now, D B P being the exterior angle of the triangle A B P, is equal to the sum of the interior and opposite, D B P = D A P + A P B, whence D B P − D A P = A P B.

(71.) It follows from what has been said that the amount of parallactic motion arising from any given change of our point of view is, *cæteris paribus*, less, as the distance of an object viewed is greater ; and when that distance is extremely great in comparison with the change in our point of view, the parallax becomes insensible ; or, in other words, objects do not appear to vary in situation at all. It is on this principle, that in alpine regions visited for the first time we are surprised and confounded at the little progress we appear to make by a considerable change of place. An hour's walk, for instance, produces but a small parallactic change in the relative situations of the vast and distant masses which surround us. Whether we walk round a circle of a hundred yards in diameter, or merely turn ourselves round in its centre, the distant panorama presents almost exactly the same aspect, — we hardly seem to have changed our point of view.

(72.) Whatever notion, in other respects, we may form of the stars, it is quite clear they must be immensely distant. Were it not so, the apparent angular interval between any two of them seen over head would be much greater than

when seen near the horizon, and the constellations, instead of
preserving the same appearances and dimensions during their
whole diurnal course, would appear to enlarge as they rise
higher in the sky, as we see a small cloud in the horizon
swell into a great overshadowing canopy when drifted by the
wind across our zenith, or as may be seen in the annexed
figure, where $a b$, A B, $a b$, are three different positions of
the same stars, as they would, if near the earth, be seen from

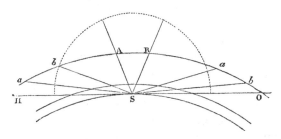

a spectator S, under the visual angles a S b, A S B.   No such
change of apparent dimension, however, is observed.   The
nicest measurements of the apparent angular distance of any
two *stars inter se*, taken in any parts of their diurnal course,
(after allowing for the unequal effects of refraction, or when
taken at such times that this cause of distortion shall act
equally on both,) manifest *not the slightest* perceptible variation.   Not only this, but at whatever point of the earth's
surface the measurement is performed, the results are *absolutely identical.*   No instruments ever yet invented by man
are delicate enough to indicate, by an increase or diminution
of the angle subtended, that one point of the earth is nearer
to or further from the *stars* than another.

(73.) The necessary conclusion from this is, that the
dimensions of the earth, large as it is, are comparatively
nothing, absolutely imperceptible, when compared with the
interval which separates the stars from the earth.   If an
observer walk round a circle not more than a few yards in
diameter, and from different points in its circumference
measure with a sextant or other more exact instrument
adapted for the purpose, the angles P A Q, P B Q, P C Q, sub-

tended at those stations by two well-defined points in his visible horizon, P Q, he will at once be advertised, by the difference of the results, of his change of distance from them arising from his change of place, although that difference may be so small as to produce no change in their *general* aspect to his unassisted sight. This is one of the innumerable instances where accurate measurement obtained by instrumental means places us in a totally different situation in respect to matters of fact, and conclusions thence deducible, from what we should

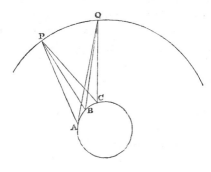

hold, were we to rely in all cases on the mere judgment of the eye. To so great a nicety have such observations been carried by the aid of an instrument called a theodolite, that a circle of the diameter above mentioned may thus be rendered *sensible*, may thus be detected to have *a size*, and an ascertainable *place*, by reference to objects distant by fully 100,000 times its own dimensions. Observations, differing, it is true, somewhat in method, but identical in principle, and executed with quite as much exactness, have been applied to the stars, and with a result such as has been already stated. Hence it follows, incontrovertibly, that the distance of the stars from the earth cannot be *so small* as 100,000 of the earth's diameters. It is, indeed, incomparably greater; for we shall hereafter find it fully demonstrated that the distance just named, immense as it may appear, is yet much underrated.

(74.) From such a distance, to a spectator with our faculties, and furnished with our instruments, the earth would

be imperceptible; and, reciprocally, an object of the earth's
size, placed at the distance of the stars, would be equally un-
discernible. If, therefore, at the point on which a spectator
stands, we draw a plane touching the globe, and prolong it in
imagination till it attain the region of the stars, and through
the centre of the earth conceive another plane parallel to the
former, and co-extensive with it, to pass; these, although
separated throughout their whole extent by the same interval,
viz. a semidiameter of the earth, will yet, on account of the
vast distance at which that interval is seen, be confounded
together, and undistinguishable from each other in the region
of the stars, when viewed by a spectator on the earth.   The
zone they there include will be of evanescent breadth to his
eye, and will only mark out a great circle in the heavens, one
and the same for both the stations.   This great circle, when
spoken of as a circle of the sphere, is called the *celestial
horizon* or simply *the* horizon, and the two planes just described are also spoken of as the *sensible* and the *rational*
horizon of the observer's station.

(75.) From what has been said (art. 73.) of the distance of
the stars, it follows, that if we suppose a spectator at the
centre of the earth to have his view bounded by the *rational*
horizon, in exactly the same manner as that of a corresponding
spectator on the surface is by his *sensible* horizon, the two
observers will see the same stars in the same relative situations, each beholding that entire hemisphere of the heavens
which is above the celestial horizon, corresponding to their
common zenith.   Now, so far as appearances go, it is clearly
the same thing whether the heavens, that is, all space, with
its contents, revolve round a spectator at rest in the earth's
centre, or whether that spectator simply turn round in the
opposite direction in his place, and view them in succession.
The aspect of the heavens, at every instant, as referred to his
horizon (which must be supposed to turn with him), will be
the same in both suppositions.   And since, as has been shown,
appearances are also, so far as the stars are concerned, the
same to a spectator on the surface as to one at the centre, it
follows that, whether we suppose the heavens to revolve

without the earth, or the earth within the heavens, *in the opposite direction*, the diurnal phenomena, to all its inhabitants, will be no way different.

(76.) The Copernican astronomy adopts the latter as the true explanation of these phenomena, avoiding thereby the necessity of otherwise resorting to the cumbrous mechanism of a solid but invisible sphere, to which the stars must be supposed attached, in order that they may be carried round the earth without derangement of their relative situations *inter se*. Such a contrivance would, indeed, suffice to explain the diurnal revolution of the stars, so as to " save appearances ; " but the movements of the sun and moon, as well as those of the planets, are incompatible with such a supposition, as will appear when we come to treat of these bodies. On the other hand, that a spherical mass of moderate dimensions (or, rather, when compared with the surrounding and visible universe, of evanescent magnitude), held by no tie, and free to move and to revolve, should do so, in conformity with those general laws which, so far as we know, regulate the motions of all material bodies, is so far from being a postulate difficult to be conceded, that the wonder would rather be should the fact prove otherwise. As a postulate, therefore, we shall henceforth regard it ; and as, in the progress of our work, analogies offer themselves in its support from what we observe of other celestial bodies, we shall not fail to point them out to the reader's notice.

(77.) The earth's rotation on its axis so admitted, explaining, as it evidently does, the apparent motion of the stars in a completely satisfactory manner, prepares us for the further admission of its motion, bodily, in space, should such a motion enable us to explain, in a manner equally so, the apparently complex and enigmatical motions of the sun, moon, and planets. The Copernican astronomy adopts this idea in its full extent, ascribing to the earth, in addition to its motion of *rotation* about an axis, also one of *translation* or transference through space, in such a course or *orbit*, and so regulated in direction and celerity, as, taken in conjunction with the motions of the other bodies of the universe, shall render a rational account of the appearances they successively present, — that is to say, an account of which the several parts, postulates, propositions, deductions, intelligibly cohere, without contradicting each other or the nature of things as concluded from experience. In this view of the Copernican doctrine it is rather a geometrical conception than a physical theory, inasmuch it simply assumes the requisite motions, without attempting to explain their mechanical origin, or assign them any dependence on physical causes. The Newtonian theory of gravitation supplies this deficiency, and, by showing that all the motions required by the Copernican conception *must*, and that no others can, result from a single, intelligible, and very simple dynamical law, has given a degree of certainty to this conception, as a matter of fact, which attaches to no other creation of the human mind.

(78.) To understand this conception in its further developments, the reader must bear steadily in mind the distinction between *relative* and *absolute* motion. Nothing is easier to perceive than that, if a spectator at rest view a certain number of moving objects, they will group and arrange themselves *to his eye*, at each successive moment, in a very different way from what they would do were he in active motion among them, — if he formed one of them, for instance, and joined in their dance. This is evident from what has been said before of parallactic motion; but it will be asked, How is such a spectator to disentangle from each other the two parts of the apparent motions of these external objects,— that which arises from the effect of his own change of place, and which is therefore only apparent (or, as a German metaphysician would say, *subjective* — having reference only to him as perceiving it), — and that which is real (or *objective* — having a positive existence, whether perceived by him or not)? By what rule is he to ascertain, from the appearances presented to him while himself in motion, what *would be* the appearances were he at rest? It by no means follows, indeed, that he would even then at once obtain a clear conception of all the motions of all the objects. The appearances so presented to him would have still something *subjective* about them.

They would be still *appearances,* not geometrical realities. They would still have a reference to the point of view, which might be very unfavourably situated (as, indeed, is the case in our system) for affording a clear notion of the real movement of each object. No geometrical figure, or curve, is seen by the eye as it is conceived by the mind to exist in reality. The laws of perspective interfere and alter the apparent directions and foreshorten the dimensions of its several parts. If the spectator be unfavourably situated, as, for instance, nearly in the plane of the figure (which is the case we have to deal with), they may do so to such an extent, as to make a considerable effort of imagination necessary to pass from the sensible to the real form.

(79.) Still, preparatory to this ultimate step, it is first necessary that the spectator should free or clear the appearances from the disturbing influence of his own change of place. And this he can always do by the following general rule or proposition : —

The relative motion of two bodies is the same as if either of them were at rest, and all its motion communicated to the other in an opposite direction. *

Hence, if two bodies move alike, they will, when seen from each other (without reference to other near bodies, but only to the starry sphere), appear at rest. Hence, also, if the absolute motions of two bodies be uniform and rectilinear, their relative motion is so also.

(80.) The stars are so distant, that as we have seen it is absolutely indifferent from what point of the earth's surface we view them. Their configurations *inter se* are identically the same. It is otherwise with the sun, moon, and planets, which are near enough (especially the moon) to be *parallactically* displaced by change of station from place to place on one globe. In order that astronomers residing on different points

---

* This proposition is equivalent to the following, which precisely meets the case proposed, but requires somewhat more thought for its clear apprehension than can perhaps be expected from a beginner : —

PROP. — *If two bodies, A and B, be in motion independently of each other, the motion which B seen from A would appear to have if A were at rest is the same with that which it would appear to have, A being in motion, if, in addition to its own motion, a motion equal to A's and in the same direction were communicated to it.*

of the earth's surface should be able to compare their observations with effect, it is necessary that they should clearly
understand and take account of this effect of the difference
of their stations on the appearance of the outward universe
as seen from each. As an exterior object seen from one
would appear to have shifted its place were the spectator
suddenly transported to the other, so two spectators, viewing
it from the two stations at the same instant, do not see it in
the same *direction*. Hence arises a necessity for the adoption
of a conventional centre of reference, or imaginary station
of observation common to all the world, to which each observer, wherever situated, may refer (or, as it is called,
reduce) his observations, by calculating and allowing for the
effect of his local position with respect to that common centre
(supposing him to possess the necessary data). If there were
only two observers, in fixed stations, one might agree to refer
his observations to the other station; but, as every locality
on the globe may be a station of observation, it is far more
convenient and natural to fix upon a point equally related to
all, as the common point of reference; and this can be no
other than the centre of the globe itself. The parallactic
change of apparent place which would arise in an object,
could any observer suddenly transport himself to the centre
of the earth, is evidently the angle C S P, subtended at the
object S by that radius C P of the earth which joins its
centre and the place P of observation.

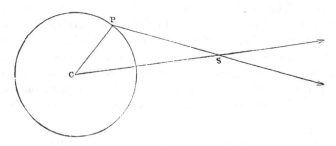

CHAPTER II.

TERMINOLOGY AND ELEMENTARY GEOMETRICAL CONCEPTIONS AND RELATIONS. — TERMINOLOGY RELATING TO THE GLOBE OF THE EARTH — TO THE CELESTIAL SPHERE. — CELESTIAL PERSPECTIVE.

(81.) SEVERAL of the terms in use among astronomers have been explained in the preceding chapter, and others used anticipatively. But the technical language of every subject requires to be formally stated, both for consistency of usage and definiteness of conception. We shall therefore proceed, in the first place, to define a number of terms in perpetual use, having relation to the globe of the earth and the celestial sphere.

(82.) DEFINITION 1. The *axis* of the earth is that diameter about which it revolves, with a uniform motion, *from west to east ;* performing one revolution in the interval which elapses between any star leaving a certain point in the heavens, and returning to the same point again.

(83.) DEF. 2. The *poles* of the earth are the points where its axis meets its surface. The North Pole is that nearest to Europe ; the South Pole that most remote from it.

(84.) DEF. 3. The *earth's equator* is a great circle on its surface, equidistant from its poles, dividing it into two hemispheres — a northern and a southern ; in the midst of which are situated the respective poles of the earth of those names. The *plane* of the equator is, therefore, a plane perpendicular to the earth's axis, and passing through its centre.

(85.) DEF. 4. The *terrestrial meridian* of a station on the earth's surface, is a great circle of the globe passing through both poles and through the plane. The plane of the meridian is the plane in which that circle lies.

(86.) Def. 5. The *sensible* and the *rational horizon* of any station have been already defined in art. 74.

(87.) Def. 6. *A meridian line* is the line of intersection of the plane of the meridian of any station with the plane of the sensible horizon, and therefore marks the north and south points of the horizon, or the directions in which a spectator must set out if he would travel directly towards the north or south pole.

(88.) Def. 7. The *latitude* of a place on the earth's surface is its angular distance from the equator, measured on its own terrestrial meridian: it is reckoned in degrees, minutes, and seconds, from 0 up to 90°, and northwards or southwards according to the hemisphere the place lies in. Thus, the observatory at Greenwich is situated in 51° 28' 40" north latitude. This definition of latitude, it will be observed, is to be considered as only temporary. A more exact knowledge of the physical structure and figure of the earth, and a better acquaintance with the niceties of astronomy, will render some modification of its terms, or a different manner of considering it, necessary.

(89.) Def. 8. *Parallels of latitude* are small circles on the earth's surface parallel to the equator. Every point in such a circle has the same latitude. Thus, Greenwich is said to be situated *in the parallel of* 51° 28' 40".

(90.) Def. 9. The *longitude* of a place on the earth's surface is the inclination of its meridian to that of some fixed station referred to as a point to reckon from. English astronomers and geographers use the observatory at Greenwich for this station; foreigners, the principal observatories of their respective nations. Some geographers have adopted the island of Ferro. Hereafter, when we speak of longitude, we reckon from Greenwich. The longitude of a place is, therefore, measured by the arc of the equator intercepted between the meridian of the place and that of Greenwich; or, which is the same thing, by the spherical angle at the pole included between these meridians.

(91.) As *latitude* is reckoned north or south, so *longitude* is usually said to be reckoned west or east. It would add greatly, however, to systematic regularity, and tend much to avoid confusion and ambiguity in computations, were this mode of expression abandoned, and longitudes reckoned invariably *westward* from their origin round the whole circle from 0 to 360°. Thus, the longitude of Paris is, in common parlance, either 2° 20′ 22″ east, or 357° 39′ 38″ west of Greenwich. But, in the sense in which we shall henceforth use and recommend others to use the term, the latter is its proper designation. Longitude is also reckoned in time at the rate of 24 h. for 360°, or 15° per hour. In this system the longitude of Paris is 23 h. 50 m. 38½ s.*

(92.) Knowing the longitude and latitude of a place, it may be laid down on an artificial globe; and thus a map of the earth may be constructed. Maps of particular countries are detached portions of this general map, extended into planes; or, rather, they are representations on planes of such portions, executed according to certain conventional systems of rules, called *projections*, the object of which is either to distort as little as possible the outlines of countries from what they are on the globe — or to establish easy means of ascertaining, by inspection or graphical measurement, the latitudes and longitudes of places which occur in them, without referring to the globe or to books — or for other peculiar uses. See Chap. IV.

(93.) Def. 10. The *Tropics* are two parallels of latitude, one on the north and the other on the south side of the equator, over every point of which respectively, the sun in its diurnal course passes vertically on the 21st of March and the 21st of September in every year. Their latitudes are about 23° 28′ respectively, north and south.

(94.) Def. 11. The Arctic and Antarctic circles are two small circles or parallels of latitude as distant from the north and south poles as the tropics are from the equator, that is to say, about 23° 28′; their latitudes, therefore, are about

---

* To distinguish minutes and seconds of time from those of angular measure we shall invariably adhere to the distinct system of notation here adopted (° ′ ″, and h. m. s.). Great confusion sometimes arises from the practice of using the same marks for both.

66° 32′. We say *about*, for the places of these circles and of the tropics are continually shifting on the earth's surface, though with extreme slowness, as will be explained in its proper place.

(95.) DEF. 12. The sphere of the heavens or of the stars is an imaginary spherical surface of infinite radius, having the eye of any spectator for its centre, and which may be conceived as a ground on which the stars, planets, &c., the visible contents of the universe, are seen projected as in a vast picture. *

(96.) DEF. 13. The *poles* of the celestial sphere are the points of that imaginary sphere towards which the earth's axis is directed.

(97.) DEF. 14. The celestial equator, or, as it is often called by astronomers, the *equinoctial*, is a great circle of the celestial sphere, marked out by the indefinite extension of the plane of the terrestrial equator.

(98.) DEF. 15. The celestial horizon of any place is a great circle of the sphere marked out by the indefinite extension of the plane of any spectator's *sensible* or (which comes to the same thing as will presently be shown), his *rational* horizon, as in the case of the equator.

(99.) DEF. 16. The *zenith* and *nadir*† of a spectator are the two points of the sphere of the heavens, vertically over his head, and vertically under his feet, or the *poles* of

---

* The ideal sphere without us, to which we refer the places of objects, and which we carry along with us wherever we go, is no doubt intimately connected by association, if not entirely dependent on that obscure perception of sensation in the retinæ of our eyes, of which, even when closed and unexcited, we cannot entirely divest them.   We have a real spherical surface within our eyes, the seat of sensation and vision, corresponding, point for point, to the external sphere.   On this the stars, &c. are really mapped down, as we have supposed them in the text to be, on the imaginary concave of the heavens.   When the whole surface of the retina is excited by light, habit leads us to associate it with the idea of a real surface existing without us.   Thus we become impressed with the notion of *a sky* and *a heaven*, but the concave surface of the retina itself is the true seat of all *visible* angular dimension and angular motion.   The substitution of the *retina* for the *heavens* would be awkward and inconvenient in language, but it may always be mentally made.   (See Schiller's pretty enigma on the eye in his Turandot.)

† From Arabic words.   Nadir corresponds evidently to the German *nieder* (down), whence our *nether*.

the celestial horizon; that is to say, points 90° distant from every point in it.

(100.) Def. 17. *Vertical circles* of the sphere are great circles passing through the zenith and nadir, or great circles perpendicular to the horizon. On these are measured the *altitudes* of objects above the horizon — the complements to which are their *zenith distances*.

(101.) Def. 18. The *celestial meridian* of a spectator is the great circle marked out on the sphere by the prolongation of the plane of his terrestrial meridian. If the earth be supposed at rest, this is a fixed circle, and all the stars are carried across it in their diurnal courses from east to west. If the stars rest and the earth rotate, the spectator's meridian, like his horizon (art. 52.), sweeps daily across the stars from west to east. Whenever in future we speak of the meridian of a spectator or observer, we intend the celestial meridian, which being a circle passing through the poles of the heavens and the zenith of the observer, is necessarily a vertical circle, and passes through the north and south points of the horizon.

(102.) Def. 19. The *prime vertical* is a vertical circle perpendicular to the meridian, and which therefore passes through the east and west points of the horizon.

(103.) Def. 20. *Azimuth* is the angular distance of a celestial object from the north or south point of the horizon (according as it is the north or south pole which is *elevated*), when the object is referred to the horizon by a vertical circle; or it is the angle comprised between two vertical planes — one passing through the elevated pole, the other through the object. Azimuth *may* be reckoned eastwards or westwards, from the north or south point, and is usually so reckoned only to 180° either way. But to avoid confusion, and to preserve continuity of interpretation when algebraic symbols are used (a point of essential importance, hitherto too little insisted on), we shall always reckon azimuth *from* the points of the horizon *most remote* from the *elevated* pole, *westward* (so as to agree in general directions with the apparent diurnal motion of the stars), and carry its reckoning from 0° to 360° if always reckoned positive, considering the eastward reckoning as negative.

(104.) DEF. 21. The *altitude* of a heavenly body is its apparent angular elevation above the horizon. It is the complement to 90°, therefore, of its zenith distance. The altitude and azimuth of an object being known, its place in the visible heavens is determined.

(105.) DEF. 22. The *declination* of a heavenly body is its angular distance from the equinoctial or celestial equator, or the complement to 90° of its angular distance from the nearest pole, which latter distance is called its *Polar distance.* Declinations are reckoned *plus* or *minus,* according as the object is situated in the northern or southern celestial hemisphere. *Polar distances* are always reckoned from the North Pole, from 0° up to 180°, by which all doubt or ambiguity of expression with respect to sign is avoided.

(106.) DEF. 23. Hour circles of the sphere, or circles of declination, are great circles passing through the poles, and of course perpendicular to the equinoctial. The hour circle, passing through any particular heavenly body, serves to refer it to a point in the equinoctial, as a vertical circle does to a point in the horizon.

(107.) DEF. 24. The hour angle of a heavenly body is the angle at the pole included between the hour circle passing through the body, and the celestial meridian of the place of observation. We shall always reckon it *positively* from the *upper* culmination (art. 125.) *westwards,* or in conformity with the *apparent* diurnal motion, completely round the circle from 0° to 360°. *Hour angles,* generally, are angles included at the pole between different hour circles.

(108.) DEF. 25. The *right ascension* of a heavenly body is the arc of the equinoctial included between a certain point in that circle called the *Vernal Equinox,* and the point in the same circle to which it is referred by the circle of declination passing through it. Or it is the angle included between two hour circles, one of which passes through the vernal equinox (and is called the equinoctial colure), the other through the body. How the place of this initial point on the equinoctial is determined, will be explained further on.

(109.) The right ascensions of celestial objects are always reckoned *eastwards* from the equinox, and are estimated either in degrees, minutes, and seconds, as in the case of terrestrial longitudes, from 0° to 360°, which completes the circle; or, in time, in hours, minutes, and seconds, from 0h. to 24h. The apparent diurnal motion of the heavens being contrary to the real motion of the earth, this is in conformity with the westward reckoning of longitudes. (Art. 91.)

(110.) *Sidereal time* is reckoned by the diurnal motion of the stars, or rather of that point in the equinoctial from which right ascensions are reckoned. This point may be considered as a star, though no star is, in fact, there; and, moreover, the point itself is liable to a certain slow variation, —so slow however, as not to affect, perceptibly, the interval, of any two of its successive returns to the meridian. This interval is called a sidereal day, and is divided into 24 sidereal hours, and these again into minutes and seconds. A clock which marks sidereal time, *i. e.* which goes at such a rate as always to show 0h. 0m. 0s. when the equinox comes on the meridian, is called a sidereal clock, and is an indispensable piece of furniture in every observatory. Hence the hour angle of an object reduced to time at the rate of 15° per hour, expresses the interval of sidereal time by which (if its reckoning be positive) it has past the meridian; or, if negative, the time it wants of arriving at the meridian of the place of observation. So also the right ascension of an object, if converted into time at the same rate (since 360° being described uniformly in 24 hours, 15° must be so described in 1 hour), will express the interval of sidereal time which elapses from the passage of the vernal equinox across the meridian to that of the object next subsequent.

(111.) As a globe or maps may be made of the whole or particular regions of the surface of the earth, so also a globe, or general map of the heavens, as well as charts of particular parts, may be constructed, and the stars laid down in their proper situations relative to each other, and to the poles of the heavens and the celestial equator. Such a representation, once made, will exhibit a true appearance of the stars as they present themselves in succession to every spectator on the surface, or as they may be conceived to be seen at once by one at the centre of the globe. It is, therefore, independent of all *geographical* localities. There will occur in such a representation neither zenith, nadir, nor horizon — neither east nor west points; and although great circles may be drawn on it from pole to pole, corresponding to terrestrial meridians, they can no longer, in this point of view, be regarded as the celestial meridians of fixed points on the earth's surface, since, in the course of one diurnal revolution, every point in it passes beneath each of them. It is on account of this change of conception, and with a view to establish a complete distinction between the two branches of *Geography* and *Uranography**, that astronomers have adopted different terms, (viz. *declination* and *right ascension*) to represent those arcs in the heavens which correspond to *latitudes* and *longitudes* on the earth. It is for this reason that they term the equator of the heavens the *equinoctial ;* that what are meridians on the earth are called *hour circles* in the heavens, and the angles they include between them at the poles are called *hour angles.* All this is convenient and intelligible ; and had they been content with this nomenclature, no confusion could ever have arisen. Unluckily, the early astronomers have employed *also* the words latitude and longitude in their uranography, in speaking of arcs of circles not corresponding to those meant by the same words on the earth, but having reference to the motion of the sun and planets among the stars. It is now too late to remedy this confusion, which is ingrafted into every existing work on astronomy : we can only regret, and warn the reader of it, that he may be on his guard when, at a more advanced period of our work, we shall have occasion to define and use the terms in their *celestial sense,* at the same time urgently recommending to future writers the adoption of others in their places.

* Γη, the earth ; γραφειν, to describe or represent ; ουρανος, the heaven.

(112.) It remains to illustrate these descriptions by reference to a figure. Let C be the centre of the earth, N C S

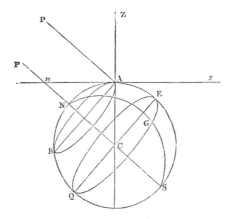

its axis; then are N and S its *poles*; E Q its *equator ;* A B
the *parallel* of latitude of the station A on its surface; A P
parallel to S C N, the direction in which an observer at A will
see the *elevated* pole of the heavens; and A Z, the prolongation of the terrestrial radius C A, that of his zenith. N A E S
will be his meridian; N G S that of some fixed station, as
Greenwich; and G E, or the spherical angle G N E, his longitude, and E A his latitude. Moreover, if *n s* be a plane
touching the surface in A, this will be his sensible horizon;
n A *s* marked on that plane by its intersection with his meridian will be his meridian line, and *n* and *s* the north and
south points of his horizon.

(113.) Again, neglecting the size of the earth, or conceiving
him stationed at its centre, and referring every thing to his
rational horizon; let the annexed figure represent the sphere
of the *heavens*; C the spectator; Z his zenith; and N his
nadir: then will H A O a great circle of the sphere, whose
poles are Z N, be his *celestial horizon ;* P *p* the *elevated* and
depressed POLES of the heavens; H P the *altitude of the pole,*
and H P Z E O his *meridian ;* E T Q, a great circle perpendicular to P *p*, will be the *equinoctial ;* and if ♈ represent the
equinox, ♈ T will be the *right ascension,* T S the *declination,*

and P S the *polar distance* of any star or object S, referred
to the equinoctial by the *hour circle* P S T *p;* and B S D

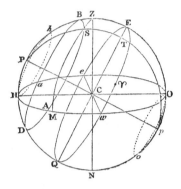

will be the diurnal circle it will appear to describe about the
pole.   Again, if we refer it to the horizon by the *vertical
circle* Z S M, O M will be its azimuth, M S its altitude, and
Z S its zenith distance.   H and O are the north and south,
e w the east and west points of his horizon, or of the heavens.
Moreover, if H *h,* O *o,* be small circles, or *parallels of declination,* touching the horizon in its north and south points,
H *h* will be the circle of *perpetual apparition,* between which
and the elevated pole the stars never set ; O *o* that of *perpetual occultation,* between which and the depressed pole they
never rise.   In all the zone of the heavens between H *h* and
O *o,* they rise and set; any one of them, as S, remaining
above the horizon in that part of its diurnal circle represented by *a* B A, and below it throughout all the part represented by A D *a.*   It will exercise the reader to construct
this figure for several different *elevations of the pole,* and for a
variety of positions of the star S in each.

    (114.) Celestial perspective is that branch of the general
science of perspective which teaches us to conclude, from a
knowledge of the real situation and forms of objects, lines,
angles, motions, &c. with respect to the spectator, their apparent aspects, as seen by him projected on the imaginary
concave of the heavens; and, *vice versâ,* from the apparent
configurations and movements of objects so seen projected,

to conclude, so far as they can be thence concluded, their real geometrical relations to each other and to the spectator. It agrees with ordinary perspective when only a small visual area is contemplated, because the concave ground of the celestial sphere, for a small extent, may be regarded as a plane surface, on which objects are seen projected or depicted as in common perspective. But when large amplitudes of the visual area are considered, or when the whole contents of space are regarded as projected on the whole interior surface of the sphere, it becomes necessary to use a different phraseology, and to resort to a different form of conception. In common perspective there is a single " point of sight," or " centre of the picture," the visual line from the eye to which is perpendicular to the " plane of the picture," and all straight lines are represented by straight lines. In celestial perspective, every point to which the view is for the moment directed, is equally entitled to be considered as the " centre of the picture," every portion of the surface of the sphere being similarly related to the eye. Moreover, every straight line (supposed to be indefinitely prolonged) is projected into a semicircle of the sphere, that, namely, in which a plane passing through the line and the eye cuts its surface. And every system of parallel straight lines, in whatever direction, is projected into a system of semicircles of the sphere, meeting in two common apexes, or vanishing points, diametrically opposite to each other, one of which corresponds to the vanishing point of parallels in ordinary perspective; the other, in such perspective has no existence. In other words, every point in the sphere to which the eye is directed may be regarded as one of the vanishing points, or one apex of a system of straight lines, parallel to that radius of the sphere which passes through it, or to the direction of the line of sight, seen in perspective from the earth, and the points diametrically opposite, or that from which he is looking, as the other. And any great circle of the sphere may similarly be regarded as the *vanishing circle* of a system of planes, parallel to its own.

(115.) A familiar illustration of this is often to be had by attending to the lines of light seen in the air, when the sun's rays are darted through apertures in clouds, the sun itself being at the time obscured behind them. These lines which, marking the course of rays emanating from a point almost infinitely distant, are to be considered as parallel straight lines, are thrown into great circles of the sphere, having two apexes or points of common intersection — one in the place where the sun itself (if not obscured) would be seen. The other diametrically opposite. The first only is most commonly suggested when the spectator's view is towards the sun. But in mountainous countries, the phenomenon of sunbeams converging towards a point diametrically opposite to the sun, and as much depressed below the horizon as the sun is elevated above it, is not unfrequently noticed, the back of the spectator being turned to the sun's place. Occasionally, but much more rarely, the whole course of such a system of sunbeams, stretching in semicircles across the hemisphere from horizon to horizon (the sun being near setting), may be seen.* Thus again, the streamers of the Aurora Borealis, which are doubtless electrical rays, parallel, or nearly parallel to each other, and to the dipping needle, usually appear to diverge from the point towards which the needle, freely suspended, would dip northwards (*i. e.* about 70° below the horizon and 23° west of north from London), and in their upward progress pursue the course of great circles till they again converge (in appearance) towards the point diametrically opposite (*i. e.* 70° above the horizon, and 23° to the eastward of south), forming a sort of canopy over head, having that point for its centre. So also in the phenomenon of shooting stars, the lines of direction which they appear to take on certain remarkable occasions of periodical recurrence, are observed, if

---

* It is in such cases only that we conceive them as circles, the ordinary conventions of plane perspective becoming untenable. The author had the good fortune to witness on one occasion the phenomenon described in the text under circumstances of more than usual grandeur. Approaching Lyons from the south on Sept. 30. 1826, about 5¼ h. P. M., the sun was seen nearly setting behind broken masses of stormy cloud, from whose apertures streamed forth beams of rose-coloured light, traceable all across the hemisphere almost to their opposite point of convergence behind the snowy precipices of Mont Blanc, conspicuously visible at nearly 100 miles to the eastward. The impression produced was that of another but feebler sun about to rise from behind the mountain, and darting forth precursory beams to meet those of the real one opposite.

prolonged backwards, apparently to meet nearly in one point of the sphere; a certain indication of a general near approach to parallelism in the real directions of their motions on those occasions. On which subject more hereafter.

(116.) In relation to this idea of celestial perspective, we may conceive the north and south poles of the sphere as the two vanishing points of a system of lines parallel to the axis of the earth; and the zenith and nadir of those of a system of perpendiculars to its surface at the place of observation, &c. It will be shown that the direction of a *plumb-line*, at every place is perpendicular to the surface of still water at that place which is the true horizon, and though mathematically speaking no two plumb-lines are exactly parallel (since they converge to the earth's centre), yet over very small tracts, such as the area of a building—in one and the same town, &c., the difference from exact parallelism is so small that it may be practically disregarded.* To a spectator looking upwards such a system of plumb-lines will appear to converge to his zenith; downwards, to his nadir.

(117.) So also the celestial equator, or the equinoctial, must be conceived as the vanishing circle of a system of planes parallel to the earth's equator, or perpendicular to its axis. The celestial horizon of any spectator is in like manner the vanishing circle of all planes parallel to his true horizon, of which planes his *rational* horizon (passing through the earth's centre) is one, and his *sensible* horizon (the tangent plane of his station) another.

(118.) Owing, however, to the absence of all the ordinary indications of distance which influence our judgment in respect of terrestrial objects, owing to the want of determinate figure and magnitude in the stars and planets as commonly seen—the projection of the celestial bodies on the ground of the heavenly concave is not usually regarded in this its true light of *a perspective representation or picture*, and it even requires an effort of imagination to conceive them in their true relations, as at vastly different distances, one behind the other,

* An interval of a mile corresponds to a convergence of plumb-lines amounting to somewhat less space than a minute.

F 3

and forming with one another lines of junction violently fore-shortened, and including angles altogether differing from those which their projected representations appear to make. To do so at all with effect presupposes a knowledge of their actual situations in space, which it is the business of astronomy to arrive at by appropriate considerations. But the connections which subsist among the several parts *of the picture*, the purely geometrical relations among the angles and sides of the spherical triangles of which it consists, constitute, under the name of Uranometry*, a preliminary and subordinate branch of the general science, with which it is necessary to be familiar before any further progress can be made. Some of the most elementary and frequently occurring of these relations we proceed to explain. And first, as immediate consequences of the above definitions, the following propositions will be borne in mind.

(119.) *The altitude of the elevated pole is equal to the latitude of the spectator's geographical station.*

For it appears, see *fig.* art. 112., that the angle P A Z between the pole and the zenith is equal to N C A, and the angles Z A n and N C E being right angles, we have P A n = A C E. Now the former of these is the elevation of the pole as seen from E, the latter is the angle at the earth's centre subtended by the arc E A, or the latitude of the place.

(120.) Hence to a spectator at the north pole of the earth, the north pole of the heavens is in his zenith. As he travels southward it becomes less and less elevated till he reaches the equator, when both poles are in his horizon — south of the equator the north pole becomes depressed below, while the south rises above his horizon, and continues to do so till the south pole of the globe is reached, when that of the heavens will be in the zenith.

(121.) The same stars, in their diurnal revolution, come to the meridian, *successively*, of every place on the globe once in twenty-four sidereal hours. And, since the diurnal rotation is uniform, the interval, in sidereal time, which elapses

---

* Ουρανος, the heavens ; μετρειν, to measure : the measurement of the heavens.

between the same star coming upon the meridians of two different places is measured by the difference of longitudes of the places.

(122.) *Vice versâ*—the interval elapsing between two *different stars* coming on the meridian of *one and the same place*, expressed in sidereal time, is the measure of the difference of right ascensions of the stars.

(123.) The equinoctial intersects the horizon in the east and west points, and the meridian in a point whose altitude is equal to the co-latitude of the place. Thus, at Greenwich, of which the latitude is 51° 28′ 40″, the altitude of the intersection of the equinoctial and meridian is 38° 31′ 20″. The north and south poles of the heavens are the poles of the equinoctial. The east and west points of the horizon of a spectator are the poles of his celestial meridian. The north and south points of his horizon are the poles of his prime vertical, and his zenith and nadir are the poles of his horizon.

(124.) All the heavenly bodies *culminate* (*i. e.* come to their greatest altitudes) on the meridian; which is, therefore, the best situation to observe them, being least confused by the inequalities and vapours of the atmosphere, as well as least displaced by refraction.

(125.) All celestial objects within the circle of perpetual apparition come twice on the meridian, above the horizon, in every diurnal revolution; once *above* and once *below* the pole. These are called their *upper* and *lower culminations*.

(126.) The problems of uranometry, as we have described it, consist in the solution of a variety of spherical triangles, both right and oblique angled, according to the rules, and by the formulæ of spherical trigonometry, which we suppose known to the reader, or for which he will consult appropriate treatises. We shall only here observe generally, that in all problems in which spherical geometry is concerned, the student will find it a useful practical maxim rather to consider the poles of the great circles which the question before him refers to than the circles themselves. To use, for example, in the relations he has to consider, polar distances rather than declinations, zenith distances rather than altitudes, &c. Bear-

ing this in mind, there are few problems in uranometry which
will offer any difficulty. The following are the combinations
which most commonly occur for solution *when the place of
one celestial object only* on the sphere is concerned.

(127.) In the triangle Z P S, Z is the zenith, P the
elevated pole, and S the star, sun, or other celestial object.
In this triangle occur, 1st, P Z, which being the complement of P H (the altitude of the pole), is obviously the complement of the latitude (or the *co-latitude*, as it is called) of
the place; 2d, P S, the *polar distance*, or the complement of
the declination (*co-declination*) of the star; 3d, Z S, the
zenith distance or *co-altitude* of the star. If P S be greater
than 90°, the object is situated on the side of the equinoctial
opposite to that of the elevated pole. If Z S be so, the object is below the horizon.

In the same triangle the angles are, 1st, Z P S the lower
angle ; 2d, P Z S (the supplement of S Z O, which latter
is the azimuth of the star or other heavenly body), 3d, P S Z,
an angle which, from the infrequency of any practical reference to it, has not acquired a name.*

The following five astronomical magnitudes, then, occur
among the sides and angles of this most useful triangle : viz.
1st, The co-latitude of the place of observation ; 2d, the
polar distance ; 3d, the zenith distance ; 4th, the hour angle ;
and 5th, the sub-azimuth (supplement of azimuth) of a given
celestial object ; and by its solution therefore may all problems be resolved, in which three of these magnitudes are
directly or indirectly given, and the other two required to be
found.

(128.) For example, suppose the time of rising or setting
of the sun or of a star were required, having given its right
ascension and polar distance. The star rises when *apparently*
on the horizon, or *really* about 34' below it (owing to refraction), so that, at the moment of its apparent rising, its zenith

---

* In the practical discussion of the measures of double stars and other objects
by the aid of the position micrometer, this angle is sometimes required to be
known ; and, when so required, it will be not inconveniently referred to as " the
angle of position of the zenith."

distance is 90° 34′ = Z S.   Its polar distance P S being also
given, and the co-latitude Z P of the place, we have given

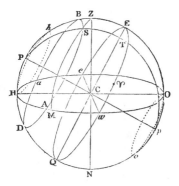

the three sides of the triangle, to find the hour angle Z P S,
which, being known, is to be added to or subtracted from the
star's right ascension, to give the sidereal time of setting or
rising, which, if we please, may be converted into solar time
by the proper rules and tables.

(129.) As another example of the use of the same triangle,
we may propose to find the local sidereal time, and the latitude
of the place of observation, by observing equal altitudes of the
same star east and west of the meridian, and noting the interval
of the observations in sidereal time.

The hour angles corresponding to equal altitudes of a fixed
star being equal, the hour angle east or west will be measured
by half the observed interval of the observations.   In our
triangle, then, we have given this hour angle Z P S, the polar
distance P S of the star, and Z S, its co-altitude at the moment
of observation.   Hence we may find P Z, the co-latitude of
the place.   Moreover, the hour angle of the star being known,
and also its right ascension, the point of the equinoctial is
known, which is on the meridian at the moment of observa-
ation; and, therefore, the local sidereal time at that moment.
This is a very useful observation for determining the latitude
and-time at an unknown station.

CHAPTER III. *

(130.) OUR first chapters have been devoted to the acquisition
chiefly of preliminary notions respecting the globe we inhabit,
its relation to the celestial objects which surround it, and the
physical circumstances under which all astronomical observations must be made, as well as to provide ourselves with a
stock of *technical words* and elementary ideas of most frequent
and familiar use in the sequel.   We might now proceed to a
more exact and detailed statement of the facts and theories
of astronomy; but, in order to do this with full effect, it will
be desirable that the reader be made acquainted with the

* The student who is anxious to become acquainted with the chief subject
matter of this work, may defer the reading of that part of this chapter which is
devoted to the description of particular instruments, or content himself with a
cursory perusal of it, until farther advanced, when it will be necessary to return
to it.

principal means which astronomers possess, of determining, with the degree of nicety their theories require, the data on which they ground their conclusions; in other words, of ascertaining by measurement the apparent and real magnitudes with which they are conversant.   It is only when in possession of this knowledge that he can fully appretiate either the truth of the theories themselves, or the degree of reliance to be placed on any of their conclusions antecedent to trial: since it is only by knowing what amount of error can certainly be perceived and distinctly measured, that he can satisfy himself whether any theory offers so close an approximation, in its numerical results, to actual phenomena, as will justify him in receiving it as a true representation of nature.

(131.) Astronomical instrument-making may be justly regarded as the most refined of the mechanical arts, and that in which the nearest approach to geometrical precision is required, and has been attained.   It may be thought an easy thing, by one unacquainted with the niceties required, to turn a circle in metal, to divide its circumference into 360 equal parts, and these again into smaller subdivisions, — to place it accurately on its centre, and to adjust it in a given position; but practically it is found to be one of the most difficult. Nor will this appear extraordinary, when it is considered that, owing to the application of telescopes to the purposes of angular measurement, every imperfection of structure of division becomes magnified by the whole optical power of that instrument; and that thus, not only direct errors of workmanship, arising from unsteadiness of hand or imperfection of tools, but those inaccuracies which originate in far more uncontrollable causes, such as the unequal expansion and contraction of metallic masses, by a change of temperature, and their unavoidable flexure or bending by their own weight, become perceptible and measurable.   An angle of one minute occupies, on the circumference of a circle of 10 inches in radius, only about $\frac{1}{350}$th part of an inch, a quantity too small to be *certainly* dealt with without the use of magnifying glasses; yet one minute is a gross quantity in the astronomical measurement of an angle.   With the instruments now employed in observatories, a single second, or the 60th part of a minute, is rendered a distinctly visible and appretiable quantity. Now, the arc of a circle, subtended by one second, is less than the 200,000th part of the radius, so that on a circle of 6 feet in diameter it would occupy no greater linear extent than $\frac{1}{3700}$th part of an inch; a quantity requiring a powerful microscope to be *discerned* at all. Let any one figure to himself, therefore, the difficulty of placing on the circumference of a metallic circle of such dimensions (supposing the difficulty of its construction surmounted), 360 marks, dots, or cognizable divisions, which shall all be true to their places within such narrow limits; to say nothing of the subdivision of the degrees so marked off into minutes, and of these again into seconds. Such a work has probably baffled, and will probably for ever continue to baffle, the utmost stretch of human skill and industry; nor, if executed, could it endure. The ever varying fluctuations of heat and cold have a tendency to produce not merely temporary and transient, but permanent, uncompensated changes of form in all considerable masses of those metals which alone are applicable to such uses; and their own weight, however symmetrically formed, must always be unequally sustained, since it is impossible to apply the sustaining power to *every part* separately: even could this be done, at all events force must be used to move and to fix them; which can never be done without producing temporary and risking permanent change of form. It is true, by dividing them on their centres, and in the identical places they are destined to occupy, and by a thousand ingenious and delicate contrivances, wonders have been accomplished in this department of art, and a degree of perfection has been given, not merely to *chefs d'œuvre*, but to instruments of moderate prices and dimensions, and in ordinary use, which, on due consideration, must appear very surprising. But though we are entitled to look for *wonders* at the hands of scientific artists, we are not to expect *miracles*. The demands of the astronomer will always surpass the power of the artist; and it must, therefore, be constantly the aim of the former to make himself, as far as possible, independent of the imperfections incident to every work the latter can place in his hands. He must, therefore, endeavour so to combine his observations, so to choose his opportunities, and so to familiarize himself with all the causes which may produce instrumental derangement, and with all the peculiarities of structure and material of each instrument he possesses, as not to allow himself to be misled by their errors, but to extract from their indications, as far as possible, all that is *true*, and reject all that is erroneous. It is in this that the art of the practical astronomer consists, — an art of itself of a curious and intricate nature, and of which we can here only notice some of the leading and general features.

(132.) The great aim of the practical astronomer being numerical correctness in the results of instrumental measurement, his constant care and vigilance must be directed to the detection and compensation of errors, either by annihilating, or by taking account of, and allowing for them. Now, if we examine the sources from which errors may arise in any instrumental determination, we shall find them chiefly reducible to three principal heads : —

(133.) 1st, External or incidental causes of error ; comprehending such as depend on external, uncontrollable circumstances : such as, fluctuations of weather, which disturb the amount of refraction from its tabulated value, and, being reducible to no fixed law, induce uncertainty to the extent of their own possible magnitude ; such as, by varying the temperature of the air, vary also the form and position of the instruments used, by altering the relative magnitudes and the tension of their parts ; and others of the like nature.

(134.) 2dly, *Errors of observation:* such as arise, for example, from inexpertness, defective vision, slowness in seizing the exact *instant* of occurrence of a phenomenon, or precipitancy in anticipating it, &c. ; from atmospheric indistinctness ; insufficient optical power in the instrument, and the like. Under this head may also be classed all errors arising from momentary instrumental derangement, — slips in clamping, looseness of screws, &c.

(135.) 3dly, The third, and by far the most numerous class

of errors to which astronomical measurements are liable, arise
from causes which may be deemed instrumental, and which
may be subdivided into two principal classes.   The *first* comprehends those which arise from an instrument not *being* what
it professes to be, which is *error* of *workmanship*.   Thus, if
a pivot or axis, instead of being, as it ought, exactly cylindrical, be slightly flattened, or elliptical, — if it be not exactly
(as it is intended it should) concentric with the circle it
carries ; — if this circle (so called) be in reality *not* exactly
circular, or not in one plane ; — if its divisions, intended to
be precisely equidistant, should be placed in reality at unequal intervals, — and a hundred other things of the same
sort.   These are not mere speculative sources of error, but
practical annoyances, which every observer has to contend
with.

(136.) The *other* subdivision of instrumental errors comprehends such as arise from an instrument not being placed
in the *position* it ought to have ; and from those of its parts,
which are made purposely moveable, not being properly disposed *inter se*.   These are *errors of adjustment*.   Some are
unavoidable, as they arise from a general unsteadiness of the
soil or building in which the instruments are placed ; which,
though too minute to be noticed in any other way, become
appretiable in delicate astronomical observations : others,
again, are consequences of imperfect workmanship, as where
an instrument once well adjusted will not remain so, but
keeps deviating and shifting.   But the most important of this
class of errors arise from the non-existence of natural indications, *other* than those afforded by astronomical observations
themselves, whether an instrument has or has not the exact
position, with respect to the horizon and its cardinal points,
the axis of the earth, or to other principal astronomical lines
and circles, which it ought to have to fulfil properly its objects.

(137.) Now, with respect to the first two classes of error,
it must be observed, that, in so far as they cannot be reduced
to known laws, and thereby become subjects of calculation
and due allowance, they actually vitiate, to their full extent,
the results of any observations in which they subsist.   Being,

however, in their nature casual and accidental, their effects necessarily lie sometimes one way, sometimes the other; sometimes diminishing, sometimes tending to increase the results. Hence, by greatly multiplying observations, under varied circumstances, by avoiding unfavourable, and taking advantage of favourable circumstances of weather, or otherwise using opportunity to advantage — and finally, by taking the *mean or average* of the results obtained, this class of errors may be so far *subdued*, by setting them to destroy one another, as no longer sensibly to vitiate any theoretical or practical conclusion. This is the great and indeed only resource against such errors, not merely to the astronomer, but to the investigator of numerical results in every department of physical research.

(138.) With regard to errors of adjustment and workmanship, not only the *possibility*, but the *certainty* of their existence, in every imaginable form, in all instruments, must be contemplated. Human hands or machines never formed a circle, drew a straight line, or erected a perpendicular, nor ever placed an instrument in *perfect* adjustment, unless accidentally; and then only during an instant of time. This does not prevent, however, that a great approximation to all these desiderata should be attained. But it is the peculiarity of astronomical observation to be the *ultimate means of detection* of all mechanical defects which elude by their minuteness every other mode of detection. What the eye cannot discern nor the touch perceive, a course of astronomical observations will make distinctly evident. The imperfect products of man's hands are here tested by being brought into comparison under very great magnifying powers (corresponding in effect to a great increase in acuteness of perception) with the perfect workmanship of nature; and there is none which will bear the trial. Now, it may seem like arguing in a vicious circle, to deduce theoretical conclusions and laws from observation, and then to turn round upon the instruments with which those observations were made, accuse them of imperfection, and attempt to detect and rectify their errors by means of the very laws and theories which they have helped us to a knowledge of. A little consideration, however, will suffice to show that such a course of proceeding is perfectly legitimate.

(139.) The steps by which we arrive at the laws of natural phenomena, and especially those which depend for their verification on numerical determinations, are necessarily successive. Gross results and palpable laws are arrived at by rude observation with coarse instruments, or without any instruments at all, and are expressed in language which is not to be considered as absolute, but is to be interpreted with a degree of latitude commensurate to the imperfection of the observations themselves. These results are corrected and refined by nicer scrutiny, and with more delicate means. The first rude expressions of the laws which embody them are perceived to be inexact. The language used in their expression is corrected, its terms more rigidly defined, or fresh terms introduced, until the new state of language and terminology is brought to fit the improved state of knowledge of facts. In the progress of this scrutiny subordinate laws are brought into view which still further modify both the verbal statement and numerical results of those which first offered themselves to our notice; and when these are traced out and reduced to certainty, others, again, subordinate to them, make their appearance, and become subjects of further inquiry. Now, it invariably happens (and the reason is evident) that the first glimpse we catch of such subordinate laws — the first form in which they are dimly shadowed out to our minds — is that of *errors*. We perceive a discordance between what we *expect*, and what we *find*. The first occurrence of such a discordance we attribute to accident. It happens again and again; and we begin to suspect our instruments. We then inquire, to what amount of error their determinations can, *by possibility*, be liable. If their *limit of possible error* exceed the observed deviation, we at once condemn the instrument, and set about improving its construction or adjustments. Still the same deviations occur, and, so far from being palliated, are more marked and better defined than before. We are now sure that we are on the traces of a law of nature, and we pursue it till we have reduced it to a definite statement, and verified it by repeated observation, under every variety of circumstances.

(140.) Now, in the course of this inquiry, it will not fail to happen that other discordances will strike us. Taught by experience, we suspect the existence of some natural law, before unknown; we tabulate (*i. e.* draw out in order) the results of our observations; and we perceive, in this synoptic statement of them, distinct indications of a regular progression. Again we improve or vary our instruments, and we now lose sight of this supposed new law of nature altogether, or find it replaced by some other, of a totally different character. Thus we are led to suspect an instrumental cause for what we have noticed. We examine, therefore, the *theory* of our instrument; we suppose defects in its structure, and, by the aid of geometry, we trace their influence in introducing *actual errors* into its indications. These errors have *their laws*, which, so long as we have no knowledge of causes to guide us, may be confounded with laws of nature, as they are mixed up with them in their effects. They are not fortuitous, like errors of observation, but, as they arise from sources inherent in the instrument, and unchangeable while it and its adjustments remain unchanged, they are reducible to fixed and ascertainable forms; each particular defect, whether of structure or adjustment, producing its own appropriate *form* of error. When these are thoroughly investigated, we recognize among them one which coincides in its nature and progression with that of our observed discordances. The mystery is at once solved. We have detected, by direct observation, an instrumental defect.

(141.) It is, therefore, a chief requisite for the practical astronomer to make himself completely familiar with the *theory* of his instruments. By this alone is he enabled at once to decide what *effect* on his observations any given imperfection of structure or adjustment will produce in any given circumstances under which an observation can be made. This alone also can place him in a condition to derive available and practical means of destroying and eliminating altogether

G

the influence of such imperfections, by so arranging his observations, that it shall affect their results in opposite ways, and that its influence shall thus disappear from their mean, which is one of the chief modes by which precision is attained in practical astronomy. Suppose, for example, the principle of an instrument required that a circle should be concentric with the axis on which it is made to turn. As this is a condition which no workmanship can *exactly* fulfil, it becomes necessary to inquire what errors will be produced in observations made and registered on the faith of such an instrument, by any assigned deviation in this respect; that is to say, what would be the disagreement between observations made with it and with one absolutely perfect, could such be obtained. Now, simple geometrical considerations suffice to show — 1st. that if the axis be excentric by a given fraction (say one thousandth part) of the radius of the circle, all angles read off on that part of the circle *towards* which the excentricity lies, will appear by that fractional amount too small, and all on the opposite side too large. And, 2dly, that *whatever* be the amount of the excentricity, and on *whatever* part of the circle any proposed angle is measured, the effect of the error in question on the result of observations depending on the graduation of its circumference (or limb, as it is technically called) will be completely annihilated by the very easy method of always reading off the divisions on two diametrically opposite points of the circle, and taking a mean; for the effect of excentricity is always to increase the arc representing the angle in question on one side of the circle, by just the same quantity by which it diminishes that on the other. Again, suppose that the proper use of the instrument required that this axis should be exactly parallel to that of the earth. As it never can be *placed* or remain so, it becomes a question, what amount of error will arise, in its use, from any assigned deviation, whether in a horizontal or vertical plane, from this precise position. Such inquiries constitute the theory of instrumental errors; a theory of the utmost importance to practice, and one of which a complete knowledge will enable an observer, with moderate instrumental means, often to

attain a degree of precision which might seem to belong only
to the most refined and costly.   This theory, as will readily
be apprehended, turns almost entirely on considerations of
pure geometry, and those for the most part not difficult.   In
the present work, however, we have no further concern with
it.   The astronomical instruments we propose briefly to describe in this chapter will be considered as perfect both in
construction and adjustment.*

(142.) As the above remarks are very essential to a right
understanding of the philosophy of our subject and the spirit
of astronomical methods, we shall elucidate them by taking
one or two special cases.   Observant persons, before the invention of astronomical instruments, had already concluded
the apparent diurnal motions of the stars to be performed in
circles about fixed poles in the heavens, as shown in the
foregoing chapter.   In drawing this conclusion, however,
refraction was entirely overlooked, or, if forced on their notice
by its great magnitude in the immediate neighbourhood
of the horizon, was regarded as a local irregularity, and, as
such, neglected, or slurred over.   As soon, however, as the
diurnal paths of the stars were attempted to be traced by instruments, even of the coarsest kind, it became evident that
the notion of exact circles described about one and the
same pole would not represent the phenomena correctly,
but that, owing to some cause or other, the apparent diurnal
orbit of every star is distorted from a circular into an oval
form, its lower segment being *flatter* than its upper; and the
deviation being greater the nearer the star approached the
horizon, the effect being the same as if the circle had been
squeezed upwards from below, and the lower parts more than
the higher.   For such an effect, as it was soon found to arise
from no casual or instrumental cause, it became necessary to
seek a natural one; and refraction readily occurred, to solve
the difficulty.   In fact, it is a case precisely analogous to

* The principle on which the chief adjustments of two or three of the most
useful and common instruments, such as the transit, the equatorial, and the
sextant, are performed, are, however, noticed, for the convenience of readers who
may use such instruments without going farther into the arcana of practical
astronomy.

what we have already noticed (art. 47.), of the apparent distortion of the sun near the horizon, only on a larger scale, and traced up to greater altitudes. This new law once established, it became necessary to modify the expression of that anciently received, by inserting in it a *salvo* for the effect of refraction, or by making a distinction between the *apparent* diurnal orbits, as affected by refraction, and the *true* ones cleared of that effect. This distinction between the *apparent* and the *true*—between the *uncorrected* and *corrected*—between the *rough* and *obvious*, and the *refined* and *ultimate*—is of perpetual occurrence in every part of astronomy.

(143.) Again. The first impression produced by a view of the diurnal movement of the heavens is, that *all* the heavenly bodies perform this revolution in one common period, viz. *a day*, or 24 hours. But no sooner do we come to examine the matter *instrumentally*, i. e. by noting, by timekeepers, their successive arrivals on the meridian, than we find differences which cannot be accounted for by any error of observation. All the *stars*, it is true, occupy the same interval of time between their successive appulses to the meridian, or to any vertical circle; but *this* is a very different one from that occupied by the sun. It is palpably shorter; being, in fact, only 23^h 56' 4·09'', instead of 24 hours, such hours as our common clocks mark. Here, then, we have already *two different days*, a *sidereal* and a *solar;* and if, instead of the sun, we observe the moon, we find a third, much longer than either, — a *lunar* day, whose average duration is 24^h 54^m of our ordinary time, which last is *solar* time, being of necessity conformable to the *sun*'s successive re-appearances, on which all the business of life depends.

(144.) Now, all the *stars* are found to be unanimous in giving the same exact duration of 23^h 56' 4·09'', for the *sidereal day;* which, therefore, we cannot hesitate to receive as the period in which the earth makes one revolution on its axis. We are, therefore, compelled to look on the sun and moon as exceptions to the general law; as having a different nature, or at least a different relation to us, from the stars; and as having motions, real or apparent, of their own, independent of the rotation of the earth on its axis. Thus a great and most important distinction is disclosed to us.

(145.) To establish these facts, almost no apparatus is required. An observer need only station himself to the north of some well-defined vertical object, as the angle of a building, and, placing his eye exactly at a certain fixed point (such as a small hole in a plate of metal nailed to some immoveable support), notice the successive disappearances of any star behind the building, by a watch.* When he observes the sun, he must shade his eye with a dark-coloured or smoked glass, and notice the moments when its western and eastern edges successively come up to the wall, from which, by taking half the interval, he will ascertain (what he cannot directly *observe*) the moment of disappearance of its centre.

(146.) When, in pursuing and establishing this general fact, we are led to attend more nicely to the times of the daily arrival of the sun on the meridian, irregularities (such they first seem to be) begin to make their appearance. The intervals between two successive arrivals are not the same at all times of the year. They are sometimes greater, sometimes less, than 24 hours, as shown by the clock; that is to say, the *solar day* is not always of the same length. About the 21st of December, for example, it is half a minute *longer*, and about the same day of September nearly as much *shorter*, than its *average duration*. And thus a distinction is again pressed upon our notice betwen the *actual* solar day, which is never two days in succession alike, and the *mean solar day* of 24 hours, which is an average of all the solar days throughout the year. Here, then, a new source of inquiry opens to us. The sun's apparent motion is not only not the

* This is an excellent practical method of ascertaining the rate of a clock or watch, being exceedingly accurate if a few precautions are attended to; the chief of which is, to take care that that part of the edge behind which the star (a bright one, *not a planet*) disappears shall be quite smooth; as otherwise variable refraction may transfer the point of disappearance from a protuberance to a notch, and thus vary the moment of observation unduly. This is easily secured, by nailing up a smooth-edged board. The verticality of its edge should be ensured by the use of a plumb-line.

same with that of the stars, but it is not (as the latter is)
uniform. It is subject to fluctuations, whose laws become
matter of investigation. But to pursue these laws, we require nicer means of observation than what we have described, and are obliged to call in to our aid an instrument
called the *transit instrument*, especially destined for such
observations, and to attend minutely to all the causes of
irregularity in the going of clocks and watches which may
affect our reckoning of time. Thus we become involved by
degrees in more and more delicate instrumental inquiries;
and we speedily find that, in proportion as we ascertain the
amount and law of one great or leading fluctuation, or inequality, as it is called, of the sun's diurnal motion, we bring
into view others continually smaller and smaller, which were
before obscured, or mixed up with errors of observation and
instrumental imperfections. In short, we may not inaptly
compare the *mean* length of the solar day to the mean or
average height of water in a harbour, or the general level
of the sea unagitated by tide or waves. The great annual
fluctuation above noticed may be compared to the daily variations of level produced by the tides, which are nothing but
enormous waves extending over the whole ocean, while the
smaller subordinate inequalities may be assimilated to waves
ordinarily so called, on which, when large, we perceive lesser
undulations to ride, and on these, again, minuter ripplings,
to the series of whose subordination we can perceive no end.

(147.) With the causes of these irregularities in the solar
motion we have no concern at present; their explanation belongs to a more advanced part of our subject: but the distinction between the solar and sidereal days, as it pervades
every part of astronomy, requires to be early introduced, and
never lost sight of. It is, as already observed, the *mean* or
average length of the solar day, which is used in the civil
reckoning of time. It commences at midnight, but astronomers, even when they use mean solar time, depart from
the civil reckoning, commencing their day at noon, and
reckoning the hours from 0 round to 24. Thus, 11 o'clock
in the forenoon of the second of January, in the civil reckon-

ing of time, corresponds to January 1 day 23 hours in the astronomical reckoning; and 1 o'clock in the afternoon of the former, to January 2 days 1 hour of the latter reckoning. This usage has its advantages and disadvantages, but the latter seem to preponderate; and it would be well if, in consequence, it could be broken through, and the civil reckoning substituted. *Uniformity in nomenclature and modes of reckoning in all matters relating to time, space, weight, measure, &c., is of such vast and paramount importance in every relation of life as to outweigh every consideration of technical convenience or custom.* *

(148.) Both astronomers and civilians, however, who inhabit different points of the earth's surface, differ from each other in their reckoning of time; as it is obvious they must, if we consider that, when it is noon at one place, it is midnight at a place diametrically opposite; sunrise at another; and sunset, again, at a fourth. Hence arises considerable inconvenience, especially as respects places differing very widely in situation, and which may even in some critical cases involve the mistake of a whole day. To obviate this inconvenience, there has lately been introduced a system of reckoning time by mean solar days and parts of a day counted from a fixed instant, common to all the world, and determined by no *local* circumstance, such as noon or midnight, but by the motion of the sun among the stars. Time, so reckoned, is called equinoctial time; and is numerically the same, at the same instant, in every part of the globe. Its origin will be explained more fully at a more advanced stage of our work.

(149.) Time is an essential element in astronomical observation, in a twofold point of view: — 1st, As the represen-

---

* The only disadvantage to astronomers of using the civil reckoning is this — that their observations being chiefly carried on during the night, the day of their date will, in this reckoning, always have to be changed at midnight, and the former and latter portion of every night's observations will belong to two differently numbered civil days of the month. There is no denying this to be an inconvenience. Habit, however, would alleviate it; and *some* inconveniences must be cheerfully submitted to by all who resolve to act on general principles. All other classes of men, whose occupation extends to the night as well as day, submit to it, and find their advantage in doing so.

tative of angular motion. The earth's diurnal motion being
uniform, every star describes its diurnal circle uniformly ;
and the time elapsing between the passage of the stars in
succession across the meridian of any observer becomes,
therefore, a direct measure of their differences of right ascension. 2dly, As the fundamental element (or natural *independent variable*, to use the language of geometers) in all
dynamical theories. The great object of astronomy is the
determination of the laws of the celestial motions, and their
reference to their proximate or remote causes. Now, the
statement of the *law* of any observed motion in a celestial
object can be no other than a proposition declaring what has
been, is, and will be, the real or apparent situation of that
object *at any time*, past, present, or future. To compare
such laws, therefore, with observation, we must possess a
register of the observed situations of the object in question,
and of the *times when* they were observed.

(150.) The measurement of time is performed by clocks,
chronometers, clepsydras, and hour-glasses. The two former
are alone used in modern astronomy. The hour-glass is a
coarse and rude contrivance for measuring, or rather counting
out, fixed portions of time, and is entirely disused. The
clepsydra, which measured time by the gradual emptying of
a large vessel of water through a determinate orifice, is susceptible of considerable exactness, and was the only dependence of astronomers before the invention of clocks and
watches. At present it is abandoned, owing to the greater
convenience and exactness of the latter instruments. In one
case only has the revival of its use been proposed; viz. for
the accurate measurement of very small portions of time, by
the flowing out of mercury from a small orifice in the bottom
of a vessel, kept constantly full to a fixed height. The stream
is intercepted at the moment of noting any event, and
directed aside into a receiver, into which it continues to run,
till the moment of noting any other event, when the intercepting cause is suddenly removed, the stream flows in its
original course, and ceases to run into the receiver. The
weight of mercury received, compared with the weight re-

ceived in an interval of time observed by the clock, gives the
interval between the events observed. This ingenious and
simple method of resolving, with all possible precision, a problem of much importance in many physical inquiries, is due
to the late Captain Kater.

(151.) The pendulum clock, however, and the balance
watch, with those improvements and refinements in its structure which constitute it emphatically a *chronometer* *, are the
instruments on which the astronomer depends for his knowledge of the lapse of time. These instruments are now
brought to such perfection, that an habitual irregularity in
the *rate* of going, to the extent of a single second in twenty-
four hours in two consecutive days, is not tolerated in one of
good character ; so that any interval of time less than twenty-
four hours may be certainly ascertained within a few tenths
of a second, by their use. In proportion as intervals are
longer, the risk of error, as well as the amount of error
risked, becomes greater, because the accidental errors of many
days may accumulate ; and causes producing a slow progressive change in the rate of going may subsist unperceived. It
is not safe, therefore, to trust the determination of time to
clocks, or watches, for many days in succession, without
checking them, and ascertaining their errors by reference to
natural events which we know to happen, day after day, at
equal intervals. But if this be done, the longest intervals
may be fixed with the same precision as the shortest ; since,
in fact, it is then only the times intervening between the first
and the last moments of such long intervals, and such of those
periodically recurring events adopted for our points of reckoning, as occur within twenty-four hours respectively of either,
that we measure by artificial means. The whole days are
counted out for us by nature ; the fractional parts only, at
either end, are measured by our clocks. To keep the reckoning of the integer days correct, so that none shall be lost or
counted twice, is the object of the calendar. Chronology
marks out the order of succession of events, and refers them

---

* Χρονος, time ; μετρειν, to measure.

to their proper years and days; while chronometry, grounding its determinations on the precise observation of such
regularly periodical events as can be conveniently and exactly
subdivided, enables us to fix the moments in which phenomena
occur, with the last degree of precision.

(152.) In the *culmination* or *transit* (*i. e.* the passage across
the meridian of an observer,) of every star in the heavens, he
is furnished with such a regularly periodical natural event as
we allude to. Accordingly, it is to the *transits* of the
brightest and most conveniently situated fixed stars that
astronomers resort to ascertain their exact time, or, which
comes to the same thing, to determine the exact amount of
error of their clocks.

(153.) Before we describe the instrument destined for the
purpose of observing such culminations, however, or those intended for the measurement of angular intervals in the sphere,
it is requisite to place clearly before the reader the principle
on which the telescope is applied in astronomy to the precise
determination of a direction in space, — that, namely of the
visual ray by which we see a star or any other distant object.

(154.) The telescope most commonly used in astronomy
for these purposes is the refracting telescope, which consists
of an object-glass (either single, or as is now almost universal,
double, forming what is called in optics, an achromatic combination) A ; a tube A B, into which the brass cell of the

object-glass is firmly screwed, and an eye-lens C, for which is
often substituted a combination of glasses designed to increase
the magnifying power of the telescope, or otherwise give
more distinctness of vision according to optical principles
which we have no occasion here to refer to. This also is
fitted into a cell, which is screwed firmly into the end B of
the tube, so that object-glass, tube, and eye-glass may be
considered as forming one piece, invariable in the relative
position of its parts.

(155.) The line P Q joining the centres of the object and eye-glasses and produced, is called the *axis* or *line of collimation* of the telescope. And it is evident, that the situation of this line holds a fixed relation to the tube and its appendages, so long as the object and eye-glasses maintain *their* fixity in this respect.

(156.) Whatever distant object E, this line is directed to, an inverted picture or image of that object F is formed (according to the principles of optics), in the focus of the object-glass, and may there be viewed as if it *were a real object*, through the eye-lens C, which (if of short focus) enables us to *magnify* it just as such a lens would magnify a material object in the same place.

(157.) Now as this image is formed and viewed in the air, being itself immaterial and impalpable — nothing prevents our placing in that very place F in the axis of the telescope, a real, substantial object of very definite form and delicate make, such as a fine metallic point, as of a needle — or better still, a cross formed by two very fine threads (spider-lines), thin metallic wires, or lines drawn on glass intersecting each other at right angles — and whose intersection is all but a mathematical point. If such a point, wire, or cross be carefully placed and firmly fixed in the exact focus F, both of the object and eye-glass, it will be seen through the latter *at the same time*, and occupying *the same precise place* as the image of the distant star E. The magnifying power of the lens renders perceptible the smallest deviation from perfect coincidence, which, should it exist, is a proof, that the axis Q P is not directed rigorously towards E. In that case, a fine motion (by means of a screw duly applied), communicated to the telescope, will be necessary to vary the direction of the axis till the coincidence is rendered perfect. So precise is this mode of pointing found in practice, that the axis of a telescope may be directed towards a star or other definite celestial object without an error of more than a few tenths of a second of angular measure.

(158.) This application of the telescope may be considered as completely annihilating that part of the error of observation which might otherwise arise from an erroneous estimation of the direction in which an object lies from the observer's eye, or from the centre of the instrument. It is, in fact, the grand source of all the precision of modern astronomy, without which all other refinements in instrumental workmanship would be thrown away; the errors capable of being committed in pointing to an object, without such assistance, being far greater than what could arise from any but the very coarsest graduation. * In fact, the telescope thus applied becomes, with respect to angular, what the microscope is with respect to linear dimension. By concentrating attention on its smallest parts, and magnifying into palpable intervals the minutest differences, it enables us not only to scrutinise the form and structure of the objects to which it is pointed, but to refer their apparent places, with all but geometrical precision, to the parts of any scale with which we propose to compare them.

(159.) We now return to our subject, the determination of time by the transits or culminations of celestial objects. The instrument with which such culminations are observed is called a *transit instrument*. It consists of a telescope firmly fastened on a horizontal axis directed to the east and west points of the horizon, or at right angles to the plane of the

---

* The honour of this capital improvement has been successfully vindicated by Derham (Phil. Trans. xxx. 603.) to our young, talented, and unfortunate countryman Gascoigne, from his correspondence with Crabtree and Horrockes, in his (Derham's) possession. The passages cited by Derham from these letters leave no doubt that, so early as 1640, Gascoigne had applied telescopes to his quadrants and sextants, *with threads in the common focus of the glasses ;* and had even carried the invention so far as to illuminate the field of view by artificial light, which he found "*very helpful when the moon appeareth not, or it is not otherwise light enough.*" These inventions were freely communicated by him to Crabtree, and through him to his friend Horrockes, the pride and boast of British astronomy; both of whom expressed their unbounded admiration of this and many other of his delicate and admirable improvements in the art of observation. Gascoigne, however, perished, at the age of twenty-three, at the battle of Marston Moor ; and the premature and sudden death of Horrockes, at a yet earlier age, will account for the temporary oblivion of the invention. It was revived, or re-invented, in 1667, by Picard and Auzout (Lalande, Astron. 2310.), after which its use became universal. Morin, even earlier than Gascoigne (in 1635), had proposed to substitute the telescope for plain sights ; but it is the thread or wire stretched in the focus which the image of a star can be brought to exact coincidence, which gives the telescope its advantage in practice ; and the idea of this does not seem to have occurred to Morin. See Lalande, *ubi suprà*)

meridian of the place of observation. The extremities of the axis are formed into cylindrical pivots of exactly equal diameters, which rest in notches formed in metallic supports, bedded (in the case of large instruments) on strong pieces of stone, and susceptible of nice adjustment by screws, both in a vertical and horizontal direction. By the former adjustment, the axis can be rendered precisely horizontal, by *levelling*  it with a *level* made to rest on the pivots. By the latter adjustment the axis is brought precisely into the east and west direction, the criterion of which is furnished by the observations themselves made with the instrument, in a manner presently to be explained, or by a well-defined object, called a *meridian mark*, originally determined by such observations, and then, for convenience of ready reference, permanently established, at a great distance, exactly in a *meridian line* passing through the central point of the whole instrument. It is evident, from this description, that, if the axis, or line of collimation of the telescope be once well adjusted at right angles to the axis of the transit, it will never quit the plane of the meridian, when the instrument is turned round on its axis of rotation.

(160.) In the focus of the eye-piece, and at right angles to the length of the telescope, is placed, not a single cross, as in our general explanation in art. 157., but a system of one horizontal and several equidistant vertical threads or wires, (five or seven are more usually employed,) as represented in the annexed figure, which always appear in *the field of view*, when properly illuminated, by day by the light of the sky, by night by that of a lamp introduced by a contrivance not necessary here to explain. The place of this system of wires may be altered by adjusting screws, giving it a lateral (horizontal) motion; and it is by this means brought to such a position, that the middle one of the vertical wires shall intersect *the line of collimation* of the telescope, where it is arrested and permanently fastened.* In this situation it is evident that the middle thread will be a visible representation of that portion of the celestial meridian to which the telescope is pointed; and when a star is seen to cross this wire in the telescope, it is in the act of culminating, or passing the celestial meridian. The instant of this event is noted by the clock or chronometer, which forms an indispensable accompaniment of the transit instrument. For greater precision, the moments of its crossing all the vertical threads is noted, and a mean taken, which (since the threads are equidistant) would give exactly the same result, were all the observations perfect, and will, of course, tend to subdivide and destroy their errors in an average of the whole in the contrary case.

(161.) For the mode of executing the adjustments, and allowing for the errors unavoidable in the use of this simple and elegant instrument, the reader must consult works especially devoted to this department of practical astronomy.† We shall here only mention one important verification of its correctness, which consists in *reversing* the ends of the axis, or turning it east for west. If this be done, and it continue to give the same results, and intersect the same point on the meridian mark, we may be sure that the line of collimation of the telescope is truly at right angles to the axis, and describes strictly a plane, *i. e.* marks out in the heavens a *great circle*. In good transit observations, an error of two or three tenths of a second of time in the moment of a star's culmination is the utmost which need be apprehended, exclusive of the error of the clock: in other words, a clock may be compared with the earth's diurnal motion by a single observation, without risk of greater error. By multiplying observations, of course, a yet greater degree of precision may be obtained.

(162.) The plane described by the line of collimation of

---

* There is no way of bringing the *true optic axis* of the object glass to coincide *exactly* with the line of collimation, but, so long as the object glass does not shift or shake in its cell, any line *holding an invariable position* with respect to that *axis*, may be taken for the *conventional* or astronomical axis with equal effect.

† See Dr. Pearson's Treatise on Practical Astronomy. Also Bianchi Sopra lo Stromento de' Passagi. Ephem. di Milano, 1824.

a transit ought to be that of the meridian of the place of observation. To ascertain whether it is so or not, celestial
observation must be resorted to. Now, as the meridian is a
great circle passing through the pole, it necessarily bisects
the diurnal circles described by all the stars, all which describe
the two semicircles so arising in equal intervals of 12 sidereal
hours each. Hence, if we choose a star whose whole diurnal
circle is above the horizon, or which never sets, and observe
the moments of its upper and lower transits across the
middle wire of the telescope, if we find the two semidiurnal
portions east and west of the plane described by the telescope
to be described in *precisely* equal times, we may be sure that
plane is the meridian.

(163.) The angular intervals measured by means of the
transit instrument and clock are arcs of the equinoctial, intercepted between circles of declination passing through the
objects observed; and their measurement, in this case, is performed by no artificial graduation of circles, but by the help
of the earth's diurnal motion, which carries equal arcs of the
equinoctial across the meridian, in equal times, at the rate of
15° per sidereal hour. In all other cases, when we would
measure angular intervals, it is necessary to have recourse to
circles, or portions of circles, constructed of metal or other
firm and durable material, and mechanically subdivided into
equal parts, such as degrees, minutes, &c. The simplest and
most obvious mode in which the measurement of the angular
interval between two directions in space can be performed
is as follows. Let A B C D be a circle, divided into 360
degrees, (numbered in order from any point 0° in the circumference, round to the same point again,) and connected with
its centre by spokes or rays, *x, y, z*, firmly united to its circumference or *limb*. At the centre let a circular hole be pierced,
in which shall move a pivot exactly fitting it, carrying a tube,
whose axis, *a b*, is exactly parallel to the plane of the circle,
or perpendicular to the pivot; and also two arms, *m, n*, at
right angles to it, and forming one piece with the tube and
the axis; so that the motion of the axis on the centre shall
carry the tube and arms smoothly round the circle, to be

arrested and fixed at any point we please, by a contrivance
called a *clamp*. Suppose, now, we would measure the angular interval between two fixed objects, S, T. The plane of

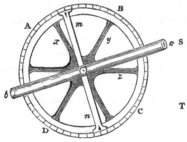

the circle must first be adjusted so as to pass through
them both, and immoveably
fixed and maintained in that
position. This done, let the
axis a b of the tube be
directed to one of them, S,
and *clamped*. Then will a
mark on the arm *m* point
either exactly to some one of the divisions on the limb,
or between two of them adjacent. In the former case,
the division must be noted as *the reading* of the arm *m*. In
the latter, the fractional part of one whole interval between
the consecutive divisions by which the mark on *m surpasses*
the last inferior division must be estimated or measured by
some mechanical or optical means. (See art. 165.) The
division and fractional part thus noted, and reduced into
degrees, minutes, and seconds, is to be set down as the *reading of the limb* corresponding to that position of the *tube a b*,
where it points to the object S. The same must then be
done for the object T; the tube pointed to it, and the *limb*
" *read off*," the position of the circle remaining meanwhile
unaltered. It is manifest, then, that, if the lesser of these
readings be subtracted from the greater, *their difference* will
be the angular interval between S and T, as seen from the
centre of the circle, at whatever point of the limb the commencement of the graduations or the point 0° be situated.

(164.) The very same result will be obtained, if, instead
of making the tube moveable upon the circle, we connect it
invariably with the latter, and make both revolve together
on an axis concentric with the circle, and forming one piece
with it, working in a hollow formed to receive and fit it in
some fixed support. Such a combination is represented in
section in the annexed sketch. T is the tube or sight,
fastened, at *p p*, on the circle A B, whose axis, D, works in

the solid metallic centring E, from which originates an arm, F, carrying at its extremity an index, or other proper mark,

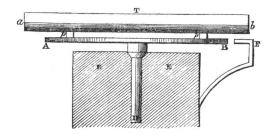

to point out and read off the exact division of the circle at B, the point close to it. It is evident that, as the telescope and circle revolve through any angle, the part of the limb of the latter, which by such revolution is carried past the index F, will measure the angle described. This is the most usual mode of applying divided circles in astronomy.

(165.) The index F may either be a simple pointer, like a clock hand (*fig. a*); or a vernier (*fig. b*); or, lastly, a com-

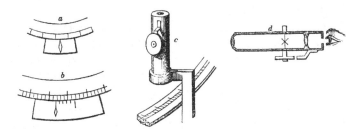

pound microscope (*fig. c*), represented in section in *fig. d*, and furnished with a cross in the common focus of its object and eye-glass, moveable by a fine-threaded screw, by which the intersection of the cross may be brought to exact coincidence with the image of the nearest of the divisions of the circle formed in the focus of the object lens upon the very same principle with that explained, art. 157. for the pointing of the telescope, only that here the fiducial cross is made moveable; and by the turns and parts of a turn of the screw required for this purpose the distance of that division from the original or zero point of the microscope may be estimated. This simple but delicate contrivance gives to the reading off of a circle a degree of accuracy only limited by the power of the microscope, and the perfection with which a screw can be executed, and places the subdivision of angles on the same footing of optical certainty which is introduced into their measurement by the use of the telescope.

(166.) The exactness of the result thus obtained must depend, 1st, on the precision with which the tube $a\ b$ can be pointed to the objects; 2dly, on the accuracy of graduation of the limb; 3dly, on the accuracy with which the subdivision of the intervals between any two consecutive graduations can be performed. The mode of accomplishing the latter object with any required exactness has been explained in the last article. With regard to the graduation of the limb, being merely of a mechanical nature, we shall pass it without remark, further than this, that, in the present state of instrument-making, the amount of error from this source of inaccuracy is reduced within very narrow limits indeed. * With regard to the first, it must be obvious that, if the sights $a\ b$ be nothing more than simple crosses, or pin-holes at the ends of a hollow tube, or an eye-hole at one end, and a cross at the other, no greater nicety in pointing can be expected than what simple vision with the naked eye can command. But if, in place of these simple but coarse contrivances, the tube itself be converted into *a telescope*, having an object-glass at b, an eye-piece at a, and a fiducial cross in their common focus, as explained in art. 157.; and if the motion of the tube on the limb of the circle be arrested when the object is brought just into coincidence with the intersectional point of that cross, it is evident that a greater degree of exactness may be attained in the pointing of the tube than by the unassisted eye, in proportion to the magnifying power and distinctness of the telescope used.

* In the great Ertel circle at Pulkova, the probable amount of the *accidental* error of division is stated by M. Struve not to exceed 0″·264. Desc. de l'Obs. centrale de Pulkova, p. 147.

(167.) The simplest mode in which the measurement of an angular interval can be executed, is what we have just described; but, in strictness, this mode is applicable only to terrestrial angles, such as those occupied on the sensible horizon by the objects which surround our station, — because these only remain stationary during the interval while the telescope is shifted on the limb from one object to the other. But the diurnal motion of the heavens, by destroying this essential condition, renders the direct measurement of an-gular distance from *object* to *object* by this means impossible. The same objection, however, does not apply if we seek only to determine the interval between the *diurnal circles* de-scribed by any two celestial objects. Suppose every star, in its diurnal revolution, were to leave behind it a visible trace in the heavens, — a fine line of light, for instance, — then a telescope once pointed to a star, so as to have its image brought to coincidence with the intersection of the wires, would constantly remain pointed to some portion or other of this line, which would therefore continue to appear in its field as a luminous line, permanently intersecting the same point, till the star came round again. From one such line to another the telescope might be shifted, at leisure, without error; and then the angular interval between the two diurnal circles, *in the plane of the telescope's rotation,* might be mea-sured. Now, though we cannot *see* the path of a star in the heavens, we can *wait* till the star itself crosses the field of view, and seize the moment of its passage to place the inter-section of its wires so that the star shall traverse it; by which, when the telescope is well clamped, we equally well secure the position of its diurnal circle as if we continued to *see* it ever so long. The reading off of the limb may then be performed at leisure; and when another star comes round *into the plane* of the circle, we may unclamp the telescope, and a similar observation will enable us to assign the place of *its* diurnal circle on the limb: and the observations may be repeated alternately, every day, as the stars pass, till we are satisfied with their result.

(168.) This is the principle of the mural circle, which is nothing more than such a circle as we have described in art. 163., firmly supported, in the plane of the meridian, on a long and powerful horizontal axis. This axis is let into a massive pier, or wall, of stone (whence the name of the instrument), and so secured by screws as to be capable of adjustment both in a vertical and horizontal direction; so that, like the axis of the transit, it can be maintained in the exact direction of the east and west points of the horizon, the plane of the circle being consequently truly meridional.

(169.) The meridian, being at right angles to all the diurnal circles described by the stars, its arc intercepted between any two of them will measure the least distance between these circles, and will be equal to the difference of the declinations, as also to the difference of the *meridian altitudes* of the objects — at least when corrected for refraction. These differences, then, are the angular intervals *directly* measured by the mural circle. But from these, supposing the law and amount of refraction known, it is easy to conclude, not their differences only, but the quantities themselves, as we shall now explain.

(170.) The declination of a heavenly body is the complement of its distance from the pole. The pole, being a point in the meridian, might be directly observed on the limb of the circle, if any star stood *exactly* therein; and thence the *polar distances*, and, of course, the declinations of all the rest, might be at once determined. But this not being the case, a bright star as near the pole as can be found is selected, and observed in its *upper* and *lower* culminations; that is, when it passes the meridian *above* and *below* the pole. Now, as its distance from the pole remains the same, the difference of reading off the circle in the two cases is, of course (when corrected for refraction), equal to twice the polar distance of the star; the arc intercepted on the limb of the circle being, in this case, equal to the angular diameter of the star's diurnal circle. In the annexed diagram, H P O represents the celestial meridian, P the pole, B R, A Q, C D the diurnal circles of stars which arrive on the meridian at B, A, and C in their upper and at R, Q, D in their lower culminations, of which D and Q happen above the horizon H O. P is the pole; and if we suppose *h p o* to be the mural circle, having S for its centre, *b a c p d* will be the points on its circumference corresponding to B A C P D in the heavens. Now the arcs *b a*, *b c*, *b d*, and *c d* are given immediately by observation;

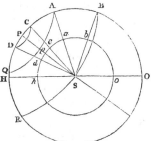

and since C P = P D, we have also *c p* = *p d*, and each of them = $\frac{1}{2}$ *c d*, consequently the place of *the polar point*, as it is called, upon the limb of the circle becomes known, and the arcs *p b*, *p a*, *p c*, which represent on the circle the *polar distances* required, become also known.

(171.) The situation of the pole star, which is a very brilliant one, is eminently favourable for this purpose, being only about a degree and half from the pole; it is, therefore, the star usually and almost solely chosen for this important purpose; the more especially because, both its culminations taking place at great and not very different altitudes, the refractions by which they are affected are of small amount, and differ but slightly from each other, so that their correction is easily and safely applied. The brightness of the pole star, too, allows it to be easily observed in the daytime. In consequence of these peculiarities, this star is one of constant resort with astronomers for the adjustment and verification of instruments of almost every description. In the case of the transit, for instance, it furnishes an excellent object for the application of the method of testing the meridional situation of the instrument described in art. 162., in fact, the most advantageous of any for that purpose, owing to its being the most remote from the zenith, at its upper culmination, of all bright stars observable both above and below the pole.

(172.) The place of the *polar point* on the limb of the mural circle once determined, becomes an origin, or zero point, from which the polar distances of all objects, referred to other points on the same limb, reckon. It matters not whether the actual commencement 0° of the graduations stand there, or not; since it is only by the *differences* of the readings that the arcs on the limb are determined; and hence a great advantage is obtained in the power of commencing anew a fresh series of observations, in which a different part of the circumference of the circle shall be employed, and different graduations brought into use, by which inequalities of division may be detected and neutralized. This is accomplished practically by detaching the telescope from its old bearings on the circle, and fixing it afresh, by screws or clamps, on a different part of the circumference.

(173.) A point on the limb of the mural circle, not less important than the *polar point*, is the *horizontal point*, which, being once known, becomes in like manner an origin, or zero point, from which altitudes are reckoned. The principle of its determination is ultimately nearly the same with that of the polar point. As no star exists in the celestial horizon, the observer must seek to determine two points on the limb, the one of which shall be precisely as far *below* the horizontal point as the other is above it. For this purpose, a star is observed at its culmination on one night, by pointing the telescope directly to it, and the next, by pointing to *the image of the same star reflected* in the still, unruffled surface of a fluid at perfect rest. Mercury, as the most reflective fluid known, is generally chosen for that use. As the surface of a fluid at rest is necessarily horizontal, and as the angle of reflection, by the laws of optics, is equal to that of incidence, this image will be just as much depressed below the horizon as the star itself is above it (allowing for the difference of refraction at the moments of observation). The arc intercepted on the limb of the circle between the star and its reflected image thus consecutively observed, when corrected for refraction, is the double altitude of the star, and its point of bisection the horizontal point. The reflecting surface of a fluid so used for the determination of the altitudes of objects is called an *artificial horizon*.*

(174.) The mural circle is, in fact, at the same time, a transit instrument; and, if furnished with a proper system of vertical wires in the focus of its telescope, may be used as such. As the axis, however, is only supported at one end, it has not the strength and permanence necessary for the more delicate purposes of a transit; nor can it be verified, as a transit may, by the *reversal* of the two ends of its axis, east for west. Nothing, however, prevents a divided circle being permanently fastened on the axis of a transit instrument, either near to one of its extremities, or close to the telescope, so as to revolve with it, the reading off being performed by one or more microscopes fixed on one of its piers. Such an instrument is called a TRANSIT CIRCLE, or a MERIDIAN CIRCLE, and serves for the simultaneous determination of the right ascensions and polar distances of objects observed with it; the time of transit being noted by the clock, and the circle being read off by the lateral microscopes. There is much advantage, when extensive catalogues of small stars have to be formed, in this simultaneous determination of both their celestial co-ordinates: to which may be added the facility of applying to the meridian circle a telescope of any length and optical power. The construction of the mural circle renders this highly inconvenient, and indeed impracticable beyond very moderate limits.

(175.) The determination of the horizontal point on the limb of an instrument is of such essential importance in astronomy, that the student should be made acquainted with every means employed for this purpose. These are, the artificial horizon, the plumb-line, the level, and the collimator. The artificial horizon has been already explained. The plumb-

---

* By a peculiar and delicate manipulation and management of the setting, bisection, and reading off of the circle, aided by the use of a moveable horizontal micrometic wire in the focus of the object-glass, it is found practicable to observe a slow moving star (as the pole star) *on one and the same night*, both by reflection and direct vision, sufficiently near to either culmination to give the horizontal point, without risking the change of refraction in twenty-four hours; so that this source of error is thus completely eliminated.

line is a fine thread or wire, to which is suspended a weight, whose oscillations are impeded and quickly reduced to rest by plunging it in water. The direction ultimately assumed by such a line, *admitting its perfect flexibility,* is that of gravity, or perpendicular to the surface of still water. Its application to the purposes of astronomy is, however, so delicate, and difficult, and liable to error, unless extraordinary precautions are taken in its use, that it is at present almost universally abandoned, for the more convenient, and equally exact instrument *the level.*

(176.) The level is a glass tube nearly filled with a liquid, (spirit of wine, or sulphuric ether, being those now generally

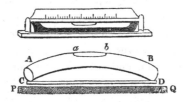

used, on account of their extreme *mobility,* and not being liable to freeze,) the bubble in which, when the tube is placed horizontally, would rest indifferently in any part if *the tube* could be mathematically *straight.* But that being impossible to execute, and every tube having some slight curvature; if the convex side be placed upwards the bubble will occupy the higher part, as in the figure (where the curvature is purposely exaggerated). Suppose such a tube, as A B, firmly fastened on a straight bar, C D, and marked at *a b,* two points distant by the length of the bubble; then, if the instrument be so placed that the bubble shall occupy this interval, it is clear that C D can have no other than one definite inclination to the horizon; because, were it ever so little moved one way or other, the bubble would shift its place, and run towards the elevated side. Suppose, now, that we would ascertain whether any given line P Q be horizontal; let the base of the level C D be set upon it, and note the points *a b,* between which the bubble is exactly contained; then turn the level end for end, so that C shall rest on Q, and D on P. If then the bubble continue to occupy the same place between *a* and *b*, it is evident that P Q can be no otherwise than horizontal. If not, the side towards which the bubble runs is highest, and must be lowered. Astronomical levels are furnished with a divided scale, by which the places of the ends of the bubble can be nicely marked; and it is said that they can be executed with such delicacy, as to indicate a single second of angular deviation from exact horizontality. In such levels accident is not trusted to to give the requisite curvature. They are ground and polished internally by peculiar mechanical processes of great delicacy.

(177.) The mode in which a level may be applied to find the horizontal point on the limb of a vertical divided circle may be thus explained: Let A B be a telescope firmly fixed to such a circle, D E F, and moveable in one with it on a

horizontal axis C, which must be like that of a transit, susceptible of reversal (see art. 161.), and with which the circle is inseparably connected. Direct the telescope on some distant well-defined object S, and bisect it by its horizontal wire, and in this position clamp it fast. Let L be a level fastened at right angles to an arm, L E F, furnished with a micro-

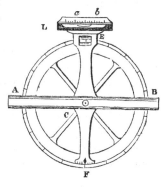

scope, or vernier at F, and, if we please, another at E. Let this arm be fitted by grinding on the axis C, but capable of moving smoothly on it without carrying it round, and also of being clamped fast on it, so as to prevent it from moving until required. While the telescope is kept fixed on the object S, let the level be set so as to bring its bubble to the marks *a b*, and clamp it there. Then will the arm L C F have some certain determinate inclination (no matter what) to the horizon. In this position let the circle be read off at F, and then let the whole apparatus be *reversed* by turning its horizontal axis end for end, *without unclamping the level arm* from the axis. This done, by the motion of the whole instrument (level and all) on its axis, restore the *level* to its horizontal position with the bubble at *a b*. Then we are sure that the telescope has now the same inclination to the horizon *the other way*, that it had when pointed to S, and the reading off at F will not have been changed. Now unclamp the level, and, keeping it nearly horizontal, turn round the circle on the axis, so as to carry back the telescope *through the zenith* to S, and in that position clamp the circle and telescope fast. Then it is evident that an angle equal to twice the zenith distance of S has been moved over by the axis of the telescope from its last position. Lastly, without unclamping the telescope and circle, let the level be once more rectified. Then will the arm L E F once more assume the same definite position with respect to the horizon ; and, consequently, if the circle be again read off, the difference between this and the previous reading must measure the arc of its circumference which has passed under the point F, which may be considered as having all the while retained an invariable position. This difference, then, will be the double zenith distance of S, and its half will be the zenith distance simply, the complement of which is its altitude. Thus the altitude corresponding to a given reading of the limb becomes known, or, in other words, the horizontal point on the limb is ascertained. Circuitous as this process may appear, there is no other mode of employing the level for this purpose which does not in the end come to the same thing. Most commonly, however, the level is used as a mere *fiducial* reference, to preserve a horizontal point once well determined by other means, which is done by adjusting it so as to stand level when the telescope is truly horizontal, and thus leaving it, depending on the permanence of its adjustment.

(178.) The last, but probably not the least exact, as it certainly is, in innumerable cases, the most convenient means of ascertaining the *horizontal point*, is that afforded by the floating collimator, an invention of Captain Kater, but of which the optical principle was first employed by Rittenhouse, in 1785, for the purpose of fixing a definite direction in space by the emergence of parallel rays from a material

object placed in the focus of a fixed lens. This elegant instrument is nothing more than a small telescope furnished
with a cross-wire in its focus, and fastened horizontally, or
as nearly so as may be, on a flat iron *float*, which is made to
swim on mercury, and which, of course, will, when left to
itself, assume always one and the same invariable inclination
to the horizon. If the cross-wires of the collimator be illu-

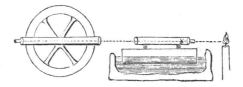

minated by a lamp, being in the focus of its object-glass, the
rays from them will issue parallel, and will therefore be in a
fit state to be brought to a focus by the object-glass of any
other telescope, in which they will form an image *as if they
came from a celestial object in their direction*, i. e. at an altitude equal to their inclination. Thus the intersection of the
cross of the collimator may be observed *as if it were a star*,
and that, however near the two telescopes are to each other.
By transferring then, the collimator *still floating* on a vessel
of mercury from the one side to the other of a circle, we are
furnished with two *quasi-celestial* objects, at precisely equal
altitudes, on opposite sides of the centre; and if these be
observed in succession with the telescope of the circle, bringing its cross to bisect the image of the cross of the collimator
(for which end the wires of the latter cross are purposely set
45° inclined to the horizon), the difference of the readings on
its limb will be twice the zenith distance of either; whence,
as in the last article, the horizontal or zenith point is immediately determined. Another, and, in many respects, preferable form of the floating collimator, in which the telescope is
vertical, and whereby the *zenith* point is directly ascertained,
is described in the Phil. Trans. 1828, p. 257., by the same
author.

(179.) By far the neatest and most delicate application of

the *principle of collimation* of Rittenhouse, however, is suggested by Benzenberg, which affords at once, and by a single
observation, an exact knowledge of the *nadir* point of an
astronomical circle. In this combination, the telescope of the
circle is its own collimator. The
object observed is the central intersectional cross of the wires in its
own focus reflected in mercury.
A strong illumination being thrown
upon the system of wires (art. 160.)
by a lateral lamp, the telescope of
the instrument is directed vertically
downwards towards the surface of
the mercury, as in the figure annexed. The rays diverging from
the wires issue in parallel pencils
from the object-glass, are incident
on the mercury, and are thence reflected back (without losing their
parallel character) to the object-
glass, which is therefore enabled to collect them again in its
focus.   Thus is formed a reflected image of the system of
cross-wires, which, when brought by the slow motion of
the telescope to exact coincidence (intersection upon intersection) with the real system as seen in the eye-piece of the
instrument, indicates the precise and rigorous verticality of
the optical axis of the telescope when directed to the nadir
point.

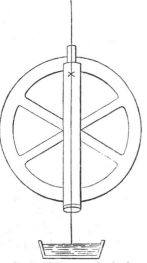

(180.) The transit and mural circle are essentially meridian
instruments, being used only to observe the stars at the moment of their meridian passage.   Independent of this being
the most favourable moment for seeing them, it is that in
which their diurnal motion is parallel to the horizon.   It is
therefore easier at this time than it could be at any other, to
place the telescope exactly in their true direction ; since their
apparent course in the field of view being parallel to the
horizontal thread of the system of wires therein, they may,
by giving a fine motion to the telescope, be brought to exact

coincidence with it, and time may be allowed to examine and correct this coincidence, if not at first accurately hit, which is the case in no other situation. Generally speaking, all angular magnitudes which it is of importance to ascertain exactly, should, if possible, be observed at their maxima or minima of increase or diminution; because at these points they remain not perceptibly changed during a time long enough to complete, and even, in many cases, to repeat and verify, our observations in a careful and leisurely manner. The angle which, in the case before us, is in this predicament, is the altitude of the star, which attains its maximum or minimum on the meridian, and which is measured on the limb of the mural circle.

(181.) The purposes of astronomy, however, require that an observer should possess the means of observing any object not directly on the meridian, but at any point of its diurnal course, or wherever it may present itself in the heavens. Now, a point in the sphere is determined by reference to two great circles at right angles to each other; or of two circles, one of which passes through the pole of the other. These, in the language of geometry, are *co-ordinates* by which its situation is ascertained: for instance, — on the earth, a place is known if we know its longitude and latitude; — in the starry heavens, if we know its right ascension and declination; — in the visible hemisphere, if we know its azimuth and altitude, &c.

(182.) To observe an object at any point of its diurnal course, we must possess the means of directing a telescope to it; which, therefore, must be capable of motion in two planes at right angles to each other; and the amount of its angular motion in each must be measured on two circles *co-ordinate* to each other, whose planes must be parallel to those in which the telescope moves. The practical accomplishment of this condition is effected by making the axis of one of the circles penetrate that of the other at right angles. The pierced axis turns on fixed supports, while the other has no connection with any external support, but is sustained entirely by that which it penetrates, which is strengthened and enlarged at the point of penetration to receive it. The annexed figure exhibits the simplest form of such a combination, though very far indeed from the best in point of mechanism. The two circles are *read off* by verniers, or microscopes; the one attached to the fixed support which carries the principal axis, the other to an arm projecting from that axis. Both circles also are susceptible of being clamped, the clamps being attached to the same ultimate bearing with which the apparatus for reading off is connected.

(183.) It is manifest that such a combination, however its principal axis be pointed (provided that its direction be invariable), will enable us to ascertain the situation of any object with respect to the observer's station, by angles reckoned upon two great circles in the visible hemisphere, one of which has for its poles the prolongations of the principal axis or the vanishing points of a system of lines parallel to it, and the other passes always through these poles: for the former great circle is the vanishing line of all planes parallel to the circle A B, while the latter, in any position of the instrument, is the vanishing line of all the

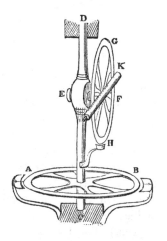

planes parallel to the circle G H; and these two planes being, by the construction of the instrument, at right angles, the great circles, which are their vanishing lines, must be so too. Now, if two great circles of a sphere be at right angles to each other, the one will always pass through the other's poles.

(184.) There are, however, but two positions in which such an apparatus can be mounted so as to be of any practical utility in astronomy. The first is, when the principal axis C D is parallel to the earth's axis, and therefore points to the poles of the heavens which are the vanishing points of all lines in this system of parallels; and when, of course, the plane of the circle A B is parallel to the earth's equator, and therefore has the equinoctial for its vanishing circle, and measures, by its arcs read off, hour angles, or differences of right ascension. In this case, the great circles in the heavens, corresponding to the various positions, which the circle G H can be made to assume, by the rotation of the instrument round its axis C D, are all hour-circles; and the arcs read off on this circle will be declinations, or polar distances, or their differences.

(185.) In this position the apparatus assumes the name of an *equatorial*, or, as it was formerly called, a *parallactic* instrument. It is a most convenient instrument for all such observations as require an object to be kept long in view, because, being once set upon the object, it can be followed as long as we please by a *single motion*, i. e. by merely turning the whole apparatus round on its polar axis. For since, when the telescope is set on a star, the angle between its direction and that of the polar axis is equal to the polar distance of the star, it follows, that when turned about its axis, without altering the position of the telescope on the circle G H, the point to which it is directed will always lie in the small circle of the heavens coincident with the star's diurnal path. In many observations this is an inestimable advantage, and one which belongs to no other instrument. The equatorial is also used for determining the place of an unknown by comparison with that of a known object, in a manner to be described in the fifth chapter. The adjustments of the equatorial are somewhat complicated and difficult. They are best performed in this manner: — 1st, Follow the pole star round its whole diurnal course, by which it will become evident whether the polar axis is directed above or below, to the right or to the left, of the true pole, — and correct it accordingly (without any attempt, during this process, to correct the errors, if any, in the position of the declination axis). 2dly, after the polar axis is thus brought into adjustment, place the plane of the declination circle in or near the meridian; and, having there secured it, observe the transits of several known stars of widely different declinations. If the intervals between these transits correspond to the known

differences of right ascensions of the stars, we may be sure
that the telescope describes a true meridian, and that, therefore, the declination axis is truly perpendicular to the polar
one; — if not, the deviation of the intervals from this law
will indicate the direction and amount of the deviation of the
axis in question, and enable us to correct it. *

(186.) A very great improvement has, within a few years
from the present time, been introduced into the construction
of the equatorial instrument. It consists in applying a clock-
work movement to turn the whole instrument round upon its
polar axis, and so to follow the diurnal motion of any celestial
object, without the necessity of the observer's manual intervention. The driving power is the descent of a weight which
communicates motion to a train of wheelwork, and thus,
ultimately, to the polar axis, while, at the same time, its *too
swift* descent is controlled and regulated to the exact and
uniform rate required to give that axis one turn in 24 hours,
by connecting it with a regulating clock, or (which is found
preferable in practice) by exhausting all the superfluous
energy of the driving power, by causing it to overcome
a regulated friction. Artists have thus succeeded in obtaining a perfectly smooth, uniform, and *regulable* motion, which,
when so applied, serves to retain any object on which the
telescope may be set, commodiously, in the centre of the field
of view for whole hours in succession, leaving the attention
of the observer undistracted by having a mechanical movement to direct, and with both his hands at liberty.

(187.) The other position in which such a compound
apparatus as we have described in art. 182. may be advantageously mounted, is that in which the principal axis
occupies a vertical position, and the one circle, A B, consequently corresponds to the celestial horizon, and the other,
G H, to a vertical circle of the heavens. The angles measured on the former are therefore *azimuths*, or differences

* See Littrow on the Adjustment of the Equatorial (Mem. Ast. Soc. vol. ii.
p. 45.), where formulæ are given for ascertaining the amount and direction of
all the misadjustments simultaneously. But the practical observer, who wishes
to avoid bewildering himself by doing two things at once, had better proceed as
recommended in the text.

of azimuth, and those of the latter zenith distances, or altitudes, according as the graduation commences from the upper point of its limb, or from one 90° distant from it. It is therefore known by the name of an *azimuth and altitude instrument*. The vertical position of its principal axis is secured either by a plumb-line suspended from the upper end, which, however it be turned round, should continue always to intersect one and the same fiducial mark near its lower extremity, or by a level fixed directly across it, whose bubble ought not to shift its place, on moving the instrument in azimuth. The north or south point on the horizontal circle is ascertained by bringing the vertical circle to coincide with the plane of the meridian, by the same criterion by which the azimuthal adjustment of the transit is performed (art. 162.), and noting, in this position, the reading off of the lower circle; or by the following process.

(188.) Let a bright star be observed at a considerable distance to the *east* of the meridian, by bringing it on the cross wires of the telescope. In this position let the horizontal circle be read off, and the telescope securely clamped on the vertical one. When the star has passed the meridian, and is in the descending point of its daily course, let it be followed by moving the whole instrument round to the west, without, however, unclamping the telescope, until it comes into the field of view; and until, by continuing the horizontal motion, the star and the cross of the wires come once more to coincide. In this position it is evident the star must have the same precise altitude above the *western* horizon, that it had at the moment of the first observation above the *eastern*. At this point let the motion be arrested, and the horizontal circle be again read off. The difference of the readings will be the azimuthal arc described in the interval. Now, it is evident that when the altitudes of any star are equal on either side of the meridian, its *azimuths*, whether reckoned both from the north or both from the south point of the horizon, must also be equal, — consequently the north or south point of the horizon must bisect the azimuthal arc thus determined, and will therefore become known.

I

(189.) This method of determining the north and south points of a horizontal circle is called the "method of equal altitudes," and is of great and constant use in practical astronomy. If we note, at the moments of the two observations, the time, by a clock or chronometer, the instant halfway between them will be the moment of the star's meridian passage, which may thus be determined without a transit; and, *vice versâ*, the error of a clock or chronometer may by this process be discovered. For this last purpose, it is not necessary that our instrument should be provided with a horizontal circle at all. Any means by which altitudes can be measured will enable us to determine the moments when the same star arrives at *equal* altitudes in the eastern and western halves of its diurnal course; and, these once known, the instant of meridian passage and the error of the clock become also known.

(190.) Thus also a meridian line may be drawn and a *meridian mark* erected. For the readings of the north and south points on the limb of the horizontal circle being known, the vertical circle may be brought exactly into the plane of the meridian, by setting it to that precise reading. This done, let the telescope be depressed to the north horizon, and let the point intersected there by its cross-wires be noted, and a mark erected there, and let the same be done for the south horizon. The line joining these points is a meridian line, passing through the centre of the horizontal circle. The marks may be made secure and permanent if required.

(191.) One of the chief purposes to which the altitude and azimuth circle is applicable is the investigation of the amount and laws of refraction. For, by following with it a circumpolar star which passes the zenith, and another which grazes the horizon, through their whole diurnal course, the exact *apparent* form of their diurnal orbits, or the ovals into which their circles are distorted by refraction, can be traced; and their deviation from circles, being at every moment given by the nature of the observation *in the direction in which the refraction itself takes place* (i. e. in altitude), is made a matter of direct observation.

(192.) The *zenith sector* and the *theodolite* are peculiar modifications of the altitude and azimuth instrument. The former is adapted for the very exact observation of stars in or near the zenith, by giving a great length to the vertical axis, and suppressing all the circumference of the vertical circle, except a few degrees of its lower part, by which a great length of radius, and a consequent proportional enlargement of the divisions of its arc, is obtained. The latter is especially devoted to the measures of horizontal angles between terrestrial objects, in which the telescope never requires to be elevated more than a few degrees, and in which, therefore, the vertical circle is either dispensed with, or executed on a smaller scale, and with less delicacy; while, on the other hand, great care is bestowed on securing the exact perpendicularity of the plane of the telescope's motion, by resting its horizontal axis on two supports like the piers of a transit-instrument, which themselves are firmly bedded on the spokes of the horizontal circle, and turn with it.

(193.) The next instrument we shall describe is one by whose aid the angular distance of any two objects may be measured, or the altitude of a single one determined, either by measuring its distance from the visible horizon (such as the sea-offing, allowing for its dip), or from its own reflection on the surface of mercury. It is the sextant, or quadrant, commonly called *Hadley's*, from its reputed inventor, though the priority of invention belongs undoubtedly to Newton, whose claims to the gratitude of the navigator are thus doubled, by his having furnished at once the only theory by which his vessel can be securely guided, and the only instrument which has ever been found to avail, in applying that theory to its nautical uses.*

(194.) The principle of this instrument is the optical property of reflected rays, thus announced: — " The angle be-

---

* Newton communicated it to Dr. Halley, who suppressed it. The description of the instrument was found, after the death of Halley, among his papers, in Newton's own handwriting, by his executor, who communicated the papers to the Royal Society, twenty-five years after Newton's death, and eleven after the publication of Hadley's invention, which might be, and probably was, independent of any knowledge of Newton's, though Hutton insinuates the contrary.

tween the first and last directions of a ray which has suffered
two reflections in one plane is equal to twice the inclination
of the reflecting surfaces to each other." Let A B be the
limb, or graduated arc, of a por-
tion of a circle 60° in extent,
but divided into 120 equal parts.
On the radius C B let a silvered plane glass D be fixed, at
right angles to the plane of the
circle, and on the moveable radius C E let another such silvered glass, C, be fixed. The
glass D is permanently fixed

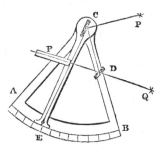

parallel to A C, and only one half of it is silvered, the other
half allowing objects to be seen through it. The glass C is
wholly silvered, and its plane is parallel to the length of the
moveable radius C E, at the extremity E of which a vernier
is placed to read off the divisions of the limb. On the radius
A C is set a telescope F, through which any object, Q, may
be seen by *direct* rays which pass through the unsilvered portion of the glass D, while another object, P, is seen through
the same telescope by rays, which, after reflection at C, have
been thrown upon the silvered part of D, and are thence
directed by a second reflection into the telescope. The two
images so formed will both be seen in the field of view at
once, and by moving the radius C E will (if the reflectors be
truly perpendicular to the plane of the circle) meet and pass
over, without obliterating each other. The motion, however,
is arrested when they meet, and at this point the angle included between the direction C P of one object, and F Q of
the other, is twice the angle E C A included between the
fixed and moveable radii C A, C E. Now, the graduations
of the limb being purposely made only half as distant as would
correspond to degrees, the arc A E, when read off, as if the
graduations were *whole* degrees, will, in fact, read double its
real amount, and therefore the numbers so read off will express not the angle E C A, but its double, the angle subtended by the objects.

(195.) To determine the exact distances between the stars by direct observation is comparatively of little service; but in nautical astronomy the measurement of their distances from the moon, and of their altitudes, is of essential import- ance; and as the sextant requires no fixed support, but can be held in the hand, and used on ship-board, the utility of the instrument becomes at once obvious. For altitudes at sea, as no level, plumb-line, or artificial horizon can be used, the sea-offing affords the only resource; and the image of the star observed, seen by reflection, is brought to coin- cide with the boundary of the sea seen by direct rays. Thus the altitude above the sea-line is found; and this corrected for the *dip of the horizon* (art. 23.) gives the true altitude of the star. On land, an artificial horizon may be used (art. 173.), and the consideration of dip is rendered unnecessary.

(196.) The adjustments of the sextant are simple. They consist in fixing the two reflectors, the one on the revolving radius C E, the other on the fixed one C B, so as to have their planes perpendicular to the plane of the circle, and parallel to each other, when the reading of the instrument is zero. This adjustment in the latter respect is of little moment, as its effect is to produce a *constant* error, whose amount is readily ascertained by bringing the two images of one and the same star or other distant object to coincidence; when the instru- ment *ought* to read zero, and if it does not, the angle which it *does* read is the zero correction and must be subtracted from *all* angles measured with the sextant. The former ad- justments are essential to be maintained, and are performed by small screws, by whose aid either or both the glasses may be *tilted* a little one way or another until the direct and re- flected images of a vertical *line* (a plumb-line) can be brought to coincidence over *their whole extent*, so as to form a single unbroken straight line, whatever be the position of the move- able arm, in the middle of the field of view of the telescope, whose axis is carefully adjusted by the optician to parallelism with the plane of the limb. In practice it is usual to leave only the reflector C on the moveable radius adjustable, that on the fixed being set to great nicety by the maker. In this case the best way of making the adjustment is to view a pair of lines crossing each other at right angles (one being horizontal the other vertical) through the telescope of the instrument, holding the plane of its limb vertical, — then having brought the horizontal line and its reflected image to coincidence by the motion of the radius, the two images of the vertical arm must be brought to coincidence by tilting one way or other the fixed reflector D by means of an adjusting screw, with which every sextant is provided for that purpose. When both lines coincide *in the centre of the field* the adjustment is correct.

(197.) The reflecting circle is an instrument destined for the same uses as the sextant, but more complete, the circle being entire, and the divisions carried all round. It is usually furnished with three verniers, so as to admit of three distinct readings off, by the average of which the error of graduation and of reading is reduced. This is altogether a very refined and elegant instrument.

(198.) We must not conclude this part of our subject without mention of the "principle of repetition;" an invention of Borda, by which the error of graduation may be diminished to any degree, and, practically speaking, annihilated. Let P Q be two objects which we may suppose fixed, for purposes of mere explanation, and let K L be a telescope moveable on O, the common axis of two circles, A M L and *a b c*, of which

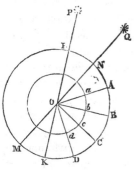

the former, A M L, is absolutely fixed in the plane of the objects, and carries the graduations, and the latter is freely moveable on the axis. The telescope is attached permanently to the latter circle, and moves with it. An arm O *a* A carries the index, or vernier, which reads off the graduated limb of the fixed circle. This arm is provided with two clamps, by which it can be temporarily connected with either circle, and detached at pleasure. Suppose, now, the telescope directed to P. Clamp the index arm O A to the *inner* circle, and unclamp it from the outer, and read off. Then carry the telescope round to the other object Q. In so doing, the inner circle, and the index-arm which is clamped to it, will also be carried round, over an arc A B, on the graduated limb of the outer, equal to the angle P O Q. Now clamp the index to the outer circle, and unclamp the inner, and read off: the difference of readings will of course measure the angle P O Q; but the result will be liable to two sources of error — that of *graduation* and that of observation, both which it is our object to get rid of. To this end transfer the telescope back to P, *without* unclamping the arm from the outer circle; *then*, having made the bisection of P, clamp the arm to *b*, and unclamp it from B, and again transfer the telescope to Q, by which the arm will now be carried with it to C, over a second arc, B C, equal to the angle P O Q. Now again read off; then will the difference between this reading and the *original* one measure *twice* the angle P O Q, affected with *both* errors of observation, but only with *the same error of graduation as before*. Let this process be repeated as often as we please (suppose ten times); then will the final arc A B C D read off on the circle be ten times the required angle, affected by the joint errors of all the ten observations, but only by the same constant error of graduation, which depends on the initial and final readings off alone. Now the errors of observation, when numerous, tend to balance and destroy one another; so that, if sufficiently multiplied, their influence will disappear from the result. There remains, then, only the constant error of graduation, which comes to be divided in the final result by the number of observations, and is therefore diminished in its influence to one tenth of its possible amount, or to less if need be. The abstract beauty and advantage of this principle seem to be counterbalanced in practice by some unknown cause, which, probably, must be sought for in imperfect clamping.

(199.) Micrometers are instruments (as the name imports*) for measuring, with great precision, small angles, not exceeding a few minutes, or at most a whole degree. They are very various in construction and principle, nearly all, however, depending on the exceeding delicacy with which space can be subdivided by the turns and parts of a turn of fine screws. Thus — in the *parallel wire micrometer,* two parallel threads (spider's lines are generally used) stretched on sliding frames, one or both moveable by screws in a direction perpendicular to that of the threads, are placed in the common focus of the object and eye-glasses of a telescope, and brought by the motion of the screws

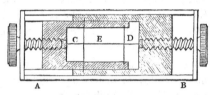

exactly to cover the two extremities of the image of any small object seen in the telescope, as the diameter of a planet, &c., the angular distance between which it is required to measure. This done, the threads are closed up by turning one of the screws till they exactly cover each other, and the number of turns and parts of a turn required gives the interval of the threads, which must be converted into angular measure, either by actual calculation from the linear measure of the threads of the screw and the focal length of the object-glass, or experimentally, by measuring the image of a known object placed at a known distance (as a foot-rule at a hundred yards, &c.) and therefore subtending a known angle.

(200.) The *duplication of the image* of an object by optical means furnishes a valuable and fertile resource in micrometry. Suppose by any optical contrivance the single image A of any object can be converted into two, exactly equal and similar, A B, at a distance from one another, dependent (by

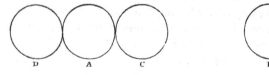

* Μικρος, small; μετρειν, to measure.

some mechanical movement) on the will of the observer, and in any required direction from one another. As these can, therefore, be made to approach to or recede from each other at pleasure, they may be brought in the first place to approach till they touch one another on one side, as at A C, and then being made by continuing the motion to cross and touch on the opposite side, as A D, it is evident that the quantity of movement required to produce the change from one contact to the other, *if uniform*, will *measure* the double diameter of the object A.

(201.) Innumerable optical combinations may be devised to operate such duplication. The chief and most important (from its recent applications), is the *heliometer*, in which the image is divided by bisecting *the object-glass of the telescope*, and making its two halves, set in separate brass frames, slide laterally on each other, as A B, the motion being produced and measured by a screw. Each half, by the laws of optics, forms its own image (somewhat blurred, it is true, by diffraction *), in its own axis ; and thus two equal and similar images are formed side by side in the focus of the eye-piece, which may be made to approach and recede by the motion of the screw, and thus afford the means of measurement as above described.

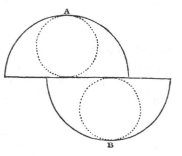

(202.) Double refraction through crystallized media affords another means of accomplishing the same end. Without going into the intricacies of this difficult branch of optics, it will suffice to state that objects viewed through certain crystals (as Iceland spar, or quartz) appear double, two images equally distinct being formed, whose angular distance from each other varies from nothing (or perfect coincidence), up to

* This might be cured, though at an expense of light, by limiting each half to a circular space by diaphragms, as represented by the dotted lines.

a certain limit, *according to the direction with respect to a certain fixed line in the crystal,* called its optical axis. Suppose, then, to take the simplest case, that the eye-lens of a telescope, instead of glass, were formed of such a crystal (say of quartz, which may be worked as well or better than glass), and *of a spherical form,* so as to offer no difference when turned about on its centre, other than the inclination of its optical axis to the visual ray. Then when that axis coincides with the line of collimation of the object-glass, one image only will be seen, but when made to revolve on an axis perpendicular to that line, two will arise, opening gradually out from each other, and thus originating the desired duplication. In this contrivance, the angular amount of the rotation of the sphere affords the necessary datum for determining the separation of the images.

(203.) Of all methods which have been proposed, however, the simplest and most unobjectionable would appear to be the following. It is well known to every optical student, that two prisms of glass, a flint and a crown, may be opposed to each other, so as to produce a *colourless deflection* of parallel rays. An object seen through such a compound or achromatic prism, will be seen simply deviated in direction, but in no way otherwise altered or distorted. Let such a prism be constructed with its surfaces so nearly parallel that the *total deviation* produced in traversing them shall not exceed a small amount (say 5′). Let this be cut in half, and from each half let a circular disc be formed, and cemented on a circular plate of parallel glass, or otherwise sustained, close to and concentric with the other by a framework of metal so light as to intercept but a small portion of the light which passes on the outside (as in the annexed figure), where the dotted lines represent the radii sustaining one, and the un-

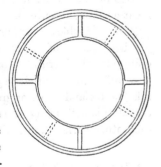

dotted those carrying the other disc. The whole must be
so mounted as to allow one disc to revolve in its own plane
behind the other, fixed, and to allow the amount of rotation
to be read off. It is evident, then, that when the deviations
produced by the two discs conspire, a total deviation of 10′
will be effected on all the light which has passed through
them; that when they oppose each other, the rays will
emerge undeviated, and that in intermediate positions a deviation varying from 0 to 10′, and calculable from the angular rotation of the one disc on the other, will arise. Now,
let this combination be applied at such a point of the cone
of rays, between the object-glass and its focus, that the discs
shall occupy exactly half the area of its section. Then will
half the light of the object lens pass undeviated—the other
half deviated, as above described; and thus a duplication of
image, variable and measureable (as required for micrometric
measurement) will occur. If the object-glass be not very
large, the most convenient point of its application will be externally before it, in which case the diameter of the discs
will be to that of the object-glass as 707 : 1000; or (allowing for the spokes) about as 7 to 10.

(204.) The *Position Micrometer* is simply a straight thread
or wire, which is carried round by a smooth revolving motion,
in the common focus of the object and eye-glasses, in a plane
perpendicular to the axis of the telescope. It serves to determine the situation with respect to some fixed line in the
field of view, of the line joining any two objects or points of
an object seen in that field — as two stars, for instance, near
enough to be seen at once. For this purpose the moveable
thread is placed so as to cover both of them, or stand, as may
best be judged, parallel to their line of junction. And its
angle, with the fixed one, is then read off upon a small
divided circle exterior to the instrument. When such a
micrometer is applied (as it most commonly is) to an equatorially mounted telescope, the *zero* of its position corresponds
to a direction of the wire, such as, prolonged, will represent
a circle of declination in the heavens — and the "angles
of position" so read off are reckoned invariably from one

point, and in one direction, viz., *north, following, south, preceding;* so that 0° position corresponds to the situation of an object exactly *north* of that assumed as a centre of reference, — 90° to a situation exactly *eastward* or *following;* 180° exactly *south;* and 270° exactly *west,* or *preceding* in the order of diurnal movement.

CHAPTER IV.

OF GEOGRAPHY.

OF THE FIGURE OF THE EARTH. — ITS EXACT DIMENSIONS. — ITS
FORM THAT OF EQUILIBRIUM MODIFIED BY CENTRIFUGAL FORCE.
— VARIATION OF GRAVITY ON ITS SURFACE. — STATICAL AND
DYNAMICAL MEASURES OF GRAVITY. — THE PENDULUM. — GRAVITY
TO A SPHEROID. — OTHER EFFECTS OF THE EARTH'S ROTATION.
— TRADE WINDS. — DETERMINATION OF GEOGRAPHICAL POSITIONS.
— OF LATITUDES. — OF LONGITUDES. — CONDUCT OF A TRIGO-
NOMETRICAL SURVEY.—OF MAPS. — PROJECTIONS OF THE SPHERE.
—MEASUREMENT OF HEIGHTS BY THE BAROMETER.

(205.) GEOGRAPHY is not only the most important of the
practical branches of knowledge to which astronomy is
applied, but it is also, theoretically speaking, an essential
part of the latter science. The earth being the general station
from which we view the heavens, a knowledge of the local
situation of particular stations on its surface is of great consequence, when we come to inquire the distances of the nearer
heavenly bodies from us, as concluded from observations of
their parallax as well as on all other occasions, where a
difference of locality can be supposed to influence astronomical
results. We propose, therefore, in this chapter, to explain
the principles, by which astronomical observation is applied
to geographical determinations, and to give at the same time
an outline of geography so far as it is to be considered a part
of astronomy.

(206.) Geography, as the word imports, is a delineation or
description of the earth. In its widest sense, this comprehends not only the delineation of the form of its continents
and seas, its rivers and mountains, but their physical condition,
climates, and products, and their appropriation by communities of men. With physical and political geography, however,
we have no concern here. Astronomical geography has for

its objects the exact knowledge of the form and dimensions
of the earth, the parts of its surface occupied by sea and land,
and the configuration of the surface of the latter, regarded as
protuberant above the ocean, and broken into the various
forms of mountain, table land, and valley; neither should the
form of the bed of the ocean, regarded as a continuation of
the surface of the land beneath the water, be left out of consideration: we know, it is true, very little of it; but this is
an ignorance rather to be lamented, and, if possible, remedied,
than acquiesced in, inasmuch as there are many very important branches of inquiry which would be greatly advanced
by a better acquaintance with it.

(207.) With regard to the figure of the earth *as a whole*,
we have already shown that, speaking loosely, it may be
regarded as spherical; but the reader who has duly appreciated
the remarks in art. 22. will not be at a loss to perceive that
this result, concluded from observations not susceptible of
much exactness, and embracing very small portions of the
surface at once, can only be regarded as a first approximation,
and may require to be materially modified by entering into
minutiæ before neglected, or by increasing the delicacy of our
observations, or by including in their extent larger areas of
its surface. For instance, if it should turn out (as it will),
on minuter inquiry, that the true figure is somewhat elliptical, or flattened, in the manner of an orange, having the
diameter which coincides with the axis about $\frac{1}{300}$th part shorter
than the diameter of its equatorial circle; — this is so trifling
a deviation from the spherical form that, if a model of such
proportions were turned in wood, and laid before us on a
table, the nicest eye or hand would not detect the flattening,
since the difference of diameters, in a globe of fifteen inches,
would amount only to $\frac{1}{20}$th of an inch. In all common
parlance, and for all ordinary purposes, then, it would still be
called a globe; while, nevertheless, by careful measurement,
the difference would not fail to be noticed; and, speaking
strictly, it would be termed, not a globe, but an oblate
ellipsoid, or spheroid, which is the name appropriated by
geometers to the form above described.

(208.) The sections of such a figure by a plane are not circles, but ellipses; so that, on such a shaped earth, the horizon of a spectator would nowhere (except at the poles) be exactly circular, but somewhat elliptical. It is easy to demonstrate, however, that its deviation from the circular form, arising from so very slight an *"ellipticity"* as above supposed, would be quite imperceptible, not only to our eye-sight, but to the test of the dip-sector; so that by that mode of observation we should never be led to notice so small a deviation from perfect sphericity. How we are led to this conclusion, as a practical result, will appear, when we have explained the means of determining with accuracy the dimensions of the whole, or any part of the earth.

(209.) As we cannot grasp the earth, nor recede from it far enough to view it at once as a whole, and compare it with a known standard of measure in any degree commensurate to its own size, but can only creep about upon it, and apply our diminutive measures to comparatively small parts of its vast surface in succession, it becomes necessary to supply, by geometrical reasoning, the defect of our physical powers, and from a delicate and careful measurement of such small parts to conclude the form and dimensions of the whole mass. This would present little difficulty, if we were sure the earth were strictly a sphere, for the proportion of the circumference of a circle to its diameter being known (viz. that of 3·1415926 to 1·0000000), we have only to ascertain the length of the entire circumference of any great circle, such as a meridian, in miles, feet, or any other standard units, to know the diameter in units of the same kind. Now, the circumference of the whole circle is known as soon as we know the exact length of any aliquot part of it, such as 1° or $\frac{1}{360}$th part; and this, being not more than about seventy miles in length, is not beyond the limits of very exact measurement, and could, in fact, be measured (if we knew its exact termination at each extremity) within a very few feet, or, indeed, inches, by methods presently to be particularized.

(210.) Supposing, then, we were to begin measuring with all due nicety from any station, in the exact direction of a meridian, and go measuring on, till by some indication we were informed that we had accomplished an exact *degree* from the point we set out from, our problem would then be at once resolved. It only remains, therefore, to inquire by what indications we can be sure, 1st, that we *have* advanced *an exact degree;* and, 2dly, that we have been measuring in the *exact direction of a great circle.*

(211.) Now, the earth has no landmarks on it to indicate degrees, nor traces inscribed on its surface to guide us in such a course. The compass, though it affords a tolerable guide to the mariner or the traveller, is far too uncertain in its indications, and too little known in its laws, to be of any use in *such* an operation. We must, therefore, look outwards, and refer our situation on the surface of our globe to natural marks, *external* to it, and which are of equal permanence and stability with the earth itself. Such marks are afforded by the stars. By observations of their meridian altitudes, performed at any station, and from their known polar distances, we conclude the height of the pole; and since the altitude of the pole is equal to the latitude of the place (art. 119.), the same observations give the latitudes of any stations where we may establish the requisite instruments. When our latitude, then, is found to have diminished a degree, we know that, *provided we have kept to the meridian*, we have described one three hundred and sixtieth part of the earth's circumference.

(212.) The direction of the meridian may be secured at every instant by the observations described in art. 162. 188.; and although local difficulties may oblige us to deviate in our measurement from this exact direction, yet if we keep a strict account of the amount of this deviation, a very simple calculation will enable us to *reduce* our observed measure to its *meridional* value.

(213.) Such is the principle of that most important geographical operation, the measurement of an arc of the meridian. In its detail, however, a somewhat modified course must be followed. An observatory cannot be mounted and dismounted at every step; so that we cannot identify and measure an exact degree *neither more nor less.* But this is of no consequence, provided we know with equal precision *how much*, more or less, we have measured. In place, then, of measuring this precise aliquot part, we take the more convenient method of measuring from one good observing station to another, *about* a degree, or two or three degrees, as the case may be, or indeed any determinate angular interval apart, and determining by astronomical observation the precise difference of latitudes between the stations.

(214.) Again, it is of great consequence to avoid in this operation every source of uncertainty, because an error committed in the length of a single degree will be multiplied 360 times in the circumference, and nearly 115 times in the diameter of the earth concluded from it. Any error which may affect the astronomical determination of a star's altitude will be especially influential. Now, there is still too much uncertainty and fluctuation in the amount of refraction at moderate altitudes, not to make it especially desirable to avoid this source of error. To effect this, we take care to select for observation, at the extreme stations, some star which passes through or near the zeniths of both. The amount of refraction, within a few degrees of the zenith, is very small, and its fluctuations and uncertainty, in point of quantity, so excessively minute as to be utterly inappretiable. Now, it is the same thing whether we observe the *pole* to be raised or depressed a degree, or the *zenith distance* of a star when on the meridian to have changed by the same quantity (fig. art. 128.). If at one station we observe any star to pass through the zenith, and at the other to pass one degree south or north of the zenith, we are sure that the geographical latitudes, or the altitudes of the pole at the two stations, must differ by the same amount.

(215.) Granting that the terminal points of one degree can be ascertained, its *length* may be measured by the methods which will be presently described, as we have before remarked, to within a very few feet. Now, the error which may be committed in fixing each of these terminal points cannot exceed that which may be committed in the observation of the zenith distance of a star properly situated for the purpose in question.   This error, with proper care, can hardly exceed half a second.   Supposing we grant the possibility of ten feet of error in the length of each degree in a measured arc of five degrees, and of half a second in each of the zenith distances of one star, observed at the northern and southern stations, and, lastly, suppose all these errors to conspire, so as to tend all of them to give a result greater, or all less, than the truth, it will appear, by a very easy proportion, that the whole amount of error which would be thus entailed on an estimate of the earth's diameter, as concluded from such a measure, would not exceed 1147 yards, or about two thirds of a mile, and this is ample allowance.

(216.) This, however, supposes that the form of the earth is that of a perfect sphere, and, in consequence, the lengths of its degrees in all parts precisely equal.   But, when we come to compare the measures of meridional arcs made in various parts of the globe, the results obtained, although they agree sufficiently to show that the supposition of a spherical figure is not *very* remote from the truth, yet exhibit discordances far greater than what we have shown to be attributable to error of observation, and which render it evident that the hypothesis, in strictness of its wording, is untenable.   The following table exhibits the lengths of a degree of the meridian (astronomically determined as above described), expressed in British standard feet, as resulting from actual measurement made with all possible care and precision, by commissioners of various nations, men of the first eminence, supplied by their respective governments with the best instruments, and furnished with every facility which could tend to ensure a successful result of their important labours.

Country.	Latitude of Middle of Arc.	Arc measured.	Measured Length in Feet.	Mean Length of the Degree at the Middle Latitude in Feet.
Sweden*, B - -	+66° 20′ 10″·0	1° 37′ 19″·6	593277	365744
Sweden, A - -	+ 66 19 37	0 57 30·4	351832	365782
Russia, A - -	+ 58 17 37	3 35 5·2	1309742	365368
Russia, B - -	+ 56 3 55·5	8 2 28·9	2937439	365291
Prussia, B - -	+ 54 58 26·0	1 30 29·0	551073	365420
Denmark, B - -	+ 54 8 13·7	1 31 53·3	559121	365087
Hanover, A B - -	+ 52 32 16·6	2 0 57·4	736425	365300
England, A - -	+ 52 35 45	3 57 13·1	1442953	364971
England, B - -	+ 52 2 19·4	2 50 23·5	1036409	364951
France, A - -	+ 46 52 2	8 20 0·3	3040605	364872
France, A B - -	+ 44 51 2·5	12 22 12·7	4509832	364572
Rome, A - -	+ 42 59 —	2 9 47	787919	364262
America, A - -	+ 39 12 —	1 28 45·0	538100	363786
India, A B - -	+ 16 8 21·5	15 57 40·7	5794598	363044
India, A B - -	+ 12 32 20·8	1 34 56·4	574318	362956
Peru, A B - -	— 1 31 0·4	3 7 3·5	1131050	363626
Cape of Good Hope, A	— 33 18 30	1 13 17·5	445506	364713
Cape of Good Hope, B	— 35 43 20·0	3 34 34·7	1301993	364060

It is evident from a mere inspection of the second and fifth columns of this table, that *the measured length of a degree increases with the latitude,* being greatest near the poles, and least near the equator. Let us now consider what interpretation is to be put upon this conclusion, as regards the form of the earth.

(217.) Suppose we held in our hands a model of the earth smoothly turned in wood, it would be, as already observed, so nearly spherical, that neither by the eye nor the touch, unassisted by instruments, could we detect any deviation from that form. Suppose, too, we were debarred from measuring directly across from surface to surface in different directions

* The astronomers by whom these measurements were executed were as follows : —

Sweden, A B — Svanberg.
Sweden, A — Maupertuis.
Russia, A — Struve.
Russia, B — Struve, Tenner.
Prussia — Bessel, Bayer.
Denmark — Schumacher.
Hanover — Gauss.
England — Ray, Kater.
France, A — Lacaille, Cassini.

France, A B — Delambre, Mechain.
Rome — Boscovich.
America — Mason and Dixon.
India, 1st — Lambton.
India, 2d — Lambton, Everest.
Peru — Lacondamine, Bouguer.
Cape of Good Hope, A — Lacaille.
Cape of Good Hope, B — Maclear.
— *Astr. Nachr.* 574.

with any instrument, by which we might at once ascertain
whether one diameter were longer than another; how, then,
we may ask, are we to ascertain whether it is a true sphere or
not? It is clear that we have no resource, but to endeavour
to discover, by some nicer means than simple inspection or
feeling, whether the convexity of its surface is the same in
every part; and if not, where it is greatest, and where least.

Suppose, then, a thin plate of
metal to be cut into a concavity at its edge, so as exactly to fit the surface at A:
let this now be removed from
A, and applied successively
to several other parts of the
surface, taking care to keep its
plane always on a great circle

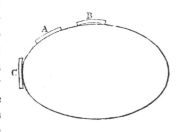

of the globe, as here represented. If, then, we find any
position, B, in which the light can enter in the middle between the globe and plate, or any other, C, where the latter
tilts by pressure, or admits the light under its edges, we
are sure that the *curvature* of the surface at B is less, and
at C greater, than at A.

(218.) What we here do by the application of a metal plate
of determinate length and curvature, we do on the earth by
the measurement of a degree of variation in the altitude of
the pole. Curvature of a surface is nothing but the continual
deflection of its tangent from one fixed direction as we advance along it. When, in the *same measured distance of
advance* we find the tangent (which answers to our horizon)
to have shifted its position with respect to a fixed direction
in space, (such as the axis of the heavens, or the line joining
the earth's centre and some given star,) *more* in one part of
the earth's meridian than in another, we conclude, of necessity,
that the curvature of the surface at the former spot is greater
than at the latter; and *vice versâ*, when, in order to produce
the same change of horizon with respect to the pole (suppose 1°) we require to travel over a *longer* measured space at
one point than at another, we assign to that point a less cur-

vature. Hence we conclude that *the curvature of a meridional section of the earth is sensibly greater at the equator than towards the poles;* or, in other words, that the earth is not spherical, but *flattened* at the poles, or, which comes to the same, protuberant at the equator.

(219.) Let N A B D E F represent a meridional section of the earth, C its centre, and N A, B D, G E, arcs of a meridian,

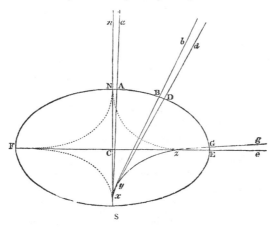

each corresponding to one degree of difference of latitude, or to one degree of variation in the meridian altitude of a star, as referred to the horizon of a spectator travelling along the meridian. Let *n* N, *a* A, *b* B, *d* D, *g* G, *e* E, be the respective directions of the *plumb-line* at the stations N, A, B, D, G, E, of which we will suppose N to be at the pole and E at the equator; then will the tangents to the surface at these points respectively be perpendicular to these directions; and, consequently, if each pair, viz. *n* N and *a* A, *b* B and *d* D, *g* G and *e* E, be prolonged till they intersect each other (at the points x, y, z), the angles N x A, B y D, G z E, will each be one degree, and, therefore, all equal; so that the small curvilinear arcs N A, B D, G E, may be regarded as arcs of circles of one degree each, described about x, y, z, as centres. These are what in geometry are called *centres of curvature,* and the radii x N or x A, y B or y D, z G or z E, represent *radii of curvature,* by which the curvatures at those points are determined and measured. Now, as the arcs of different circles, which subtend equal angles at their respective centres, are in the direct proportion of their radii, and as the arc N A is greater than B D, and that again than G E, it follows that the radius N x must be greater than B y, and B y than E z. Thus it appears that the mutual intersections of the plumb-lines will not, as in the sphere, all coincide in one point C, the centre, but will be arranged along a certain curve, $x\,y\,z$ (which will be rendered more evident by considering a number of intermediate stations). To this curve geometers have given the name of the *evolute* of the curve N A B D G E, from whose centres of curvature it is constructed.

(220). In the flattening of a round figure at two opposite points, and its protuberance at points rectangularly situated to the former, we recognize the distinguishing feature of the elliptic form. Accordingly, the next and simplest supposition that we can make respecting the nature of the meridian, since it is proved not to be a circle, is, that it is an ellipse, or nearly so, having N S, the axis of the earth, for its shorter, and E F, the equatorial diameter, for its longer axis; and that the form of the earth's surface is that which would arise from making such a curve revolve about its shorter axis N S. This agrees well with the general course of the increase of the degree in going from the equator to the pole. In the ellipse, the radius of curvature at E, the extremity of the longer axis is the least, and at that of the shorter axis, the greatest it admits, and the form of its *evolute* agrees with that here represented.* Assuming, then, that it is an ellipse, the geometrical properties of that curve enable us to assign the proportion between the lengths of its axes which shall correspond to any proposed rate of variation in its curvature, as well as to fix upon their absolute lengths, corresponding to any assigned length of the degree in a given latitude. Without troubling the reader with the investigation, (which may be found in any work on the conic sections,) it will be suf-

---

* The dotted lines are the portions of the evolute belonging to the other quadrants.

ficient to state the results which have been arrived at by the most systematic combinations of the measured arcs which have hitherto been made by geometers. The most recent is that of Bessel *, who by a combination of the ten arcs, marked B in our table, has concluded the dimensions of the terrestrial spheroid to be as follows : —

		Feet.	Miles.
Greater or equatorial diameter	- - -	= 41,847,192	= 7925·604
Lesser or polar diameter	- - -	= 41,707,324	= 7899·114
Difference of diameters, or polar compression	=	139,768	= 26·471
Proportion of diameters as 299·15 to 298·15.			

The other combination whose results we shall state, is that of Mr. Airy †, who concludes as follows : —

		Feet.	Miles.
Equatorial diameter	- - - - -	= 41,847,426	= 7925·648
Polar diameter	- - - - - -	= 41,707,620	= 7899·170
Polar compression	- - - - -	=	139,806	= 26·478
Proportion of diameters as 299·33 to 298·33.			

These conclusions are based on the consideration of those 13 arcs, to which the letter A is annexed ‡, and of one other arc of 1° 7′ 31″·1, measured in Piedmont by Plana and Carlini, whose discordance with the rest, owing to local causes hereafter to be explained, arising from the exceedingly mountainous nature of the country, render the propriety of so employing it very doubtful. Be that as it may, the strikingly near accordance of the two sets of dimensions is such as to inspire the greatest confidence in both. The measurement at the Cape of Good Hope by Lacaille, also used in this determination, has always been regarded as unsatisfactory, and has recently been demonstrated by Mr. Maclear to be erroneous to a considerable extent. The omission of the former, and the substitution for the latter, of the far preferable result of Mr. Maclear's second measurement would induce, however, but a trifling change in the final result.

(221.) Thus we see that the rough diameter of 8000 miles we have hitherto used, is rather too great, the excess being

---

* Schumacher's Astronomische Nachrichten, Nos. 333, 334, 335. 438.
† Encyclopædia Metropolitana, " Figure of the Earth " (1831).
‡ In those which have both A and B, the numbers used by Mr. Airy differ slightly from Bessel's, which are those we have preferred.

about 100 miles, or $\frac{1}{80}$th part.   As convenient numbers to remember, the reader may bear in mind, that in our latitude there are just as many thousands of feet in a degree of the meridian as there are days in the year (365): that, speaking loosely, a degree is about 70 British statute miles, and a second about 100 feet; that the equatorial circumference of the earth is a little less than 25,000 miles (24,899), and the ellipticity or polar flattening amounts to one 300th part of the diameter.

(222.) The two sets of results above stated are placed in juxtaposition, and the particulars given more in detail than may at first sight appear consonant, either with the general plan of this work, or the state of the reader's presumed acquaintance with the subject.   But it is of importance that he should early be made to see how, in astronomy, results in admirable concordance emerge from data accumulated from totally different quarters, and how local and accidental irregularities in the data themselves become neutralized and obliterated by their impartial geometrical treatment.   In the cases before us, the modes of calculation followed are widely different, and in each the mass of figures to be gone through to arrive at the result, enormous.

(223.) The supposition of an elliptic form of the earth's section through the axis is recommended by its simplicity, and confirmed by comparing the numerical results we have just set down with those of actual measurement.   When this comparison is executed, discordances, it is true, are observed, which, although still too great to be referred to error of measurement, are yet so small, compared to the errors which would result from the spherical hypothesis, as completely to justify our regarding the earth as an ellipsoid, and referring the observed deviations to either local or, if general, to comparatively small causes.

(224.) Now, it is highly satisfactory to find that the general elliptical figure thus *practically* proved to exist, is precisely what *ought theoretically* to result from the rotation of the earth on its axis.   For, let us suppose the earth a sphere, at rest, of uniform materials throughout, and externally covered with an ocean of equal depth in every part. Under such circumstances it would obviously be in a state of *equilibrium* ; and the water on its surface would have no tendency to run one way or the other. Suppose, now, a quantity of its materials were taken from the polar regions, and piled up all around the equator, so as to produce that difference of the polar and equatorial diameters of 26 miles which we know to exist. It is not less evident that a mountain ridge or equatorial *continent, only,* would be thus formed, from which the water would run down the excavated part at the poles. However solid matter might rest where it was placed, the liquid part, at least, would not remain there, any more than if it were thrown on the side of a hill. The consequence therefore, would be the formation of two great polar seas, hemmed in all round by equatorial land. Now, this is by no means the case in nature. The ocean occupies, indifferently, all latitudes, with no more partiality to the polar than to the equatorial. Since, then, as we see, the water occupies an elevation above the centre no less than 13 miles greater at the equator than at the poles, and yet manifests no tendency to leave the former and run towards the latter, it is evident that it must be *retained* in that situation by some adequate *power.* No such power, however, would exist in the case we have supposed, which is therefore not conformable to nature. In other words, the spherical form is *not* the *figure of equilibrium* ; and *therefore* the earth is either not at rest, or is so internally constituted as to *attract* the water to its equatorial regions, and retain it there. For the latter supposition there is no *primâ facie* probability, nor any analogy to lead us to such an idea. The former is in accordance with all the phenomena of the apparent diurnal motion of the heavens ; and therefore, if it will furnish us with the *power* in question, we can have no hesitation in adopting it as the true one.

(225.) Now, every body knows that when a weight is whirled round, it acquires thereby a tendency to recede from the centre of its motion ; which is called the centrifugal force. A stone whirled round in a sling is a common illustration ; but a better, for our present purpose, will be a pail of water,

suspended by a cord, and made to *spin round*, while the cord
hangs perpendicularly.   The surface of the water, instead of
remaining horizontal, will become concave,
as in the figure.   The centrifugal force generates a tendency in *all* the water to leave
the axis, and press towards the circumference ; it is, therefore, urged against the
pail, and forced up its sides, till the excess
of height, and consequent increase of pre-
sure downwards, just counterbalances its
centrifugal force, and a *state of equilibrium*
is attained.   The experiment is a very easy
and instructive one, and is admirably cal-
calculated to show how the *form of equili-*
brium accommodates itself to varying circumstances.   If, for example, we allow the
rotation to cease by degrees, as it becomes
slower we shall see the concavity of the
water regularly diminish; the elevated out-

ward portion will descend, and the depressed central rise,
while all the time a perfectly *smooth* surface is maintained,
till the rotation is exhausted, when the water resumes its
horizontal state.

(226.) Suppose, then, a globe, of the size of the earth, at
rest, and covered with a uniform ocean, were to be set in rotation about a certain axis, at first very slowly, but by degrees
more rapidly, till it turned round once in twenty-four hours;
a centrifugal force would be thus generated, whose general
tendency would be to urge the water at every point of the
surface to *recede* from the *axis*.   A rotation might, indeed,
be conceived so swift as to *flirt* the whole ocean from the
surface, like water from a mop.   But this would require a
far greater velocity than what we now speak of.   In the case
supposed, the *weight* of the water would still keep it *on* the
earth; and the tendency to recede from the axis *could* only
be satisfied, therefore, by the water leaving the poles, and
flowing towards the equator; there heaping itself up in a
ridge, just as the water in our pail accumulates against the

side; and being retained in opposition to its weight, or natural tendency towards the centre, by the pressure thus caused. This, however, could not take place without laying dry the polar portions of the land in the form of immensely protuberant continents; and the difference of our supposed cases, therefore, is this:—in the former, a great equatorial continent and polar seas would be formed; in the latter, protuberant land would appear at the poles, and a zone of ocean be disposed around the equator. This would be the first or immediate effect. Let us now see what would afterwards happen, in the two cases, if things were allowed to take their natural course.

(227.) The sea is constantly beating on the land, grinding it down, and scattering its worn off particles and fragments, in the state of mud and pebbles, over its bed. Geological facts afford abundant proof that the existing continents have all of them undergone this process, even more than once, and been entirely torn in fragments, or reduced to powder, and submerged and reconstructed. Land, in this view of the subject, loses its attribute of fixity. As a mass it might hold together in opposition to forces which the water freely obeys; but in its state of successive or simultaneous degradation, when disseminated through the water, in the state of sand or mud, it is subject to all the impulses of that fluid. In the lapse of time, then, the protuberant land in both cases would be destroyed, and spread over the bottom of the ocean, filling up the lower parts, and tending continually to remodel the surface of the solid nucleus, in correspondence with the *form of equilibrium* in both cases. Thus, after a sufficient lapse of time, in the case of an earth at rest, the equatorial continent, thus forcibly constructed, would again be levelled and transferred to the polar excavations, and the spherical figure be so at length restored. In that of an earth in rotation, the polar protuberances would gradually be cut down and disappear, being transferred to the equator (as being *then* the *deepest sea*), till the earth would assume by degrees the form we observe it to have—that of a flattened or *oblate* ellipsoid.

(228.) We are far from meaning here to trace the process *by which* the earth really assumed its actual form; all we intend is, to show that this is the form to which, under the conditions of a rotation on its axis, it must *tend;* and which it would attain, even if originally and (so to speak) perversely constituted otherwise.

(229.) But, further, the dimensions of the earth and the time of its rotation being known, it is easy thence to calculate the exact amount of the centrifugal force *, which, at the equator, appears to be $\frac{1}{289}$th part of the force or weight by which all bodies, whether solid or liquid, tend to fall towards the earth. By this fraction of its weight, then, the sea at the equator is *lightened,* and thereby rendered susceptible of being supported on a higher level, or more remote from the centre than at the poles, where no such counteracting force exists; and where, in consequence, the water may be considered as *specifically heavier.* Taking this principle as a guide, and combining it with the laws of gravity (as developed by Newton, and as hereafter to be more fully explained), mathematicians have been enabled to investigate, *à priori,* what would be the figure of equilibrium of such a body, constituted internally as we have reason to believe the earth to be; covered wholly or partially with a fluid; and revolving uniformly in twenty-four hours; and the result of this inquiry is found to agree very satisfactorily with what experience shows to be the case. From their investigations it appears that the form of equilibrium is, in fact, no other than an oblate ellipsoid, of a degree of ellipticity very nearly identical with what is observed, and which would be no doubt accurately so, did we know, with precision, the internal constitution and materials of the earth.

(230.) The confirmation thus incidentally furnished, of the hypothesis of the earth's rotation on its axis, cannot fail to strike the reader. A deviation of its figure from that of a sphere was not contemplated among the original reasons for adopting that hypothesis, which was assumed solely on account of the easy explanation it offers of the apparent diurnal motion of the heavens. Yet we see that, once admitted,

* Newton's Principia, iii. Prop. 19.

it draws with it, as a necessary consequence, this other remarkable phenomenon, of which no other satisfactory account could be rendered. Indeed, so direct is their connection, that the ellipticity of the earth's figure was discovered and demonstrated by Newton to be a consequence of its rotation, and its amount actually calculated by him, long before any measurement had suggested such a conclusion. As we advance with our subject, we shall find the same simple principle branching out into a whole train of singular and important consequences, some obvious enough, others which at first seem entirely unconnected with it, and which, until traced by Newton up to this their origin, had ranked among the most inscrutable arcana of astronomy, as well as among its grandest phenomena.

(231.) Of its more obvious consequences, we may here mention one which falls naturally within our present subject. If the earth really revolve on its axis, this rotation must generate a centrifugal force (see art. 225.), the effect of which must of course be to counteract a certain portion of the *weight* of every body situated at the equator, as compared with its weight at the poles, or in any intermediate latitudes. Now, this is fully confirmed by experience. There is actually observed to exist a difference in the *gravity*, or downward tendency, of one and the same body, when conveyed successively to stations in different latitudes. Experiments made with the greatest care, and in every accessible part of the globe, have fully demonstrated the fact of a regular and progressive increase in the weights of bodies corresponding to the increase of latitude, and fixed its amount and the law of its progression. From these it appears, that the extreme amount of this variation of gravity, or the difference between the equatorial and polar weights of one and the same mass of matter, is 1 part in 194 of its whole weight, the rate of increase in travelling from the equator to the pole being *as the square of the sine of the latitude.*

(232.) The reader will here naturally inquire, what is *meant* by speaking of the same body as having different weights at different stations; and, how such a fact, if true, can be as-

certained. When we weigh a body by a balance or a steelyard we do but counteract its weight by the equal weight of
another body under the very same circumstances; and if both
the body weighed and its counterpoise be removed to another
station, their gravity, if changed at all, will be changed
equally, so that they will still continue to counterbalance
each other. A difference in the intensity of gravity could,
therefore, never be detected by these means; nor is it in *this*
sense that we assert that a body weighing 194 pounds at the
equator will weigh 195 at the pole. If counterbalanced in a
scale or steelyard at the former station, an additional pound
placed in one or other scale at the latter would inevitably
sink the beam.

(233.) The meaning of the proposition may be thus explained: — Conceive a weight x suspended at the equator by
a string without weight passing over a
pulley, A, and conducted (supposing
such a thing possible) over other pulleys, such as B, round the earth's convexity, till the other end hung down
at the pole, and there sustained the
weight y. If, then, the weights x and
y were such as, at any one station,
equatorial or polar, would exactly counterpoise each other on
a balance, or when suspended side by side over a single
pulley, they would not counterbalance each other in this
supposed situation, but the polar weight y would preponderate; and to restore the equipoise the weight x must be
increased by $\frac{1}{194}$th part of its quantity.

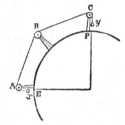

(234.) The means by which this variation of gravity may
be shown to exist, and its amount measured, are twofold (like
all estimations of mechanical power), statical and dynamical.
The former consists in putting the gravity of a weight in
equilibrium, not with that of another weight, but with a
natural power of a different kind not liable to be affected by
local situation. Such a power is the elastic force of a spring.
Let A B C be a strong support of brass standing on the foot
A E D cast in one piece with it, into which is let a smooth

plate of agate, D, which can be adjusted to perfect horizontality by a level.   At C let a spiral spring G be attached,
which carries at its lower end a weight F,
polished and convex below.   The length
and strength of the spring must be so
adjusted that the weight F shall be sustained by it just to swing clear of contact
with the agate plate in the highest latitude at which it is intended to use the
instrument.   Then, if small weights be
added cautiously, it may be made to descend till it *just grazes* the agate, a contact which can be made with the utmost
imaginable delicacy.   Let these weights
be noted; the weight F detached; the
spring G carefully lifted off its hook, and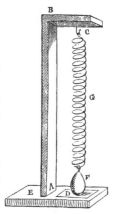
secured, for travelling, from rust, strain, or disturbance, and
the whole apparatus conveyed to a station in a lower latitude.
It will then be found, on remounting it, that, although loaded
with the same additional weights as before, the weight F will no
longer have power enough to stretch the spring to the extent
required for producing a similar contact.   More weights will
require to be added; and the additional quantity necessary
will, it is evident, measure the difference of gravity between
the two stations, as exerted on the whole quantity of pendent
matter, *i.e.* the sum of the weight of F and *half* that of the
spiral spring itself.   Granting that a spiral spring can be constructed of such strength and dimensions that a weight of
10,000 grains, including its own, shall produce an elongation
of 10 inches without permanently straining it*, one additional
grain will produce a further extension of $\frac{1}{1000}$th of an inch, a
quantity which cannot possibly be mistaken in such a contact as that in question.   Thus we should be provided with

* Whether the process above described could ever be so far perfected and refined as to become a substitute for the use of the pendulum must depend on the
degree of permanence and uniformity of action of springs, on the constancy or va-
riability of the effect of temperature on their elastic force, on the possibility of transporting them, absolutely unaltered, from place to place, &c. The great advantages,
however, which such an apparatus and mode of observation would possess, in point
of convenience, cheapness, portability, and expedition, over the present laborious,
tedious, and expensive process, render the attempt well worth making.

the means of measuring the power of gravity at any station to within $\frac{1}{10000}$th of its whole quantity.

(235.) The other, or dynamical process, by which the force urging any given weight to the earth may be determined, consists in ascertaining the velocity imparted by it to the weight when suffered to fall freely in a given time, as one second. This velocity cannot, indeed, be directly measured; but indirectly, the principles of mechanics furnish an easy and certain means of deducing it, and, consequently, the intensity of gravity, by observing the oscillations of a pendulum. It is proved from mechanical principles[*], that, if one and the same pendulum be made to oscillate at different stations, or under the influence of different forces, and the numbers of oscillations made in the same time in each case be counted, the intensities of the forces will be to each other as the squares of the numbers of oscillations made, and thus their proportion becomes known. For instance, it is found that, under the equator, a pendulum of a certain form and length makes 86,400 vibrations in a mean solar day; and that, when transported to London, the same pendulum makes 86,535 vibrations in the same time. Hence we conclude, that the intensity of the force urging the pendulum downwards at the equator is to that at London as $(86,400)^2$ to $(86,535)^2$, or as 1 to 1.00315; or, in other words, that a mass of matter weighing in London 100,000 pounds, exerts the same pressure on the ground, or the same effort to crush a body placed below it, that 100,315 *of the same pounds* transported to the equator would exert there.

(236.) Experiments of this kind have been made, as above stated, with the utmost care and minutest precaution to ensure exactness in all accessible latitudes; and their general and final result has been, to give $\frac{1}{194}$ for the fraction expressing the difference of gravity at the equator and poles. Now, it will not fail to be noticed by the reader, and will, probably, occur to him as an objection against the explanation here given of the fact by the earth's rotation, that this differs materially from the fraction $\frac{1}{289}$ expressing the centrifugal force at the equator. The difference by which the former fraction exceeds the latter is $\frac{1}{590}$, a small quantity in itself,

---

[*] Newton's Principia, ii. Prop. 24. Cor. 3.

but still far too large, compared with the others in question, not to be distinctly accounted for, and not to prove fatal to this explanation if it will not render a strict account of it.

(237.) The mode in which this difference arises affords a curious and instructive example of the indirect influence which mechanical causes often exercise, and of which astronomy furnishes innumerable instances. The rotation of the earth gives rise to the centrifugal force; the centrifugal force produces an ellipticity in the form of the earth itself; and this very ellipticity of form modifies its power of attraction on bodies placed at its surface, and thus gives rise to the difference in question. Here, then, we have the same cause exercising at once a direct and an indirect influence. The amount of the former is easily calculated, that of the latter with far more difficulty, by an intricate and profound application of geometry, whose steps we cannot pretend to trace in a work like the present, and can only state its nature and result.

(238.) The weight of a body (considered as undiminished by a centrifugal force) is the effect of the earth's attraction on it. This attraction, as Newton has demonstrated, consists, not in a tendency of all matter to any one particular centre, but in a disposition of every particle of matter in the universe to press towards, and if not opposed to approach to, every other. The attraction of the earth, then, on a body placed on its surface, is not a simple but a complex force, resulting from the separate attractions of all its parts. Now, it is evident, that if the earth were a perfect sphere, the attraction exerted by it on a body any where placed on its surface, whether at its equator or pole, must be exactly alike, — for the simple reason of the exact symmetry of the sphere in every direction. It is not less evident that, the earth being elliptical, and this symmetry or similitude of all its parts not existing, the same result cannot be expected. A body placed at the equator, and a similar one at the pole of a flattened ellipsoid, stand in a different geometrical relation to the mass as a whole. This difference, without entering further into particulars, may be expected to draw with it a difference in

L

its forces of attraction on the two bodies. Calculation confirms this idea. It is a question of purely mathematical investigation, and has been treated with perfect clearness and
precision by Newton, Maclaurin, Clairaut, and many other
eminent geometers; and the result of their investigations is
to show that, owing to the elliptic form of the earth alone,
and independent of the centrifugal force, its attraction ought
to increase the weight of a body in going from the equator to
the pole by almost exactly $\frac{1}{590}$th part; which, together with
$\frac{1}{289}$th due to the centrifugal force, make up the whole
quantity, $\frac{1}{194}$th, observed.

(239.) Another great geographical phenomenon, which
owes its existence to the earth's rotation, is that of the trade-
winds. These mighty currents in our atmosphere, on which
so important a part of navigation depends, arise from, 1st,
the unequal exposure of the earth's surface to the sun's rays,
by which it is unequally heated in different latitudes; and,
2dly, from that general law in the constitution of all fluids, in
virtue of which they occupy a larger bulk, and become specifically lighter when hot than when cold. These causes,
combined with the earth's rotation from west to east, afford an
easy and satisfactory explanation of the magnificent phenomena in question.

(240.) It is a matter of observed fact, of which we shall
give the explanation farther on, that the sun is constantly
vertical over some one or other part of the earth between
two parallels of latitude, called the tropics, respectively $23\frac{1}{2}°$
north, and as much south of the equator; and that the whole
of that zone or belt of the earth's surface included between
the tropics, and equally divided by the equator, is, in consequence of the great altitude attained by the sun in its
diurnal course, maintained at a much higher temperature than
those regions to the north and south which lie nearer the
poles. Now, the heat thus acquired by the earth's surface
is communicated to the incumbent air, which is thereby
expanded, and rendered specifically lighter than the air incumbent on the rest of the globe. It is therefore, in obedience
to the general laws of hydrostatics, displaced and buoyed up

from the surface, and its place occupied by colder, and therefore heavier air, which glides in, on both sides, along the
surface, from the regions beyond the tropics; while the displaced air, thus raised above its due level, and unsustained by
any lateral pressure, flows over, as it were, and forms an upper
current in the contrary direction, or towards the poles; which,
being cooled in its course, and also sucked down to supply the
deficiency in the extra-tropical regions, keeps up thus a
continual circulation.

(241.) Since the earth revolves about an axis passing
through the poles, the equatorial portion of its surface has
the greatest velocity of rotation, and all other parts less in
the proportion of the radii of the circles of latitude to which
they correspond. But as the air, when relatively and apparently at rest on any part of the earth's surface, is only so
because in reality it participates in the motion of rotation
proper to that part, it follows that when a mass of air near
the poles is transferred to the region near the equator by any
impulse urging it directly towards that circle, in every point
of its progress towards its new situation it must be found
deficient in rotatory velocity, and therefore unable to keep
up with the speed of the new surface over which it is brought.
Hence, the currents of air which set in towards the equator
from the north and south must, as they glide along the surface, at the same time lag, or hang back, and *drag upon* it in
the direction *opposite* to the earth's rotation, *i. e.* from east to
west. Thus these currents, which but for the rotation would
be simply northerly and southerly winds, acquire, from this
cause, a *relative* direction towards the west, and assume the
character of permanent north-easterly and south-easterly
winds.

(242.) Were any considerable mass of air to be *suddenly*
transferred from beyond the tropics to the equator, the difference of the rotatory velocities proper to the two situations
would be so great as to produce not merely a wind, but a
tempest of the most destructive violence. But this is not
the case : the advance of the air from the north and south is
gradual, and all the while the earth is continually acting on,

and by the friction of its surface accelerating its rotatory velocity. Supposing its progress towards the equator to cease at any point, this cause would almost immediately communicate to it the deficient motion of rotation, after which it would revolve quietly with the earth, and be at relative rest. We have only to call to mind the comparative *thinness* of the coating which the atmosphere forms around the globe (art. 35.), and the immense *mass* of the latter, compared with the former (which it exceeds at least 100,000,000 times), to appreciate fully the absolute *command* of any extensive territory of the earth over the atmosphere immediately incumbent on it, in point of motion.

(243.) It follows from this, then, that as the winds on both sides approach the equator, their easterly tendency must diminish. * The lengths of the diurnal circles increase very slowly in the immediate vicinity of the equator, and for several degrees on either side of it hardly change at all. Thus the friction of the surface has more time to act in accelerating the velocity of the air, bringing it towards a state of *relative* rest, and diminishing thereby the relative set of the currents from east to west, which, on the other hand, is feebly, and, at length, not at all reinforced by the cause which originally produced it. Arrived, then, at the equator, the trades must be expected to lose their easterly character altogether. But not only this but the northern and southern currents here meeting and opposing, will mutually destroy each other, leaving only such preponderancy as may be due to a difference of local causes acting in the two hemispheres, — which in some regions around the equator may lie one way, in some another.

(244.) The result, then, must be the production of two great tropical belts, in the northern of which a constant north-easterly, and in the southern a south-easterly, wind must prevail, while the winds in the equatorial belt, which separates the two former, should be comparatively calm and

* See Captain Hall's " Fragments of Voyages and Travels," 2d series, vol. i. p. 162., where this is very distinctly, and, so far as I am aware, for the first time, reasoned out.

free from any steady prevalence of easterly character. All these consequences are agreeable to observed fact, and the system of aërial currents above described constitutes in reality what is understood by the regular *trade winds.*

(245.) The constant friction thus produced between the earth and atmosphere in the regions near the equator must (it may be objected) by degrees reduce and at length destroy the rotation of the whole mass. The laws of dynamics, however, render such a consequence, generally, impossible; and it is easy to see, in the present case, where and how the compensation takes place. The heated equatorial air, while it rises and flows over towards the poles, carries with it the rotatory velocity due to its equatorial situation into a higher latitude, where the earth's surface has less motion. Hence, as it travels northward or southward, it will *gain* continually more and more on the surface of the earth in its diurnal motion, and assume constantly more and more a *westerly* relative direction; and when at length it returns to the surface, in its circulation, which it must do more or less in all the interval between the tropics and the poles, it will act on it by its friction as a powerful south-west wind in the northern hemisphere, and a north-west in the southern, and restore to it the impulse taken up from it at the equator. We have here the origin of the south-west and westerly gales so prevalent in our latitudes, and of the almost universal westerly winds in the North Atlantic, which are, in fact, nothing else than a part of the general system of the re-action of the trades, and of the process by which the equilibrium of the earth's motion is maintained under their action. *

* As it is our object merely to illustrate the mode in which the earth's rotation affects the atmosphere on the great scale, we omit all consideration of local periodical winds, such as monsoons, &c.

It seems worth inquiry, whether hurricanes in tropical climates may not arise from portions of the upper currents prematurely diverted downwards before their relative velocity has been sufficiently reduced by friction on, and gradual mixing with, the lower strata; and so dashing upon the earth with that tremendous velocity which gives them their destructive character, and of which hardly any rational account has yet been given. But it by no means follows that this must always be the case. In general, a rapid transfer, either way, in latitude, of any mass of air which local or temporary causes might carry *above the immediate reach of the friction of the earth's surface,* would give a fearful exaggeration to its velocity. Wherever such a mass should strike the earth, a hurricane might arise; and should two such masses encounter in mid air, a tornado of any degree of intensity on record might easily result from their combination.

(246.) In order to construct a map or model of the earth, and obtain a knowledge of the distribution of sea and land over its surface, the forms of the outlines of its continents and islands, the courses of its rivers and mountain chains, and the relative situations, with respect to each other, of those points which chiefly interest us, as centres of human habitation, or from other causes, it is necessary to possess the means of determining correctly the situation of any proposed station on its surface. For this two elements require to be known, the latitude and longitude, the former assigning its distance from the poles or the equator, the latter, the meridian on which that distance is to be reckoned. To these, in strictness, should be added, its height above the sea level; but the consideration of this had better be deferred, to avoid complicating the subject.

(247.) The latitude of a station on a sphere would be merely the length of an arc of the meridian, intercepted between the station and the nearest point of the equator, reduced into degrees. (See art. 88.) But as the earth is elliptic, this mode of conceiving latitudes becomes inapplicable, and we are compelled to resort for our definition of latitude to a generalization of that property (art. 119.), which affords the readiest means of determining it by observation, and which has the advantage of being independent of the figure of the earth, which, after all, is not *exactly* an ellipsoid, or any known geometrical solid. The latitude of a station, then, is the altitude of the elevated pole, and is, therefore, astronomically determined by those methods already explained for ascertaining that important element. In consequence, it will be remembered that, to make a *perfectly* correct map of the whole, or any part of the earth's surface, equal differences of latitude are not represented by exactly equal intervals of surface.

(248.) For the purposes of geodesical* measurements and trigonometrical surveys, an exceedingly correct determination of the latitudes of the most important stations is required.

---

* Γη, the earth ; δεσις (from δεω, to bind), a joining or connexion (of parts).

For this purpose, therefore, the zenith sector (an instrument capable of great precision) is most commonly used to observe stars passing the meridian near the zenith, whose declinations have become known by previous long series of observations at fixed observatories, and which are therefore called standard or fundamental stars. Recently a method* has been employed

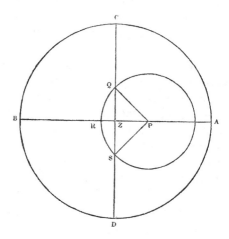

with great success, which consists in the use of an instrument similar in every respect to the transit instrument, but having the plane of motion of the telescope not coincident with the meridian, but with the prime vertical, so that its axis of rotation prolonged passes through the north and south points of the horizon. Let A B C D be the celestial hemisphere projected on the horizon, P the pole, Z the zenith, A B the meridian, C D the prime vertical, Q R S part of the diurnal circle of a star passing near the zenith, whose polar distance P R is but little greater than the co-latitude of the place, or the arc P Z, between the zenith and pole (art. 112.). Then the moments of this star's arrival on the prime vertical at Q and S will, if the instrument be correctly adjusted, be those of its crossing the middle wire in the field of view of the

---

* Devised originally by Römer.   Revived or re-invented by Bessel. — *Astr. Nachr.* No. 40.

telescope (art. 160.). Consequently the interval between these moments will be the time of the star passing from Q to S, or the measure of the diurnal arc Q R S, which corresponds to the angle Q P S at the pole. This angle, therefore, becomes known *by the mere observation of an interval of time*, in which it is not even necessary to know the error of the clock, and in which, when the star passes near the zenith, so that the interval in question is small, even the *rate* of the clock, or its gain or loss on true sidereal time, may be neglected. Now the angle Q P S, or its half Q P R, and P Q the polar distance of the star, being known, P Z the zenith distance of the pole can be calculated by the resolution of the right-angled spherical triangle P Z Q, and thus the co-latitude (and of course the latitude) of the place of observation becomes known. The advantages gained by this mode of observation are, 1st, that no readings of a divided arc are needed, so that errors of graduation and reading are avoided: 2dly, that the arc Q R S is very much greater than its versed sine R Z, so that the difference R Z between the latitude of the place and the declination of the star is given by the observation of a magnitude very much greater than itself, or is, as it were, observed on a greatly enlarged scale. In consequence, a very minute error is entailed on R Z by the commission of even a considerable one in Q R S: 3dly, that in this mode of observation all the merely instrumental errors which affect the ordinary use of the transit instrument are either uninfluential or eliminated by simply reversing the axis.

(249.) To determine the latitude of a station, then, is easy. It is otherwise with its longitude, whose exact determination is a matter of more difficulty. The reason is this: — as there are no meridians marked upon the earth, any more than parallels of latitude, we are obliged in this case, as in the case of the latitude, to resort to marks external to the earth, *i. e.* to the heavenly bodies, for the objects of our measurement; but with this difference in the two cases — to observers situated at stations on the same *meridian* (*i. e.* differing in latitude) the heavens present different aspects at *all* moments.

The portions of them which become visible in a complete diurnal rotation are not the same, and stars which are common to both describe circles differently inclined to their horizons, and differently divided by them, and attain different altitudes. On the other hand, to observers situated on the same *parallel* (*i. e.* differing only in longitude) the heavens present the same aspects. Their visible portions are the same; and the same stars describe circles equally inclined, and similarly divided by their horizons, and attain the same altitudes. In the former case there *is*, in the latter there is *not*, any thing in the appearance of the heavens, watched through a whole diurnal rotation, which indicates a difference of locality in the observer.

(250.) But no two observers, at different points of the earth's surface, can have at *the same instant* the same celestial hemisphere visible. Suppose, to fix our ideas, an observer stationed at a given point of the equator, and that at the moment when he noticed some bright star to be in his zenith, and therefore on his meridian, he should be suddenly transported, in an instant of time, round one quarter of the globe in a *westerly* direction, it is evident that he will no longer have the same star vertically above him : it will now appear to him to be just rising, and he will have to wait six hours before it again comes to his zenith, *i. e.* before the earth's rotation from west to east *carries him back again* to the line joining the star and the earth's centre from which he set out.

(251.) The difference of the cases, then, may be thus stated, so as to afford a key to the astronomical solution of the problem of the longitude. In the case of stations differing only in latitude, the same star comes to the meridian at the same *time*, but at different *altitudes*. In that of stations differing only in longitude, it comes to the meridian at the same *altitude*, but at different *times*. Supposing, then, that an observer is in possession of any means by which he can certainly ascertain the *time* of a known star's transit across his meridian, he knows his longitude; or if he knows the difference between its time of transit across his meridian and across that of any other station, he knows their difference of longitudes. For instance, if the same star pass the meridian of a place A at a certain moment, and that of B exactly one hour of sidereal time, or one twenty-fourth part of the earth's diurnal period, later, then the difference of longitude between A and B is one hour of time or 15° of arc, and B is so much west of A.

(252.) In order to a perfectly clear understanding of the principle on which the problem of finding the longitude by astronomical observations is resolved, the reader must learn to distinguish between time, in the abstract, as common to the whole universe, and therefore reckoned from an epoch independent of local situation, and *local time*, which reckons, at each particular place, from an epoch, or initial instant, determined by local convenience. Of time reckoned in the former, or abstract manner, we have an example in what we have before defined as equinoctial time, which dates from an epoch determined by the sun's motion among the stars. Of the latter, or *local* reckoning, we have instances in every sidereal clock in an observatory, and in every town clock for common use. Every astronomer regulates, or aims at regulating, his sidereal clock, so that it shall indicate 0^h 0^m 0^s, when a certain point in the heavens, called the equinox, is on the meridian of his station. This is the *epoch* of his sidereal time; which is, therefore, entirely a *local* reckoning. It gives no information to say that an event happened at such and such an hour of sidereal time, unless we particularize the station to which the sidereal time meant appertains. Just so it is with mean or common time. This is also a local reckoning, having for its epoch *mean noon*, or the average of all the times throughout the year, when the sun is on the meridian *of that particular place to which it belongs;* and, therefore, in like manner, when we date any event by mean time, it is necessary to name the place, or particularize *what* mean time we intend. On the other hand, a date by equinoctial time is absolute, and requires no such explanatory addition.

(253.) The astronomer sets and regulates his sidereal clock by observing the meridian passages of the more conspicuous and well-known stars. Each of these holds in the heavens a certain determinate and known place with respect to that imaginary point called the equinox, and by noting the times of their passage in succession by his clock he knows when the equinox passed. At that moment his clock ought to have marked 0^h 0^m 0^s; and if it did not, he knows and can correct its error, and by the agreement or disagreement of the errors assigned by each star he can ascertain whether his clock is correctly regulated to go twenty-four hours in one diurnal period, and if not, can ascertain and allow for its rate. Thus, although his clock may not, and indeed cannot, either be set correctly, or go truly, yet by applying its *error* and *rate* (as they are technically termed), he can correct its indications, and ascertain the exact sidereal times corresponding to them, and proper to his locality. This indispensable operation is called getting his *local time*. For simplicity of explanation, however, we shall suppose the clock a perfect instrument; or, which comes to the same thing, its error and rate applied at every moment it is consulted, and included in its indications.

(254.) Suppose, now, of two observers, at distant stations, A and B, each, independently of the other, to set and regulate his clock to the true sidereal time of his station. It is evident that if one of these clocks could be taken up without deranging its going, and set down by the side of the other, they would be found, on comparison, to differ by the exact difference of their local epochs; that is, by the time occupied by the equinox, or by any star, in passing from the meridian of A to that of B; in other words, by their difference of longitude, expressed in sidereal hours, minutes, and seconds.

(255.) A pendulum clock cannot be thus taken up and transported from place to place without derangement, but a chronometer may. Suppose, then, the observer at B to use a chronometer instead of a clock, he may, by bodily transfer of the instrument to the other station, procure a direct comparison of sidereal times, and thus obtain his longitude from A. And even if he employ a clock, yet by comparing it first with a good chronometer, and then transferring the latter instrument for comparison with the other clock, the same end will be accomplished, provided the going of the chronometer can be depended on.

(256.) Were chronometers perfect, nothing more complete and convenient than this mode of ascertaining differences of longitude could be desired. An observer, provided with such an instrument, and with a portable transit, or some equivalent method of determining the local time at any given station, might, by journeying from place to place, and observing the meridian passages of stars at each, (taking care not to alter his chronometer, or let it run down,) ascertain their differences of longitude with any required precision. In this case, the same time-keeper being used at every station, if, at one of them, A, it mark true sidereal time, at any other, B, it will be just so much sidereal time in error as the difference of longitudes of A and B is equivalent to: in other words, the longitude of B from A will appear as the error of the time-keeper on the local time of B. If he travel westward, then his chronometer will appear continually to gain, although it really goes correctly. Suppose, for instance, he set out from A, when the equinox was on the meridian, or his chronometer at 0^h, and in twenty-four hours (sid. time) had travelled $15°$ westward to B. At the moment of arrival there, his chronometer will again point to 0^h; but the equinox will be, not on his new meridian, but on that of A, and he must wait one hour more for its arrival at that of B. When it does arrive there, then his watch will point not to 0^h but to 1^h, and will therefore be 1^h *fast* on the local time of B. If he travel eastward, the reverse will happen.

(257.) Suppose an observer now to set out from any station as above described, and constantly travelling westward to make the tour of the globe, and return to the point he set out from. A singular consequence will happen: he will have lost a day in his reckoning of time. He will enter the day of his arrival in his diary, as Monday, for instance, when, in fact, it is Tuesday. The reason is obvious. Days and nights are caused by the alternate appearance of the sun and stars, as the rotation of the earth carries the spectator round to view them in succession. So many turns as he makes absolutely round the centre, so often will he pass through the earth's shadow, and emerge into light, and so many nights and days will he experience. But if he travel once round the globe in the direction of its motion, he will, on his arrival, have really made one turn *more* round its centre; and if in the opposite direction, one turn *less* than if he had remained upon one point of its surface: in the former case, then, he will have witnessed one alternation of day and night more, in the latter one less, than if he had trusted to the rotation of the earth alone to carry him round. As the earth revolves from west to east, it follows that a westward direction of his journey, by which he counteracts its rotation, will cause him to lose a day, and an eastward direction, by which he conspires with it, to gain one. In the former case, all his days will be longer; in the latter, shorter than those of a stationary observer. This contingency has actually happened to circumnavigators. Hence, also, it must necessarily happen that distant settlements, *on the same meridian*, will differ a day in their usual reckoning of time, according as they have been colonized by settlers arriving in an eastward or in a westward direction, — a circumstance which may produce strange confusion when they come to communicate with each other. The only mode of correcting the ambiguity, and settling the disputes which such a difference may give rise to, consists in having recourse to the equinoctial date, which can never be ambiguous.

(258.) Unfortunately for geography and navigation, the chronometer, though greatly and indeed wonderfully improved by the skill of modern artists, is yet far too imperfect an instrument to be relied on implicitly. However such an instrument may preserve its uniformity of rate for a few hours, or even days, yet in long absences from home the chances of error and accident become so multiplied, as to destroy all security of reliance on even the best. To a certain extent this may, indeed, be remedied by carrying out several, and using them as checks on each other; but, besides the expense and trouble, this is only a palliation of the evil — the great and fundamental, — as it is the only one to which *the determination of longitudes by time-keepers* is liable. It becomes necessary, therefore, to resort to other means of communicating from one station to another a knowledge of its local time, or of propagating from some principal station, as a centre, its local time as a universal standard with which the local time at any other, however situated, may be at once compared, and thus the longitudes of all places be referred to the meridian of such central point.

(259.) The simplest and most accurate method by which this object can be accomplished, when circumstances admit of its adoption, is that by telegraphic signal. Let A and B be two observatories, or other stations, provided with accurate means of determining *their respective local times,* and let us first suppose them visible from each other. Their clocks being regulated, and their errors and rates ascertained and applied, let a signal be made at A, of some sudden and definite kind, such as the flash of gunpowder, the explosion of a rocket, the sudden extinction of a bright light, or any other which admits of no mistake, and can be seen at great distances. The moment of the signal being made must be noted by *each* observer at his respective clock or watch, as if it were the transit of a star, or any astronomical phenomenon, and the error and rate of the clock at each station being applied, the local time of the signal at each is determined. Consequently, when the observers communicate their observations of the signal to each other, since (owing to the almost instantaneous transmission of light) it must have been seen at the same *absolute* instant by both, the difference of their local times, and therefore of their longitudes, becomes known. For example, at A the signal is observed to happen at $5^h\ 0^m\ 0^s$ sid. time at A, as obtained by applying the error and rate to the time shown by the clock at A, when the signal was seen there. At B the same signal was seen at $5^h\ 4^m\ 0^s$, sid. time at B, similarly deduced from the time noted by the clock at B, by applying *its* error and rate. Consequently, the difference of their local epochs is $4^m\ 0^s$, which is also their difference of longitudes in time, or $1°\ 0'\ 0''$ in hour angle.

(260.) The accuracy of the final determination may be increased by making and observing several signals at stated intervals, each of which affords a comparison of times, and the mean of all which is, of course, more to be depended on than the result of any single comparison. By this means, the error introduced by the comparison of clocks may be regarded as altogether destroyed.

(261.) The distances at which signals can be rendered visible must of course depend on the nature of the interposed country. Over sea the explosion of rockets may easily be seen at fifty or sixty miles; and in mountainous countries the flash of gunpowder in an open spoon may be seen, if a proper station be chosen for its exhibition, at much greater distances.

(262.) When the direct light of the flash can no longer be perceived, either owing to the convexity of the interposed segment of the earth, or to intervening obstacles, the sudden illumination cast on the under surface of the clouds by the explosion of considerable quantities of powder may often be observed with success; and in this way signals have been made at very much greater distances. Whatever means can be devised of exciting in two distant observers the same sensation, whether of sound, light, or visible motion, at *precisely the same instant of time,* may be employed as a longitude signal. Wherever, for instance, an unbroken line of electro-telegraphic connection has been, or hereafter may be, established, the means exist of making as complete a comparison of clocks or watches as if they stood side by side, so that no method more complete for the determination of differences of longitude can be desired. The differences of longitude between the observatories of New York, Washington and Philadelphia have been very recently determined in this manner by the astronomers at those observatories.

(263.) Where no such electric communication exists, however, the interval between observing stations may be increased by causing the signals to be made not at one of them, but at an intermediate point; for, provided they are seen by both parties, it is a matter of indifference where they are exhibited. Still the interval which could be thus embraced would be very limited, and the method in consequence of little use, but for the following ingenious contrivance, by which it can be extended to any distance, and carried over any tract of country, however difficult.

(264.) This contrivance consists in establishing, between the extreme stations, whose difference of longitude is to be ascertained, and at which the local times are observed, a chain of intermediate stations, alternately destined for signals and for observers. Thus, let A and Z be the extreme stations. At B let a signal station be established, at which rockets, &c. are fired at stated intervals. At C let an observer be placed, provided with a chronometer; at D, another signal station; at E, another observer and chronometer; till the whole line is occupied by stations so arranged, that the signal at B can be seen from A and C; those at D, from C and E; and so on. Matters being thus arranged, and the errors and rates of the clocks at A and Z ascertained by astronomical observation, let a signal be made at B, and observed at A and C, and the times noted. Thus the difference between A's clock and C's chronometer becomes known. After a short interval (five minutes for instance) let a signal be made at D, and observed by C and E. Then will the difference between their respective chronometers be determined; and the difference between the former and the clock at A being already ascertained, the difference between the clock A and chronometer E is therefore known. This, however, supposes that the intermediate chronometer C has kept true sidereal time, or at least a known rate, in the interval between the signals.

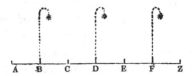

Now this interval is purposely made so very short, that no instrument of any pretensions to character can possibly produce an appretiable amount of error in its lapse by deviations from its usual *rate*. Thus the time propagated from A to C may be considered as handed over, without gain or loss (save from error of observation), to E. Similarly, by the signal made at F, and observed at E and Z, the time so transmitted to E is forwarded on to Z; and thus at length the clocks at A and Z are compared. The process may be repeated as often as is necessary to destroy error by a mean of results; and when the line of stations is numerous, by keeping up a succession of signals, so as to allow each observer to note alternately those on either side, which is easily pre-arranged, many comparisons may be kept running along the line at once, by which time is saved, and other advantages obtained.* In important cases the process is usually repeated on several nights in succession.

(265.) In place of artificial signals, natural ones, when they occur sufficiently definite for observation, may be equally employed. In a clear night the number of those singular meteors, called shooting stars, which may be observed, is often very great, especially on the 9th and 10th of August, and some other days, as November 12 and 13; and as they are sudden in their appearance and disappearance, and from the great height at which they have been ascertained to take place are visible over extensive regions of the earth's surface, there is no doubt that they may be resorted to with advantage, by previous concert and agreement between distant observers to watch and note them.† Those sudden disturbances of the magnetic needle, to which the name of magnetic shocks has been given, have been satisfactorily ascertained to be, very often at least, simultaneous over whole continents, and in some, perhaps, over the whole globe. These, if observed at magnetic observatories with precise attention to astronomical time, may become the means of determining their differences of longitude with more precision, possibly, than by any other method, if a sufficient number of remarkable shocks be observed to ascertain their *identity*, about which the intervals of time between their occurrence (exactly alike at both stations) will leave no doubt.

* For a complete account of this method, and the mode of deducing the most advantageous result from a combination of all the observations, see a paper on the difference of longitudes of Greenwich and Paris, Phil. Trans. 1826; by the Author of this volume.

† This idea was first suggested by the late Dr. Maskelyne, to whom, however, the practically useful fact of their periodic recurrence was unknown. Mr. Cooper has thus employed the meteors of the 10th and 12th August, 1847, to determine the difference of longitudes of Markree and Mount Eagle, in Ireland. Those of the same epoch have also been used in Germany for ascertaining the longitudes of several stations, and with very satisfactory results.

(266.) Another species of natural signal, visible at once over a whole terrestrial hemisphere, is afforded by the eclipses of Jupiter's satellites, of which we shall speak more at large when we come to treat of those bodies. Every such eclipse is an event which possesses one great advantage in its applicability to the purpose in question, viz. that the time of its happening, at any fixed station, such as Greenwich, can be *predicted* from a long course of previous recorded observation and calculation thereon founded, and that this prediction is sufficiently precise and certain, to stand in the place of a corresponding observation. So that an observer at any other station wherever, who shall have observed one or more of these eclipses, and ascertained his local time, instead of waiting for a communication with Greenwich, to inform him at what moment the eclipse took place there, may use the *predicted Greenwich time* instead, and thence, at once, and on the spot, determine his longitude. This mode of ascertaining longitudes is, however, as will hereafter appear, not susceptible of great exactness, and should only be resorted to when others cannot be had. The nature of the observation also is such that it cannot be made at sea*; so that, however useful to the geographer, it is of no advantage to navigation.

(267.) But such phenomena as these are of only occasional occurrence; and in their intervals, and when cut off from all communication with any fixed station, it is indispensable to possess some means of determining longitudes, on which not only the geographer may rely for a knowledge of the exact position of important stations on land in remote regions, but on which the navigator can securely stake, at every instant of his adventurous course, the lives of himself and comrades, the interests of his country, and the fortunes of his employers. Such a method is afforded by LUNAR OBSERVATIONS. Though

---

* To accomplish this is still a desideratum. Observing chairs, suspended with studious precaution for ensuring freedom of motion, have been resorted to, under the vain hope of mitigating the effect of the ship's oscillation. The opposite course seems more promising, viz. to merely deaden the motion by a somewhat stiff suspension (as by a coarse and rough cable), and by friction strings attached to weights running through loops (not pulleys) fixed in the wood-work of the vessel. At least, such means have been found by the author of singular efficacy in increasing personal comfort in the suspension of a cot.

we have not yet introduced the reader to the phenomena of the moon's motion, this will not prevent us from giving here the exposition of the principle of the lunar method; on the contrary, it will be highly advantageous to do so, since by this course we shall have to deal with the naked principle, apart from all the peculiar sources of difficulty with which the lunar theory is encumbered, but which are, in fact, completely extraneous to the *principle* of its application to the problem of the longitudes, which is quite elementary.

(268.) If there were in the heavens a clock furnished with a dial-plate and hands, which always marked Greenwich time, the longitude of any station would be at once determined, so soon as the *local time* was known, by comparing it with this clock. Now, the offices of the dial-plate and hands of a clock are these : — the former carries a set of marks upon it, whose position is known; the latter, by passing over and among these marks, informs us, by the place it holds with respect to them, what it is o'clock, or what time has elapsed since a certain moment when it stood at one particular spot.

(269.) In a clock the marks on the dial-plate are uniformly distributed all around the circumference of a circle, whose centre is that on which the hands revolve with a uniform motion. But it is clear that we should, with equal certainty, though with much more trouble, tell what o'clock it were, if the marks on the dial-plate were *un*equally distributed, — if the hands were *ex*centric, and their motion *not* uniform, — provided we knew, 1st, the exact intervals round the circle at which the hour and minute marks were placed; which would be the case if we had them all registered in a table, from the results of previous careful measurement : — 2dly, if we knew the exact amount and direction of excentricity of the centre of motion of the hands; — and, 3dly, if we were fully acquainted with all the mechanism which put the hands in motion, so as to be able to say at every instant what were their velocity of movement, and so as to be able to calculate, without fear of error, HOW MUCH *time* should correspond to SO MUCH *angular movement.*

(270.) The visible surface of the starry heavens is the

dial-plate of our clock, the stars are the fixed marks distributed around its circuit, the moon is the moveable hand, which, with a motion that, superficially considered, seems uniform, but which, when carefully examined, is found to be far otherwise, and which, regulated by mechanical laws of astonishing complexity and intricacy in result, though beautifully simple in principle and design, performs a monthly circuit among them, passing visibly over and hiding, or, as it is called, occulting some, and gliding beside and between others; and whose position among them can, at any moment when it is visible, be exactly measured by the help of a sextant, just as we might measure the place of our clock-hand among the marks on its dial-plate with a pair of compasses, and thence, from the known and calculated laws of its motion, deduce the time. That the moon *does* so move *among the stars,* while the latter hold constantly, with respect to each other, the same relative position, the notice of a few nights, or even hours, will satisfy the commencing student, and this is all that at present we require.

(271.) There is only one circumstance wanting to make our analogy complete. Suppose the hands of our clock, instead of moving *quite close* to the dial-plate, were considerably elevated above, or distant in front of it. Unless, then, in viewing it, we kept our eye just in the line of their centre, we should not see them exactly thrown or *projected* upon their proper places on the dial. And if we were either unaware of this cause of optical change of place, this *parallax*—or negligent in not taking it into account—we might make great mistakes in reading the time, by referring the hand to the wrong mark, or incorrectly appreciating its distance from the right. On the other hand, if we took care to note, in every case when we had occasion to observe the time, the exact position of the eye, there would be no difficulty in ascertaining and allowing for the precise influence of this cause of apparent displacement. Now, this is just what obtains with the apparent motion of the moon among the stars. The former (as will appear) is comparatively near to the earth — the latter immensely distant; and in consequence of our not occupying the centre of the earth, but being carried about on its surface, and constantly changing place, there arises a *parallax*, which displaces the moon apparently among the stars, and must be allowed for before we can tell the true place she would occupy if seen from the centre.

(272.) Such a clock as we have described might, no doubt, be considered a very bad one; but if it were our *only* one, and if incalculable interests were at stake on a perfect know-ledge of time, we should justly regard it as most precious, and think no pains ill bestowed in studying the laws of its movements, or in facilitating the means of *reading* it correctly. Such, in the parallel we are drawing, is the lunar theory, whose object is to reduce to regularity, the indications of this strangely irregular-going clock, to enable us to predict, long beforehand, and with absolute certainty, *whereabouts* among the stars, at every hour, minute, and second, in every day of every year, in Greenwich local time, the moon *would* be seen from the earth's centre, and *will* be seen from every accessible point of its surface; and such is the *lunar method* of longi-tudes. The moon's apparent angular distance from all those principal and conspicuous stars which lie in its course, as seen from the earth's centre, are computed and tabulated with the utmost care and precision in almanacks published under national control. No sooner does an observer, in any part of the globe, at sea or on land, measure its actual distance from any one of those *standard stars* (whose places in the heavens have been ascertained for the purpose with the most anxious solicitude), than he has, in fact, performed that com-parison of his local time with the local times of every ob-servatory in the world, which enables him to ascertain his difference of longitude from one or all of them.

(273.) The latitudes and longitudes of any number of points on the earth's surface may be ascertained by the methods above described; and by thus laying down a sufficient number of principal points, and filling in the intermediate spaces by local surveys, might maps of countries be constructed. In prac-tice, however, it is found simpler and easier to divide each particular nation into a series of great triangles, the angles of which are stations conspicuously visible from each other. Of these triangles, the *angles* only are measured by means of the *theodolite*, with the exception of *one side* only of *one triangle*, which is called *a base*, and which is measured with every refinement which ingenuity can devise or expense command. This *base* is of moderate extent, rarely surpassing six or seven miles, and purposely selected in a perfectly horizontal plane, otherwise conveniently adapted to the purposes of measurement. Its length between its two extreme points (which are dots on plates of gold or platina let into massive blocks of stone, and which are, or at least *ought to be*, in all cases preserved with almost religious care, as monumental records of the highest importance), is then measured, with every precaution to ensure precision*, and its position with respect to the meridian, as well as the geographical positions of its extremities, carefully ascertained.

(274.) The annexed figure represents such a chain of

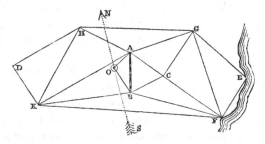

triangles. A B is the base, O, C, stations visible from both its extremities (one of which, O, we will suppose to be a national observatory, with which it is a principal object that the base should be as closely and immediately connected as possible); and D, E, F, G, H, K, other stations, remarkable points in the country, by whose connection its whole surface may be covered, as it were, with a network of triangles. Now, it is evident that the angles of the triangle A, B, C being observed, and one of its sides, A B, measured, the other two sides, A C, B C, may be calculated by the rules of trigonometry; and thus each of the sides A C and B C

---

* The greatest *possible* error in the Irish base of between seven and eight miles, near Londonderry, is supposed not to exceed two inches.

becomes in its turn a *base* capable of being employed as known sides of other triangles. For instance, the angles of the triangles A C G and B C F being known by observation, and their sides A C and B C, we can thence calculate the lengths A G, C G, and B F, C F. Again, C G and C F being known, and the included angle G C F, G F may be calculated, and so on. Thus may all the stations be accurately determined and laid down, and as this process may be carried on to any extent, a map of the whole country may be thus constructed, and filled in to any degree of detail we please.

(275.) Now, on this process there are two important remarks to be made. The first is, that it is necessary to be careful in the selection of stations, so as to form triangles free from any *very* great inequality in their angles. For instance, the triangle K B F would be a very improper one to determine the situation of F from observations at B and K, because the angle F being very acute, a small error in the angle K would produce a great one in the place of F *upon the line B F*. Such *ill-conditioned* triangles, therefore, must be avoided. But if this be attended to, the accuracy of the determination of the calculated sides will not be much short of that which would be obtained by actual measurement (were it practicable); and, therefore, as we recede from the base on all sides as a centre, it will speedily become practicable to use *as bases*, the sides of *much larger* triangles, such as G F, G H, H K, &c.; by which means the next step of the operation will come to be carried on on a much larger scale, and embrace far greater intervals, than it would have been safe to do (for the above reason) in the immediate neighbourhood of the base. Thus it becomes easy to divide the whole face of a country into *great triangles* of from 30 to 100 miles in their sides (according to the nature of the ground), which, being once well determined, may be afterwards, by a second series of subordinate operations, broken up into smaller ones, and these again into others of a still minuter order, till the final filling in is brought within the limits of personal survey and draftsmanship, and till a map is constructed, with any required degree of detail.

M 4

(276.) The next remark we have to make is, that all the triangles in question are not, rigorously speaking, *plane*, but *spherical* — existing on the surface of a sphere, or rather, to speak correctly, of an ellipsoid.   In very small triangles, of six or seven miles in the side, this may be neglected, as the difference is imperceptible ; but in the larger ones it must be taken into consideration.   It is evident that, as every object used for pointing the telescope of a theodolite has some certain *elevation*, not only above the *soil*, but above the level of the *sea*, and as, moreover, these elevations differ in every instance, a *reduction to the horizon* of all the measured angles would appear to be required.   But, in fact, by the construction of the theodolite (art. 192.), which is nothing more than an altitude and azimuth instrument, this reduction is *made* in the very act of reading off the horizontal angles.   Let E be the centre of the earth; A, B, C, the places on its *spherical surface*, to which three stations, A, P, Q, in a country are referred by radii E A, E B P, E C Q. If a theodolite be stationed at A, the axis of its horizontal circle will point to E when truly adjusted, and its plane will be a tangent to the sphere at A, intersecting the radii E B P, E C Q, at M and N, *above* the spherical surface.   The telescope of the theodolite, it is true, is pointed in succession to P, and Q ; but the readings off of its azimuth circle give — *not* the angle P A Q between the directions of the telescope, or between the objects P, Q, as seen from A ; *but the azimuthal angle* M A N, which is the measure of the angle A of the spherical triangle B A C.   Hence arises this remarkable circumstance, — that the sum of the three observed angles of any of the great triangles in geodesical operations is always found to be rather *more* than 180°.   Were the earth's surface a *plane*, it ought to be exactly 180° ; and this *excess*, which is called the *spherical excess*, is so far from being a proof of incorrectness in the work, that it is essential to its accuracy, and offers at the same time another palpable proof of the earth's sphericity.

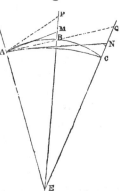

(277.) The true way, then, of conceiving the subject of a trigonometrical survey, when the spherical form of the earth is taken into consideration, is to regard the network of triangles with which the country is covered, as the bases of an assemblage of pyramids converging to the centre of the earth. The theodolite gives *us the true measures of the angles included by the planes of these pyramids;* and the surface of an imaginary sphere on the level of the sea intersects them in an assemblage of spherical triangles, above whose angles, in the radii prolonged, the real stations of observation are raised, by the superficial inequalities of mountain and valley. The operose calculations of spherical trigonometry which this consideration would seem to render necessary for the reductions of a survey, are dispensed with in practice by a very simple and easy rule, called the *rule for the spherical excess,* which is to be found in most works on trigonometry. If we would take into account the ellipticity of the earth, it may also be done by appropriate processes of calculation, which, however, are too abstruse to dwell upon in a work like the present.

(278.) Whatever process of calculation we adopt, the result will be a reduction to the level of the sea, of all the triangles, and the consequent determination of the geographical latitude and longitude of every station observed. Thus we are at length enabled to construct maps of countries; to lay down the outlines of continents and islands; the courses of rivers; the places of cities, towns and villages; the direction of mountain ridges, and the places of their principal summits; and all those details which, as they belong to physical and statistical, rather than to astronomical geography, we need not here dilate on. A few words, however, will be necessary respecting maps, which are used as well in astronomy as in geography.

(279.) A map is nothing more than a representation, upon a plane, of some portion of the surface of a sphere, on which are traced the particulars intended to be expressed, whether they be continuous outlines or points. Now, as a spherical surface *

---

* We here neglect the ellipticity of the earth, which, for such a purpose as map-making, is too trifling to have any material influence.

can by no contrivance be extended or projected into a
plane, without undue enlargement or contraction of some
parts in proportion to others; and as the system adopted in
so extending or projecting it will decide *what* parts shall be
enlarged or relatively contracted, and in what proportions;
it follows, that when large portions of the sphere are to be
mapped down, a great difference in their representations may
subsist, according to the system of projection adopted.

(280.) The projections chiefly used in maps, are the *orthographic*, *stereographic*, and *Mercator's*.   In the *orthographic*
projection, every point of the hemi-
sphere is referred to its diametral
plane or base, by a perpendicular let
fall on it, so that the representation
of the hemisphere thus mapped on
its base, is such as would actually
appear to an eye placed at an infinite
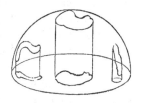
distance from it.   It is obvious, from the annexed figure,
that in this projection only the central portions are represented of their true forms, while all the exterior is more
and more distorted and crowded together as we approach the
edges of the map.   Owing to this cause, the orthographic
projection, though very good for small portions of the globe,
is of little service for large ones.

(281.) The *stereographic*
projection is in great mea
sure free from this defect.
To understand this projection, we must conceive an
eye to be placed at E, one
extremity of a diameter,
E C B, of the sphere, and
to view the concave surface
of the sphere, every point
of which, as P, is referred to
the diametral plane A D F,
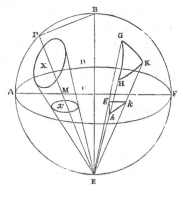
perpendicular to E B by the visual line P M E.   The
stereographic projection of a sphere, then, is a true perspec-

tive representation of its concavity on a diametral plane; and, as such, it possesses some singularly elegant geometrical properties, of which we shall state one or two of the principal.

(282.) And first, then, all circles on the sphere are represented by circles in the projection. Thus the circle X is projected into *x*. Only great circles passing through the vertex B are projected into straight lines traversing the centre C : thus, B P A is projected into C A.

2dly. Every very small triangle, G H K, on the sphere, is represented by a *similar* triangle, *g h k*, in the projection. This is a very valuable property, as it insures a general similarity of appearance in the map to the reality in all its parts, and enables us to project at least a hemisphere in a single map, without any violent distortion of the configurations on the surface from their real forms. As in the orthographic projection, the borders of the hemisphere are unduly crowded together; in the stereographic, their projected dimensions are, on the contrary, somewhat enlarged in receding from the centre.

(283.) Both these projections may be considered *natural* ones, inasmuch as they are really perspective representations of the surface on a plane. Mercator's is entirely an artificial one, representing the sphere as it cannot be seen from any one point, but as it might be seen by an eye carried successively over every part of it. In it, the degrees of *longitude*,

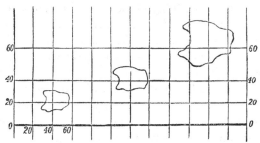

and those of *latitude*, bear always to each other their due proportion : the equator is conceived to be extended out into a straight line, and the meridians are straight lines at right angles to it, as in the figure. Altogether, the general cha-

racter of maps on this projection is not very dissimilar to
what would be produced by referring every point in the globe
to a circumscribing cylinder, by lines drawn from the centre,
and then unrolling the cylinder into a plane. Like the
stereographic projection, it gives a true representation, as to
form, of every particular small part, but varies greatly in
point of *scale* in its different regions; the polar portions in
particular being extravagantly enlarged; and the whole map,
even of a single hemisphere, not being comprizable within
any finite limits.

(284.) We shall not, of course, enter here into any
geographical details; but one result of maritime discovery on
the great scale is, so to speak, *massive* enough to call for
mention as an astronomical feature. When the continents
and seas are laid down on a globe (and since the discovery of
Australia and the recent addition to our antarctic knowledge
of Victoria Land by Sir J. C. Ross, we are sure that no
very extensive tracts of land remain unknown), we find that
it is possible so to divide the globe into two hemispheres, that
one shall contain *nearly all the land;* the other being almost
entirely sea. It is a fact, not a little interesting to English-
men, and, combined with our insular station in that great
highway of nations, the Atlantic, not a little explanatory of
our commercial eminence, that London* occupies nearly the
centre of the terrestrial hemisphere. Astronomically speaking,
the fact of this divisibility of the globe into an oceanic and a
terrestrial hemisphere is important, as demonstrative of a
want of absolute equality in the density of the solid material
of the two hemispheres. Considering the whole mass of land
and water as in a state of *equilibrium,* it is evident that the
half which protrudes must of necessity be *buoyant;* not, of
course, that we mean to assert it to be lighter than *water,*
but, as compared with the whole globe, *in a less degree heavier*
than that fluid. We leave to geologists to draw from these
premises their own conclusions (and we think them obvious

* More exactly, Falmouth. The central point of the hemisphere which
contains the maximum of land falls very nearly indeed upon this port. The
land in the opposite hemisphere, with exception of the tapering extremity of
South America and the slender peninsula of Malacca, is wholly insular, and
were it not for New Holland would be quite insignificant in amount.

enough) as to the internal constitution of the globe, and the immediate nature of the forces which sustain its continents at their actual elevation; but in any future investigations which may have for their object to explain the local deviations of the intensity of gravity, from what the hypothesis of an exact elliptic figure would require, this, as a general fact, ought not to be lost sight of.

(285.) Our knowledge of the surface of our globe is incomplete, unless it include the heights above the sea level of every part of the land, and the depression of the bed of the ocean below the surface over all its extent.   The latter object is attainable (with whatever difficulty and howsoever slowly) by direct sounding; the former by two distinct methods: the one consisting in trigonometrical measurement of the differences of level of all the stations of a survey; the other, by the use of the barometer, the principle of which is, in fact, identical with that of the sounding line.   In both cases we measure the distance of the point whose level we would know from the surface of an equilibrated ocean: only in the one case it is an ocean of water; in the other, of air.   In the one case our sounding line is real and tangible; in the other, an imaginary one, measured by the length of the column of quicksilver the superincumbent air is capable of counterbalancing.

(286.) Suppose that instead of air, the earth and ocean were covered with oil, and that human life could subsist under such circumstances.   Let A B C D E be a continent, of which the portion A B C projects above the water, but is

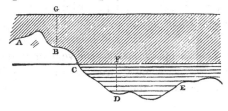

covered by the oil, which also floats at an uniform depth on the whole ocean.   Then if we would know the depth of any point D below the sea level, we let down a plummet from F. But, if we would know the height of B above the same level,

we have only to send up a float from B to the surface of the
oil; and *having done the same at C, a point at the sea level, the
difference of the two float lines gives the height in question.*

(287.) Now, though the atmosphere differs from oil in not
having a positive *surface* equally definite, and in not being
capable of carrying up any float adequate to such an use, yet
it possesses all the properties of a fluid really essential to the
purpose in view, and this in particular, — that, over the whole
surface of the globe, its *strata of equal density* supposed in a
state of equilibrium, are parallel to the surface of equilibrium,
or to what *would be* the surface of the sea, if *prolonged under
the continents,* and therefore each or any of them has all the
characters of a definite surface to measure from, provided it
can be ascertained and identified. Now, the height at which,
at any station B, the mercury in a barometer is supported,
informs us *at once* how much of the atmosphere is incumbent
on B, or, in other words, *in what stratum* of the general
atmosphere (indicated by its density) B is situated: whence
we are enabled finally to conclude, by mechanical reasoning*,
at what height above the sea-level *that degree of density* is to
be found over the whole surface of the globe. Such is the
principle of the application of the barometer to the measurement of heights. For details, the reader is referred to other
works. †

(288.) We will content ourselves here with a general caution against an implicit dependence on barometric measurements, except as a differential process, at stations not too
remote from each other. They rely in their application on
the assumption of a state of equilibrium in the atmospheric
strata over the whole globe — which is very far from being
their actual state (art. 37.). Winds, especially steady and
general currents sweeping over extensive continents, undoubtedly *tend* to produce some degree of conformity in the curvature of these strata to the *general* form of the land-surface,

---

* Newton's Princip. ii. Prop. 22.
† Biot, Astronomie Physique, vol. iii. For tables, see she work of Biot cited.
Also those of Oltmann, annually published by the French board of longitudes
in their Annuaire; and Mr. Baily's Collection of Astronomical Tables and
Formulæ.

and therefore to give an undue elevation to the mercurial column at some points.   On the other hand, the existence of localities on the earth's surface where a permanent depression of the barometer prevails to the astonishing extent of nearly an inch, has been clearly proved by the observations of Ermann in Siberia and of Ross in the Antarctic Seas, and is probably a result of the same cause, and may be conceived as complementary to an undue habitual elevation in other regions.

(289.) Possessed of a knowledge of the heights of stations above the sea, we may connect all stations at the same altitude by level lines, the lowest of which will be the outline of the sea-coast; and the rest will mark out the successive coast-lines which would take place were the sea to rise by regular and equal accessions of level over the whole world, till the highest mountains were submerged.   The bottoms of valleys and the ridge-lines of hills are determined by their property of intersecting all these level lines at right angles, and being, subject to that condition, the shortest and longest, that is to say, the steepest, and the most gently sloping courses respectively which can be pursued from the summit to the sea.   The former constitute " the water courses" of a country; the latter its lines of " water-shed" by which it is divided into distinct basins of drainage.   Thus originate natural districts of the most ineffaceable character, on which the distribution, limits, and peculiarities of human communities are in great measure dependent.   The mean height of the continent of Europe, or that height which its surface would have were all inequalities levelled and the mountains spread equally over the plains, is according to Humboldt 671 English feet; that of Asia, 1137; of North America, 748; and of South America, 1151.

CHAPTER V.

OF URANOGRAPHY.

CONSTRUCTION OF CELESTIAL MAPS AND GLOBES BY OBSERVATIONS
OF RIGHT ASCENSION AND DECLINATION. — CELESTIAL OBJECTS
DISTINGUISHED INTO FIXED AND ERRATIC. — OF THE CONSTEL-
LATIONS. — NATURAL REGIONS IN THE HEAVENS. — THE MILKY
WAY.—THE ZODIAC. — OF THE ECLIPTIC. — CELESTIAL LATITUDES
AND LONGITUDES.— PRECESSION OF THE EQUINOXES.—NUTATION.
— ABERRATION.—REFRACTION.—PARALLAX.—SUMMARY VIEW OF
THE URANOGRAPHICAL CORRECTIONS.

(290.) THE determination of the relative situations of objects
in the heavens, and the construction of maps and globes
which shall truly represent their mutual configurations as
well as of catalogues which shall preserve a more precise
numerical record of the position of each, is a task at once
simpler and less laborious than that by which the surface of
the earth is mapped and measured. Every star in the great
constellation which appears to revolve above us, constitutes,
so to speak, a celestial station; and among these stations we
may, as upon the earth, triangulate, by measuring with
proper instruments their angular distances from each other,
which, cleared of the effect of refraction, are then in a state
for laying down on charts, as we would the towns and villages
of a country: and this without moving from our place, at
least for all the stars which rise above our horizon.

(291.) Great exactness might, no doubt, be attained by
this means, and excellent celestial charts constructed; but
there is a far simpler and easier, and at the same time, infinitely more accurate course laid open to us if we take advantage of the earth's rotation on its axis, and by observing each
celestial object as it passes our meridian, refer it separately
and independently to the celestial equator, and thus ascertain

its place on the surface of an imaginary sphere, which may be conceived to revolve with it, and on which it may be considered as projected.

(292.) The right ascension and declination of a point in the heavens correspond to the longitude and latitude of a station on the earth ; and the place of a star on a celestial sphere is determined, when the former elements are known, just as that of a town on a map, by knowing the latter. The great advantages which the method of meridian observation possesses over that of triangulation from star to star, are, then, 1st, That in it every star is observed in that point of its diurnal course, when it is best seen and least displaced by refraction. 2dly, That the instruments required (the transit and meridian circle) are the simplest and least liable to error or derangement of any used by astronomers. 3dly, That all the observations can be made systematically, in regular succession, and with equal advantages ; there being here no question about advantageous or disadvantageous triangles, &c. And, lastly, That, by adopting this course, the very quantities which we should otherwise have to calculate by long and tedious operations of spherical trigonometry, and which are essential to the formation of a catalogue, are made the objects of immediate measurement. It is almost needless to state, then, that this is the course adopted by astronomers.

(293.) To determine the right ascension of a celestial object, all that is necessary is to observe the moment of its meridian passage with a transit instrument, by a clock regulated to exact sidereal time, or reduced to such by applying its known error and rate. The *rate* may be obtained by repeated observations of the same star at its successive meridian passages. The *error*, however, requires a knowledge of the *equinox*, or initial point from which all right ascensions in the heavens reckon, as longitudes do on the earth from a first meridian.

(294.) The nature of this point will be explained presently ; but for the purposes of uranography, in so far as they concern only the actual configurations of the stars *inter*

N

se, a knowledge of the equinox is not necessary. The choice of the equinox, as a zero point of right ascensions, is purely artificial, and a matter of convenience : but as on the earth, any station (as a national observatory) may be chosen for an origin of longitudes ; so in uranography, any conspicuous star might be selected as an initial point from which hour angles might be reckoned, and from which, by merely observing *differences* or *intervals* of time, the situation of all others might be deduced.   In practice, these intervals are affected by certain minute causes of inequality, which must be allowed for, and which will be explained in their proper places.

(295.) The declinations of celestial objects are obtained, 1st, By observation of their *meridian altitudes,* with the mural or meridian circle, or other proper instruments.   This requires a knowledge of the geographical latitude of the station of observation, which itself is only to be obtained by celestial observation. 2dly, And more directly, by observation of their *polar distances* on the mural circle, as explained in art. 170., which is independent of any previous determination of the latitude of the station ; neither, however, in this case, does observation give directly and immediately the *exact* declinations.   The observations require to be corrected, first for refraction, and moreover for those minute causes of inequality which have been just alluded to in the case of right ascensions.

(296.) In this manner, then, may the places, one among the other, of all celestial objects be ascertained, and maps and globes constructed.   Now here arises a very important question.   How far are these places permanent?   Do these stars and the greater luminaries of heaven preserve for ever one invariable connection and relation of place *inter se,* as if they formed part of a solid though invisible firmament ; and, like the great natural land-marks on the earth, preserve inmutably the same distances and bearings each from the other ?   If so, the most rational idea we could form of the universe would be that of an earth at absolute rest in the centre, and a hollow crystalline sphere circulating round it, and carrying sun, moon, and stars along in its diurnal motion. If not, we must dismiss all such notions, and inquire individually into the distinct history of each object, with a view to discovering the laws of its peculiar motions, and whether any and what other connection subsists between them.

(297.) So far is this, however, from being the case, that observations, even of the most cursory nature, are sufficient to show that some, at least, of the celestial bodies, and those the most conspicuous, are in a state of continual change of place among the rest. In the case of the moon, indeed, the change is so rapid and remarkable, that its alteration of situation with respect to such bright stars as may happen to be near it may be noticed any fine night in a few hours; and if noticed on two successive nights, cannot fail to strike the most careless observer. With the sun, too, the change of place among the stars is constant and rapid; though, from the invisibility of stars to the naked eye in the day-time, it is not so readily recognized, and requires either the use of telescopes and angular instruments to measure it, or a longer continuance of observation to be struck with it. Nevertheless, it is only necessary to call to mind its greater meridian altitude in summer than in winter, and the fact that the stars which come into view at night (and which are therefore situated in an hemisphere opposite to that occupied by the sun, and having that luminary for its centre) vary with the season of the year, to perceive that a great change must have taken place in that interval in its relative situation with respect to all the stars. Besides the sun and moon, too, there are several other bodies, called planets, which, for the most part, appear to the naked eye only as the largest and most brilliant stars, and which offer the same phenomenon of a constant change of place among the stars; now approaching, and now receding from, such of them as we may refer them to as marks; and, some in longer, some in shorter periods, making, like the sun and moon, the complete tour of the heavens.

(298.) These, however, are exceptions to the general rule. The innumerable multitude of the stars which are distributed over the vault of the heavens form a constellation, which preserves, not only to the eye of the casual observer, but to the nice examination of the astronomer, a uniformity of aspect which, when contrasted with the perpetual change in the configurations of the sun, moon, and planets, may well be termed invariable. It is true, indeed, that, by the refinement of exact measurements prosecuted from age to age, some small changes of apparent place, attributable to no illusion and to no *terrestrial* cause, have been detected in many of them. Such are called, in astronomy, the *proper motions* of the stars. But these are so excessively slow, that their accumulated amount (even in those stars for which they are greatest) has been insufficient, in the whole duration of astronomical history, to produce any obvious or material alteration in the appearance of the starry heavens.

(299.) This circumstance, then, establishes a broad distinction of the heavenly bodies into two great classes; — the fixed, among which (unless in a course of observations continued for many years) no change of mutual situation can be detected; and the erratic, or wandering — (which is implied in the word planet*) — including the sun, moon, and planets, as well as the singular class of bodies termed comets, in whose apparent places among the stars, and among each other, the observation of a few days, or even hours, is sufficient to exhibit an indisputable alteration.

(300.) Uranography, then, as it concerns the fixed celestial bodies (or, as they are usually called, the *fixed stars*), is reduced to a simple marking down of their relative places on a globe or on maps; to the insertion on that globe, in its due place in the great constellation of the stars, of the pole of the heavens, or the vanishing point of parallels to the earth's axis; and of the equator and place of the equinox: points and circles these, which, though artificial, and having reference entirely to our earth, and therefore subject to all changes (if any) to which the earth's axis may be liable, are yet so convenient in practice, that they have obtained an

---

* Πλανητης, a wanderer.

admission (with some other circles and lines), sanctioned by usage, in all globes and planispheres. The reader, however, will take care to keep them separate in his mind, and to familiarize himself with the idea rather of *two* or more celestial globes, superposed and fitting on each other, on one of which — a real one — are inscribed the stars ; on the others those imaginary points, lines, and circles, which astronomers have devised for their own uses, and to aid their calculations; and to accustom himself to conceive in the latter or artificial spheres a capability of being shifted in any manner upon the surface of the other; so that, should experience demonstrate (as it does) that these artificial points and lines are brought, by a slow motion of the earth's axis, or by other *secular variations* (as they are called), to coincide, at very distant intervals of time, with different stars, he may not be unprepared for the change, and may have no confusion to correct in his notions.

(301.) Of course we do not here speak of those uncouth figures and outlines of men and monsters, which are usually scribbled over celestial globes and maps, and serve, in a rude and barbarous way, to enable us to talk of groups of stars, or districts in the heavens, by names which, though absurd or puerile in their origin, have obtained a currency from which it would be difficult to dislodge them. In so far as they have really (as some have) any slight resemblance to the figures called up in imagination by a view of the more splendid " constellations," they have a certain convenience ; but as they are otherwise entirely arbitrary, and correspond to no *natural* subdivisions or groupings of the stars, astronomers treat them lightly, or altogether disregard them*, except for briefly *naming* remarkable stars, as *a* Leonis, *β* Scorpii, &c. &c., by letters of the Greek alphabet attached

---

* This disregard is neither supercilious nor causeless. The constellations seem to have been almost purposely named and delineated to cause as much confusion and inconvenience as possible. Innumerable snakes twine through long and contorted areas of the heavens, where no memory can follow them ; bears, lions, and fishes, large and small, northern and southern, confuse all nomenclature, &c. A better system of constellations might have been a material help as an artificial memory.

to them. The reader will find them on any celestial charts
or globes, and may compare them with the heavens, and
there learn for himself their position.

(302.) There are not wanting, however, *natural* districts
in the heavens, which offer great peculiarities of character,
and strike every observer: such is the *milky way*, that great
luminous band, which stretches, every evening, all across the
sky, from horizon to horizon, and which, when traced with
diligence, and mapped down, is found to form a zone completely encircling *the whole sphere*, almost in a great circle,
which is neither an *hour* circle, nor coincident with any other
of our astronomical *grammata*. It is divided in one part of
its course, sending off a kind of branch, which unites again
with the main body, after remaining distinct for about 150
degrees, within which it suffers an interruption in its continuity. This remarkable belt has maintained, from the
earliest ages, the same relative situation among the stars;
and, when examined through powerful telescopes, is found
(wonderful to relate!) *to consist entirely of stars scattered by
millions*, like glittering dust, on the black ground of the
general heavens. It will be described more particularly in
the subsequent portion of this work.

(303.) Another remarkable region in the heavens is the
zodiac, not from any thing peculiar in its own constitution,
but from its being the area within which the apparent
motions of the sun, moon, and all the greater planets are confined. To trace the path of any one of these, it is only
necessary to ascertain, by continued observation, its places
at successive epochs, and entering these upon our map or
sphere in sufficient number to form a series, not too far
disjoined, to connect them by lines from point to point, as we
mark out the course of a vessel at sea by mapping down its
place from day to day. Now when this is done, it is found,
first, that the apparent path, or track, of the sun on the surface of the heavens, is no other than an exact great circle of
the sphere which is called the *ecliptic*, and which is inclined
to the equinoctial at an angle of about 23° 28′, intersecting
it at two opposite points, called the equinoctial points, or

equinoxes, and which are distinguished from each other by the epithets vernal and autumnal; the vernal being that at which the sun crosses the equinoctial from south to north; the autumnal, when it quits the northern and enters the southern hemisphere. Secondly, that the moon and all the planets pursue paths which, in like manner, encircle the whole heavens, but are not, like that of the sun, great circles exactly returning into themselves and bisecting the sphere, but rather spiral curves of much complexity, and described with very unequal velocities in their different parts. They have all, however, this in common, that the *general direction* of their motions is the same with that of the sun, viz. from *west to east*, that is to say, the contrary to that in which both they and the stars appear to be carried by the diurnal motion of the heavens; and, moreover, that they never deviate far from the ecliptic on either side, crossing and recrossing it at regular and equal intervals of time, and confining themselves within a zone, or belt (the *zodiac* already spoken of), extending (with one or two exceptions among the smaller planets) not further than 9° on either side of the ecliptic.

(304.) It would manifestly be useless to map down on globes or charts the apparent paths of any of those bodies which never retrace the same course, and which, therefore, demonstrably, must occupy at some one moment or other of their history, every point in the area of that zone of the heavens within which they are circumscribed. The apparent complication of their movements arises (that of the moon excepted) from our viewing them from a station which is itself in motion, and would disappear, could we shift our point of view and observe them from the sun. On the other hand the apparent motion of the sun is presented to us under its least involved form, and is studied, from the station we occupy, to the greatest advantage. So that, independent of the importance of that luminary to us in other respects, it is by the investigation of the laws of its motions in the first instance that we must rise to a knowledge of those of all the other bodies of our system.

(305.) The ecliptic, which is its apparent path among the

stars, is traversed by it in the period called the *sidereal year*, which consists of 365^d 6^h 9^m $9 \cdot 6^s$, reckoned in mean solar time or 366^d 6^h 9^m $9 \cdot 6^s$ reckoned in sidereal time. The reason of this difference (and it is this which constitutes the origin of the difference between solar and sidereal time) is, that as the sun's apparent annual motion *among* the stars is performed in a contrary direction to the apparent *diurnal* motion of both sun and stars, it comes to the same thing as if the diurnal motion of the sun were so much *slower* than that of the stars, or as if the sun lagged behind them in its daily course. When this has gone on for a whole year, the sun will have fallen behind the stars by a whole circumference of the heavens — or, in other words — in a year the sun will have made fewer diurnal revolutions, by one, than the stars. So that the same interval of time which is measured by 366^d 6^h, &c. of sidereal time, will be called 365 days, 6 hours, &c., if reckoned in mean solar time. Thus, then, is the proportion between the mean solar and sidereal day established, which, reduced into a decimal fraction, is that of $1 \cdot 00273791$ to 1. The measurement of time by these different standards may be compared to that of space by the standard feet, or ells of two different nations; the proportion of which, once settled and borne in mind, can never become a source of error.

(306.) The position of the ecliptic among the stars may, for our present purpose, be regarded as invariable. It is true that this is not strictly the case; and on comparing together its position at present with that which it held at the most distant epoch at which we possess observations, we find evidences of a small change, which theory accounts for, and whose nature will be hereafter explained; but that change is so excessively slow, that for a great many successive years, or even for whole centuries, this circle may be regarded, for most ordinary purposes, as holding the same position in the sidereal heavens.

(307.) The *poles of the ecliptic*, like those of any other great circle of the sphere, are opposite points on its surface, equidistant from the ecliptic in every direction. They are of course not coincident with those of the equinoctial, but

removed from it by an angular interval equal to the inclination of the ecliptic to the equinoctial (23° 28′), which is called
the *obliquity of the ecliptic.*   In the next figure, if P *p* represent the north and south poles (by which when used without
qualification we always mean the poles of *the equinoctial*),
and E A Q V the equinoctial, V S A W the ecliptic, and K *k*,
its poles — the spherical angle Q V S is the obliquity of the
ecliptic, and is equal in angular measure to P K or S Q.
If we suppose the sun's apparent motion to be in the direction
V S A W, V will be the *vernal* and A the *autumnal equinox.*
S and W, the two points at which the ecliptic is most distant
from the equinoctial, are termed *solstices,* because, when
arrived there, the sun ceases to recede from the equator, and
(in that sense, so far as its motion in declination is concerned)
to stand still in the heavens.   S, the point where the sun
has the greatest *northern* declination, is called the *summer,*
and W, that where it is farthest south, the *winter* solstice.
These epithets obviously have their origin in the dependence
of the seasons on the sun's declination, which will be explained
in the next chapter.   The circle E K P Q *k p,* which passes
through the poles of the ecliptic and equinoctial, is called the
solstitial colure ; and a meridian drawn through the equinoxes,
P V *p* A, the *equinoctial colure.*

(308.) Since the ecliptic holds a determinate situation in
the starry heavens, it may be employed, like the equinoctial,
to refer the positions of the stars to, by circles drawn through
them from *its* poles, and therefore perpendicular to it.   Such
circles are termed, in astronomy,
circles of latitude — the distance
of a star from the ecliptic, reckoned on the circle of latitude
passing through it, is called the
latitude of the stars — and the
arc of the ecliptic intercepted
between the vernal equinox and
this circle, its *longitude.*   In the
figure, X is a star, P X R a
circle of declination drawn

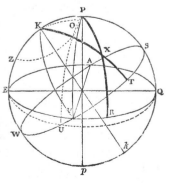

through it, by which it is referred to the equinoctial, and
K X T a circle of latitude referring it to the ecliptic —
then, as V R is the right ascension, and R X the declination,
of X, so also is V T its longitude, and T X its latitude. The
use of the terms longitude and latitude, in this sense, seems
to have originated in considering the ecliptic as forming a
kind of natural equator to the heavens, as the terrestrial
equator does to the earth — the former holding an invariable
position with respect to the stars, as the latter does with
respect to stations on the earth's surface. The force of this
observation will presently become apparent.

(309.) Knowing the right ascension and declination of an
object, we may find its longitude and latitude, and *vice versâ*
This is a problem of great use in physical astronomy — the
following is its solution : — In our last figure, E K P Q, the
solstitial colure is of course 90° distant from V, the vernal
equinox, which is one of its poles — so that V R (the right
ascension) being given, and also V E, the arc E R, and its
measure, the spherical angle E P R, or K P X, is known.
In the spherical triangle K P X, then, we have given, 1st,
The side P K, which, being the distance of the poles of the
ecliptic and equinoctial, is equal to the obliquity of the
ecliptic; 2d, The side P X, the *polar distance*, or the complement of the declination R X; and, 3d, the included angle
K P X; and therefore, by spherical trigonometry, it is easy
to find the other side K X, and the remaining angles. Now
K X is the complement of the required latitude X T, and the
angle P K X being known, and P K V being a right angle
(because S V is 90°), the angle X K V becomes known.
Now this is no other than the measure of the longitude V T
of the object. The inverse problem is resolved by the same
triangle, and by a process exactly similar.

(310.) It is often of use to know the situation of the
ecliptic in the visible heavens at any instant; that is to say,
the points where it cuts the horizon, and the altitude of its
highest point, or, as it is sometimes called, the *nonagesimal*
point of the ecliptic, as well as the longitude of this point on
the ecliptic itself from the equinox. These, and all questions

referable to the same data and quæsita, are resolved by the spherical triangle Z P E, formed by the zenith Z (considered as the pole of the horizon), the pole of the equinoctial P, and the pole of the ecliptic E. The

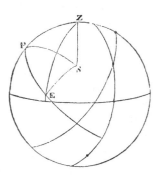

sidereal time being given, and also the right ascension of the pole of the ecliptic (which is always the same, viz. $18^h\ 0^m\ 0^s$), the hour angle Z P E of that point is known. Then, in this triangle we have given P Z, the colatitude ; P E, the polar distance of the pole of the ecliptic, $23°\ 28'$, and the angle Z P E from which we may find, 1st, the side Z E, which is easily seen to be equal to the altitude of the nonagesimal point sought; and 2dly, the angle Z P E, which is the azimuth of the pole of the ecliptic, and which, therefore, being added to and subtracted from $90°$, gives the azimuth of the eastern and western intersections of the ecliptic with the horizon. Lastly, the longitude of the nonagesimal point may be had, by calculating in the same triangle the angle P E Z, which is its complement.

(311.) The *angle of situation* of a star is the angle included between circles of latitude and of declination passing through it. To determine it in any proposed case, we must resolve the triangle P S E, in which are given P S, P E, and the angle S P E, which is the difference between the star's right ascension and 18 hours , from which it is easy to find the angle P S E required. This angle is of use in many inquiries in physical astronomy. It is called in most books on astronomy, the angle of *position*, but this expression has become otherwise and more conveniently appropriated. (See Art. 204.)

(312.) The same course of observations by which the path of the sun among the fixed stars is traced, and the ecliptic marked out among them, determines, of course, the place of the equinox V (Fig. art. 308.) upon the starry sphere, at that time—a point of great importance in practical astronomy, as it is the origin or zero point of right ascension. Now, when this process is repeated at considerably distant intervals of time, a very remarkable phenomenon is observed; viz. that the equinox does *not* preserve a constant place among the stars, but shifts its position, travelling continually and regularly, although with extreme slowness, *backwards*, along the ecliptic, in the direction V W from east to west, or the *contrary* to that in which the sun appears to move in that circle. As the ecliptic and equinoctial are not very much inclined, this motion of the equinox from east to west along the former, conspires (speaking generally) with the diurnal motion, and carries it, with reference to that motion, continually in advance upon the stars: hence it has acquired the name of the *precession of the equinoxes*, because the place of the equinox among the stars, at every subsequent moment, *precedes* (with reference to the diurnal motion) that which it held the moment before. The amount of this motion by which the equinox travels backward, or retrogrades (as it is called), on the ecliptic, is 0° 0′ 50·10″ *per annum*, an extremely minute quantity, but which, by its continual accumulation from year to year, at last makes itself very palpable, and that in a way highly inconvenient to practical astronomers, by destroying, in the lapse of a moderate number of years, the arrangement of their catalogues of stars, and making it necessary to reconstruct them. Since the formation of the earliest catalogue on record, the place of the equinox has retrograded already about 30°. The period in which it performs a complete tour of the ecliptic, is 25,868 years. *

(313.) The immediate uranographical effect of the precession of the equinoxes is to produce a uniform *increase of longitude* in all the heavenly bodies, whether fixed or erratic. For the vernal equinox being the initial point of longitudes, as well as of right ascension, a retreat of this point on the ecliptic *tells* upon the longitudes of all alike, whether at rest or in motion, and produces, so far as its amount extends, the

---

* Incipiunt magni procedere menses. — VIRGIL, *Pollio.*

appearance of a motion in longitude common to all, *as if* the whole heavens had a slow rotation round the poles of the ecliptic in the long period above mentioned, similar to what they have in twenty-four hours round those of the equinoctial. This increase of longitude, the reader will of course observe and bear in mind, is, properly speaking, neither a real nor an apparent *movement of the stars*. It is a purely technical result, arising from the gradual shifting of the zero point from which longitudes are reckoned. Had a fixed star been chosen as the origin of longitudes, they would have been invariable.

(314.) To form a just idea of this curious astronomical phenomenon, however, we must abandon, for a time, the consideration of the ecliptic, as tending to produce confusion in our ideas; for this reason, that the stability of the ecliptic itself among the stars is (as already hinted, art. 306.) only approximate, and that in consequence its intersection with the equinoctial is liable to a certain amount of change, arising from *its* fluctuation, which mixes itself with what is due to the principal uranographical cause of the phenomenon. This cause will become at once apparent, if, instead of regarding the equinox, we fix our attention on the pole of the equinoctial, or the vanishing point of the earth's axis.

(315.) The place of this point among the stars is easily determined at any epoch, by the most direct of all astronomical observations, — those with the meridian or mural circle. By this instrument we are enabled to ascertain at every moment the exact distance of the polar point from any three or more stars, and therefore to lay it down, by triangulating from these stars, with unerring precision, on a chart or globe, without the least reference to the position of the ecliptic, or to any other circle not naturally connected with it. Now, when this is done with proper diligence and exactness, it results that, although for short intervals of time, such as a few days, the place of the pole may be regarded as not sensibly variable, yet in reality it is in a state of constant, although extremely slow motion; and, what is still more remarkable, this motion is not uniform, but compounded of one principal, uniform, or nearly uniform, part, and other smaller and subordinate periodical fluctuations: the former giving rise to the phenomena of *precession;* the latter to another distinct phenomenon called *nutation.* These two phenomena, it is true, belong, theoretically speaking, to one and the same general head, and are intimately connected together, forming part of a great and complicated chain of consequences flowing from the earth's rotation on its axis: but it will be conducive to clearness at present to consider them separately.

(316.) It is found, then, that in virtue of the uniform part of the motion of the pole, it describes a circle in the heavens around the pole of the ecliptic as a centre, keeping constantly at the same distance of 23° 28′ from it in a direction from east to west, and with such a velocity, that the annual angle described by it, in this its imaginary orbit, is 50·10″; so that the whole circle would be described by it in the above-mentioned period of 25,868 years. It is easy to perceive how such a motion of the pole will give rise to the retrograde motion of the equinoxes; for in the figure, art. 308, suppose the pole P in the progress of its motion in the small circle P O Z round K to come to O, then, as the situation of the equinoctial E V Q is determined by that of the pole, this, it is evident, must cause a displacement of the equinoctial, which will take a new situation, E U Q, 90° distant in every part from the new position O of the pole. The point U, therefore, in which the displaced equinoctial will intersect the ecliptic, *i. e.* the displaced equinox, will lie on that side of V, its original position, towards which the motion of the pole is directed, or to the westward.

(317.) The precession of the equinoxes thus conceived, consists, then, in a real but very slow motion of the pole of the heavens among the stars, in a small circle round the pole of the ecliptic. Now this cannot happen without producing corresponding changes in the apparent diurnal motion of the sphere, and the aspect which the heavens must present at very remote periods of history. The pole is nothing more than the vanishing point of the earth's axis. As this point, then,

has such a motion as we have described, it necessarily follows
that the earth's *axis* must have a conical motion, in virtue of
which it points successively to every part of the small circle
in question.   We may form the best idea of such a motion
by noticing a child's peg-top, when it spins not upright, or
that amusing toy the te-to-tum, which, when delicately executed, and nicely balanced, becomes an elegant philosophical
instrument, and exhibits, in the most beautiful manner, the
whole phenomenon.   The reader will take care not to confound the variation of the *position of the earth*'s axis *in space*
with a mere shifting of the imaginary line about which it
revolves, in its interior.   The whole earth participates in the
motion, and goes along with the axis as if it were really a
bar of iron driven through it.   That such is the case is proved
by the two great facts: 1st, that the latitudes of places on
the earth, or their geographical situation with respect to the
poles, have undergone no perceptible change from the earliest
ages.   2dly, that the sea maintains its level, which could not
be the case if the motion of the axis were not accompanied
with a motion of the whole mass of the earth.*

(318.) The visible effect of precession on the aspect of the
heavens consists in the *apparent* approach of some stars and
constellations to the pole and recess of others.   The bright
star of the Lesser Bear, which we call the pole star, has not
always been, nor will always continue to be, our cynosure:
at the time of the construction of the earliest catalogues it
was 12° from the pole — it is now only 1° 24′, and will
approach yet nearer, to within half a degree, after which it
will again recede, and slowly give place to others, which will
succeed in its companionship to the pole.   After a lapse of
about 12,000 years, the star *a* Lyræ, the brightest in the
northern hemisphere, will occupy the remarkable situation of
a pole star approaching within about 5° of the pole.

(319.)   At the date of the erection of the Great Pyramid
of Gizeh, which precedes by 3970 years (say 4000) the pre-

* Local changes of the sea level, arising from purely geological causes, are
easily distinguished from that general and systematic alteration which a shifting
of the axis of rotation would give rise to.

sent epoch, the longitudes of all the stars were less by 55° 45' than at present. Calculating from this datum* the place of the pole of the heavens among the stars, it will be found to fall near *a* Draconis; its distance from that star being 3° 44' 25". This being the most conspicuous star in the immediate neighbourhood was therefore the pole star at that epoch. And the latitude of Gizeh being just 30° north, and consequently the altitude of the north pole there also 30°, it follows that the star in question must have had at its lower culmination, at Gizeh, an altitude of 26° 15' 35". Now it is a remarkable fact, ascertained by the late researches of Col. Vyse, that of the nine pyramids still existing at Gizeh, six (including all the largest) have the narrow passages by which alone they can be entered, (all which open out on the northern faces of their respective pyramids) inclined to the horizon downwards at angles as follows.

1st, or Pyramid of Cheops	-	-	-	26° 41'
2d, or Pyramid of Cephren	-	-	-	25 55
3d, or Pyramid of Mycerinus	-	-	-	26 2
4th,	-	-	-	-	27 0
5th,	-	-	-	-	27 12
9th,	-	-	-	-	28 0
			Mean -	26 47

Of the two pyramids at Abousseir also, which alone exist in a state of sufficient preservation to admit of the inclinations of their entrance passages being determined, one has the angle 27° 5', the other 26°.

(320.) At the bottom of every one of these passages therefore, the *then* pole star must have been visible at its lower culmination, a circumstance which can hardly be supposed to have been unintentional, and was doubtless connected (perhaps superstitiously) with the astronomical observation of that star, of whose proximity to the pole at the epoch of the erec-

* On this calculation the diminution of the obliquity of the ecliptic in the 4000 years elapsed has no influence. That diminution arises from a change in the plane of the earth's *orbit*, and has nothing to do with the change in the position of its *axis*, as referred to the starry sphere.

tion of these wonderful structures, we are thus furnished with a monumental record of the most imperishable nature.

(321.) The *nutation* of the earth's axis is a small and slow subordinate gyratory movement, by which, if subsisting alone, the pole would describe among the stars, in a period of about nineteen years, a minute ellipsis, having its longer axis equal to 18″·5, and its shorter to 13″·74 ; the longer being directed towards the pole of the ecliptic, and the shorter, of course, at right angles to it. The consequence of this real motion of the pole is an *apparent* approach and recess of all the stars in the heavens to the pole in the same period. Since, also, the place of the equinox on the ecliptic is determined by the place of the pole in the heavens, the same cause will give rise to a small alternate advance and recess of the equinoctial points, by which, in the same period, both the longitudes and the right ascensions of the stars will be also alternately increased and diminished.

(322.) Both these motions, however, although here considered separately, subsist jointly ; and since, while in virtue of the nutation, the pole is describing its little ellipse of 18″·5 in diameter, it is carried by the greater and regularly progressive motion of precession over so much of its circle round the pole of the ecliptic as corresponds to nineteen years, — that is to say, over an angle of nineteen times 50″·1 round the centre (which, in a small circle of 23° 28′ in diameter, corresponds to 6′ 20″, as seen from the centre of the sphere): the path which it will pursue in virtue of the two motions, subsisting jointly, will be neither an ellipse nor an exact circle, but a gently undulated ring like that in the figure (where, however, the undulations are much exaggerated). (See *fig.* to art. 325.)

(323.) These movements of precession and nutation are common to all the celestial bodies, both fixed and erratic ; and this circumstance makes it impossible to attribute them to any other cause than a real motion of the earth's axis such as we have described. Did they only affect the stars, they might, with equal plausibility, be urged to arise from a *real* rotation of the starry heavens, as a solid shell, round an axis passing through the poles of the ecliptic in 25,868 years, and a real elliptic gyration of *that* axis in nineteen years : but since they also affect the sun, moon, and planets, which, having motions independent of the general body of the stars, cannot without extravagance be supposed *attached to* the celestial concave*, this idea falls to the ground; and there only remains, then, a real motion in the earth by which they *can* be accounted for. It will be shown in a subsequent chapter that they are necessary consequences of the rotation of the earth, combined with its elliptical figure, and the unequal attraction of the sun and moon on its polar and equatorial regions.

(324.) Uranographically considered, as affecting the apparent places of the stars, they are of the utmost importance in practical astronomy. When we speak of the right ascension and declination of a celestial object, it becomes necessary to state what *epoch* we intend, and whether we mean the *mean* right ascension — cleared, that is, of the periodical fluctuation in its amount, which arises from nutation, or the *apparent* right ascension, which, being reckoned from the actual place of the vernal equinox, is affected by the periodical advance and recess of the equinoctial point produced by nutation — and so of the other elements. It is the practice of astronomers to *reduce*, as it is termed, all their observations, both of right ascension and declination, to some common and convenient epoch — such as the beginning of the year for temporary purposes, or of the decade, or the century for more permanent uses, by subtracting from them the whole effect of precession in the interval ; and, moreover, to divest them of the influence of nutation by investigating and subducting the amount of change, both in right ascension and declination, due to the displacement of the pole from the centre to the circumference of the little ellipse above mentioned. This last process is technically termed correcting

---

* This argument, cogent as it is, acquires additional and decisive force from the law of nutation, which is dependent on the position, for the time, of the lunar orbit. If we attribute it to a real motion of the celestial sphere, we must then maintain that sphere to be kept in a constant state of tremor by the motion of the moon !

or *equating* the observation for nutation; by which latter
word is always understood, in astronomy, the getting rid of
a periodical cause of fluctuation, and presenting a result, not
as it *was* observed, but as it would have been observed, had
that cause of fluctuation had no existence.

(325.) For these purposes, in the present case, very convenient formulæ have been derived, and tables constructed.
They are, however, of too technical a character for this
work; we shall, however, point out the manner in which
the investigation is conducted. It has been shown in art.
309. by what means the right ascension and declination of an
object are derived from its longitude and latitude. Referring
to the figure of that article, and supposing the triangle
K P X orthographically projected on the plane of the ecliptic
as in the annexed figure: in the triangle K P X, K P is the
obliquity of the ecliptic, K X the *co-latitude* (or complement
of latitude), and the angle P K X the *co-longitude* of the
object X. These are the *data* of our question, of which the
second is constant, and the other two are varied by the effect
of precession and nutation: and their variations (considering
the minuteness of the latter effect generally, and the small
number of years in comparison of the whole period of 25,868,
for which we ever require to estimate the effect of the
former,) are of that order
which may be regarded as
infinitesimal in geometry,
and treated as such without
fear of error. The whole
question, then, is reduced
to this: — In a spherical
triangle K P X, in which
one side K X is constant,
and an angle K, and ad-
jacent side K P vary by
given infinitesimal changes
of the position of P: re-

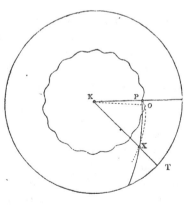

quired the changes thence arising in the other side P X, and
the angle K P X. This is a very simple and easy problem of

spherical geometry, and being resolved, it gives at once the
reductions we are seeking; for P X being the polar distance
of the object, and the angle K P X its right ascension *plus*
90°, their variations are the very quantities we seek. It only
remains, then, to express in proper form the amount of the
precession and nutation in *longitude* and *latitude*, when their
amount in right ascension and declination will immediately
be obtained.

(326.) The precession in *latitude* is zero, since the latitudes
of objects are not changed by it : that in longitude is a quantity proportional to the time at the rate of $50''\cdot10$ per annum.
With regard to the nutation in *longitude* and *latitude*, these
are no other than the abscissa and ordinate of the little
ellipse in which the pole moves. The law of its motion,
however, therein, cannot be understood till the reader has
been made acquainted with the principal features of the
moon's motion on which it depends.

(327.) Another consequence of what has been shown
respecting precession and nutation is, that *sidereal time*, as
astronomers use it, *i. e.* as reckoned from the transit of the
equinoctial point, is *not a mean or uniformly flowing quantity*,
being affected by nutation ; and, moreover, that *so* reckoned,
even when cleared of the periodical fluctuation of nutation,
it does not *strictly* correspond to the earth's diurnal rotation.
As the sun *loses* one day in the year on the stars, by its
direct motion in longitude ; so the equinox *gains* one day in
25,868 years on them by its *retrogradation*. We ought,
therefore, as carefully to distinguish between mean and
apparent sidereal as between mean and apparent solar time.

(328.) Neither precession nor nutation change the apparent
places of celestial objects *inter se*. We see them, so far as
these causes go, *as they are*, though from a station more or
less unstable, as we see distant land objects correctly formed,
though appearing to rise and fall when viewed from the
heaving deck of a ship in the act of pitching and rolling.
But there is an optical cause, independent of refraction or of
perspective, which displaces them *one among the other*, and
causes us to view the heavens under an aspect always to a

certain slight extent false; and whose influence must be estimated and allowed for before we can obtain a precise knowledge of the place of any object. This cause is what is called the aberration of light; a singular and surprising effect arising from this, that we occupy a station not at rest but in rapid motion; and that the apparent directions of the rays of light are not the same to a spectator in motion as to one at rest. As the estimation of its effect belongs to uranography, we must explain it here, though, in so doing, we must anticipate some of the results to be detailed in subsequent chapters.

(329.) Suppose a shower of rain to fall perpendicularly in a dead calm; a person exposed to the shower, who should stand quite still and upright, would receive the drops on his hat, which would thus shelter him, but if he ran forward in any direction they would strike him in the face. The effect would be the same as if he remained still, and a wind should arise of the same velocity, and drift them against him. Suppose a ball let fall from a point A above a horizontal line E F, and that at B were placed to receive it the open mouth of an inclined hollow tube P Q; if the tube were held im-

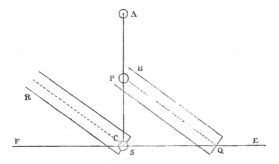

moveable the ball would strike on its lower side, but if the tube were carried forward in the direction E F, with a velocity properly adjusted at every instant to that of the ball, while *preserving its inclination* to the horizon, so that when the ball in its natural descent reached C, the tube should have been carried into the position R S, it is evident that the ball would, throughout its whole descent, be found in the axis of the tube; and a spectator referring to the tube the motion of the ball, and carried along with the former, unconscious of its motion, would fancy that the ball had been moving in the inclined direction R S of the tube's axis.

(330.) Our eyes and telescopes are such tubes. In whatever manner we consider light, whether as an advancing wave in a motionless ether, or a shower of atoms traversing space, (provided that in both cases we regard it as absolutely incapable of suffering resistance or corporeal obstruction from the particles of transparent media traversed by it*,) if in the interval between the rays traversing the object glass of the one or the cornea of the other (*at which moment* they acquire that convergence which directs them to a certain point *in fixed space*), and their arrival at their focus, the cross wires of the one or the retina of the other be *slipped aside*, the point of convergence (which remains unchanged) will no longer correspond to the intersection of the wires or the central point of our visual area. The object then will *appear* displaced; and the amount of this displacement is *aberration*.

(331.) The earth is moving through space with a velocity of about 19 miles per second, in an elliptic path round the sun, and is therefore changing the direction of its motion at every instant. Light travels with a velocity of 192,000 miles per second, which, although much greater than that of the earth, is yet not *infinitely so*. Time is occupied by it in traversing any space, and in that time the earth describes a space which is to the former as 19 to 192,000, or as the tangent of 20″·5 to radius. Suppose now A P S to represent a ray of light from a star at A, and let the tube P Q be that of a telescope so inclined forward that the focus *formed* by

---

* This condition is indispensable. Without it we fall into all those difficulties which M. Doppler has so well pointed out in his paper on Aberration (Abhandlungen der k. boemischen Gesellschaft der Wissenschaften. Folge V. vol. iii.). If light itself, or the luminiferous ether, be corporeal, the condition insisted on amounts to a formal surrender of the dogma, either of the extension or of the impenetrability of matter; at least in the sense in which those terms have been hitherto used by metaphysicians. At the point to which science is arrived, probably few will be found disposed to maintain either the one or the other.

its object glass shall be *received* upon its cross wire, it is evident from what has been said, that the inclination of the tube must be such as to make P S : S Q : : velocity of light : velocity of the earth : : 1 : tan. 20″·5 ; and, therefore, the angle S P Q, or P S R, by which the axis of the telescope must deviate from the true direction of the star, must be 20″·5.

(332.) A similar reasoning will hold good when the direction of the earth's motion is not perpendicular to the visual ray. If S B be the true direction of the visual ray, and A C the position in which the telescope requires to be held in the apparent direction, we must still have the proportion B C : B A : : velocity of light : velocity of the earth : : rad. : sine of 20″·5 (for in such small angles it matters not whether we use the sines or tangents). But we have, also, by trigonometry, B C : B A : : sine of B A C : sine of A C B or C B D, which last is the apparent displacement caused by aberration. Thus it appears that the sine of the aberration, or (since the angle is extremely small) the aberration itself, is proportional to the sine of the angle made by the earth's motion in space with the visual ray, and is therefore a maximum when the line of sight is perpendicular to the direction of the earth's motion.

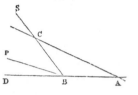

(333.) The uranographical effect of aberration, then, is to distort the aspect of the heavens, causing all the stars to crowd as it were directly towards that point in the heavens which is the vanishing point of all lines parallel to that in which the earth is for the moment moving. As the earth moves round the sun in the plane of the ecliptic, this point must lie in that plane, 90° in advance of the earth's longitude, or 90° *behind* the sun's, and shifts of course continually, describing the circumference of the ecliptic in a year. It is easy to demonstrate that the effect on each particular star will be to make it apparently describe a small ellipse in the heavens, having for its centre the point in which the star would be seen if the earth were at rest.

(334.) Aberration then affects the apparent right ascensions and declinations of all the stars, and that by quantities easily calculable. The formulæ most convenient for that purpose, and which, systematically embracing at the same time the corrections for precession and nutation, enable the observer, with the utmost readiness, to disencumber his observations of right ascension and declination of their influence, have been constructed by Prof. Bessel, and tabulated in the appendix to the first volume of the Transactions of the Astronomical Society, where they will be found accompanied with an extensive catalogue of the places, for 1830, of the principal fixed stars, one of the most useful and best arranged works of the kind which has ever appeared.

(335.) When the body from which the visual ray emanates is itself in motion, an effect arises which is not properly speaking aberration, though it is usually treated under that head in astronomical books, and indeed confounded with it, to the production of some confusion in the mind of the student. The effect in question (which is independent of any theoretical views respecting the nature of light*) may be explained as follows. The ray by which we see any object is not that which it emits at the moment we look at it, but that which it *did* emit some time before, *viz.* the time occupied by light in traversing the interval which separates it from us. The aberration of such a body then arising from the earth's velocity must be applied as a correction, not to the line joining the earth's place at the moment of observation with that occupied by the body *at the same moment,* but at that antecedent instant when the ray quitted it. Hence it is easy to derive the rule given by astronomical writers for the case

---

* The results of the undulatory and corpuscular theories of light, in the matter of aberration are, in the main, the same. We say *in the main.* There is, however, a minute difference even of numerical results. In the undulatory doctrine, the propagation of light takes place with equal velocity in all directions, whether the luminary be at rest or in motion. In the corpuscular, with an excess of velocity in the direction of the motion over that in the contrary equal to twice the velocity of the body's motion. In the cases, then, of a body moving with equal velocity directly to and directly from the earth, the aberrations will be *alike* on the undulatory, but different on the corpuscular hypothesis. The utmost difference which can arise from this cause *in our system* cannot amount to above six thousandths of a second.

of a moving object. *From the known laws of its motion and the earth's, calculate its apparent or relative angular motion in the time taken by light to traverse its distance from the earth. This is the total amount of its apparent misplacement.* Its effect is to displace the body observed in a direction contrary to its apparent motion in the heavens. And it is a compound or aggregate effect consisting of two parts, one of which is the aberration, properly so called, resulting from the composition of the earth's motion with that of light, the other being what is not inaptly termed the *Equation of light,* being the allowance to be made for the *time* occupied by the light in traversing a variable space.

(336.) The complete *Reduction,* as it is called, of an astronomical observation consists in applying to the place of the observed heavenly body as read off on the instruments (supposed perfect and in perfect adjustment) five distinct and independent corrections, viz. those for refraction, parallax, aberration, precession, and nutation. Of these the correction for refraction enables us to declare what would have been the observed place, were there no atmosphere to displace it. That for parallax enables us to say from its place observed at the surface of the earth, where it would have been seen if observed from the centre. That for aberration, where it would have been observed from a motionless, instead of a moving station : while the corrections for precession and nutation refer it to fixed and determinate instead of constantly varying celestial circles. The great importance of these corrections, which pervade all astronomy, and have to be applied to every observation before it can be employed for any practical or theoretical purpose, renders this recapitulation far from superfluous.

(337.) Refraction has been already sufficiently explained, Art. 40. and it is only, therefore, necessary here to add that in its use as an astronomical correction its amount must be applied in a contrary sense to that in which it affects the observation ; a remark equally applicable to all other corrections.

(338.) The general nature of parallax or rather of parallactic motion has also been explained in Art. 80. But parallax in the uranographical sense of the word has a more technical meaning. It is understood to express that optical displacement of a body observed which is due to its being observed, not from that point which we have fixed upon as a conventional central station (from which we conceive the apparent motion would be more simple in its laws), but from some other station remote from such conventional centre: not from the centre of the earth, for instance, but from its surface: not from the centre of the sun (which, as we shall hereafter see, is for some purposes a preferable conventional station), but from that of the earth. In the former case this optical displacement is called the *diurnal* or geocentric parallax; in the latter the *annual* or heliocentric. In either case parallax is the *correction* to be applied to the apparent place of the heavenly body, as actually seen from the station of observation, to *reduce* it to its place as it would have been seen at that instant from the conventional station.

(339.) The diurnal or geocentric parallax at any place of the earth's surface is easily calculated if we know the distance of the body, and, *vice versâ*, if we know the diurnal parallax that distance may be calculated. For supposing S the object, C the centre of the earth, A the station of observation at its surface, and C A Z the direction of a perpendicular to the surface at A, then will the object be seen from A in the direction A S, and its apparent zenith distance will be Z A S; whereas, if seen from the centre, it will appear in the direction C S, with an angular distance from the zenith of A equal to Z C S; 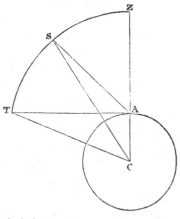 so that Z A S—Z C S or A S C is the parallax. Now since by trigonometry C S : C A :: sin C A S = sin

Z A S : sin A S C, it follows that the sine of the parallax

$$= \frac{\text{Radius of earth}}{\text{Distance of body}} \times \sin \text{Z A S}.$$

(340.) The diurnal or geocentric parallax, therefore, at a given place, and for a given distance of the body observed, is proportional to the sine of its apparent zenith distance, and is, therefore, the greatest when the body is observed in the act of rising or setting, in which case its parallax is called its *horizontal* parallax, so that at any other zenith distance, parallax = horizontal parallax × sine of apparent zenith distance, and since A C S is always less than Z A S it appears that the application of the reduction or correction for parallax always acts in diminution of the apparent zenith distance or increase of the apparent altitude or distance from the Nadir, *i. e.* in a contrary sense to that for refraction.

(341.) In precisely the same manner as the geocentric or diurnal parallax refers itself to the zenith of the observer for its direction and quantitative rule, so the heliocentric or annual parallax refers itself for its law to the point in the heavens diametrically opposite to the place of the sun as seen from the earth. Applied as a correction, its effect takes place in a plane passing through the sun, the earth, and the observed body. Its effect is always to decrease its observed distance from that point or to increase its angular distance from the sun. And its sine is given by the relation, Distance of the observed body from the sun : distance of the earth from the sun :: sine of apparent angular distance of the body from the sun (or its apparent *elongation*) : sine of heliocentric parallax. *

(342.) On a summary view of the whole of the uranographical corrections, they divide themselves into two classes, those which *do*, and those which *do not*, alter the apparent configurations of the heavenly bodies *inter se*. The former are *real*, the latter *technical* corrections. The real corrections are refraction, aberration and parallax. The technical are

---

* This account of the law of heliocentric parallax is in anticipation of what follows in a subsequent chapter, and will be better understood by the student when somewhat farther advanced.

precession and nutation, unless, indeed, we choose to consider parallax as a technical correction introduced with a view to simplification by a better choice of our point of sight.

(343.) The corrections of the first of these classes have one peculiarity in respect of their law, common to them all, which the student of practical astronomy will do well to fix in his memory. *They all refer themselves to definite apexes or points of convergence in the sphere.*   Thus, refraction in its apparent effect causes all celestial objects to draw together or converge towards the zenith of the observer : geocentric parallax, towards his Nadir : heliocentric, towards the place of the sun in the heavens : aberration towards that point in the celestial sphere which is the vanishing point of all lines parallel to the direction of the earth's motion at the moment, or (as will be hereafter explained) towards a point in the great circle called the ecliptic, 90° behind the sun's place in that circle.   When applied as corrections to an observation, these directions are of course to be reversed.

(344.) In the quantitative law, too, which this class of corrections follow, a like agreement takes place, at least as regards the geocentric and heliocentric parallax and aberration, in all three of which the amount of the correction (or more strictly its sine) increases in the direct proportion of the sine of the apparent distance of the observed body from the *apex* appropriate to the particular correction in question.   In the case of refraction the law is less simple, agreeing more nearly with the tangent than the sine of that distance, but agreeing with the others in placing the maximum at 90° from its apex.

(345.) As respects the *order* in which these corrections are to be applied to any observation, it is as follows: 1. Refraction; 2. Aberration; 3. Geocentric Parallax; 4. Heliocentric Parallax; 5. Nutation; 6. Precession.   Such, at least, is the order in theoretical strictness.   But as the amount of aberration and nutation is in all cases a very minute quantity, it matters not in what order they are applied ; so that for practical convenience they are always thrown together with the precession, and applied after the others.

CHAPTER VI.

OF THE SUN'S MOTION.

APPARENT MOTION OF THE SUN NOT UNIFORM. — ITS APPARENT DIAMETER ALSO VARIABLE. — VARIATION OF ITS DISTANCE CONCLUDED. — ITS APPARENT ORBIT AN ELLIPSE ABOUT THE FOCUS. — LAW OF THE ANGULAR VELOCITY. — EQUABLE DESCRIPTION OF AREAS. — PARALLAX OF THE SUN. — ITS DISTANCE AND MAGNITUDE. — COPERNICAN EXPLANATION OF THE SUN'S APPARENT MOTION. — PARALLELISM OF THE EARTH'S AXIS. — THE SEASONS. — HEAT RECEIVED FROM THE SUN IN DIFFERENT PARTS OF THE ORBIT. — MEAN AND TRUE LONGITUDES OF THE SUN. — EQUATION OF THE CENTRE. — SIDEREAL, TROPICAL, AND ANOMALISTIC YEARS. — PHYSICAL CONSTITUTION OF THE SUN — ITS SPOTS. — FACULÆ. — PROBABLE NATURE AND CAUSE OF THE SPOTS. — ATMOSPHERE OF THE SUN — ITS SUPPOSED CLOUDS — TEMPERATURE AT ITS SURFACE — ITS EXPENDITURE OF HEAT. — TERRESTRIAL EFFECTS OF SOLAR RADIATION.

(346.) IN the foregoing chapters, it has been shown that the apparent path of the sun is a great circle of the sphere, which it performs in a period of one sidereal year. From this it follows, that the line joining the earth and sun lies constantly *in one plane ;* and that, therefore, whatever be the real motion from which this apparent motion arises, it must be confined to one plane, which is called the *plane of the ecliptic.*

(347.) We have already seen (art. 146.) that the sun's motion in right ascension among the stars is not uniform. This is partly accounted for by the obliquity of the ecliptic, in consequence of which equal variations in longitude do not correspond to equal changes of right ascension. But if we observe the place of the sun daily throughout the year, by the transit and circle, and from these calculate the longitude for each day, it will still be found that, even in its own proper path, its apparent angular motion is far from uniform. The change of longitude in twenty-four mean solar hours *averages* 0° 59′ 8″·33 ; but about the 31st of December it amounts to 1° 1′ 9″·9, and about the 1st of July is only 0° 57′ 11″·5. Such are the extreme limits, and such the mean value of the sun's apparent angular velocity in its annual orbit.

(348.) This variation of its angular velocity is accompanied with a corresponding change of its distance from us. The change of distance is recognized by a variation observed to take place in its apparent diameter, when measured at different seasons of the year, with an instrument adapted for that purpose, called the *heliometer**, or, by calculating from the time which its disc takes to traverse the meridian in the transit instrument. The greatest apparent diameter corresponds to the 1st of December, or to the greatest angular velocity, and measures 32′ 35″·6, the least is 31′ 31″·0 ; and corresponds to the 1st of July; at which epochs, as we have seen, the angular motion is also at its extreme limit either way. Now, as we cannot suppose the sun to alter its real size periodically, the observed change, of its apparent size can only arise from an actual change of distance. And the sines or tangents of such small arcs being proportional to the arcs themselves, its distances from us, at the above-named epoch, must be in the inverse proportion of the apparent diameters. It appears, therefore, that the greatest, the mean, and the least distances of the sun from us are in the respective proportions of the numbers 1·01679, 1·00000, and 0·98321; and that its apparent angular velocity diminishes as the distance increases, and *vice versâ*.

(349.) It follows from this, that the real orbit of the sun, as referred to the earth supposed at rest, is not a circle with the earth in the centre. The situation of the earth within it is *excentric*, the *excentricity* amounting to 0·01679 of the mean distance, which may be regarded as our unit of measure in this inquiry. But besides this, the *form* of the orbit is not circular, but elliptic. If from any point O, taken to represent the earth, we draw a line, O A, in some fixed

---

* Ἥλιος the sun, and μετρειν to measure.

direction, from which we then set off a series of angles, A O B, A O C, &c. equal to the observed longitudes of the sun throughout the year, and in these respective directions mea- sure off from O the distances O A, O B, O C, &c. representing the distances deduced from the observed diameter, and then con- nect all the extremities A, B, C,

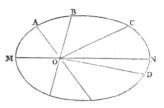

&c. of these lines by a continuous curve, it is evident this will be a correct representation of the relative orbit of the sun about the earth.   Now, when this is done, a deviation from the cir- cular figure in the resulting curve becomes apparent ; it is found to be evidently longer than it is broad — that is to say, ellip- tic, and the point O to occupy, not the *centre*, but one of the foci of the ellipse.   The graphical process here described is sufficient to point out the general figure of the curve in ques- tion ; but for the purposes of exact verification, it is necessary to recur to the properties of the ellipse *, and to express the distance of any one of its points in terms of the angular situ- ation of that point with respect to the longer axis, or diameter of the ellipse.   This, however, is readily done; and when nu- merically calculated, on the supposition of the excentricity being such as above stated, a perfect coincidence is found to subsist between the distances thus computed, and those de- rived from the measurement of the apparent diameter.

(350.) The mean distance of the earth and sun being taken for unity, the extremes are 1·01679 and 0·98321.   But if we compare, in like manner, the mean or average angular velocity with the extremes, greatest and least, we shall find these to be in the proportions of 1·03386, 1·00000, and 0·96670. The variation of the sun's *angular velocity*, then, is much greater in proportion than that of its distance — fully twice as great ; and if we examine its numerical expressions at dif- ferent periods, comparing them with the mean value, and also with the corresponding distances, it will be found, that, by

---

* See Conic Sections, by the Rev. H. P. Hamilton, or any other of the very numerous works on this subject.

whatever fraction of its mean value the distance exceeds the mean, the angular velocity will fall short of *its* mean or average quantity by very nearly *twice* as great a fraction of the latter, and *vice versa*. Hence we are led to conclude that the *angular velocity* is in the inverse proportion, not of the distance simply, but of its *square;* so that, to compare the daily motion in longitude of the sun, at one point, A, of its path, with that at B, we must state the proportion thus : —

O B² : O A² :: daily motion at A : daily motion at B. And this is found to be exactly verified in every part of the orbit.

(351.) Hence we deduce another remarkable conclusion — viz. that if the sun be supposed really to move around the circumference of this ellipse, its actual speed cannot be uniform, but must be greatest at its least distance and less at its greatest. For, were it uniform, the apparent angular velocity would be, of course, inversely proportional to the distance; simply because the same linear change of place, being produced in the same time at different distances from the eye, must, by the laws of perspective, correspond to apparent angular displacements inversely as those distances. Since, then, observation indicates a more rapid law of variation in the angular velocities, it is evident that mere change of distance, unaccompanied with a change of actual speed, is insufficient to account for it; and that the increased proximity of the sun to the earth must be accompanied with an actual increase of its real velocity of motion along its path.

(352.) This elliptic form of the sun's path, the excentric position of the earth within it, and the unequal speed with which it is actually traversed by the sun itself, all tend to render the calculation of its longitude from theory (*i. e.* from a knowledge of the causes and nature of its motion) difficult; and indeed impossible, so long as the *law* of its actual velocity continues unknown. This *law*, however, is not immediately apparent. It does not come forward, as it were, and present itself at once, like the elliptic form of the orbit, by a direct comparison of angles and distances, but requires an attentive consideration of the whole series of observations registered during an entire period. It was not, therefore, without much painful and laborious calculation, that it was discovered by Kepler (who was also the first to ascertain the elliptic form of the orbit), and announced in the following terms : — Let a line be always supposed to connect the sun, supposed in motion, with the earth, supposed at rest; then, as the sun moves along its ellipse, this line (which is called in astronomy the *radius vector*) will *describe* or *sweep over* that portion of the whole *area* or *surface* of the ellipse which is included between its consecutive positions : and the motion of the sun will be such that *equal areas are* thus *swept over* by the revolving radius vector *in equal times,* in whatever part of the circumference of the ellipse the sun may be moving.

(353.) From this it necessarily follows, that in *un*equal times, the areas described must be proportional to the times. Thus, in the figure of art. 349. the time in which the sun moves from A to B, is the time in which it moves from C to D, as the area of the elliptic sector A O B is to the area of the sector D O C.

(354.) The circumstances of the sun's apparent annual motion may, therefore, be summed up as follows : — It is performed in an orbit lying in one plane passing through the earth's centre, called the plane of the ecliptic, and whose projection on the heavens is the great circle so called. In this plane, however, the actual path is not circular, but elliptical; having the earth, not in its centre, but in one focus. The excentricity of this ellipse is 0·01679, in parts of a unit equal to the *mean distance,* or *half the longer diameter of the ellipse;* i. e. about one sixtieth part of that semi-diameter; and the motion of the sun in its circumference is so regulated, that equal areas of the ellipse are passed over by the radius vector in equal times.

(355.) What we have here stated supposes no knowledge of the sun's actual distance from the earth, nor, consequently, of the actual dimensions of its orbit, nor of the body of the sun itself. To come to any conclusions on these points, we must first consider by what means we can arrive at any knowledge of the distance of an object to which we have no

access. Now, it is obvious, that its *parallax* alone can afford
us any information on this subject. Suppose P A B Q to
represent the earth, C its centre, and S the sun, and A, B

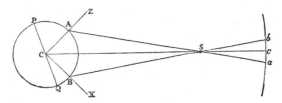

two situations of a spectator, or, which comes to the same
thing, the stations of two spectators, both observing the sun
S at the same instant. The spectator A will see it in the
direction A S *a*, and will refer it to a point *a* in the infinitely
distant sphere of the fixed stars, while the spectator B will
see it in the direction B S *b*, and refer it to *b*. The angle
included between these directions, or the measure of the
celestial arc *a b*, by which it is *displaced*, is equal to the angle
A S B; and if this angle be known, and the local situations
of A and B, with the part of the earth's surface A B included
between them, it is evident that the distance C S may be
calculated. Now, since A S C (art. 339.) is the parallax of
the sun as seen from A, and B S C as seen from B, the angle
A S B, or the total apparent displacement is the sum of the
two parallaxes. Suppose, then, two observers — one in the
northern, the other in the southern hemisphere — at stations
on the same meridian, to observe on the same day the meridian
altitudes of the sun's centre. Having thence derived the
apparent zenith distances, and cleared them of the effects of
refraction, if the distance of the sun were equal to that of the
fixed stars, the sum of the zenith distances thus found would
be precisely equal to the sum of the latitudes north and south
of the places of observation. For the sum in question would
then be equal to the angle Z C X, which is the meridional
distance of the stations across the equator. But the effect
of parallax being in both cases to increase the apparent zenith
distances, their observed sum will be greater than the sum of
the latitudes, by the sum of the two parallaxes, or by the

angle A S B. This angle, then, is obtained by subducting the sum of the north and south latitudes from that of the zenith distances; and this once determined, the horizontal parallax is easily found, by dividing the angle so determined by the sum of the sines of the two latitudes.

(356.) If the two stations be not exactly on the same meridian (a condition very difficult to fulfil), the same process will apply, if we take care to allow for the change of the sun's actual zenith distance in the interval of time elapsing between its arrival on the meridians of the stations. This change is readily ascertained, either from tables of the sun's motion, grounded on the experience of a long course of observations, or by actual observation of its meridional altitude on several days before and after that on which the observations for parallax are taken. Of course, the nearer the stations are to each other in longitude, the less is this interval of time, and, consequently, the smaller the amount of this correction; and, therefore, the less injurious to the accuracy of the final result is any uncertainty in the daily change of zenith distance which may arise from imperfection in the solar tables, or in the observations made to determine it.

(357.) The horizontal parallax of the sun has been concluded from observations of the nature above described, performed in stations the most remote from each other in latitude, at which observatories have been instituted. It has also been deduced from other methods of a more refined nature, and susceptible of much greater exactness, to be hereafter described. Its amount so obtained, is about $8''\cdot6$. Minute as this quantity is, there can be no doubt that it is a tolerably correct approximation to the truth; and in conformity with it, we must admit the sun to be situated at a mean distance from us, of no less than 23984 times the length of the earth's radius, or about 95000000 miles.

(358.) That at so vast a distance the sun should appear to us of the size it does, and should so powerfully influence our condition by its heat and light, requires us to form a very grand conception of its actual magnitude, and of the scale on which those important processes are carried on within it, by

which it is enabled to keep up its liberal and unceasing supply
of these elements. As to its actual magnitude we can be at
no loss, knowing its distance, and the angles under which its
diameter appears to us. An object, placed at the distance of
95000000 miles, and subtending an angle of 32′ 3″, must
have a real diameter of 882000 miles. Such, then, is the
diameter of this stupendous globe. If we compare it with
what we have already ascertained of the dimensions of our
own, we shall find that in linear magnitude it exceeds the
earth in the proportion $111\frac{1}{2}$ to 1, and in bulk in that of
1384472 to 1.

(359.) It is hardly possible to avoid associating our conception of an object of definite globular figure, and of such
enormous dimensions, with some corresponding attribute of
massiveness and material solidity. That the sun is not a
mere phantom, but a body having its own peculiar structure
and economy, our telescopes distinctly inform us. They show
us dark spots on its surface, which slowly change their places
and forms, and by attending to whose situation, at different
times, astronomers have ascertained that the sun revolves
about an axis nearly perpendicular to the plane of the
ecliptic, performing one rotation in a period of about 25 days,
and in the same direction with the diurnal rotation of
the earth, *i. e.* from west to east. Here, then, we have an
analogy with our own globe; the slower and more majestic
movement only corresponding with the greater dimensions of
the machinery, and impressing us with the prevalence of
similar mechanical laws, and of, at least, such a community
of nature as the existence of inertia and obedience to force
may argue. Now, in the exact proportion in which we invest
our idea of this immense bulk with the attribute of inertia, or
weight, it becomes difficult to conceive its circulation round
so comparatively small a body as the earth, without, on the
one hand, dragging it along, and displacing it, if bound to it
by some invisible tie; or, on the other hand, if not so held to
it, pursuing its course alone in space, and leaving the earth
behind. If we connect two solid masses by a rod, and fling
them aloft, we see them circulate about a point between them,

which is their common centre of gravity ; but if one of them be greatly more ponderous than the other, this common centre will be proportionally nearer to that one, and even within its surface ; so that the smaller one will circulate, in fact, about the larger, which will be comparatively but little disturbed from its place.

(360.) Whether the earth move round the sun, the sun round the earth, or both round their common centre of gravity, will make no difference, so far as appearances are concerned, provided the stars be supposed sufficiently distant to undergo no sensible apparent *parallactic* displacement by the motion so attributed to the earth. Whether they are so or not must still be a matter of enquiry; and from the absence of any measureable amount of such displacement, we can conclude nothing but this, that the scale of the sidereal universe is so great, that the mutual orbit of the earth and sun may be regarded as an imperceptible point in comparison with the distance of its nearest members. Admitting, then, in conformity with the laws of dynamics, that two bodies connected with and revolving about each other in free space do, in fact, revolve about their common centre of gravity, which remains immoveable by their mutual action, it becomes a matter of further enquiry, *whereabouts* between them this centre is situated. Mechanics teach us that its place will divide their mutual distance in the inverse ratio of their *weights* or *masses* ;* and calculations grounded on phenomena, of which an account will be given further on, inform us that this ratio, in the case of the sun and earth, is actually that of 354936 to 1, — the sun being, in that proportion, more ponderous than the earth. From this it will follow that the common point about which they both circulate is only 267 miles from the sun's centre, or about $\frac{1}{3300}$th part of its own diameter.

(361.) Henceforward, then, in conformity with the above statements, and with the Copernican view of our system, we must learn to look upon the sun as the comparatively motionless centre about which the earth performs an annual elliptic orbit of the dimensions and excentricity, and with a velocity,

* Principia, lib. i. lex. iii. cor. 14.

regulated according to the law above assigned; the sun occupying one of the foci of the ellipse, and from that station quietly disseminating on all sides its light and heat; while the earth travelling round it, and presenting itself differently to it at different times of the year and day, passes through the varieties of day and night, summer and winter, which we enjoy.

(362.) In this annual motion of the earth, its axis preserves, at all times, the same direction as if the orbital movement had no existence; and is carried round parallel to itself, and pointing always to the same vanishing point in the sphere of the fixed stars. This it is which gives rise to the variety of seasons, as we shall now explain. In so doing, we shall neglect (for a reason which will be presently explained) the ellipticity of the orbit, and suppose it a circle, with the sun in the centre.

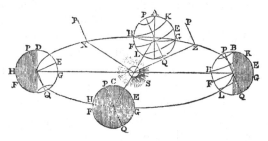

(363.) Let, then, S represent the sun, and A, B, C, D, four positions of the earth in its orbit 90° apart, viz. A that which it has on the 21st of March, or at the time of the vernal equinox; B that of the 21st of June, or the summer solstice; C that of the 21st of September, or the autumnal equinox; and D that of the 21st of December, or the winter solstice. In each of these positions let P Q represent the axis of the earth, about which its diurnal rotation is performed without interfering with its annual motion in its orbit. Then, since the sun can only enlighten one half of the surface at once, viz. that turned towards it, the shaded portions of the globe in its several positions will represent the dark, and the bright, the enlightened halves of the earth's surface in these positions. Now, 1st, in the position A, the sun is vertically over the intersection of the equinoctial F E and the ecliptic H G. It is, therefore, in the equinox; and in this position the poles P Q, both fall on the extreme confines of the enlightened side. In this position, therefore, it is day over half the northern and half the southern hemisphere at once; and as the earth revolves on its axis, every point of its surface describes half its diurnal course in light, and half in darkness; in other words, the duration of day and night is here equal over the whole globe: hence the term *equinox.* The same holds good at the autumnal equinox on the position C.

(364.) B is the position of the earth at the time of the *northern summer* solstice. Here the north pole P, and a considerable portion of the earth's surface in its neighbourhood, as far as B, are situated *within* the enlightened half. As the earth turns on its axis in this position, therefore, the whole of that part remains constantly enlightened; therefore, at this point of its orbit, or at this season of the year, it is continual day at the north pole, and in all that region of the earth which encircles this pole as far as B — that is, to the distance of 23° 28′ from the pole, or within what is called in geography, the *arctic circle.* On the other hand, the opposite or south pole Q, with all the region comprised within the *antarctic* circle, as far as 23° 28′ from the south pole, are immersed at this season in darkness during the entire diurnal rotation, so that it is here continual night.

(365.) With regard to that portion of the surface comprehended between the arctic and antarctic circles, it is no less evident that the nearer any point is to the north pole, the larger will be the portion of its diurnal course comprised within the bright, and the smaller within the dark hemisphere; that is to say, the longer will be its day, and the shorter its night. Every station north of the equator will have a day of more and a night of less than twelve hours' duration, and *vice versâ.* All these phenomena are exactly inverted when the earth comes to the opposite point D of its orbit.

(366.) Now, the temperature of any part of the earth's surface depends mainly on its exposure to the sun's rays. Whenever the sun is above the horizon of any place, that place is receiving heat ; when below, parting with it, by the process called radiation ; and the whole quantities received and parted with in the year (secondary causes apart) must balance each other at every station, or the equilibrium of temperature (that is to say, the constancy which is observed to prevail in the annual averages of temperature as indicated by the thermometer) would not be supported.   Whenever, then, the sun remains more than twelve hours above the horizon of any place, and less beneath, the general temperature of that place will be above the average; when the reverse, below. As the earth, then, moves from A to B, the days growing longer, and the nights shorter, in the northern hemisphere, the temperature of every part of that hemisphere increases, and we pass from spring to summer; while, at the same time, the reverse obtains in the southern hemisphere.   As the earth passes from B to C, the days and nights again approach to equality—the excess of temperature in the northern hemisphere above the mean state grows less, as well as its defect in the southern ; and at the autumnal equinox C, the mean state is once more attained.   From thence to D, and, finally, round again to A, all the same phenomena, it is obvious, must again occur, but reversed,—it being now winter in the northern and summer in the southern hemisphere.

(367.) All this is exactly consonant to observed fact. The continual day within the polar circles in summer, and night in winter, the general increase of temperature and length of day as the sun approaches the elevated pole, and the reversal of the seasons in the northern and southern hemispheres, are all facts too well known to require further comment.   The positions A, C of the earth correspond, as we have said, to the equinoxes ; those at B, D to the *solstices*. This term must be explained.   If, at any point, X, of the orbit, we draw X P the earth's axis, and X S to the sun, it is evident that the angle P X S will be the sun's *polar distance.* Now, this angle is at its maximum in the position D, and at its minimum at B ; being in the former case = 90° + 23° 28′ = 103° 28′, and in the latter 90°—23° 28′ = 66° 32′. At these points the sun ceases to approach to or to recede from the pole, and hence the name solstice.

(368.) The elliptic form of the earth's orbit has but a very trifling share in producing the variation of temperature corresponding to the difference of seasons. This assertion may at first sight seem incompatible with what we know of the laws of the communication of heat from a luminary placed at a variable distance. Heat, like light, being equally dispersed from the sun in all directions, and being spread over the surface of a sphere continually enlarging as we recede from the centre, must, of course, diminish in intensity according to the inverse proportion of the surface of the sphere over which it is spread; that is, in the inverse proportion of the square of the distance. But we have seen (art. 350.) that this is also the proportion in which the *angular velocity* of the earth about the sun varies. Hence it appears, that the *momentary supply of heat* received by the earth from the sun varies in the exact proportion of the angular velocity, *i. e.* of the *momentary increase of longitude:* and from this it follows, that equal amounts of heat are received from the sun in passing over equal angles round it, in whatever part of the ellipse those angles may be situated. Let, then, S represent the sun; A Q M P the earth's orbit; A its nearest point to the sun, or, as it is called, the *perihelion* of its orbit ; M the farthest, or the *aphelion* ; and therefore A S M the *axis* of the ellipse. Now, suppose the orbit divided into two segments by a straight line P S Q, drawn through the sun, and anyhow situated as to direction; then, if we

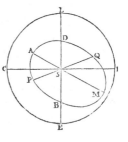

suppose the earth to circulate in the direction P A Q M P, it will have passed over 180° of longitude in moving from P to Q, and as many in moving from Q to P. It appears, therefore, from what has been shown, that the supplies of heat received from the sun will be equal in the two segments, in whatever direction the line P S Q be drawn. They will, indeed, be described in unequal times; that in which the perihelion A lies in a shorter, and the other in a longer, in proportion to their unequal area: but the greater proximity of the sun in the smaller segment compensates exactly for its more rapid description, and thus an equilibrium of heat is, as it were, maintained. Were it not for this, the excentricity of the orbit would materially influence the transition of seasons. The fluctuation of distance amounts to nearly $\frac{1}{30}$th of its mean quantity, and, consequently, the fluctuation in the sun's direct heating power to double this, or $\frac{1}{15}$th of the whole. Now, the perihelion of the orbit is situated nearly at the place of the northern winter solstice; so that, were it not for the compensation we have just described, the effect would be to exaggerate the difference of summer and winter in the southern hemisphere, and to moderate it in the northern; thus producing a more violent alternation of climate in the one hemisphere, and an approach to perpetual spring in the other. As it is, however, no such inequality subsists, but an equal and impartial distribution of heat and light is accorded to both.

(369.) This does not prevent, however, the direct impression of the solar heat in the height of summer, — the glow and ardour of his rays, under a perfectly clear sky, at noon, in equal latitudes and under equal circumstances of exposure, — from being very materially greater in the southern hemisphere than in the northern. One fifteenth is tooconsiderable a fraction of the whole intensity of sunshine not to aggravate in a serious degree the sufferings of those who are exposed to it in thirsty deserts, without shelter. The accounts of these sufferings in the interior of Australia, for instance, are of the most frightful kind, and would seem far to exceed what have ever been undergone by travellers in the northern deserts of Africa. *

* See the account of Captain Sturt's exploration in Athenæum, No. 1012. " The ground was almost a molten surface, and if a match accidentally fell upon it, it immediately ignited." The author has observed the temperature of the surface soil in South Africa as high as 159° Fahrenheit. An ordinary

(370.) A conclusion of a very remarkable kind, recently drawn by Professor Dove from the comparison of thermometric observations at different seasons in very remote regions of the globe, may appear on first sight at variance with what is above stated. That eminent meteorologist has shown, by taking at all seasons the mean of the temperatures of points diametrically opposite to each other, that the mean temperature *of the whole earth's surface* in June considerably exceeds that in December. This result, which is at variance with the greater proximity of the sun in December, is, however, due to a totally different and very powerful cause, — the greater amount of land in that hemisphere which has its summer solstice in June (*i. e.* the northern, see art. 362.); and the fact is so explained by him. The effect of land under sunshine is to throw heat into the general atmosphere, and so distribute it by the carrying power of the latter over the whole earth. Water is much less effective in this respect, the heat penetrating its depths, and being there absorbed; so that the surface never acquires a very elevated temperature even under the equator.

(371.) The great key to simplicity of conception in astronomy, and, indeed, in all sciences where motion is concerned, consists in contemplating every movement as referred to points which are either permanently fixed, or so nearly so, as that their motions shall be too small to interfere materially with and confuse our notions. In the choice of these primary points of reference, too, we must endeavour, as far as possible, to select such as have simple and symmetrical geometrical relations of situation with respect to the curves described by the moving parts of the system, and which are thereby fitted to perform the office of natural centres — advantageous stations for the eye of reason and theory. Having learned to attribute an orbitual motion to the earth, it loses this advantage, which is transferred to the sun, as the fixed centre about which its orbit is performed. Precisely as, when embarrassed

lucifer match does not ignite when simply pressed upon a smooth surface at 212°, but *in the act of withdrawing it*, it takes fire, and the slightest friction upon such a surface of course ignites it.

by the earth's diurnal motion, we have learned to transfer, in imagination, our station of observation from its surface to its centre, by the application of the diurnal parallax; so, when we come to inquire into the movements of the planets, we shall find ourselves continually embarrassed by the orbitual motion of our point of view, unless, by the consideration of the *annual or heliocentric parallax*, we consent to refer all our observations on them to the centre of the sun, or rather to the common centre of gravity of the sun, and the other bodies which are connected with it in our system. Hence arises the distinction between the *geocentric* and *heliocentric* place of an object. The former refers its situation in space to an imaginary sphere of infinite radius, having the centre of the earth for its centre — the latter to one concentric with the sun. Thus, when we speak of the *heliocentric longitudes* and *latitudes* of objects, we suppose the spectator situated in the sun, and referring them by circles perpendicular to the plane of the ecliptic, to the great circle marked out in the heavens by the infinite prolongation of that plane.

(372.) The point in the imaginary concave of an infinite heaven, to which a spectator in the sun refers the earth, must, of course, be diametrically opposite to that to which a spectator on the earth refers the sun's centre; consequently the heliocentric *latitude* of the earth is always nothing, and *its heliocentric longitude always equal to the sun's geocentric longitude* + 180°. The heliocentric equinoxes and solstices are, therefore, the same as the geocentric reversely named; and to conceive them, we have only to imagine a plane passing through the sun's centre, parallel to the earth's equator, and prolonged infinitely on all sides. The line of intersection of this plane and the plane of the ecliptic is the line of equinoxes, and the solstices are 90° distant from it.

(373.) The position of the longer axis of the earth's orbit is a point of great importance. In the figure (art. 368.) let E C L I be the ecliptic, E the vernal equinox, L the autumnal (*i. e.* the points *to which the earth is referred from the sun when its heliocentric longitudes are* 0° *and* 180° *respectively*). Supposing the earth's motion to be performed in the direction

E C L I, the angle E S A, or the longitude of the perihelion, in the year 1800 was 99° 30′ 5′ : we say in the year 1800, because, in point of fact, by the operation of causes hereafter to be explained, its position is subject to an extremely slow variation of about 12″ per annum to the eastward, and which in the progress of an immensely long period — of no less than 20984 years — carries the axis A S M of the orbit completely round the whole circumference of the ecliptic. But this motion must be disregarded for the present, as well as many other minute deviations, to be brought into view when they can be better understood.

(374.) Were the earth's orbit a circle, described with a uniform velocity about the sun placed in its centre, nothing could be easier than to calculate its position at any time with respect to the line of equinoxes, or its longitude, for we should only have to reduce to numbers the proportion following; viz. One year : the time elapsed :: 360° : the arc of longitude passed over. The longitude so calculated is called in astronomy the *mean* longitude of the earth. But since the earth's orbit is neither circular, nor uniformly described, this rule will not give us the true place in the orbit at any proposed moment. Nevertheless, as the excentricity and deviation from a circle are small, the *true place* will never deviate very far from that so determined (which, for distinction's sake, is called the *mean place*), and the former may at all times be calculated from the latter, by applying to it a *correction* or *equation* (as it is termed), whose amount is never very great, and whose computation is a question of pure geometry, depending on the equable description of areas by the earth about the sun. For since, in elliptic motion according to Kepler's law above stated, *areas* not *angles* are described uniformly, the proportion must now be stated thus ;—One year : the time elapsed :: the whole *area* of the ellipse : the *area* of the sector swept over by the radius vector in that time. This area, therefore, becomes known, and it is then, as above observed, a problem of pure geometry to ascertain the *angle* about the sun (A S P, *fig.* art. 368.), which corresponds to any proposed fractional area of the whole ellipse supposed to be

contained in the sector A P S.   Suppose we set out from A
the perihelion, then will the angle A S P at first increase more
rapidly than the *mean longitude*, and will, therefore, during
the whole semi-revolution from A to M, exceed it in amount;
or, in other words, the *true place* will be in advance of the
mean: at M, one half the year will have elapsed, and one
half the orbit have been described, whether it be circular or
elliptic.   Here, then, the *mean* and *true* places coincide; but
in all the other half of the orbit, from M to A, the true place
will fall short of the mean, since at M the angular motion is
slowest, and the true place from this point begins to lag
behind the mean — to make up with it, however, as it
approaches A, where it once more overtakes it.

(375.) The quantity by which the *true* longitude of the
earth differs from the *mean* longitude is called the equation
of the centre, and is *additive* during all the half-year, in which
the earth passes from A to M, beginning at 0° 0′ 0″, increasing
to a maximum, and again diminishing to zero at M; after
which it becomes subtractive, attains a maximum of subtractive
magnitude between M and A, and again diminishes to 0 at A.
Its maximum, both additive and substractive, is 1° 55′ 33″·3.

(376.) By applying, then, to the earth's mean longitude,
the equation of the centre corresponding to any given time
at which we would ascertain its place, the true longitude becomes known; and since the sun is always seen from the
earth in 180° more longitude than the earth from the sun, in
this way also the sun's true place in the ecliptic becomes
known.   The calculation of the equation of the centre is performed by a table constructed for that purpose, to be found in
all " Solar Tables."

(377.) The maximum value of the equation of the centre
depends only on the ellipticity of the orbit, and may be expressed in terms of the excentricity.   *Vice versâ*, therefore,
if the former quantity can be ascertained by observation, the
latter may be derived from it; because, whenever the law,
or numerical connection, between two quantities is known,
the one can always be determined from the other.   Now, by
assiduous observation of the sun's transits over the meridian,

we can ascertain, for every day, its exact right ascension, and thence conclude its longitude (art. 309.). After this, it is easy to assign the angle by which this *observed* longitude exceeds or falls short of the mean; and the greatest amount of this excess or defect which occurs in the whole year, is the maximum equation of the centre. This, as a means of ascertaining the eccentricity of the orbit, is a far more easy and accurate method than that of concluding the sun's distance by measuring its apparent diameter. The results of the two methods coincide, however, perfectly.

(378.) If the ecliptic coincided with the equinoctial, the effect of the equation of the centre, by disturbing the uniformity of the sun's apparent motion in longitude, would cause an inequality in its time of coming on the meridian on successive days. When the sun's centre comes to the meridian, it is *apparent noon,* and if its motion in longitude were uniform, and the ecliptic coincident with the equinoctial, this would always coincide with *mean noon,* or the stroke of 12 on a well-regulated solar clock. But, independent of the want of uniformity in its motion, the obliquity of the ecliptic gives rise to another inequality in this respect; in consequence of which, the sun, even supposing its motion in the ecliptic uniform, would yet alternately, in its time of attaining the meridian, anticipate and fall short of the mean noon as shown by the clock. For the right ascension of a celestial object forming a side of a right-angled spherical triangle, of which its longitude is the hypothenuse, it is clear that the uniform increase of the latter must necessitate a deviation from uniformity in the increase of the former.

(379.) These two causes, then, acting conjointly, produce, in fact, a very considerable fluctuation in the time as shown per clock, when the sun really attains the meridian. It amounts, in fact, to upwards of half an hour; apparent noon sometimes taking place as much as $16\frac{1}{4}$ min. before mean noon, and at others as much as $14\frac{1}{2}$ min. after. This difference between apparent and mean noon is called the *equation of time,* and is calculated and inserted in ephemerides for every day of the year, under that title : or else, which comes to the same thing, the moment, *in mean time,* of the sun's culmination for each day, is set down as an astronomical phænomenon to be observed.

(380.) As the sun, in its apparent annual course, is carried along the ecliptic, its declination is continually varying between the extreme limits of 23° 28′ 40″ north, and as much south, which it attains at the solstices. It is consequently always vertical over some part or other of that zone or belt of the earth's surface which lies between the north and south parallels of 23° 28′ 40″. These parallels are called in geography the *tropics ;* the northern one that of *Cancer,* and the southern, of *Capricorn ;* because the sun, at the respective solstices, is situated in the divisions, or signs of the ecliptic so denominated. Of these signs there are twelve, each occupying 30° of its circumference. They commence at the vernal equinox, and are named in order—Aries, Taurus, Gemini, Cancer, Leo, Virgo, Libra, Scorpio, Sagittarius, Capricornus, Aquarius, Pisces.* They are denoted also by the following symbols :— ♈, ♉, ♊, ♋, ♌, ♍, ♎, ♏, ♐, ♑, ♒, ♓. Longitude itself is also divided into signs, degrees, and minutes, &c. Thus 5ˢ 27° 0′ corresponds to 177° 0′.

(381.) These *Signs* are purely technical subdivisions of the ecliptic, commencing from the actual equinox, and are not to be confounded with the *constellations* so called (and sometimes so symbolized). The constellations of the zodiac, as they now stand arranged on the ecliptic, are all a full " sign " in advance or anticipation of their symbolic cognomens thereon marked. Thus the constellation Aries actually occupies the sign Taurus ♉, the constellation Taurus, the sign Gemini ♊, and so on, the *signs* having retreated† among the stars (together with the equinox their origin), by the effect of precession. The bright star Spica in the constellation Virgo (*a* Virginis), by the observations of

---

* They may be remembered by the following memorial hexameters : —
    Sunt Aries, Taurus, Gemini, Cancer, Leo, Virgo,
    Libraque, Scorpius, Arcitenens, Caper, Amphora, Pisces.

† *Retreated* is here used with reference to *longitude,* not to the apparent diurnal motion.

Hipparchus, 128 years B.C., *preceded*, or was westward of the autumnal equinox in longitude by 6°. In 1750 it followed or stood eastward of the same equinox by 20° 21'. Its place then, as referred to the ecliptic at the former epoch, would be in longitude 5ˢ 24° 0', or in the 24th degree of the *sign* ♌, whereas in the latter epoch it stood in the 21st degree of ♍, the equinox having retreated by 26° 21' in the interval, 1878 years, elapsed. To avoid this source of misunderstanding, the use of " signs " and their symbols in the reckoning of celestial longitudes is now almost entirely abandoned, and the ordinary reckoning (by degrees, &c. from 0 to 360) adopted in its place, and the names Aries, Virgo, &c. are becoming restricted to the constellations so called.*

(382.) When the sun is in either tropic, it enlightens, as we have seen, the pole on that side the equator, and shines over or beyond it to the extent of 23° 28' 40". The parallels of latitude, at this distance from either pole, are called the polar circles, and are distinguished from each other by the names *arctic* and *antarctic*. The regions within these circles are sometimes termed frigid zones, while the belt between the tropics is called the torrid zone, and the immediate belts temperate zones. These last, however, are merely names given for the sake of naming; as, in fact, owing to the different distribution of land and sea in the two hemispheres, zones of *climate* are not co-terminal with zones of *latitude*.

(383.) Our seasons are determined by the apparent passages of the sun across the equinoctial, and its alternate arrival in the northern and southern hemisphere. Were the equinox invariable, this would happen at intervals precisely equal to the duration of the sidereal year; but, in fact, owing to the slow conical motion of the earth's axis described in art. 317., the equinox retreats on the ecliptic, and *meets* the advancing sun somewhat *before* the whole sidereal circuit is completed. The annual retreat of the equinox is 50"·1, and this arc is

---

* When, however, the place of the sun is spoken of, the old usage prevails. Thus, if we say " the sun is in Aries," it would be interpreted to mean between 0° and 30° of longitude. So, also, " the first point of Aries" is still understood to mean the vernal, and " the first point of Libra," the autumnal equinox ; and so in a few other cases.

described by the sun in the ecliptic in 20ᵐ 19ˢ·9. By so much *shorter*, then, is the periodical return of our seasons than the true sidereal revolution of the earth round the sun. As the latter period, or sidereal year, is equal to 365ᵈ 6ʰ 9ᵐ 9ˢ·6, it follows, then, that the former must be only 365ᵈ 5ʰ 48ᵐ 49ˢ·7 ; and this is what is meant by the *tropical* year.

(384.) We have already mentioned that the longer axis of the ellipse described by the earth has a slow motion of 11″·8 per annum in advance. From this it results, that when the earth, setting out from the perihelion, has completed one sidereal period, the perihelion will have moved forward by 11″·8, which arc must be described by the earth before it can again reach the perihelion. In so doing, it occupies 4ᵐ 39ˢ·7 and this must therefore be added to the sidereal period, to give the interval between two consecutive returns to the perihelion. This interval, then, is 365ᵈ 6ʰ 13ᵐ 49ˢ·3 *, and is what is called the *anomalistic year*. All these periods have their uses in astronomy ; but that in which mankind in general are most interested is *the tropical year*, on which the return of the seasons depends, and which we thus perceive to be a compound phenomenon, depending chiefly and directly on the annual revolution of the earth round the sun, but subordinately also, and indirectly, on its rotation round its own axis, which is what occasions the precession of the equinoxes ; thus affording an instructive example of the way in which a motion, once admitted in any part of our system, may be traced in its influence on others with which at first sight it could not possibly be supposed to have any thing to do.

(385.) As a rough consideration of the appearance of the earth points out the general roundness of its form, and more exact enquiry has led us first to the discovery of its elliptic figure, and, in the further progress of refinement, to the perception of minuter local deviations from that figure ; so, in investigating the solar motions, the first notion we obtain is that of an orbit, generally speaking, round, and not far from

---

* These numbers, as well as all the other numerical data of our system, are taken from Mr. Baily's Astronomical Tables and Formulæ, unless the contrary is expressed.

a circle, which, on more careful and exact examination, proves to be an ellipse of small excentricity, and described in conformity with certain laws, as above stated.   Still minuter enquiry, however, detects yet smaller deviations again from this form and from these laws, of which we have a specimen in the slow motion of the axis of the orbit spoken of in art. 372.; and which are generally comprehended under the name of perturbations and secular inequalities.   Of these deviations, and their causes, we shall speak hereafter at length.   It is the triumph of physical astronomy to have rendered a complete account of them all, and to have left nothing unexplained, either in the motions of the sun or in those of any other of the bodies of our system.   But the nature of this explanation cannot be understood till we have developed the law of gravitation, and carried it into its more direct consequences. This will be the object of our three following chapters; in which we shall take advantage of the proximity of the moon, and its immediate connection with and dependence on the earth, to render it, as it were, a stepping-stone to the general explanation of the planetary movements.   We shall conclude this by describing what is known of the physical constitution of the sun.

(386.) When viewed through powerful telescopes, provided with coloured glasses, to take off the heat, which would otherwise injure our eyes, the sun is observed to have frequently large and perfectly black spots upon it, surrounded with a kind of border, less completely dark, called a penumbra.   Some of these are represented at *a, b, c, d,* in Plate I. fig. 2., at the end of this volume.   They are, however, not permanent. When watched from day to day, or even from hour to hour, they appear to enlarge or contract, to change their forms, and at length to disappear altogether, or to break out anew in parts of the surface where none were before.   In such cases of disappearance, the central dark spot always contracts into a point, and vanishes before the border.   Occasionally they break up, or divide into two or more, and in those cases offer every evidence of that extreme mobility which belongs only to the fluid state, and of that excessively violent agitation which seems only compatible with the atmospheric or gaseous state of matter. The scale on which their movements take place is immense. A single second of angular measure, as seen from the earth, corresponds on the sun's disc to 461 miles; and a circle of this diameter (containing therefore nearly 167000 square miles) is the least space which can be distinctly discerned on the sun as a *visible area*. Spots have been observed, however, whose linear diameter has been upwards of 45000 miles*; and even, if some records are to be trusted, of very much greater extent. That such a spot should close up in six weeks' time (for they seldom last much longer), its borders must approach at the rate of more than 1000 miles a day.

(387.) Many other circumstances tend to corroborate this view of the subject. The part of the sun's disc not occupied by spots is far from uniformly bright. Its *ground* is finely mottled with an appearance of minute, dark dots, or *pores*, which, when attentively watched, are found to be in a constant state of change. There is nothing which represents so faithfully this appearance as the slow subsidence of some flocculent chemical precipitates in a transparent fluid, when viewed perpendicularly from above: so faithfully, indeed, that it is hardly possible not to be impressed with the idea of a luminous medium intermixed, but not confounded, with a transparent and non-luminous atmosphere, either floating as clouds in our air, or pervading it in vast sheets and columns like flame, or the streamers of our northern lights, directed in lines perpendicular to the surface.

(388.) Lastly, in the neighbourhood of great spots, or extensive groups of them, large spaces of the surface are often observed to be covered with strongly marked curved or branching streaks, more luminous than the rest, called *faculæ*, and among these, if not already existing, spots frequently break out. They may, perhaps, be regarded with most probability, as the ridges of immense waves in the luminous regions of the sun's atmosphere, indicative of violent agitation

---

* Mayer, Obs. Mar. 15. 1758. " Ingens macula in sole conspiciebatur, cujus diameter = $\frac{1}{20}$ diam. solis."

in their neighbourhood.   They are most commonly, and best seen, towards the borders of the visible disc, and their appearance is as represented in Plate I. fig. 1.

(389.) But what *are* the spots?   Many fanciful notions have been broached on this subject, but only one seems to have any degree of physical probability, viz. that they are the dark, or at least comparatively dark, solid body of the sun itself, laid bare to our view by those immense fluctuations in the luminous regions of its atmosphere, to which it appears to be subject.   Respecting the manner in which this disclosure takes place, different ideas again have been advocated.   Lalande (art. 3240.) suggests, that eminences in the nature of mountains are actually laid bare, and project above the luminous ocean, appearing black above it, while their shoaling declivities produce the penumbræ, where the luminous fluid is less deep.   A fatal objection to this theory is the uniform shade of the penumbra and its sharp termination, both inwards, where it joins the spot, and outwards, where it borders on the bright surface.   A more probable view has been taken by Sir William Herschel*, who considers the luminous strata of the atmosphere to be sustained far above the level of the solid body by a transparent elastic medium, carrying on its upper surface (*or rather*, to avoid the former objection, *at some considerably lower level within its depth*) a cloudy stratum which, being strongly illuminated from above, reflects a considerable portion of the light to our eyes, and forms a penumbra, while the solid body shaded by the clouds, reflects none   (See *fig.*) The temporary removal of both the strata, but more of the upper than the lower, he supposes effected by powerful upward currents of the atmosphere, arising, perhaps, from spiracles in the body, or from local agitations.

* Phil. Trans. 1801.

(390.) When the spots are attentively watched, their situation on the disc of the sun is observed to change. They advance regularly towards its western limb or border, where they disappear, and are replaced by others which enter at the eastern limb, and which, pursuing their respective courses, in their turn disappear at the western. The apparent rapidity of this movement is not uniform, as it would be were the spots dark bodies passing, by an independent motion of their own, between the earth and the sun; but is swiftest in the middle of their paths across the disc, and very slow at its borders. This is precisely what would be the case supposing them to appertain to and make part of the visible surface of the sun's globe, and to be carried round by a uniform rotation of that globe on its axis, so that each spot should describe a circle parallel to the sun's equator, rendered elliptic by the effect of perspective. Their apparent paths also across the disc conform to this view of their nature, being, generally speaking, ellipses, much elongated, concentric with the sun's disc, each having one of its chords for its longer axis, and all these axes parallel to each other. At two periods of the year only do the spots appear to describe straight lines, viz. on and near to the 11th of June and the 12th of December, on which days, therefore, the plane of the circle, which a spot situated on the sun's equator describes (and consequently, the plane of that equator itself), passes through the earth. Hence it is obvious, that the plane of the sun's equator is inclined to that of the ecliptic, and intersects it in a line which passes through the place of the earth on these days. The situation of this line, or *the line of the nodes of the sun's equator* as it is called, is, therefore, defined by the longitudes of the earth as seen from the sun at those epochs, which are respectively 80° 21′ and 260° 21′ (= 80° 21′ + 180°) being, of course, diametrically opposite in direction.

(391.) The inclination of the sun's axis (that of the plane of its equator) to the ecliptic is determined by ascertaining the proportion of the longer and the shorter diameter of the apparent ellipse, described by any remarkable, well-defined spot; in order to do which, its apparent place on the sun's disc must be very precisely ascertained by micrometric measures, repeated from day to day as long as it continues visible, (usually about 12 or 13 days, according to the magnitude of the spots, which always vanish by the effect of foreshortening before they attain the actual border of the disc — but the larger spots being traceable closer to the limb than the smaller. *) The *reduction* of such observations, or the conclusion from them of the element in question, is complicated with the effect of the earth's motion in the interval of the observations, and with its situation in the ecliptic, with respect to the line of nodes. For simplicity, we will suppose the earth situated as it is on the 10th of March, in a line at right angles to that of the nodes, *i. e.* in the heliocentric longitude 170° 21', and to remain there stationary during the whole passage of a spot across the disc. In this case the axis

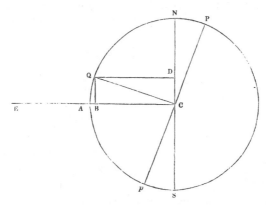

of rotation of the sun will be situated in a plane passing through the earth and at right angles to the plane of the ecliptic. Suppose C to represent the sun's centre, P C p its axis, E C the line of sight, P N Q A p S a section of the sun passing through the earth, and Q a spot situated on its equator, and in that plane, and consequently in the middle of its apparent path across the disc. If the axis of rotation were perpendicular to the ecliptic, as N S, this spot would be

* The great spot of December, 1719, is stated to have been seen as a notch in the limb of the sun.

at A, and would be seen projected on C, the centre of the sun. It is actually at Q, projected upon D, at an apparent distance C D to the *north* of the centre, which is the apparent smaller semi-axis of the ellipse described by the spot, which being known by micrometric measurement, the value of $\dfrac{C\,D}{C\,N}$ or the cosine of Q C A, the inclination of the sun's equator becomes known, C N being the apparent semi-diameter of the sun at the time. At this epoch, moreover, the northern half of the circle described by the spot is visible (the southern passing behind the body of the sun), and the south pole p of the sun is within the visible hemisphere. This is the case in the whole interval from December 11th to July 12th, during which, the visual ray falls upon the southern side of the sun's equator. The contrary happens in the other half year, from July 12th to December 11th and this is what is understood when we say that the *ascending* node (denoted ☊) of the sun's equator lies in 80° 21′ longitude—a spot on the equator passing that node being then in the act of *ascending* from the southern to the northern side of the plane of the ecliptic—such being the conventional language of astronomers in speaking of these matters.

(392.) If the observations are made at other seasons (which, however, are the less favourable for this purpose the more remote they are from the epochs here assigned); when, moreover, as in strictness is necessary, the motion of the earth in the interval of the measures is allowed for (as for a change of the point of sight); the calculations requisite to deduce the situation of the axis in space, and the duration of the revolution around it, become much more intricate, and it would be beyond the scope of this work to enter into them.* According to the best determinations we possess, the inclination of the sun's equator to the ecliptic is about 7° 20′ (its nodes being as above stated), and the period of rotation 25 days 7 hours 48 minutes. †

---

* See the theory in Lalande's Astronomy, art. 3258., and the formulæ of computation in a paper by Petersen Schumacher's Nachrichten, No. 419.

† Bianchi (Schumacher's Nach., 483.), agreeing with Laugier. Delambre makes it 25ᵈ 0ʰ 17ᵐ; Petersen, 25ᵈ 4ʰ 30ᵐ. The inclination of the axis is uncertain to half a degree, and the node to several degrees. The continual changes in the spots themselves cause this uncertainty.

(393.) The region of the spots is confined, generally speaking, within about 25° on either side of the sun's equator; beyond 30° they are very rarely seen; in the polar regions, never. The actual equator of the sun is also less frequently visited by spots than the adjacent zones on either side, and a very material difference in their frequency and magnitude subsists in its northern and southern hemisphere, those on the northern preponderating in both respects. The zone comprised between the 11th and 15th degree to the northward of the equator is particularly fertile in large and durable spots. These circumstances, as well as the frequent occurrence of a more or less regular arrangement of the spots, when numerous, in the manner of belts parallel to the equator, point evidently to physical peculiarities in certain parts of the sun's body more favourable than in others to the production of the spots, on the one hand; and on the other, to a general influence of its rotation on its axis as a determining cause of their distribution and arrangement, and would appear indicative of a system of movements in the fluids which constitute its luminous surface bearing no remote analogy to our trade winds — from whatever cause arising. (See art. 239. et seq.)

(394.) The duration of individual spots is commonly not great; some are formed and disappear within the limit of a single transit across the disc — but such are for the most part small and insignificant. Frequently they make one or two revolutions, being recognized at their reappearance by their situation with respect to the equator, their configurations *inter se*, their size, or other peculiarities, as well as by the interval elapsing between their disappearance at one limb and reappearance on the other. In a few rare cases, however, they have been watched round many revolutions. The great spot of 1779 appeared during six months, and one and the same *groupe* was observed in 1840 by Schwabe to return eight times. * It has been surmised, with considerable apparent probability, that some spots, at least, are generated again

---

* Schum. Nach. No. 418. p. 150. The recent papers of Biela, Capocci, Schwabe, Pastorff, and Schmidt, in that collection, will be found highly interesting.

and again, at distant intervals of time, over the same identical points of the sun's body (as hurricanes, for example, are known to affect given localities on the earth's surface, and to pursue definite tracks). The uncertainty which still prevails with respect to the exact duration of its rotation renders it very difficult to obtain convincing evidence of this ; nor, indeed, can it be expected, until by bringing together into one connected view the recorded state of the sun's surface during a very long period of time, and comparing together remarkable spots which have appeared *on the same parallel,* some precise periodic time shall be found which shall exactly conciliate numerous and well-characterized appearances. The enquiry is one of singular interest, as there can be no reasonable doubt that the supply of light and heat afforded to our globe stands in intimate connexion with those processes which are taking place on the solar surface, and to which the spots in some way or other owe their origin.

(395.) Above the luminous surface of the sun, and the region in which the spots reside, there are strong indications of the existence of a gaseous atmosphere having a somewhat imperfect transparency. When the whole disc of the sun is seen at once through a telescope magnifying moderately enough to allow it, and with a darkening glass such as to suffer it to be contemplated with perfect comfort, it is very evident that the borders of the disc are much less luminous than the centre. That this is no illusion is shown by projecting the sun's image undarkened and moderately magnified, so as to occupy a circle two or three inches in diameter, on a sheet of white paper, taking care to have it well in focus, when the same appearance will be observed. This can only arise from the circumferential rays having undergone the absorptive action of a much greater thickness of some imperfectly transparent envelope (due to greater obliquity of their passage through it) than the central. — But a still more convincing and indeed decisive evidence is offered by the phænomena attending a total eclipse of the sun. Such eclipses (as will be shown hereafter) are produced by the interposition of the dark body of the moon between the earth and sun, the moon being large enough to cover and surpass, by a very small breadth, the whole disc of the sun. Now when this takes place, were there no vaporous atmosphere capable of reflecting any light about the sun, the sky ought to appear totally dark, since (as will hereafter abundantly appear) there is not the smallest reason for believing the *moon* to have any atmosphere capable of doing so. So far, however, is this from being the case, that a bright ring or corona of light is seen, fading gradually away, as represented in Pl. I. fig 3., which (in cases where the moon is not *centrally* superposed on the sun) is observed to be concentric with the latter, not the former body. This corona was beautifully seen in the eclipse of July 7. 1842, and with this most remarkable addition — witnessed by every spectator in Pavia, Milan, Vienna, and elsewhere : three distinct and very conspicuous *rose-coloured* protuberances (as represented in the figure cited) were seen to project beyond the dark limb of the moon, likened by some to flames, by others to mountains, but which their enormous magnitude (for to have been seen at all by the naked eye their height must have exceeded 40,000 miles), and their faint degree of illumination, clearly prove to have been cloudy masses *of the most excessive tenuity*, and which doubtless owed their support, and probably their existence, to such an atmosphere as we are now speaking of.

(396.) That the temperature at the visible surface of the sun cannot be otherwise than very elevated, much more so than any artificial heat produced in our furnaces, or by chemical or galvanic processes, we have indications of several distinct kinds: 1st, From the law of decrease of radiant heat and light, which, being inversely as the squares of the distances, it follows, that the heat received on a given area exposed at the distance of the earth, and on an equal area at the visible surface of the sun, must be in the proportion of the area of the sky occupied by the sun's apparent disc to the whole hemisphere, or as 1 to about 300000. A far less intensity of solar radiation, collected in the focus of a burning glass, suffices to dissipate gold and platina in vapour. 2dly, From the facility with which the calorific rays of the sun traverse glass,

a property which is found to belong to the heat of artificial
fires in the direct proportion of their intensity.*   3dly, From
the fact, that the most vivid flames disappear, and the most
intensely ignited solids appear only as black spots on the disc
of the sun when held between it and the eye.†   From the
last remark it follows, that the body of the sun, however dark
it may appear when seen through its spots, *may*, nevertheless,
be in a state of most intense ignition. It does not, however,
follow of necessity that it *must* be so.   The contrary is at
least physically possible.   A *perfectly reflective* canopy would
effectually defend it from the radiation of the luminous regions
above its atmosphere, and no heat would be conducted downwards through a gaseous medium increasing rapidly in density.
That the penumbral clouds *are* highly reflective, the fact of
their visibility in such a situation can leave no doubt.

(397.) As the magnitude of the sun has been measured,
and (as we shall hereafter see) its weight, or quantity of ponderable matter, ascertained, so also attempts have been made,
and not wholly without success, from the heat actually communicated by its rays to given surfaces of material bodies
exposed to their vertical action on the earth's surface, to estimate the total expenditure of heat by that luminary in a given
time.   The result of such experiments has been thus announced.   Supposing a cylinder of ice 45 miles in diameter, to
be continually darted into the sun *with the velocity of light*, and
that the water produced by its fusion were continually carried
off, the heat now given off constantly by radiation would then
be wholly expended in its liquefaction, on the one hand, so as
to leave no radiant surplus; while, on the other, the actual
temperature at its surface would undergo no diminution.

(398.) This immense escape of heat by radiation, we may
remark, will fully explain the constant state of tumultuous

* By direct measurement with the *actinometer*, I find that out of 1000
calorific solar rays, 816 penetrate a sheet of plate glass 0·12 inch thick; and
that of 1000 rays which have passed through one such plate, 859 are capable
of passing through another.   H. 1827.

† The ball of ignited quicklime, in Lieutenant Drummond's oxy-hydrogen
lamp, gives the nearest imitation of the solar splendour which has yet been
produced.   The apppearance of this against the sun was, however, as described
in an imperfect trial at which I was present.   The experiment ought to be
repeated under favourable circumstances. — *Note to the ed. of* 1833.

agitation in which the fluids composing the visible surface are maintained, and the continual generation and filling in of *the pores*, without having recourse to internal causes. The mode of action here alluded to is perfectly represented to the eye in the disturbed subsidence of a precipitate, as described in art. 386., when the fluid from which it subsides is warm, and losing heat from its surface.

(399.) The sun's rays are the ultimate source of almost every motion which takes place on the surface of the earth. By its heat are produced all winds, and those disturbances in the electric equilibrium of the atmosphere which give rise to the phenomena of lightning, and probably also to those of terrestrial magnetism and the aurora. By their vivifying action vegetables are enabled to draw support from inorganic matter, and become, in their turn, the support of animals and of man, and the sources of those great deposits of dynamical efficiency which are laid up for human use in our coal strata.* By them the waters of the sea are made to circulate in vapour through the air, and irrigate the land, producing springs and rivers. By them are produced all disturbances of the chemical equilibrium of the elements of nature, which, by a series of compositions and decompositions, give rise to new products, and originate a transfer of materials. Even the slow degradation of the solid constituents of the surface, in which its chief geological changes consist, is almost entirely due on the one hand to the abrasion of wind and rain, and the alternation of heat and frost; on the other to the continual beating of the sea waves, agitated by winds, the results of solar radiation. Tidal action (itself partly due to the sun's agency) exercises here a comparatively slight influence. The effect of oceanic currents (mainly originating in that influence), though slight in abrasion, is powerful in diffusing and transporting the matter abraded; and when we consider the immense transfer of matter so produced, the increase of pressure over large spaces in the bed of the ocean, and diminution over corresponding portions of the land, we are not at a loss to perceive how the elastic power of subterranean fires, thus repressed on the one hand

* So in the edition of 1833.

and relieved on the other, may break forth in points when the resistance is barely adequate to their retention, and thus bring the phenomena of even volcanic activity under the general law of solar influence. *

(400.) The great mystery, however, is to conceive how so enormous a conflagration (if such it be) can be kept up. Every discovery in chemical science here leaves us completely at a loss, or rather, seems to remove farther the prospect of probable explanation. If conjecture might be hazarded, we should look rather to the known possibility of an indefinite generation of heat by friction, or to its excitement by the electric discharge, than to any actual combustion of ponderable fuel, whether solid or gaseous, for the origin of the solar radiation.†

* So in the edition of 1833.

† Electricity traversing excessively rarefied air or vapours, gives out light, and, doubtless, also heat. May not a continual current of electric matter be constantly circulating in the sun's immediate neighbourhood, or traversing the planetary spaces, and exciting, in the upper regions of its atmosphere, those phenomena of which, on however diminutive a scale, we have yet an unequivocal manifestation in our aurora borealis. The possible analogy of the solar light to that of the aurora has been distinctly insisted on by the late Sir W. Herschel, in his paper already cited. It would be a highly curious subject of experimental inquiry, how far a mere reduplication of sheets of flame, at a distance one behind the other (by which their *light* might be brought to any required intensity), would communicate to the *heat* of the resulting compound ray the *penetrating* character which distinguishes the solar calorific rays. We may also observe, that the tranquillity of the sun's polar, as compared with its equatorial regions (if its spots be really atmospheric), cannot be accounted for by its rotation on its axis only, but *must* arise from some cause *external to the luminous surface* of the sun, as we see the belts of Jupiter and Saturn, and our trade-winds, arise from a cause, external to these planets, combining itself with their rotation, which *alone* can produce no motions when once the form of equilibrium is attained.

The prismatic analysis of the solar beam exhibits in the spectrum a series of "fixed lines," totally unlike those which belong to the light of any known terrestrial flame. This may hereafter lead us to a clearer insight into its origin. But, before we can draw any conclusions from such an indication, we must recollect, that previous to reaching us it has undergone the whole absorptive action of our atmosphere, as well as of the sun's. Of the latter we know nothing, and may conjecture every thing; but of the blue colour of the former we are sure ; and if this be an inherent (*i. e.* an absorptive) colour, the air must be expected to act on the spectrum after the analogy of other coloured media, which often (and *especially light blue* media) leave unabsorbed portions separated by dark intervals. It deserves enquiry, therefore, whether some or all the fixed lines observed by Wollaston and Fraunhofer may not have their origin in our own atmosphere. Experiments made on lofty mountains, or the cars of balloons, on the one hand, and on the other with reflected beams which have been made to traverse several miles of additional air near the surface, would decide this point. The absorptive effect of the sun's atmosphere, and possibly also of the medium surrounding it (whatever it be) which resists the motions of comets, cannot be thus eliminated. — *Note to the edition of 1833.*

CHAPTER VII.

OF THE MOON.—ITS SIDEREAL PERIOD. —ITS APPARENT DIAMETER. — ITS PARALLAX, DISTANCE, AND REAL DIAMETER — FIRST APPROXIMATION TO ITS ORBIT. — AN ELLIPSE ABOUT THE EARTH IN THE FOCUS. — ITS EXCENTRICITY AND INCLINATION. —MOTION OF ITS NODES AND APSIDES. — OF OCCULTATIONS AND SOLAR ECLIPSES GENERALLY. — LIMITS WITHIN WHICH THEY ARE POSSIBLE. — THEY PROVE THE MOON TO BE AN OPAKE SOLID — ITS LIGHT DERIVED FROM THE SUN. — ITS PHASES. — SYNODIC REVOLUTION OR LUNAR MONTH.—OF ECLIPSES MORE PARTICULARLY. — THEIR PHENOMENA.—THEIR PERIODICAL RECURRENCE.—PHYSICAL CONSTITUTION OF THE MOON.— ITS MOUNTAINS AND OTHER SUPERFICIAL FEATURES. — INDICATIONS OF FORMER VOLCANIC ACTIVITY. — ITS ATMOSPHERE.—CLIMATE.— RADIATION OF HEAT FROM ITS SURFACE.— ROTATION ON ITS OWN AXIS.— LIBRATION. — APPEARANCE OF THE EARTH FROM IT.

(401.) THE moon, like the sun, appears to advance among the stars with a movement contrary to the general diurnal motion of the heavens, but much more rapid, so as to be very readily perceived (as we have before observed) by a few hours' cursory attention on any moonlight night. By this continual advance, which, though sometimes quicker, sometimes slower, is never intermitted or reversed, it makes the tour of the heavens in a mean or average period of 27^{d} 7^{h} 43^{m} $11^{s}.5$, returning, in that time, to a position among the stars nearly coincident with that it had before, and which would be exactly so, but for reasons presently to be stated.

(402.) The moon, then, like the sun, apparently describes an orbit round the earth, and this orbit cannot be *very* different from a circle, because the apparent angular diameter of the full moon is not liable to any great extent of variation.

(403.) The distance of the moon from the earth is concluded from its horizontal parallax, which may be found either directly, by observations at remote geographical stations, exactly similar to those described in art. 355., in the case of the sun, or by means of the phænomena called occultations, from which also its apparent diameter is most readily and correctly found.   From such observations it results that the mean or average distance of the center of the moon from that of the earth is 59·9643 of the earth's equatorial radii, or about 237,000 miles.   This distance, great as it is, is little more than one-fourth of the diameter of the sun's body, so that the globe of the sun would nearly twice include the whole orbit of the moon ; a consideration wonderfully calculated to raise our ideas of that stupendous luminary !

(404.) The distance of the moon's center from an observer at any station on the earth's surface, compared with its apparent angular diameter as measured from that station, will give its real or linear diameter.   Now, the former distance is easily calculated when the distance from the earth's center is known, and the apparent zenith distance of the moon also determined by observation ; for if we turn to the figure of art. 339., and suppose S the moon, A the station, and C the earth's center, the distance S C, and the earth's radius C A, two sides of the triangle A C S are given, and the angle C A S, which is the supplement of Z A S, the observed zenith distance, whence it is easy to find A S, the moon's distance from A.   From such observations and calculations it results, that the real diameter of the moon is 2160 miles, or about 0·2729 of that of the earth, whence it follows that, the bulk of the latter being considered as 1, that of the former will be 0·0204, or about $\frac{1}{49}$.   The difference of the apparent diameter of the moon, as seen from the earth's center and from any point of its surface, is technically called the *augmentation of the apparent* diameter, and its maximum occurs when the moon is in the zenith of the spectator.   Her mean angular diameter, as seen from the center, is 31′ 7″, and is always $= 0·545 \times$ her horizontal parallax.

(405.) By a series of observations, such as described in art. 403., if continued during one or more revolutions of the moon, its real distance may be ascertained at every point of its orbit ; and if at the same time its apparent places in the heavens be observed, and reduced by means of its parallax to the earth's center, their angular intervals will become known, so that the path of the moon may then be laid down on a chart supposed to represent the plane in which its orbit lies, just as was explained in the case of the solar ellipse (art. 349.) Now, when this is done, it is found that, neglecting certain small (though very perceptible) (deviations of which a satisfactory account will hereafter be rendered), the form of the apparent orbit, like that of the sun, is elliptic, but considerably more eccentric, the eccentricity amounting to 0·05484 of the mean distance, or the major semi-axis of the ellipse, and the earth's centre being situated in its focus.

(406.) The plane in which this orbit lies is not the ecliptic, however, but is inclined to it at an angle of 5° 8′ 48″, which is called the inclination of the lunar orbit, and intersects it in two opposite points, which are called its nodes — the *ascending node* being that in which the moon passes from the southern side of the ecliptic to the northern, and the *descending* the reverse. The points of the orbit at which the moon is nearest to, and farthest from, the earth, are called respectively its *perigee* and *apogee*, and the line joining them and the earth the line of *apsides*.

(407.) There are, however, several remarkable circumstances which interrupt the closeness of the analogy, which cannot fail to strike the reader, between the motion of the moon around the earth, and of the earth around the sun. In the latter case, the ellipse described remains, during a great many revolutions, unaltered in its position and dimensions; or, at least, the changes which it undergoes are not perceptible but in a course of very nice observations, which have disclosed, it is true, the existence of " perturbations," but of so minute an order, that, in ordinary parlance, and for common purposes, we may leave them unconsidered. But this cannot be done in the case of the moon. Even in a single revolution, its deviation from a perfect ellipse is very sensible. It does not return to the same exact position among the stars from which it set out, thereby indicating a continual change in the *plane* of its orbit. And, in effect,

R

if we trace by observation, from month to month, the point where it traverses the ecliptic, we shall find that the *nodes* of its orbit are in a continual state of *retreat* upon the ecliptic. Suppose O to be the earth, and A *b a d* that portion of the plane of the ecliptic which is intersected by the moon, in its alternate passages through it, from south to north, and *vice versa;* and let A B C D E F be a portion of the moon's orbit, embracing a complete sidereal revolution.   Suppose it

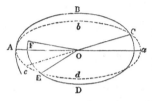

to set out from the ascending node, A; then, if the orbit lay all in one plane, passing through O, it would have *a*, the opposite point in the ecliptic, for its descending node; after passing which, it would again ascend at A.   But, in fact, its real path carries it not to *a*, but along a certain curve, A B C, to C, a point in the ecliptic less than 180° distant from A; so that the angle A O C, or the arc of longitude described between the ascending and the descending node, is somewhat less than 180°.   It then pursues its course below the ecliptic, along the curve C D E, and rises again above it, not at the point *c*, diametrically opposite to C, but at a point E, less advanced in longitude.   On the whole, then, the arc described in longitude between two consecutive passages from south to north, through the plane of the ecliptic, falls short of 360° by the angle A O E; or, in other words, the ascending node appears to have retreated in one lunation on the plane of the ecliptic by that amount. To complete a sidereal revolution, then, it must still go on to describe an arc, A⁄ F, on its orbit, which will no longer, however, bring it exactly back to A, but to a point somewhat above it, or *having north latitude.*

(408.) The actual amount of this retreat of the moon's node is about 3′ 10″ ·64 *per diem*, on an average, and in a period of 6793·39 mean solar days, or about 18·6 years, the ascending node is carried round in a direction contrary to the moon's motion in its orbit (or from east to west) over a whole circumference of the ecliptic.   Of course, in the middle of this period the position of the orbit must have been precisely reversed from what it was at the beginning. Its apparent path, then, will lie among totally different stars and constellations at different parts of this period; and this kind of spiral revolution being continually kept up, it will, at one time or other, cover with its disc every point of the heavens within that limit of latitude or distance from the ecliptic which its inclination permits; that is to say, a belt or zone of the heavens, of 10° 18′ in breadth, having the ecliptic for its middle line.   Nevertheless, it still remains true that the *actual place* of the moon, in consequence of this motion, deviates in a single revolution *very little* from what it would be were the nodes at rest.   Supposing the moon to set out from its node A, its latitude, when it comes to F, having completed a revolution in longitude, will not exceed 8′; which, though small in a single revolution, accumulates in its effect in a succession of many : it is to account for, and represent geometrically, this deviation, that the *motion of the nodes* is devised.

(409.) The moon's orbit, then, is not, strictly speaking, an ellipse returning into itself, by reason of the variation of the plane in which it lies, and the motion of its nodes.   But even laying aside this consideration, the axis of the ellipse is itself constantly changing its direction in space, as has been already stated of the solar ellipse, but much more rapidly ; making a complete revolution, in the same direction with the moon's own motion, in 3232·5753 mean solar days, or about nine years, being about 3° of angular motion in a whole revolution of the moon.   This is the phenomenon known by the name of the revolution of the moon's *apsides*.   Its cause will be hereafter explained.   Its immediate effect is to produce a variation in the moon's distance from the earth, which is not included in the laws of exact elliptic motion.   In a single revolution of the moon, this variation of distance is trifling ; but in the course of many it becomes considerable, as is easily seen, if we consider that in four years and a half the position of the axis will be completely reversed, and the apogee of the moon will occur where the perigee occurred before.

(410.) The best way to form a distinct conception of the moon's motion is to regard it as describing an ellipse about the earth in the focus, and, at the same time, to regard this ellipse itself to be in a twofold state of revolution ; 1st, in its own plane, by a continual advance of its axis in that plane ; and 2dly, by a continual *tilting* motion of the plane itself, exactly similar to, but much more rapid than, that of the earth's equator produced by the conical motion of its axis described in art 317.

(411.) As the moon is at a very moderate distance from us (astronomically speaking), and is in fact our nearest neighbour, while the sun and stars are in comparison immensely beyond it, it must of necessity happen, that at one time or other it must *pass over* and *occult* or *eclipse* every star and planet within the zone above described (and, as seen from the *surface* of earth, even somewhat beyond it, by reason of parallax, which may throw it apparently nearly a degree either way from its place as seen from the centre, according to the observer's station). Nor is the sun itself exempt from being thus hidden, whenever any part of the moon's disc, in this her tortuous course, comes to *overlap* any part of the space occupied in the heavens by that luminary. On these occasions is exhibited the most striking and impressive of all the occasional phenomena of astronomy, an *eclipse of the sun,* in which a greater or less portion, or even in some rare conjunctures the whole, of its disc is obscured, and, as it were, obliterated, by the superposition of that of the moon, which appears upon it as a circularly-terminated black spot, producing a temporary diminution of daylight, or even nocturnal darkness, so that the stars appear as if at midnight. In other cases, when, at the moment that the moon is centrally superposed on the sun, it so happens that her distance from the earth is such as to render her angular diameter less than the sun's, the very singular phenomenon of an *annular solar eclipse* takes place, when the edge of the sun appears for a few minutes as a narrow ring of light, projecting on all sides beyond the dark circle occupied by the moon in its centre.

(412.) A solar eclipse can only happen when the sun and moon are *in conjunction*, that is to say, have the *same*, or nearly the same, position in the heavens, or the same longitude. It appears by art. 409. that this condition can only be fulfilled at the time of a *new moon*, though it by no means follows, that at *every* conjunction there *must* be an eclipse of the sun. If the lunar orbit coincided with the ecliptic, this would be the case, but as it is inclined to it at an angle of upwards of 5°, it is evident that the conjunction, or equality of longitudes, may take place when the moon is in the part of her orbit too remote from the ecliptic to permit the discs to meet and overlap. It is easy, however, to assign the limits within which an eclipse is possible. To this end we must consider, that, by the effect of parallax, the moon's *apparent* edge may be thrown in *any* direction, according to a spectator's geographical station, by *any* amount not exceeding the horizontal parallax. Now, this comes to the same (so far as the possibility of an eclipse is concerned) as if the apparent diameter of the moon, seen from the earth's centre, were dilated by twice its horizontal parallax; for if, when so dilated, it can touch or overlap the sun, there *must* be an eclipse at *some* part or other of the earth's surface. If, then, at the moment of the nearest conjunction, the geocentric distance of the centres of the two luminaries do not exceed the sum of their semidiameters and of the moon's horizontal parallax, there will be an eclipse. This sum is, at its maximum, about 1° 34′ 27″. In the spherical triangle S N M, then, in which S is the sun's centre, M the moon's, S N the ecliptic, M N the moon's orbit, and N the node, we may

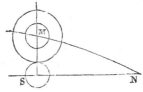

suppose the angle N S M a right angle, S M = 1° 34′ 27″, and the angle M N S = 5° 8′ 48″, the inclination of the orbit. Hence we calculate S N, which comes out 16° 58′. If, then, at the moment of the new moon, the moon's node is farther from the sun in longitude than this limit, there can be no eclipse; if within, there may, and probably will, at some part or other of the earth. To ascertain precisely whether there will or not, and, if there be, how great will be the part eclipsed, the solar and lunar tables must be consulted, the place of the node and the semidiameters exactly ascertained, and the local parallax, and apparent augmentation of the moon's diameter due to the difference of her distance from the observer and from the centre of the earth (which may amount to a sixtieth part of her horizontal diameter), determined; after which it is easy, from the above considerations, to calculate the amount overlapped of the two discs, and their moment of contact.

(413.) The calculation of the occultation of a star depends on similar considerations. An occultation is *possible*, when the moon's course, as seen from the earth's centre, carries her within a distance from the star equal to the sum of her semidiameter and horizontal parallax; and it *will happen at any particular spot*, when her apparent path, as seen from that spot, carries her centre within a distance equal to the sum of her *augmented* semidiameter and *actual* parallax. The details of these calculations, which are somewhat troublesome, must be sought elsewhere.*

(414.) The phenomenon of a solar eclipse and of an occultation are highly interesting and instructive in a physical point of view. They teach us that the moon is an opaque body, terminated by a real and sharply defined surface intercepting light like a solid. They prove to us, also, that at those times when we cannot *see* the moon, she really exists, and pursues her course, and that when we see her only as a crescent, however narrow, the whole globular body *is there*, filling up the deficient outline, though unseen. For occultations take place indifferently at the dark and bright, the visible and invisible outline, whichever happens to be towards the direction in which the moon is moving; with this only difference, that a star *occulted* by the bright limb, if the

* Woodhouse's Astronomy, vol. i.   See also Trans. Ast. Soc. vol. i. p. 325.

phenomenon be watched with a telescope, gives notice, by its gradual approach to the visible edge, when to expect its disappearance, while, if occulted at the dark limb, if the moon, at least, be more than a few days old, it is, as it were, extinguished in mid-air, without notice or visible cause for its disappearance, which, as it happens *instantaneously*, and without the slightest previous diminution of its light, is always surprising ; and, if the star be a large and bright one, even startling from its suddenness. The re-appearance of the star, too, when the moon has passed over it, takes place in those cases when the bright side of the moon is foremost, not at the concave outline of the crescent, but at the invisible outline of the complete circle, and is scarcely less surprising, from its suddenness, than its disappearance in the other case.*

(415.) The existence of the complete circle of the disc, even when the moon is not full, does not, however, rest only on the evidence of occultations and eclipses. It may be *seen*, when the moon is crescent or waning, a few days before and after the *new moon*, with the naked eye, as a pale round body, to which the crescent seems attached, and somewhat projecting beyond its outline (which is an optical illusion arising from the greater intensity of its light). The cause of this appearance will presently be explained. Meanwhile the fact is sufficient to show that the moon is not *inherently* luminous like the sun, but that her light is of an adventitious nature. And its crescent form, increasing regularly from

* There is an optical illusion of a very strange and unaccountable nature which has often been remarked in occultations. The star appears to advance actually *upon* and *within* the edge of the disc before it disappears, and that sometimes to a considerable depth. I have never myself witnessed this singular effect, but it rests on most unequivocal testimony. I have called it an optical illusion ; but it is *barely possible* that a star may shine on such occasions through deep fissures in the substance of the moon. The occultations of close double stars ought to be narrowly watched, to see whether *both* individuals are thus *projected*, as well as for other purposes connected with their theory. I will only hint at one, viz. that a double star, *too close* to be seen divided with any telescope, may yet be detected to be double by the mode of its disappearance. Should a considerable star, for instance, instead of undergoing instantaneous and complete extinction, go out by two distinct steps, following close upon each other ; first losing a portion, then the whole remainder of its light, we may be sure it is a double star, though we cannot see the individuals separately. — *Note to the edit. of* 1833.

a narrow semicircular line to a complete circular disc, corresponds to the appearance a globe would present, one
hemisphere of which was black, the other white, when
differently turned towards the eye, so as to present a greater
or less portion of each.   The obvious conclusion from this is,
that the moon is such a globe, one half of which is brightened
by the rays of some luminary sufficiently distant to enlighten
the complete hemisphere, and sufficiently intense to give it
the degree of splendour we see.   Now, the sun alone is
competent to such an effect.   Its distance and light suffice;
and, moreover, it is invariably observed that, when a crescent,
the bright edge is *towards the sun*, and that in proportion as
the moon in her monthly course becomes more and more
distant from the sun, the breadth of the crescent increases,
and *vice versâ*.

(416.) The sun's distance being 23984 radii of the earth,
and the moon's only 60, the former is nearly 400 times the
latter.   Lines, therefore, drawn from the sun to every part
of the moon's orbit may be regarded as very nearly parallel.*
Suppose, now, O to be the earth, A B C D, &c. various
positions of the moon in its orbit, and S the sun, at the vast
distance above stated; as is shown, then, in the figure, the
hemisphere of the lunar globe turned towards it (on the
right) will be bright, the opposite dark, wherever it may

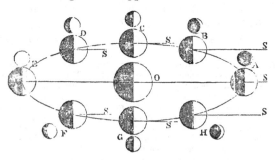

stand in its orbit.   Now, in the position A, when in conjunction with the sun, the dark part is entirely turned

* The angle subtended by the moon's orbit, as seen from the sun (in the mean
state of things), is only 17′ 12″.

towards O, and the bright from it. In this case, then, the moon is not seen, it is *new* moon. When the moon has come to C, half the bright and half the dark hemisphere are presented to O, and the same in the opposite situation G : these are the first and third quarters of the moon. Lastly, when at E, the whole bright face is towards the earth, the whole dark side from it, and it is then seen wholly bright or *full moon*. In the intermediate positions B D F H, the portions of the bright face presented to O will be at first less than half the visible surface, then greater, and finally less again, till it vanishes altogether, as it comes round again to A.

(417.) These monthly changes of appearance, or *phases*, as they are called, arise, then, from the moon, an opaque body, being illuminated on one side by the sun, and reflecting from it, in all directions, a portion of the light so received. Nor let it be thought surprising that a solid substance thus illuminated should appear to *shine* and again illuminate the earth. It is no more than a white cloud does standing off upon the clear blue sky. By day, the moon can hardly be distinguished in brightness from such a cloud ; and, in the dusk of the evening, clouds catching the last rays of the sun appear with a dazzling splendour, not inferior to the seeming brightness of the moon at night.* That the earth sends also such a light to the moon, only probably more powerful by reason of its greater apparent size†, is agreeable to optical principles, and explains the appearance of the dark portion of the young or waning moon completing its crescent (art. 413). For, when the moon is nearly new to the earth, the latter (so to speak) is nearly full to the former ; it then illuminates its dark half by strong *earth-light ;* and it is a

---

* The actual illumination of the lunar surface is not much superior to that of weathered sandstone rock in full sunshine. I have frequently compared the moon setting behind the grey perpendicular façade of the Table Mountain, illuminated by the sun just risen in the opposite quarter of the horizon, when it has been scarcely distinguishable in brightness from the rock in contact with it. The sun and moon being nearly at equal altitudes and the atmosphere perfectly free from cloud or vapour, its effect is alike on both luminaries. (H. 1848).

† The apparent diameter of the moon is 32′ from the earth ; that of the earth seen from the moon is twice her horizontal parallax, or 1° 54′. The apparent surfaces, therefore, are as $(114)^2 : (32)^2$, or as 13 : 1 nearly.

portion of this, reflected back again, which makes it visible
to us in the twilight sky.   As the moon gains age, the earth
offers it a less portion of its bright side, and the phenomenon
in question dies away.

(418.) The lunar month is determined by the recurrence
of its phases : it reckons from new moon to new moon ; that
is, from leaving its conjunction with the sun to its return to
conjunction.   If the sun stood still, like a fixed star, the
interval between two conjunctions would be the same as the
period of the moon's sidereal revolution (art. 401.) ; but, as
the sun apparently advances in the heavens in the same
direction with the moon, only slower, the latter has more
than a complete sidereal period to perform to come up with
the sun again, and will require for it a longer time, which is
the lunar month, or, as it is generally termed in astronomy, a
synodical period.   The difference is easily calculated by considering that the superfluous arc (whatever it be) is described
by the sun with the velocity of $0°·98565$ *per diem*, *in the same
time* that the moon describes that arc *plus* a complete revolution, with her velocity of $13°·17640$ *per diem;* and, the times
of description being identical, the spaces are to each other in
the proportion of the velocities.   Let V and v be the mean
angular velocities, x the superfluous arc ; then $V : v :: 1 + x : x$;

and $V - v : v :: 1 : x$, whence x is found, and $\frac{x}{v} =$ the time of

describing x, or the difference of the sidereal and synodical
periods.   From these data a slight knowledge of arithmetic
will suffice to derive the arc in question, and the time of its
description by the moon ; which being the excess of the
synodic over the sidereal period, the former will be had, and
will appear to be $29^d\ 12^h\ 44^m\ 2^s·87$.

(419.) Supposing the position of the nodes of the moon's
orbit to permit it, when the moon stands at A (or at the new
moon), it will intercept a part or the whole of the sun's rays,
and cause a solar eclipse.   On the other hand, when at E
(or at the full moon), the earth O will intercept the rays of
the sun, and *cast a shadow* on the moon, thereby causing a
lunar eclipse.   And this is perfectly consonant to fact, such

eclipses never happening but at the exact time of the full moon. But, what is still more remarkable, as confirmatory of the position of the earth's sphericity, this shadow, which we plainly see to enter upon and, as it were, eat away the disc of the moon, is always terminated by a circular outline, though, from the greater *size* of the circle, it is only partially seen at any one time. Now, a body which always casts a circular shadow must itself be spherical.

(420.) Eclipses of the sun are best understood by regarding the sun and moon as two independent luminaries, each moving according to known laws, and viewed from the earth; but it is also instructive to consider eclipses generally as arising from the shadow of one body thrown on another by a luminary *much larger than either.* Suppose, then, A B to represent the sun, and C D a spherical body, whether earth or moon, illuminated by it. If we join and prolong A C, B D; since A B is greater than C D, these lines will meet in a point E, more or less distant from the body C D, according to its size, and within the space C E D (which represents a cone,

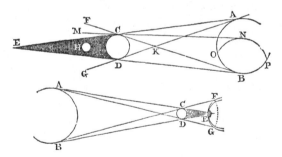

since C D and A B are spheres), there will be a total shadow. This shadow is called the *umbra,* and a spectator situated within it can see no part of the sun's disc. Beyond the umbra are two diverging spaces (or rather, a portion of a single conical space, having K for its vertex), where if a spectator be situated, as at M, he will see a portion only (A O N P) of the sun's surface, the rest (B O N P) being obscured by the earth. He will, therefore, receive only partial sunshine; and the more, the nearer he is to the

exterior borders of that cone which is called the *penumbra*.
Beyond this he will see the whole sun, and be in full illumination.   All these circumstances may be perfectly well
shown by holding a small globe up in the sun, and receiving
its shadow at different distances on a sheet of paper.

(421.) In a lunar eclipse (represented in the upper figure),
the moon is seen to enter* the *penumbra* first, and, by
degrees, get involved in the *umbra,* the former bordering
the latter like a smoky haze.   At this period of the eclipse,
and while yet a considerable part of the moon remains unobscured, the portion involved in the umbra is invisible to
the naked eye, though still perceptible in a telescope, and of
a dark grey hue.   But as the eclipse advances, and the
enlightened part diminishes in extent, and grows gradually
more and more obscured by the advance of the *penumbra,*
the eye, relieved from its glare, becomes more sensible to
feeble impressions of light and colour ; and phænomena of a
remarkable and instructive character begin to be developed.
The *umbra* is seen to be very far from totally dark ; and in
its faint illumination it exhibits a gradation of colour, being
bluish, or even (by contrast) somewhat greenish, towards
the borders for a space of about 4' or 5' of apparent angular
breadth inwards, thence passing, by delicate but rapid gradation, through rose red to a fiery or copper-coloured glow,
like that of dull red-hot iron.   As the eclipse proceeds this
glow spreads over the whole surface of the moon, which then
becomes on some occasions so strongly illuminated, as to cast
a very sensible shadow, and allow the spots on its surface to
be perfectly well distinguished through a telescope.

(422.) The cause of these singular, and sometimes very
beautiful appearances, is the refraction of the sun's light in
passing through our atmosphere, which at the same time
becomes coloured with the hues of sunset by the absorption
of more or less of the violet and blue rays, as it passes
through strata nearer or more remote from the earth's surface,

* The actual contact with the penumbra is never seen ; the defalcation of
light comes on so very gradually that it is not till when already deeply immersed,
that it is perceived to be sensibly darkened.

and, therefore, more or less loaded with vapour. To show this, let A D *a* be a section of the cone of the *umbra*, and F H *h f*

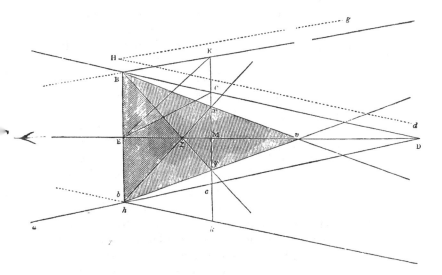

of the *penumbra,* through their common axis D E S, passing through the centres E S of the earth and sun, and let K M E be a section of these cones at a distance E M from E, equal to the radius of the moon's orbit, or 60 radii of the earth.* Taking this radius for unity, since D S, the distance of the sun, is 23984 such units, we readily calculate D E = 218, D M = 158, for the distances at which the apex of the *geometrical umbra* lies behind the earth and the moon respectively. We also find for the measure of the angle E D B, 15' 46", and therefore D B E = 89° 44' 14", whence also we get M C (the linear semidiameter of the *umbra*) = 0·725 (or in miles 2864), and the angle C E M, its apparent angular semi-diameter as seen from E = 41' 30". And instituting similar calculations for the geometrical *penumbra* we get M K = 1·005 (3970 miles), and K E M   57' 36"; and it

---

* The figure is unavoidably drawn out of all proportion, and the angles violently exaggerated. The reader should endeavour to draw the figure in its true proportions.

may be well to remember that the doubles of these angles, or the mean angular diameters of the *umbra* and *penumbra*, are described by the moon with its mean velocity in $2^h\ 43^m$, and $3^h\ 47^m$ respectively, which are therefore the respective durations of the total and partial obscuration of any one point of the moon's disc in traversing centrally the geometrical shadow.

(423.) Were the earth devoid of atmosphere, the whole of the phenomena of a lunar eclipse would consist in these partial or total obscurations. Within the space C c the whole of the light, and within K C and $c\,k$ a greater or less portion of it would be intercepted by the solid body B b of the earth. The refracting atmosphere, however, extends from B, b, to a certain unknown, but very small distance B H, $b\,h$, which, acting as a convex lens, of gradually (and very rapidly) decreasing density, disperses all that comparatively small portion of light which falls upon it over a space bounded externally by H g, parallel and very nearly coincident with B F, and internally by a line B z, the former representing the extreme exterior ray from the limb a of the sun, the latter, the extreme interior ray from the limb A. To avoid complication, however, we will trace only the courses of rays which just graze the surface at B, viz: B z from the upper border, A, and B v from the lower, a, of the sun. Each of these rays is bent inwards from its original course by *double* the amount of the horizontal refraction (33′) *i.e.* by 1° 6′, because, in passing from B out of the atmosphere, it undergoes a deviation equal to that at entering, and in the same direction. Instead, therefore, of pursuing the courses B D, B F, these rays respectively will occupy the positions B $z\,y$, B v, making, with the aforesaid lines, the angles D B b, F B v, each 1° 6′. Now we have found D B E $= 89°\ 44′\ 14″$ and therefore F B E ($=$ D B E $+$ angular diam. of $\odot$) $= 90°\ 17′\ 17″$, consequently the angles E B y and E B v will be respectively 88° 38′ 14″ and 89° 11′ 17″ from which we conclude E $z = 42\!\cdot\!03$ and E $v = 88\!\cdot\!89$ the former falling short of the moon's orbit by 17·97, and the latter surpassing it by 28·89 radii of the earth.

(424.) The penumbra, therefore, of rays refracted at B, will be spread over the space v B y, that at H over g H d, and at the intermediate points, over similar intermediate spaces, and through this compound of superposed penumbræ the moon passes during the whole of its path through the geometrical shadow, never attaining the *absolute* umbra B z b at all. Without going into detail as to the intensity of the refracted rays, it is evident that the totality of light so thrown into the shadow is to that which the earth intercepts, as the area of a circular section of the atmosphere to that of a diametrical section of the earth itself, and, therefore, at all events but feeble. And it is still further enfeebled by actual clouds suspended in that portion of the air which forms the visible border of the earth's disc as seen from the moon, as well as by the general want of transparency caused by invisible vapour, which is especially effective in the lowermost strata, within three or four miles of the surface, and which will impart to all the rays they transmit, the ruddy hue of sunset, only *of double* the depth of tint which we admire in our glowing sunsets, by reason of the rays having to traverse twice as great a thickness of atmosphere. This redness will be most intense at the points x, y, of the moon's path through the umbra, and will thence degrade very rapidly outwardly, over the spaces x c, y C, less so inwardly, over x y. And at C, c, its hue will be mingled with the bluish or greenish light which the atmosphere scatters by irregular dispersion, or in other words by our *twilight* (art. 44). Nor will the phenomenon be uniformly conspicuous at all times. Supposing a generally and deeply clouded state of the atmosphere around the edge of the earth's disc visible from the moon (*i. e.* around that great circle of the earth, in which, at the moment the sun is in the horizon,) little or no refracted light may reach the moon.* Supposing that circle partly clouded and partly clear, patches of red light corresponding to the clear portions will be thrown into the umbra, and may give rise to various and changeable distributions of light on the eclipsed

---

* As in the eclipses of June 5. 1620, April 25. 1642. Lalande, Ast. 1769.

disc*; while, if entirely clear, the eclipse will be remarkable for the conspicuousness of the moon during the whole or a part of its immersion in the umbra.†

(425.) Owing to the great size of the earth, the cone of its *umbra* always projects far beyond the moon; so that, if, at the time of a lunar eclipse, the moon's path be properly directed, it is sure to pass through the *umbra*. This is not, however, the case in solar eclipses. It so happens, from the adjustment of the size and distance of the moon, that the extremity of her *umbra* always falls *near* the earth, but sometimes attains and sometimes falls short of its surface. In the former case (represented in the lower figure art. 420.) a black spot, surrounded by a fainter shadow, is formed, beyond which there is no eclipse on any part of the earth, but within which there may be either a total or partial one, as the spectator is within the *umbra* or *penumbra*. When the apex of the umbra falls *on* the surface, the moon at that point will appear, for an instant, to *just* cover the sun; but, when it falls short, there will be no total eclipse on any part of the earth; but a spectator, situated in or near the prolongation of the axis of the cone, will see the whole of the moon on the sun, although not large enough to cover it, *i. e.* he will witness an annular eclipse.

(426.) Owing to a remarkable enough adjustment of the periods in which the moon's *synodical* revolution, and that of her nodes, are performed, eclipses return after a certain period, very nearly in the same order and of the same magnitude. For 223 of the moon's mean synodical revolutions, or *lunations,* as they are called, will be found to occupy 6585·32 days, and nineteen complete synodical revolutions of the node to occupy 6585·78. The difference in the mean position of the node, then, at the beginning and end of 223 lunations, is nearly insensible; so that a recurrence of all eclipses within that interval must take place. Accordingly,

---

* As in the eclipse of Oct. 13. 1837, observed by the author.
† As in that of March 19. 1848, when the moon is described as giving " good light " during more than an hour after its total immersion, and some persons even doubted its being eclipsed. (Notices of R. Ast. Soc. viii. p. 132.)

this period of 223 lunations, or eighteen years and ten days, is a very important one in the calculation of eclipses. It is supposed to have been known to the Chaldeans, under the name of the *Saros*; the regular return of eclipses having been known as a physical fact for ages before their exact theory was understood. In this period there occur ordinarily 70 eclipses, 29 of the moon and 41 of the sun, visible in some part of the earth. Seven eclipses of either sun or moon at most, and two at least (both of the sun), may occur in a year.

(427.) The commencement, duration, and magnitude of a lunar elipse are much more easily calculated than those of a solar, being independent of the position of the spectator on the earth's surface, and the same as if viewed from its centre. The common centre of the *umbra* and *penumbra* lies always in the ecliptic, at a point opposite to the sun, and the path described by the moon in passing through it is its true orbit as it stands at the moment of the full moon. In this orbit, its position, at every instant, is known from the lunar tables and ephemeris; and all we have, therefore, to ascertain, is, the *moment when* the distance between the moon's centre and the centre of the shadow is exactly equal to the sum of the semidiameters of the moon and *penumbra*, or of the moon and *umbra*, to know when it enters upon and leaves them respectively. No lunar eclipse can take place, if, at the moment of the full moon, the sun be at a greater angular distance from the node of the moon's orbit than 11° 21′, meaning by an *eclipse* the immersion of any part of the moon in the *umbra*, as its contact with the *penumbra* cannot be observed (see note to art. 421.).

(428.) The dimensions of the shadow, at the place where it crosses the moon's path, require us to know the distances of the sun and moon at the time. These are variable; but are calculated and set down, as well as their semidiameters, for every day, in the ephemeris, so that none of the data are wanting. The sun's distance is easily calculated from its elliptic orbit; but the moon's is a matter of more difficulty, by reason of the progressive motion of the axis of the lunar orbit. (Art 409.)

(429.) The physical constitution of the moon is better known to us than that of any other heavenly body. By the aid of telescopes, we discern inequalities in its surface which can be no other than mountains and valleys,—for this plain reason, that we see the shadows cast by the former in the exact proportion as to length which they ought to have, when we take into account the inclination of the sun's rays to that part of the moon's surface on which they stand. The convex outline of the limb turned towards the sun is always circular, and very nearly smooth; but the opposite border of the enlightened part, which (were the moon a perfect sphere) ought to be an exact and sharply defined ellipse, is always observed to be extremely ragged, and indented with deep recesses and prominent points. The mountains near this edge cast long black shadows, as they should evidently do, when we consider that the sun is in the act of rising or setting to the parts of the moon so circumstanced. But as the enlightened edge advances beyond them, i. e. as the sun to them gains altitude, their shadows shorten; and at the full moon, when all the light falls in our line of sight, no shadows are seen on any part of her surface. From micrometrical measures of the lengths of the shadows of the more conspicuous mountains, taken under the most favourable circumstances, the heights of many of them have been calculated. Messrs. Beer and Maedler in their elaborate work, entitled " Der Mond," have given a list of heights resulting from such measurements, for no less than 1095 lunar mountains, among which occur all degrees of elevation up to 3569 toises, (22823 British feet), or about 1400 feet higher than Chimborazo in the Andes. The existence of such mountains is further corroborated by their appearance, as small points or islands of light beyond the extreme edge of the enlightened part, which are their tops catching the sun-beams before the intermediate plain, and which, as the light advances, at length connect themselves with it, and appear as prominences from the general edge.

(430.) The generality of the lunar mountains present a striking uniformity and singularity of aspect. They are wonderfully numerous, especially towards the Southern portion of the disc, occupying by far the larger portion of the surface, and almost universally of an exactly circular or cup-shaped form, foreshortened, however, into ellipses towards the limb; but the larger have for the most part flat bottoms within, from which rises centrally a small, steep, conical hill. They offer, in short, in its highest perfection, the true *volcanic* character, as it may be seen in the crater of Vesuvius, and in a map of the volcanic districts of the Campi Phlegræi* or the Puy de Dôme, but with this remarkable peculiarity, viz. : that the bottoms of many of the craters are very deeply depressed below the general surface of the moon, the internal depth being often twice or three times the external height. In some of the principal ones, decisive marks of volcanic stratification, arising from successive deposits of ejected matter, and evident indications of lava currents streaming outwards in all directions, may be clearly traced with powerful tele-scopes. (See Pl. V. fig. 2.)† In Lord Rosse's magnificent reflector, the flat bottom of the crater called Albategnius is seen to be strewed with blocks not visible in inferior tele-scopes, while the exterior of another (Aristillus) is all hatched over with deep gullies radiating towards its center. What is, moreover, extremely singular in the geology of the moon is, that, although nothing having the character of seas can be traced, (for the dusky spots, which are commonly called seas, when closely examined, present appearances incom-patible with the supposition of deep water,) yet there are large regions perfectly level, and apparently of a decided alluvial character.

(431.) The moon has no clouds, nor any other decisive indications of an atmosphere. Were there any, it could not fail to be perceived in the occultations of stars and the phæ-nomena of solar eclipses, as well as in a great variety of other phænomena. The moon's diameter, for example, as measured

---

* See Breislak's map of the environs of Naples, and Desmarest's of Auvergne.
† From a drawing taken with a reflector of twenty feet focal length (*h*).

micrometrically, and as estimated by the interval between the disappearance and reappearance of a star in an occultation, ought to differ by twice the horizontal refraction at the moon's surface. No appretiable difference being perceived, we are entitled to conclude the non-existence of any atmosphere dense enough to cause a refraction of $1''$ *i. e.* having one 1980th part of the density of the earth's atmosphere. In a solar eclipse, the existence of any sensible refracting atmosphere in the moon, would enable us to trace the limb of the latter beyond the cusps, externally to the sun's disc, by *a narrow, but brilliant* line of light, extending to some distance along its edge. No such phænomenon is seen. *Very* faint stars ought to be extinguished before occultation, were any appretiable amount of vapour suspended near the surface of the moon. But such is not the case ; when occulted at the bright edge, indeed, the light of the moon extinguishes small stars, and even at the dark limb, the glare in the sky caused by the near presence of the moon, renders the occultation of *very* minute stars unobservable. But during the continuance of a total lunar eclipse, stars of the tenth and eleventh magnitude are seen to come up to the limb, and undergo *sudden* extinction as well as those of greater brightness.* Hence, the climate of the moon must be very extraordinary; the alternation being that of unmitigated and burning sunshine fiercer than an equatorial noon, continued for a whole fortnight, and the keenest severity of frost, far exceeding that of our polar winters, for an equal time. Such a disposition of things must produce a constant transfer of whatever moisture may exist on its surface, from the point beneath the sun to that opposite, by distillation *in vacuo* after the manner of the little instrument called a *cryophorus*. The consequence must be absolute aridity below the vertical sun, constant accretion of hoar frost in the opposite region, and, perhaps, a narrow zone of running water at the borders of the enlightened hemisphere.† It is possible, then, that evapo-

---

* As observed by myself in the eclipse of Oct. 13. 1837.
† So in ed. of 1833.

ration on the one hand, and condensation on the other, may to a certain extent preserve an equilibrium of temperature, and mitigate the extreme severity of both climates; but this process, which would imply the continual generation and destruction of an atmosphere of aqueous vapour, must, in conformity with what has been said above of a lunar atmosphere, be confined within very narrow limits.

(432.) Though the surface of the full moon exposed to us, must necessarily be very much heated, —*possibly* to a degree much exceeding that of boiling water, — yet we *feel* no heat from it, and even in the focus of large reflectors, it fails to affect the thermometer.   No doubt, therefore, its heat (conformably to what is observed of that of bodies heated below the point of luminosity) is much more readily absorbed in traversing transparent media than direct solar heat, and is extinguished in the upper regions of our atmosphere, never reaching the surface of the earth at all.   Some probability is given to this by the *tendency to disappearance of clouds under the full moon,* a meteorological *fact,* (for as such we think it fully entitled to rank*) for which it is necessary to seek a cause, and for which no other rational explanation seems to offer.   As for any other influence of the moon on the weather, we have no decisive evidence in its favour.

(433.) A circle of one second in diameter, as seen from the earth, on the surface of the moon, contains about a square mile.   Telescopes, therefore, must yet be greatly improved, before we could expect to see signs of inhabitants, as manifested by edifices or by changes on the surface of the soil.   It should, however, be observed, that, owing to the small density of the materials of the moon, and the comparatively feeble gravitation of bodies on her surface, muscular force would there go six times as far in overcoming the weight of materials as on the earth.   Owing to the want of air, however, it seems impossible that any form of life, analogous to those on earth, can

---

* From my own observation, made quite independently of any knowledge of such a tendency having been observed by others. Humboldt, however, in his personal narrative, speaks of it as well known to the pilots and seamen of Spanish America (*h*).

subsist there. No appearance indicating vegetation, or the slightest variation of surface, which can, in our opinion, fairly be ascribed to change of season, can any where be discerned.

(434.) The lunar summer and winter arise, in fact, from the rotation of the moon on its own axis, the period of which rotation is *exactly* equal to its sidereal revolution about the earth, and is performed in a plane 1° 30′ 11″ inclined to the ecliptic, whose *ascending* node is always precisely coincident with the *descending* node of the lunar orbit. So that the axis of rotation describes a conical surface about the pole of the ecliptic in one revolution of the node. The remarkable coincidence of the two rotations, that about the axis and that about the earth, which at first sight would seem perfectly distinct, has been asserted (but we think some-what too hastily *) to be a consequence of the general laws to be explained hereafter. Be that how it may, it is the cause why we always see the same face of the moon, and have no knowledge of the other side.

(435.) The moon's rotation on her axis is uniform; but since her motion in her orbit (like that of the sun) is not so, we are enabled to look a few degrees round the equatorial parts of her visible border, on the eastern or western side, according to circumstances; or, in other words, the line joining the centers of the earth and moon fluctuates a little in its position, from its mean or average intersection with her surface, to the east or westward. And, moreover, since the axis about which she revolves is neither exactly perpendicular to her orbit, nor holds an invariable direction in space, her poles come alternately into view for a small space at the edges of her disc. These phenomena are known by the name of *librations.* In consequence of these two distinct kinds of libration, the same identical point of the moon's surface is not always the center of her disc, and we therefore get sight of a zone of a few degrees in breadth on all sides of the border, beyond an exact hemisphere.

(436.) If there be inhabitants in the moon, the earth must

---

* See Edinburgh Review, No. 175. p. 192.

present to them the extraordinary appearance of a moon of nearly 2° in diameter, exhibiting phases complementary to those which we see the moon to do, but *immovably fixed in their sky,* (or, at least, changing its apparent place only by the small amount of the libration,) while the stars must seem to pass slowly beside and behind it. It will appear clouded with variable spots, and belted with equatorial and tropical zones corresponding to our trade-winds; and it may be doubted whether, in their perpetual change, the outlines of our continents and seas can ever be clearly discerned. During a solar eclipse, the earth's atmosphere will become visible as a narrow, but bright luminous ring of a ruddy colour, where it rests on the earth, gradually passing into faint blue, encircling the whole or part of the dark disc of the earth, the remainder being dark and rugged with clouds.

(437.) The best charts of the lunar surface are those of Cassini, of Russel (engraved from drawings, made by the aid of a seven feet reflecting telescope,) the seleno-topographical charts of Lohrmann, and the very elaborate projection of Beer and Maedler accompanying their work already cited.* Madame Witte, a Hanoverian lady, has recently succeeded in producing from her own observations, aided by Maedler's charts, more than one *complete model* of the whole visible lunar hemisphere, of the most perfect kind, the result of incredible diligence and assiduity. Single craters have also been modelled on a large scale, both by her and Mr. Nasmyth.

---

* The representations of Hevelius in his Selenographia, though not without great merit at the time, and fine specimens of his own engraving, are now become antiquated.

CHAPTER VIII.

OF TERRESTRIAL GRAVITY. — OF THE LAW OF UNIVERSAL GRA-
VITATION. — PATHS OF PROJECTILES ; APPARENT — REAL —
THE MOON RETAINED IN HER ORBIT BY GRAVITY. — ITS LAW
OF DIMINUTION. — LAWS OF ELLIPTIC MOTION. — ORBIT OF THE
EARTH ROUND THE SUN IN ACCORDANCE WITH THESE LAWS. —
MASSES OF THE EARTH AND SUN COMPARED. — DENSITY OF THE
SUN. — FORCE OF GRAVITY AT ITS SURFACE. — DISTURBING
EFFECT OF THE SUN ON THE MOON'S MOTION.

(438.) THE reader has now been made acquainted with the
chief phenomena of the motions of the earth in its orbit
round the sun, and of the moon about the earth. — We come
next to speak of the physical cause which maintains and
perpetuates these motions, and causes the massive bodies so
revolving to deviate continually from the directions they
would naturally seek to follow, in pursuance of the first law
of motion *, and bend their courses into curves concave to
their centers.

(439.) Whatever attempts may have been made by metaphysical writers to reason away the connection of cause and
effect, and fritter it down into the unsatisfactory relation of
habitual sequence †, it is certain that the conception of some
more real and intimate connection is quite as strongly impressed upon the human mind as that of the existence of an
external world, — the vindication of whose reality has (strange

---

* Princip. Lex. i.

† See Brown " On Cause and Effect,"—a work of great acuteness and subtlety
of reasoning on some points, but in which the whole train of argument is
vitiated by one enormous oversight ; the omission, namely, of a *distinct and immediate personal consciousness of causation* in his enumeration of that *sequence of
events,* by which the volition of the mind is made to terminate in the motion of
material objects.   I mean the consciousness of *effort,* accompanied with *intention
thereby* to accomplish an end, as a thing entirely distinct from mere *desire* or
volition on the one hand, and from mere spasmodic contraction of muscles on
the other.   Brown, 3d edit. Edin. 1818, p. 47.   (Note to edition of 1833.)

to say) been regarded as an achievement of no common merit in the annals of this branch of philosophy. It is our own immediate consciousness *of effort*, when we exert force to put matter in motion, or to oppose and neutralize force, which gives us this internal conviction of *power* and *causation* so far as it refers to the material world, and compels us to believe that whenever we see material objects put in motion from a state of rest, or deflected from their rectilinear paths and changed in their velocities if already in motion, it is in consequence of such an EFFORT *somehow* exerted, though not accompanied with *our* consciousness. That such an effort should be exerted with success through an interposed space, is no more difficult to conceive, than that our hand should communicate motion to a stone, with which it is *demonstrably not in contact.*

(440.) All bodies with which we are acquainted, when raised into the air and quietly abandoned, descend to the earth's surface in lines perpendicular to it. They are therefore urged thereto by a force or effort, which it is but reasonable to regard as the direct or indirect result of a *consciousness* and a *will* existing *somewhere*, though beyond our power to trace, which force we term *gravity*, and whose tendency or direction, as universal experience teaches, is towards the earth's center; or rather, to speak strictly, with reference to its spheroidal figure, perpendicular to the surface of still water. But if we cast a body obliquely into the air, this tendency, though not extinguished or diminished, is materially modified in its ultimate effect. The upward impetus we give the stone is, it is true, after a time destroyed, and a downward one communicated to it, which ultimately brings it to the surface, where it is opposed in its further progress, and brought to rest. But all the while it has been continually deflected or bent aside from its rectilinear progress, and made to describe a curved line concave to the earth's center; and having a *highest point, vertex,* or *apogee,* just as the moon has in its orbit, where the direction of its motion is perpendicular to the radius.

(441.) When the stone which we fling obliquely upwards

meets and is stopped in its descent by the earth's surface, its motion is not *towards the center*, but inclined to the earth's radius at the same angle as when it quitted our hand. As we are sure that, if not stopped by the resistance of the earth, it would continue to descend, and that *obliquely*, what presumption, we may ask, is there that it would ever reach the center towards which its motion, in no part of its visible course, was ever directed? What reason have we to believe that it might not rather circulate round it, as the moon does round the earth, returning again to the point it set out from, after completing an elliptic orbit of which the earth's center occupies the lower focus? And if so, is it not reasonable to imagine that the same force of gravity *may* (since we know that it is exerted at all accessible heights above the surface, and even in the highest regions of the atmosphere) extend as far as 60 radii of the earth, or to the moon? and may not this be the power,—for *some* power there *must* be,—which deflects *her* at every instant from the tangent of her orbit, and keeps her in the elliptic path which experience teaches us she actually pursues?

(442.) If a stone be whirled round at the end of a string it will stretch the string by a *centrifugal* force, which, if the speed of rotation be sufficiently increased, will at length break the string, and let the stone escape. However strong the string, it may, by a sufficient rotary velocity of the stone, be brought to the utmost tension it will bear without breaking; and if we know what weight it is capable of carrying, the velocity necessary for this purpose is easily calculated. Suppose, now, a string to connect the earth's center with a weight at its surface, whose strength should be just sufficient to sustain that weight suspended from it. Let us, however, for a moment imagine gravity to have no existence, and that the weight is made to revolve with the *limiting velocity* which that string can barely counteract: then will its tension be just equal to the weight of the revolving body; and any power which should continually urge the body towards the center with a force equal to its weight would perform the office, and might supply the place of the string, if divided.

Divide it then, and in its place let gravity act, and the body will circulate as before; its tendency to the center, or *its weight*, being just balanced by its centrifugal force.   Knowing the radius of the earth, we can calculate by the principles of mechanics the periodical time in which a body so balanced must circulate to keep it up; and this appears to be 1^h 23^m 22^s.

(443.) If we make the same calculation for a body at the distance of the moon, *supposing its weight* or *gravity the same as at the earth's surface*, we shall find the period required to be 10^h 45^m 30^s.   The actual period of the moon's revolution, however, is 27^d 7^h 43^m; and hence it is clear that the moon's velocity is not nearly sufficient to sustain it against *such* a power, supposing it to revolve in a circle, or neglecting (for the present) the slight ellipticity of its orbit.   In order that a body at the distance of the moon (or the moon itself) should be capable of *keeping its distance* from the earth by the outward effort of its centrifugal force, while yet its time of revolution should be what the moon's actually is, it will appear* that *gravity*, instead of being as intense as at the surface, would require to be very nearly 3600 times less energetic; or, in other words, that its intensity is so enfeebled by the remoteness of the body on which it acts, as to be capable of producing in it, in the same time, only $\frac{1}{3600}$th part of the motion which it would impart to the same mass of matter at the earth's surface.

(444.) The distance of the moon from the earth's center is a very little less than sixty times the distance from the center to the surface, and $3600 : 1 : : 60^2 : 1^2$; so that the proportion in which we must admit the earth's gravity to be enfeebled at the moon's distance, if it be really the force which retains the moon in her orbit, must be (at least in this particular instance) that of the squares of the distances at which it is compared.   Now, in such a diminution of energy with increase of distance, there is nothing *prima facie* inadmissible.   Emanations from a center, such as light and heat,

* Newton, Princip. b. i., Prop. 4., Cor. 2.

do really diminish in intensity by increase of distance, and in this identical proportion; and though we cannot certainly argue much from this analogy, yet we do see that the power of magnetic and electric attractions and repulsions is actually enfeebled by distance, and much more rapidly than in the simple proportion of the increased distances. The argument therefore, stands thus : — On the one hand, *Gravity* is a real power, of whose agency we have daily experience. We know that it extends to the greatest accessible heights, and far beyond; and we see no reason for drawing a line at any particular height, and there asserting that it must cease entirely; though we have analogies to lead us to suppose its energy may diminish as we ascend to great heights from the surface, such as that of the moon. On the other hand we are sure the moon *is* urged towards the earth by *some* power which retains her in her orbit, and that the intensity of this power is such as would correspond to a gravity, diminished in the proportion — otherwise not improbable — of the squares of the distances. If gravity be *not* that power, there must exist some other; and, besides this, gravity must cease at some inferior level, or the nature of the moon must be different from that of ponderable matter; — for if not, it would be urged by *both* powers, and therefore *too much* urged and forced inwards from her path.

(445.) It is on such an argument that Newton is understood to have rested, in the first instance, and provisionally, his law of universal gravitation, which may be thus abstractly stated :— " Every particle of matter in the universe attracts every other particle, with a force directly proportioned to the mass of the attracting particle, and inversely to the square of the distance between them." In this abstract and general form, however, the proposition is not applicable to the case before us. The earth and moon are not mere *particles*, but great spherical bodies, and to such the general law does not immediately apply; and, before we can make it applicable, it becomes necessary to enquire what will be the force with which a congeries of particles, constituting a solid mass of any assigned figure, will attract another such collection of material atoms. This problem is one purely dynamical, and, in this its general form, is of extreme difficulty. Fortunately however, for human knowledge when the attracting and attracted bodies are spheres, it admits of an easy and direct solution. Newton himself has shown (*Princip.* b. i. prop. 75.) that, in that case, the attraction is precisely the same as if the whole matter of each sphere were collected into its center, and the spheres were single particles there placed; so that, in this case, the general law applies in its strict wording. The effect of the trifling deviation of the earth from a spherical form is of too minute an order to need attention at present. It is, however, perceptible, and may be hereafter noticed.

(446.) The next step in the Newtonian argument is one which divests the law of gravitation of its provisional character, as derived from a loose and superficial consideration of the lunar orbit as a circle described with an average or mean velocity, and elevates it to the rank of a general and primordial relation by proving its applicability to the state of existing nature in all its circumstances. This step consists in demonstrating, as he has done* (*Princip.* i. 17. i. 75.), that, under the influence of such an attractive force mutually urging two spherical gravitating bodies towards each other, they will each, when moving in each other's neighbourhood, be deflected into an orbit concave towards the other, and describe, one about the other regarded as fixed, or both round their common centre of gravity, curves whose forms are limited to those figures known in geometry by the general name of conic sections. It will depend, he shows, in any assigned case, upon the particular circumstances or velocity, distance, and direction, *which* of these curves shall be described, — whether an ellipse, a circle, a parabola, or

* We refer for these fundamental propositions, as a point of duty, to the immortal work in which they were first propounded. It is impossible for us, in this volume, to go into these investigations : even did our limits permit, it would be utterly inconsistent with our plan ; a general idea, however, of their conduct will be given in the next chapter. We trust that the careful and attentive study of the Principia *in its original form* will never be laid aside, whatever be the improvements of the modern analysis as respects facility of calculation and expression. From no other quarter can a thorough and complete comprehension of the mechanism of our system, (so far as the immediate scope of that work extends,) be anything like so well, and we may add, so easily obtained

an hyperbola; but one or other it *must* be; and any one of any degree of eccentricity it *may* be, according to the circumstances of the case; and, in all cases, the point to which the motion is referred, whether it be the centre of one of the spheres, or their common centre of gravity, will of necessity be the *focus* of the conic section described. He shows, furthermore (*Princip.* i. 1.), that, in every case, the *angular velocity* with which the line joining their centres moves, must be inversely proportional to the square of their mutual distance, and that equal areas of the curves described will be swept over by their line of junction in equal times.

(447.) All this is in conformity with what we have stated of the solar and lunar movements. Their orbits are ellipses, but of different degrees of eccentricity; and this circumstance already indicates the general applicability of the principles in question.

(448.) But here we have already, by a natural and ready implication (such is always the progress of generalization), taken a further and most important step, almost unperceived. We have extended the action of gravity to the case of the earth and sun, to a distance immensely greater than that of the moon, and to a body apparently quite of a different nature than either. Are we justified in this? or, at all events, are there no modifications introduced by the change of data, if not into the general expression, at least into the particular interpretation, of the law of gravitation? Now, the moment we come to numbers, an obvious incongruity strikes us. When we calculate, as above, from the known distance of the sun (art. 357.), and from the period in which the earth circulates about it (art. 305.), what must be the centrifugal force of the latter by which the sun's attraction is balanced, (and which, therefore, becomes an exact measure of the sun's attractive energy as exerted on the earth,) we find it to be immensely greater than would suffice to counteract the *earth's* attraction on an equal body at that distance—greater in the high proportion of 354936 to 1. It is clear, then, that if the earth be retained in its orbit about the sun by *solar attraction*, conformable in its rate of diminution with the general law, this force must be no less than 354936 times more intense than what the earth would be capable of exerting, *cæteris paribus*, at an equal distance.

(449.) What, then, are we to understand from this result? Simply this,—that the sun attracts as a collection of 354936 earths occupying its place would do, or, in other words, that the sun contains 354936 times the mass or quantity of ponderable matter that the earth consists of. Nor let this conclusion startle us. We have only to recall what has been already shown in (art. 358.) of the gigantic dimensions of this magnificent body, to perceive that, in assigning to it so vast a mass, we are not outstepping a reasonable proportion. In fact, when we come to compare its *mass* with its *bulk*, we find its density* to be less than that of the earth, being no more than 0·2543. So that it must consist, in reality, of far *lighter* materials, especially when we consider the force under which its central parts must be condensed. This consideration renders it highly probable that an intense heat prevails in its interior by which its elasticity is reinforced, and rendered capable of resisting this almost inconceivable pressure without collapsing into smaller dimensions.

(450.) This will be more distinctly appretiated, if we estimate, as we are now prepared to do, the intensity of gravity at the sun's surface.

The attraction of a sphere being the same (art. 445.) as if its whole mass were collected in its centre, will, of course, be proportional to the mass directly, and the square of the distance inversely; and, in this case, the distance is the radius of the sphere. Hence we conclude†, that the intensities of solar and terrestrial gravity at the surfaces of the two globes are in the proportions of 27·9 to 1. A pound of terrestrial matter at the sun's surface, then, would exert a pressure equal to what 27·9 such pounds would do at the

---

* The density of a material body is as the *mass* directly, and the volume inversely: hence density of ⊙: density of ⊕ :: $\frac{354936}{1384472}$: 1 :: 0·2543 : 1.

† Solar gravity : terrestrial :: $\frac{354936}{(440000)^2}$: $\frac{1}{(4000)^2}$ :: 27·9 : 1; the respective radii of the sun and earth being 440000, and 4000 miles.

earth's. The efficacy of muscular power to overcome weight is therefore proportionally nearly 28 times less on the sun than on the earth. An ordinary man, for example, would not only be unable to sustain his own weight on the sun, but would literally be crushed to atoms under the load.*

(451.) Henceforward, then, we must consent to dismiss all idea of the earth's immobility, and transfer that attribute to the sun, whose ponderous mass is calculated to exhaust the feeble attractions of such comparative atoms as the earth and moon, without being perceptibly dragged from its place. Their centre of gravity lies, as we have already hinted, almost close to the centre of the solar globe, at an interval quite imperceptible from our distance; and whether we regard the earth's orbit as being performed about the one or the other center makes no appretiable difference in any one phenomenon of astronomy.

(452.) It is in consequence of the *mutual* gravitation of all the several parts of matter, which the Newtonian law supposes, that the earth and moon, while in the act of revolving, monthly, in their mutual orbits about their common center of gravity, yet continue to circulate, without parting company, in a greater annual orbit round the sun. We may conceive this motion by connecting two unequal balls by a stick, which, at their center of gravity, is tied by a long string, and whirled round. Their joint *system* will circulate as one body about the common center to which the string is attached, while yet they may go on circulating round each other in subordinate gyrations, as if the stick were quite free from any such tie, and merely hurled through the air. If the earth alone, and not the moon, gravitated to the sun, it would be dragged away, and leave the moon behind — and *vice versâ*; but, acting on both, they continue together under its attraction, just as the loose parts of the earth's surface continue to rest upon it. It is, then, in strictness, not the earth or the moon which describes an ellipse around the sun, but their common centre of gravity. The effect is to produce

* A mass weighing 12 stone or 168lbs. on the earth, would produce a pressure of 4687 lbs. on the sun.

a small, but very perceptible, monthly *equation* in the sun's apparent motion as seen from the earth, which is always taken into account in calculating the sun's place.   The moon's actual path in its compound orbit round the earth and sun is an epicycloidal curve intersecting the orbit of the earth twice in every lunar month, and alternately within and without it.   But as there are not more than twelve such months in the year, and as the total departure of the moon from it either way does not exceed one 400th part of the radius, this amounts only to a slight undulation upon the earth's ellipse, so slight, indeed, that if drawn in true propor· tion on a large sheet of paper, no eye unaided by measurement with compasses would detect it.   The real orbit of the moon is everywhere concave towards the sun.

(453.) Here moreover, *i. e.* in the attraction of the sun, we have the key to all those differences from an exact elliptic movement of the moon in her monthly orbit, which we have already noticed (arts. 407. 409.), viz. to the retrograde revolution of her nodes; to the direct circulation of the axis of her ellipse; and to all the other deviations from the laws of elliptic motion at which we have further hinted.. If the moon simply revolved about the earth under the influence of its gravity, none of these phenomena would take place.   Its orbit would be a perfect ellipse, returning into itself, and always lying in one and the same plane.   That it *is not so*, is a proof that some cause *disturbs* it, and interferes with the earth's attraction; and this cause is no other than the sun's attraction — or rather, that part of it which is not *equally* exerted on the earth.

(454.) Suppose two stones, side by side, or otherwise situated with respect to each other, to be let fall together ; then, as gravity accelerates them equally, they will retain their relative positions, and fall together as if they formed one mass.   But suppose gravity to be rather more intensely exerted on one than the other; then would that one be rather more accelerated in its fall, and would gradually leave the other ; and thus a relative motion between them would arise from the difference of action, however slight.

T

(455.) The sun is about 400 times more remote than the moon; and, in consequence, while the moon describes her monthly orbit round the earth, her distance from the sun is alternately $\frac{1}{400}$th part greater and as much less than the earth's. Small as this is, it is yet sufficient to produce a perceptible excess of attractive tendency of the moon towards

the sun, above that of the earth when in the nearer point of her orbit, M, and a corresponding defect on the opposite part, N; and, in the intermediate positions, not only will a difference of *forces* subsist, but a difference of *directions* also; since however small the lunar orbit M N, it is not a point, and, therefore, the lines drawn from the sun S to its several parts cannot be regarded as strictly parallel. If, as we have already seen, the force of the sun were equally exerted, and in parallel directions on both, no disturbance of their relative situations would take place; but from the non-verification of these conditions arises a *disturbing force,* oblique to the line joining the moon and earth, which in some situations acts to *accelerate,* in others to *retard,* her elliptic orbitual motion; in some to draw the earth from the moon, in others the moon from the earth. Again, the lunar orbit, though very nearly, is yet not quite coincident with the plane of the ecliptic; and hence the action of the sun, which is very nearly parallel to the last-mentioned plane, tends to draw her somewhat *out of the plane* of her orbit, and does actually do so — producing the revolution of her nodes, and other phenomena less striking. We are not yet prepared to go into the subject of these *perturbations,* as they are called; but they are introduced to the reader's notice as early as possible, for the purpose of re-assuring his mind, should doubts have arisen as to the logical correctness of our argument, in consequence of our temporary neglect of them while working our way upward to the law of gravity from a general consideration of the moon's orbit.

CHAPTER IX.

OF THE SOLAR SYSTEM.

APPARENT MOTIONS OF THE PLANETS.— THEIR STATIONS AND RE-
TROGRADATIONS.— THE SUN THEIR NATURAL CENTER OF MOTION.
— INFERIOR PLANETS.— THEIR PHASES, PERIODS, ETC. — DIMEN-
SIONS AND FORM OF THEIR ORBITS. — TRANSITS ACROSS THE SUN.
— SUPERIOR PLANETS.— THEIR DISTANCES, PERIODS, ETC.— KEP-
LER'S LAWS AND THEIR INTERPRETATION. — ELLIPTIC ELEMENTS
OF A PLANET'S ORBIT. — ITS HELIOCENTRIC AND GEOCENTRIC
PLACE.— EMPIRICAL LAW OF PLANETARY DISTANCES; —VIOLATED
IN THE CASE OF NEPTUNE. — THE ULTRA-ZODIACAL PLANETS. —
PHYSICAL PECULIARITIES OBSERVABLE IN EACH OF THE PLANETS.

(456.) THE sun and moon are not the only celestial objects
which appear to have a motion independent of that by which
the great constellation of the heavens is daily carried round
the earth. Among the stars there are several,—and those
among the brightest and most conspicuous, — which, when
attentively watched from night to night, are found to
change their relative situations among the rest; some rapidly,
others much more slowly. These are called *planets*. Four
of them —Venus, Mars, Jupiter, and Saturn —are remarkably large and brilliant; another, Mercury, is also visible to
the naked eye as a large star, but, for a reason which will
presently appear, is seldom conspicuous; a sixth, Uranus, is
barely discernible without a telescope; and nine others—
Neptune, Ceres, Pallas, Vesta, Juno, Astræa, Hebe, Iris,
Flora— are never visible to the naked eye. Besides these
fifteen, others yet undiscovered may exist*; and it is extremely probable that such is the case, — the multitude of

* While this sheet is passing through the press, a sixteenth, not yet named,
has been added to the list, by the observations of Mr. Graham, astronomical
assistant to E. Cooper, Esq., at his observatory at Markree, Sligo, Ireland.

telescopic stars being so great that only a small fraction of
their number has been sufficiently noticed to ascertain whether
they retain the same places or not, and the ten last-mentioned
planets having all been discovered within little more than
half a century from the present time.

(457.) The apparent motions of the planets are much more
irregular than those of the sun or moon. Generally speaking,
and comparing their places at distant times, they all advance,
though with very different *average* or *mean* velocities, in the
same direction as those luminaries, *i. e.* in opposition to the
apparent diurnal motion, or from west to east : all of them
make the entire tour of the heavens, though under very different circumstances; and all of them, with the exception of the
eight telescopic planets,— Ceres, Pallas, Juno, Vesta, Astræa,
Hebe, Iris, and Flora (which may therefore be termed *ultra-
zodiacal*), — are confined in their visible paths within very
narrow limits on either side the ecliptic, and perform their
movements within that zone of the heavens we have called,
above, the Zodiac (art. 303.).

(458.) The obvious conclusion from this is, that whatever
be, otherwise, the nature and law of their motions, they are
performed *nearly in the plane of the ecliptic*—that plane, namely,
in which our own motion about the sun is performed.   Hence
it follows, that we see their evolutions, not in *plan*, but in *section;* their real angular movements and linear distances being
all *foreshortened* and confounded undistinguishably, while
only their deviations from the ecliptic appear of their natural
magnitude, undiminished by the effect of perspective.

(459.) The apparent motions of the sun and moon, though
not uniform, do not deviate very greatly from uniformity ; a
moderate acceleration and retardation, accountable for by the
ellipticity of their orbits, being all that is remarked.   But
the case is widely different with the planets : sometimes they
advance rapidly ; then relax in their apparent speed — come
to a momentary stop; and then actually reverse their motion,
and run back upon their former course, with a rapidity at
first increasing, then diminishing, till the reversed or retrograde motion ceases altogether.   Another *station*, or moment

of apparent rest or indecision, now takes place; after which the movement is again reversed, aud resumes its original direct character.   On the whole, however, the amount of direct motion more than compensates the retrograde; and by the excess of the former over the latter, the gradual advance of the planet from west to east is maintained.   Thus, supposing the Zodiac to be unfolded into a plane surface, (or represented as in Mercator's projection, art. 283. taking the ecliptic E C for its ground line,) the track of a planet when mapped down by observation from day to day, will offer the

appearance P Q R S, &c. ; the motion from P to Q being direct, at Q stationary, from Q to R retrograde, at R again stationary, from R to S direct, and so on.

(460.) In the midst of the irregularity and fluctuation of this motion, one remarkable feature of uniformity is observed. Whenever the planet crosses the ecliptic, as at N in the figure, it is said (like the moon) to be in its node ; and as the earth necessarily lies in the plane of the ecliptic, the planet cannot be *apparently* or *uranographically* situated in the celestial circle so called, without being *really* and *locally* situated *in that plane*.   The visible passage of a planet through its node, then, is a phenomenon indicative of a circumstance in its real motion quite independent of the station from which we view it.   Now, it is easy to ascertain, by observation, when a planet passes from the north to the south side of the ecliptic : we have only to convert its right ascensions and declinations into longitudes and latitudes, and the change from north to south latitude on two successive days will advertise us on what *day* the transition took place; while a simple proportion, grounded on the observed state of its motion *in latitude* in the interval, will suffice to fix the precise hour and minute of its arrival on the ecliptic.   Now, this being done for several transitions from side to side of the ecliptic, and their dates thereby fixed, we find, universally, that the interval of time elapsing between the successive passages of each planet through *the same node* (whether it be the ascending or the descending) is always alike, whether the planet at the moment of such passage be direct or retrograde, swift or slow, in its apparent movement.

(461.) Here, then, we have a circumstance which, while it shows that the motions of the planets are in fact subject to certain laws and fixed periods, may lead us very naturally to suspect that the apparent irregularities and complexities of their movements may be owing to our not seeing them from their natural center (art. 338. 371.), and from our mixing up with their own proper motions movements of a parallactic kind, due to our own change of place, in virtue of the orbital motion of the earth about the sun.

(462.) If we abandon the earth as a center of the planetary motions, it cannot admit of a moment's hesitation where we should place that center with the greatest probability of truth. It must surely be the sun which is entitled to the first trial, as a station to which to refer to them. If it be not connected with them by any physical relation, it at least possesses the advantage, which the earth does not, of comparative immobility. But after what has been shown in art. 449., of the immense mass of that luminary, and of the office it performs to us as a quiescent center of our orbital motion, nothing can be more natural than to suppose it may perform the same to other globes which, like the earth, may be revolving round it; and these globes may be visible to us by its light reflected from them, as the moon is. Now there many facts which give a strong support to the idea that the planets are in this predicament.

(463.) In the first place, the planets really are great globes, of a size commensurate with the earth, and several of them much greater. When examined through powerful telescopes, they are seen to be round bodies, of sensible and and even of considerable apparent diameter, and offering distinct and characteristic peculiarities, which show them to be solid masses, each possessing its individual structure and mechanism; and that, in one instance at least, an exceedingly artificial and complex one. (See the representations of Mars, Jupiter, and Saturn, in Plate III.) That their distances from us are great, much greater than that of the moon, and some of them even greater than that of the sun, we infer, 1st, from their being occulted by the moon, and 2dly, from the smallness of their diurnal parallax, which, even for the nearest of them, when most favourably situated, does not exceed a few seconds, and for the remote ones is almost imperceptible. From the comparison of the diurnal parallax of a celestial body, with its apparent semidiameter, we can at once estimate its real size. For the parallax is, in fact, nothing else than the apparent semidiameter of the earth as seen from the body in question (art. 339. et seq.); and, the intervening distance being the same, the real diameters must be to each other in the proportion of the apparent ones. Without going into particulars, it will suffice to state it as a general result of that comparison, that the planets are all of them incomparably smaller than the sun, but some of them as large as the earth, and others much greater.

(464.) The next fact respecting them is, that their distances from us, as estimated from the measurement of their angular diameters, are in a continual state of change, periodically increasing and decreasing within certain limits, but by no means corresponding with the supposition of regular circular or elliptic orbits described by them about the earth as a center or focus, but maintaining a constant and obvious relation to their apparent angular distances or *elongations* from the sun. For example; the apparent diameter of Mars is greatest when in opposition (as it is called) to the sun, i. e. when in the opposite part of the ecliptic, or when it comes on the meridian at midnight, — being then about 18″, — but diminishes rapidly from that amount to about 4″, which is its apparent diameter when in *conjunction,* or when seen in nearly the same direction as that luminary. This, and facts of a similar character, observed with respect to the apparent diameters of the other planets, clearly point out the sun as having more than an accidental relation to their movements.

(465.) Lastly, certain of the planets, (Mercury, Venus,

and Mars,) when viewed through telescopes, exhibit the appearance of phases like those of the moon. This proves that they are opaque bodies, shining only by reflected light, which can be no other than that of the sun's; not only because there is no other source of light external to them sufficiently powerful, but because the appearance and succession of the phases themselves are (like their visible diameters) intimately connected with their elongations from the sun, as will presently be shown.

(466.) Accordingly it is found, that, when we refer the planetary movements to the sun as a center, all that apparent irregularity which they offer when viewed from the earth disappears at once, and resolves itself into one simple and general law, of which the earth's motion, as explained in a former chapter, is only a particular case. In order to show how this happens, let us take the case of a single planet, which we will suppose to revolve round the sun, in a plane nearly, but not quite, coincident with the ecliptic, but passing through the sun, and of course intersecting the ecliptic in a fixed line, which is the line of the planet's nodes. This line must of course divide its orbit into two segments; and it is evident that, so long as the circumstances of the planet's motion remain otherwise unchanged, the times of describing these segments must remain the same. The interval, then, between the planet's quitting either node, and returning to *the same* node again, must be that in which it describes one complete revolution round the sun, or its periodic time; and thus we are furnished with a direct method of ascertaining the periodic time of each planet.

(467.) We have said (art. 157.) that the planets make the entire tour of the heavens under very different circumstances. This must be explained. Two of them — Mercury and Venus — perform this circuit evidently as attendants upon the sun, from whose vicinity they never depart beyond a certain limit. They are seen sometimes to the east, sometimes to the west of it. In the former case they appear conspicuous over the western horizon, just after sunset, and are called evening stars: Venus, especially, appears occasionally in this situation with a dazzling lustre ; and in favourable circumstances may be observed to cast a pretty strong shadow.* When they happen to be to the west of the sun, they rise before that luminary in the morning, and appear over the eastern horizon as morning stars: they do not, however, attain the same *elongation* from the sun. Mercury never attains a greater angular distance from it than about 29°, while Venus extends her excursions on either side to about 47°. When they have receded from the sun, *eastward,* to their respective distances, they remain for a time, as it were, immovable *with respect to it,* and are carried along with it in the ecliptic with a motion equal to its own ; but presently they begin to approach it, or, which comes to the same, their motion in longitude diminishes, and the sun gains upon them. As this approach goes on, their continuance above the horizon after sunset becomes daily shorter, till at length they set before the darkness has become sufficient to allow of their being seen. For a time, then, they are not seen at all, unless on very rare occasions, when they are to be observed *passing across the sun's disc as small, round, well-defined black spots,* totally different in appearance from the solar spots (art. 386.). These phenomena are emphatically called *transits* of the respective planets across the sun, and take place when the earth happens to be passing the line of their nodes while they are in that part of their orbits, just as in the account we have given (art. 412.) of a solar eclipse. After having thus continued invisible for a time, however, they begin to appear on the other side of the sun, at first showing themselves only for a few minutes before sunrise, and gradually longer and longer as they recede from him. At this time their motion in longitude is rapidly retrograde. Before they attain their greatest elongation, however, they become stationary in the heavens; but their recess from the sun is still maintained by the advance of that luminary along the ecliptic, which continues to leave them behind, until, having reversed their

---

* It must be thrown upon a white ground. An open window in a white-washed room is the best exposure. In this situation I have observed not only the shadow, but the diffracted fringes edging its outline. — *H. Note to the edition of* 1833. Venus may often be seen with the naked eye in the daytime.

motion, and become again *direct,* they acquire sufficient speed
to commence overtaking him — at which moment they have
their greatest *western* elongation; and thus is a kind of oscillatory movement kept up, while the general advance along
the ecliptic goes on.

(468.) Suppose P Q to be the ecliptic, and A B D the
orbit of one of these planets, (for instance, Mercury,) seen
almost edgewise by an eye situated very nearly in its plane;
S, the sun, its centre; and A, B, D, S, successive positions of
the planet, of which B and S are in the nodes. If, then, the

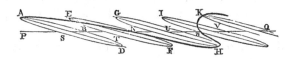

sun S stood apparently still in the ecliptic, the planets would
simply appear to oscillate backwards and forwards from A to
D, alternately passing before and behind the sun; and, if the
eye happened to lie exactly in the plane of the orbit, *transiting* his disc in the former case, and being covered by it in the
latter. But as the sun is not so stationary, but apparently
carried along the ecliptic P Q, let it be supposed to move
over the spaces S T, T U, U V, while the planet in each case
executes one quarter of its period. Then will its orbit be
apparently carried along with the sun, into the successive
positions represented in the figure; and while its real motion
round the sun brings it into the respective points, B, D, S, A,
its apparent movement in the heavens will seem to have been
along the wavy or zigzag line A N H K. In this, its motion
in longitude will have been direct in the parts A N, N H, and
retrograde in the parts H *n* K; while at the turns of the zigzag, as at H, it will have been stationary.

(469.) The only two planets — Mercury and Venus —
whose evolutions are such as above described, are called
inferior planets; their points of farthest recess from the sun
are called (as above) their *greatest* eastern and western *elongations*; and their points of nearest approach to it, their
inferior and *superior* conjunctions, — the former when the

planet passes between the earth and the sun, the latter when behind the sun.

(470.) In art. 467. we have traced the apparent path of an inferior planet, by considering its orbit in section, or as viewed from a point in the plane of the ecliptic. Let us now contemplate it *in plan*, or as viewed from a station above that plane, and projected on it. Suppose then, S to represent the sun, *a b c d* the orbit of Mercury, and A B C D a part of that of the earth — the direction of the circulation being the same in both, viz. that of the arrow.

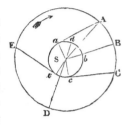

When the planet stands at *a*, let the earth be situated at A, in the direction of a tangent, *a* A, to its orbit; then it is evident that it will appear at its *greatest elongation* from the sun, — the angle *a* A S, which measures their apparent interval as seen from A, being then greater than in any other situation of *a* upon its own circle.

(471.) Now, this angle being known by observation, we are hereby furnished with a ready means of ascertaining, at least approximately, the distance of the planet from the sun, or the radius of its orbit, supposed a circle. For the triangle S A *a* is right-angled at *a*, and consequently we have S *a* : S A : : sin. S A *a* : radius, by which proportion the radii S *a*, S A of the two orbits are directly compared. If the orbits were both exact circles, this would of course be a perfectly rigorous mode of proceeding: but (as is proved by the inequality of the resulting values of S *a* obtained at different times) this is not the case ; and it becomes necessary to admit an excentricity of position, and a deviation from the exact circular form in *both* orbits, to account for this difference. Neglecting, however, at present this inequality, a mean or average value of S *a* may, at least, be obtained from the frequent repetition of this process in all varieties of situation of the two bodies. The calculations being performed, it is concluded that the mean distance of Mercury from the sun is about 36000000 miles; and that of Venus, similarly derived, about 68000000 ; the radius of the earth's orbit being 95000000.

(472.) The sidereal periods of the planets may be obtained (as before observed), with a considerable approach to accuracy, by observing their passages through the nodes of their orbits; and indeed, when a certain very minute motion of these nodes and the apsides of their orbits (similar to that of the moon's nodes and apsides, but incomparably slower) is allowed for, with a precision only limited by the imperfection of the appropriate observations. By such observation, so corrected, it appears that the sidereal period of Mercury is 87^d 23^h 15^m 43·9^s; and that of Venus, 224^d 16^h 49^m 8·0^s. These periods, however, are widely different from the intervals at which the successive appearances of the two planets at their eastern and western elongations from the sun are observed to happen. Mercury is seen at its greatest splendour as an evening star, at average intervals of about 116, and Venus at intervals of about 584 days. The difference between the *sidereal* and *synodical* revolutions (art. 418.) accounts for this. Referring again to the figure of art. 470., if the earth stood still at A, while the planet advanced in its orbit, the lapse of a sidereal period, which should bring it round again to *a*, would also produce a similar elongation from the sun. But, meanwhile, the earth has advanced in its orbit in the same direction towards E, and therefore the next greatest elongation on the same side of the sun will happen — not in the position *a* A of the two bodies, but in some more advanced position, *e* E. The determination of this position depends on a calculation exactly similar to what has been explained in the article referred to; and we need, therefore, only here state the resulting synodical revolutions of the two planets, which come out respectively 115·877^d, and 583·920^d.

(473.) In this interval, the planet will have described a whole revolution *plus* the arc *a c e*, and the earth only the arc A C E of its orbit. During its lapse, the *inferior conjunction* will happen when the earth has a certain intermediate situation, B, and the planet has reached *b*, a point between the sun and earth. The greatest elongation on the opposite side of the sun will happen when the earth has come to C, and the planet to *c*, where the line of junction C *c* is a tangent to the interior circle on the opposite side from M. Lastly, the *superior* conjunction will happen when the earth arrives at D, and the planet at *d* in the same line prolonged on the other side of the sun. The intervals at which these phænomena happen may easily be computed from a knowledge of the synodical periods and the radii of the orbits.

(474.) The circumferences of circles are in the proportion of their radii. If, then, we calculate the circumferences of the orbits of Mercury and Venus, and the earth, and compare them with the times in which their revolutions are performed, we shall find that the actual velocities with which they move in their orbits differ greatly; that of Mercury being about 109360 miles per hour, of Venus 80000, and of the earth 68040. From this it follows, that at the inferior conjunction, or at *b*, either planet is moving in the *same* direction as the earth, but with a greater velocity; it will, therefore, leave the earth *behind* it; and the apparent motion of the planet viewed from the earth, will be *as if* the planet stood still, and the earth moved in a contrary direction from what it really does. In this situation, then, the apparent motion of the planet must be contrary to the apparent motion of the sun; and, therefore, retrograde. On the other hand, at the superior conjunction, the real motion of the planet being in the opposite direction to that of the earth, the relative motion will be the same as if the planet stood still, and the earth advanced with their united velocities in its own proper direction. In this situation, then, the apparent motion will be direct. Both these results are in accordance with observed fact.

(475.) The stationary points may be determined by the following consideration. At *a* or *c*, the points of greatest elongation, the motion of the planet is directly to or from the earth, or *along* their line of junction, while that of the earth is nearly perpendicular to it. Here, then, the apparent motion must be direct. At *b*, the inferior conjunction, we have seen that it must be retrograde, owing to the planet's motion (which is there, as well as the earth's, *perpendicular* to the line of junction) surpassing the earth's. Hence, the

stationary points ought to lie, as it is found by observation
they do, between *a* and *b*, or *c* and *b*, viz. in such a position
that the obliquity of the planet's motion with respect to the
line of junction shall just compensate for the excess of its
velocity, and cause an equal advance of each extremity of
that line, by the motion of the planet at one end, and of the
earth at the other: so that, for an instant of time, the whole
line shall move parallel to itself.   The question thus proposed
is purely geometrical, and its solution on the supposition of
circular orbits is easy.   Let E *e* and P *p* represent small

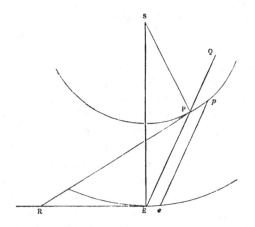

arcs of the orbits of the earth and planet described contem-
poraneously, at the moment when the latter appears stationary,
about S, the sun.   Produce *p* P and *e* E, tangents at P
and E, to meet at R, and prolong E P backwards to Q, join
e p.   Then since P E, *p e* are parallel we have by similar
triangles P *p* : E *e* :: P R : R E, and since, putting *v* and V
for the respective velocities of the planet and the earth,
P *p* : E *e* :: *v* : V; therefore

$$v : V :: \mathrm{P\,R} : \mathrm{R\,E} :: \sin. \mathrm{P\,E\,R} : \sin. \mathrm{E\,P\,R}$$
$$:: \cos. \mathrm{S\,E\,P} : \cos. \mathrm{S\,P\,Q}$$
$$:: \cos. \mathrm{S\,E\,P} : \cos. (\mathrm{S\,E\,P} + \mathrm{E\,S\,P})$$

because the angles S E R and S P R are right angles.   Moreover, if r and R be the radii of the respective orbits, we have also

$$v : R :: \sin. \; S\,E\,P : \sin. \; (S\,E\,P + E\,S\,P)$$

from which two relations it is easy to deduce the values of the two angles S E P and E S P; the former of which is the apparent elongation of the planet from the sun*, the latter the difference of heliocentric longitudes of the earth and planet.

(476.) When we regard the orbits as other than circles (which they really are), the problem becomes somewhat complex—too much so to be here entered upon. It will suffice to state the results which experience verifies, and which assigns the stationary points of Mercury at from 15° to 20° of elongation from the sun, according to circumstances; and of Venus, at an elongation never varying much from 29°. The former continues to retrograde during about 22 days; the latter, about 42.

(477.) We have said that some of the planets exhibit phases like the moon. This is the case with both Mercury and Venus; and is readily explained by a consideration of their orbits, such as we have above supposed them. In fact, it requires little more than mere inspection of the figure annexed,

to show, that to a spectator situated on the earth E, an inferior planet, illuminated by the sun, and therefore bright on the side next to him, and dark on that turned from him, will appear *full* at the superior conjunction A; *gibbous* (*i. e.* more than half full, like the moon 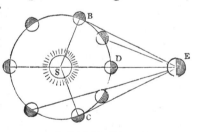 between the first and second quarter) between that point and the points B C of its greatest elongation; half-mooned at these points; and crescent-shaped, or horned, between these

---

* If $\dfrac{R}{r} = m$ and $\dfrac{V}{v} = n$, $SEP = \phi$, $ESP = \psi$, the equations to be resolved are $sin \; (\phi + \psi) = m \; sin. \; \phi$, and $cos. \; (\phi + \psi) = n \; cos. \; \psi$, which give $cos. \; \psi = \dfrac{1 + m\,n}{m + n}$

and the inferior conjunction D. As it approaches this point, the crescent ought to thin off till it vanishes altogether, rendering the planet invisible, unless in those cases where it *transits* the sun's disc, and appears on it as a black spot. All these phenomena are exactly conformable to observation.

(478.) The variation in brightness of Venus in different parts of its apparent orbit is very remarkable. This arises from two causes: 1st, the varying proportion of its visible illuminated area to its whole disc ; and, 2dly, the varying angular diameter, or whole apparent magnitude of the disc itself. As it approaches its inferior conjunction from its greater elongation, the half-moon becomes a crescent, which *thins off;* but this is more than compensated, for some time, by the increasing apparent magnitude, in consequence of its diminishing distance. Thus the total light received from it goes on increasing, till at length it attains a maximum, which takes place when the planet's elongation is about 40°.

(479.) The transits of Venus are of very rare occurrence, taking place alternately at the very unequal but regularly recurring intervals of 8, 122, 8, 105, 8, 122, &c., years in succession, and always in June or December. As astronomical phænomena, they are extremely important; since they afford the best and most exact means we possess of ascertaining the sun's distance, or its parallax. Without going into the niceties of calculation of this problem, which, owing to the great multitude of circumstances to be attended to, are extremely intricate, we shall here explain its principle, which, in the abstract, is very simple and obvious. Let E be the earth, V Venus, and S the sun, and C D the portion of Venus's relative orbit which she describes while in the act of transiting the sun's disc. Suppose A B two spectators at opposite extremities of that diameter of the earth which is perpendicular to the ecliptic, and, to avoid complicating the case, let us lay out of consideration the earth's rotation, and suppose A, B, to retain that situation during the whole time of the transit. Then, at any moment when the spectator at A sees the center of Venus projected at *a* on the sun's disc, he at B will see it projected at *b*. If then one or other spectator could suddenly transport himself from A to B, he would see Venus suddenly displaced on the disc from a to b; and if he had any means of noting accurately the place of the points on the disc, either by micrometrical measures from its edge, or by other means, he might ascertain the angular measure of $a\,b$ as seen from the earth. Now, since A V a, B V b, are straight lines, and

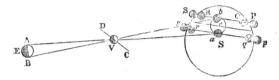

therefore make equal angles on each side V, $a\,b$ will be to A B as the distance of Venus from the sun is to its distance from the earth, or as 68 to 27, or nearly as $2\frac{1}{2}$ to 1; $a\,b$ therefore occupies on the sun's disc a space $2\frac{1}{2}$ times as great as the earth's diameter; and its angular measure is therefore equal to about $2\frac{1}{2}$ times the earth's apparent diameter at the distance of the sun, or (which is the same thing) to five times the sun's horizontal parallax (art. 298.). Any error, therefore, which may be committed in measuring $a\,b$, will entail only *one fifth* of that error on the horizontal parallax concluded from it.

(480.) The thing to be ascertained, therefore, is, in fact, neither more nor less than the breadth of the zone P Q R S, $p\,q\,r\,s$, included between the extreme apparent paths of the center of Venus across the sun's disc, from its entry on one side to its quitting it on the other. The whole business of the observers at A, B, therefore, resolves itself into this; — to ascertain, with all possible care and precision, each at his own station, this path, — where it enters, where it quits, and what segment of the sun's disc it cuts off. Now, one of the most exact ways in which (conjoined with careful micrometric measures) this can be done, is by noting the *time* occupied in the whole transit: for the relative angular motion of Venus being, in fact, very precisely known from the tables of her motion, and the apparent path being very nearly a straight line, these times give us a measure (*on a very enlarged scale*) of the lengths of the chords of the segments cut off; and the sun's diameter being known also with great precision, their versed sines, and therefore their difference, or the breadth of the zone required, becomes known. To obtain these times correctly, each observer must ascertain the instants of ingress and egress of the *center.* To do this, he must note, 1st, the instant when the first visible impression or notch on the edge of the disc at P is produced, or the *first external contact;* 2dly, when the planet is just wholly immersed, and the broken edge of the disc just closes again at Q, or the first *internal* contact; and, lastly, he must make the same observations at the egress at R, S. The mean of the internal and external contacts, corrected for the curvature of the sun's limb in the intervals of the respective points of contact, internal and external, gives the entry and egress of the planet's center.

(481.) The modifications introduced into this process by the earth's rotation on its axis, and by other geographical stations of the observers thereon than here supposed, are similar in their principles to those which enter into the calculation of a solar eclipse, or the occultation of a star by the moon, only more refined. Any consideration of them, however, here, would lead us too far ; but in the view we have taken of the subject, it affords an admirable example of the way in which minute elements in astronomy may become magnified in their effects, and, by being made subject to measurement on a greatly enlarged scale, or by substituting the measure of time for space, may be ascertained with a degree of precision adequate to every purpose, by only watching favourable opportunities, and taking advantage of nicely adjusted combinations of circumstance. So important has this observation appeared to astronomers, that at the last transit of Venus, in 1769, expeditions were fitted out, on the most efficient scale, by the British, French, Russian, and other governments, to the remotest corners of the globe, for the express purpose of performing it. The celebrated expedition of Captain Cook to Otaheite was one of them. The general result of all the observations made on this most memorable occasion gives 8″5776 for the sun's horizontal parallax. The two next occurrences of this phænomenon will happen on Dec. 8. 1874 and Dec. 6. 1882.

(482.) The orbit of Mercury is very elliptical, the excentricity being nearly one fourth of the mean distance. This appears from the inequality of the greatest elongations from the sun, as observed at different times, and which vary between the limits 16° 12' and 28° 48', and, from exact measures of such elongations, it is not difficult to show that the orbit of Venus also is slightly excentric, and that both these planets, in fact, describe ellipses, having the sun in their commom focus.

(483.) Transits of Mercury over the sun's disc occasionally occur, as in the case of Venus, but more frequently ₊those at the ascending node in November, at the descending in May. The intervals (considering each node separately) are *usually* either 13 or 7 years, and in the order 13, 13, 13, 7, &c.; but owing to the considerable inclination of the orbit of Mercury to the ecliptic, this cannot be taken as an exact expression of the said recurrence, and it requires a period of at least 217 years to bring round the transits in regular order. One will occur in the present year (1848), the next in 1861. They are of much less astronomical importance than that of Venus, on account of the proximity of Mercury to the sun, which affords a much less favourable combination for the determination of the sun's parallax.

(484.) Let us now consider the superior planets, or those whose orbits enclose on all sides that of the earth. That they do so is proved by several circumstances : — 1st, They are not, like the inferior planets, confined to certain limits of elongation from the sun, but appear at all distances from it, even in the opposite quarter of the heavens, or, as it is called, in *opposition;* which could not happen, did not the earth at such times place itself between them and the sun : 2dly, They never appear horned, like Venus or Mercury, nor even semilunar. Those, on the contrary, which, from the minuteness of their parallax, we conclude to be the most distant from us, viz. Jupiter, Saturn, Uranus, and Neptune, never appear otherwise than round; a sufficient proof, of

itself, that we see them always in a direction not very
remote from that in which the sun's rays illuminate them;
and that, therefore, we occupy a station which is never very
widely removed from the centre of their orbits, or, in other
words, that the earth's orbit is entirely enclosed within theirs,
and of comparatively small diameter.   One only of them,
Mars, exhibits any perceptible *phase*, and in its deficiency
from a circular outline, never surpasses a moderately *gibbous*
appearance, — the enlightened portion of the disc being
never less than seven-eighths of the whole.   To understand
this, we need only cast our eyes on the annexed figure, in
which E is the earth, at its apparent greatest elongation
from the sun S, as seen from Mars, M.   In this position,
the angle S M E, included between the lines
S M and E M, is at its maximum ; and therefore, in this state of things, a spectator on the
earth is enabled to see a greater portion of
the dark hemisphere of Mars than in any other
situation.   The extent of the phase, then, or
greatest observable degree of gibbosity, affords
a measure — a sure, although a coarse and rude
one — of the angle S M E, and therefore of the
proportion of the distance S M, of Mars, to
S E, that of the earth from the sun, by which
it appears that the diameter of the orbit of
Mars cannot be less than $1\frac{1}{2}$ times that of the

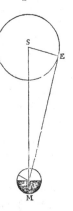

earth's.   The phases of Jupiter, Saturn, Uranus, and Neptune, being imperceptible, it follows that their orbits must
include not only that of the earth, but of Mars also.

(485.) All the superior planets are retrograde in their
apparent motions when in *opposition*, and for some time
before and after; but they differ greatly from each other,
both in the extent of their arc of retrogradation, in the
duration of their retrograde movement, and in its rapidity
when swiftest.   It is more extensive and rapid in the case
of Mars than of Jupiter, of Jupiter than of Saturn, of that
planet than of Uranus, and of Uranus again than Neptune.
The angular velocity with which a planet appears to re-

trograde is easily ascertained by observing its apparent
place in the heavens from day to day; and from such
observations, made about the time of opposition, it is easy
to conclude the relative magnitudes of their orbits, as
compared with the earth's, supposing their periodical times
known. For, from these, their mean angular velocities are
known also, being inversely as the times. Suppose, then,
E e to be a very small por-
tion of the earth's orbit,
and M m a corresponding

portion of that of a superior planet, described on the day
of opposition, about the sun S, on which day the three
bodies lie in one straight line S E M X. Then the angles
E S e and M S m are given. Now, if e m be joined and
prolonged to meet S M continued in X, the angle e X E,
which is equal to the alternate angle X e Y, is evidently
the retrogradation of Mars on that day, and is, therefore,
also given. E e, therefore, and the angle E X e, being given
in the right-angled triangle E e X, the side E X is easily
calculated, and thus S X becomes known. Consequently,
in the triangle S m X, we have given the side S X and the
two angles m S X, and m X S, whence the other sides, S m,
m X, are easily determined. Now, S m is no other than the
radius of the orbit of the superior planet required, which in
this calculation is supposed circular, as well as that of the
earth; a supposition not exact, but sufficiently so to afford a
satisfactory approximation to the dimensions of its orbit, and
which, if the process be often repeated, in every variety of
situation at which the opposition can occur, will ultimately
afford an average or mean value of its diameter fully to be
depended upon.

(486.) To apply this principle, however, to practice, it
is necessary to know the periodic times of the several planets.
These may be obtained directly, as has been already stated,
by observing the intervals of their passages through the
ecliptic; but, owing to the very small inclination of the
orbits of some of them to its plane, they cross it so obliquely
that the precise moment of their arrival on it is not ascer-

tainable, unless by very nice observations. A better method consists in determining, from the observations of several successive days, the exact moments of their arriving *in opposition* with the sun, the criterion of which is a difference of longitudes between the sun and planet of exactly 180°. The interval between successive oppositions thus obtained is nearly one *synodical* period; and would be exactly so, were the planet's orbit and that of the earth both circles, and uniformly described; but as that is found not to be the case (and the criterion is, the *inequality* of successive synodical revolutions so observed), the average of a great number, taken in all varieties of situation in which the oppositions occur, will be freed from the elliptic inequality, and may be taken as a *mean synodical* period. From this, by the considerations and by the process of calculation, indicated (art. 418.) the sidereal periods are readily obtained. The accuracy of this determination will, of course, be greatly increased by embracing a long interval between the extreme observations employed. In point of fact, that interval extends to nearly 2000 years in the cases of the planets known to the ancients, who have recorded their observations of them in a manner sufficiently careful to be made use of. Their periods may, therefore, be regarded as ascertained with the utmost exactness. Their numerical values will be found stated, as well as the mean distances, and all the other elements of the planetary orbits, in the synoptic table at the end of the volume, to which (to avoid repetition) the reader is once for all referred.

(487.) In casting our eyes down the list of the planetary distances, and comparing them with the periodic times, we cannot but be struck with a certain correspondence. The greater the distance, or the larger the orbit, evidently the longer the period. The order of the planets, beginning from the sun, is the same, whether we arrange them according to their distances, or to the time they occupy in completing their revolutions; and is as follows: — Mercury, Venus, Earth, Mars, — the ultra-zodiacal planets, or, as they are sometimes also called, Asteroids, — Jupiter, Saturn,

Uranus, and Neptune. Nevertheless, when we come to examine the numbers expressing them, we find that the relation between the two series is not that of simple *proportional* increase. The periods increase more than in proportion to the distances. Thus, the period of Mercury is about 88 days, and that of the Earth 365 — being in proportion as 1 to 4·15, while their distances are in the less proportion of 1 to 2·56; and a similar remark holds good in every instance. Still, the ratio of increase of the times is not so rapid as that of the *squares* of the distances. The square of 2·56 is 6·5536, which is considerably greater than 4·15. An intermediate rate of increase, between the simple proportion of the distances and that of their squares is therefore clearly pointed out by the sequence of the numbers; but it required no ordinary penetration in the illustrious Kepler, backed by uncommon perseverance and industry, at a period when the data themselves were involved in obscurity, and when the processes of trigonometry and of numerical calculation were encumbered with difficulties, of which the more recent invention of logarithmic tables has happily left us no conception, to perceive and demonstrate the real law of their connection. This connection is expressed in the following proposition: — "The squares of the periodic times of any two planets are to each other, in the same proportion as the cubes of their mean distances from the sun." Take, for example, the Earth and Mars *, whose periods are in the proportion of 3652564 to 6869796, and whose distance from the sun is that of 100000 to 152369; and it will be found, by any one who will take the trouble to go through the calculation, that —

$$(3652564)^2 : (6869796)^2 :: (100000)^3 : (152369)^3.$$

(488.) Of all the laws to which induction from pure observation has ever conducted man, this *third law* (as it is called) *of Kepler* may justly be regarded as the most remark-

* The expression of this law of Kepler requires a slight modification when we come to the extreme nicety of numerical calculation, for the greater planets, due to the influence of their masses. This correction is imperceptible for the Earth and Mars.

able, and the most pregnant with important consequences. When we contemplate the constituents of the planetary system from the point of view which this relation affords us, it is no longer mere analogy which strikes us — no longer a general resemblance among them, as individuals independent of each other, and circulating about the sun, each according to its own peculiar nature, and connected with it by its own peculiar tie. The resemblance is now perceived to be a true *family* likeness; they are bound up in one chain — interwoven in one web of mutual relation and harmonious agreement — subjected to one pervading influence, which extends from the centre to the farthest limits of that great system, of which all of them, the earth included, must henceforth be regarded as members.

(489.) The laws of elliptic motion about the sun as a focus, and of the equable description of areas by lines joining the sun and planets, were originally established by Kepler, from a consideration of the observed motions of Mars; and were by him extended, analogically, to all the other planets. However precarious such an extension might then have appeared, modern astronomy has completely verified it as a matter of fact, by the general coincidence of its results with entire series of observations of the apparent places of the planets. These are found to accord satisfactorily with the assumption of a particular ellipse for each planet, whose magnitude, degree of excentricity, and situation in space, are numerically assigned in the synoptic table before referred to. It is true, that when observations are carried to a high degree of precision, and when each planet is traced through many successive revolutions, and its history carried back, by the aid of calculations founded on these data, for many centuries, we learn to regard the laws of Kepler as only *first approximations* to the much more complicated ones which actually prevail; and that to bring remote observations into rigorous and mathematical accordance with each other, and at the same time to retain the extremely convenient nomenclature and relations of the ELLIPTIC SYSTEM, it becomes necessary to modify, to a certain extent, our verbal expression of the laws, and to regard the numerical data or *elliptic elements* of the planetary orbits as not absolutely permanent, but subject to a series of extremely slow and almost imperceptible changes. These changes may be neglected when we consider only a few revolutions; but going on from century to century, and continually accumulating, they at length produce material departures in the orbits from their original state. Their explanation will form the subject of a subsequent chapter; but for the present we must lay them out of consideration, as of an order too minute to affect the general conclusions with which we are now concerned. By what means astronomers are enabled to compare the results of the elliptic theory with observation, and thus satisfy themselves of its accordance with nature, will be explained presently.

(490.) It will first, however, be proper to point out what particular theoretical conclusion is involved in each of the three laws of Kepler, considered as satisfactorily established, — what indication each of them, separately, affords of the mechanical forces prevalent in our system, and the mode in which its parts are connected, — and how, when thus considered, they constitute the basis on which the Newtonian explanation of the mechanism of the heavens is mainly supported. To begin with the first law, that of the equable description of areas. — Since the planets move in curvilinear paths, they *must* (if they be bodies obeying the laws of dynamics) be deflected from their otherwise natural rectilinear progress *by force*. And from this law, taken as a matter of observed fact, it follows, that the *direction* of such force, at every point of the orbit of each planet, always *passes through the sun*. No matter from what ultimate cause the power which is called gravitation originates, — be it a virtue lodged in the sun as its receptacle, or be it pressure from without, or the resultant of many pressures or solicitations of unknown fluids, magnetic or electric ethers, or impulses, — still, when finally brought under our contemplation, and summed up into a single resultant energy — its *direction* is, *from* every point on all sides, *towards the sun's center.* As an abstract dynamical proposition, the reader will find it demonstrated by Newton, in the first proposition of the *Principia*, with an elementary simplicity to which we really could add nothing but obscurity by amplification, that any body, urged towards a certain central point by a force continually directed thereto, and thereby deflected into a curvilinear path, will describe about that center equal areas in equal times: and *vice versâ*, that such equable description of areas is itself the essential criterion of a continual direction of the acting force towards the center to which this character belongs. The first law of Kepler, then, gives us no information as to the nature or intensity of the force urging the planets to the sun; the only conclusion it involves is, that it does so urge them. It is a property of orbital rotation under the influence of central forces *generally*, and, as such, we daily see it exemplified in a thousand familiar instances. A simple experimental illustration of it is to tie a bullet to a thin string, and, having whirled it round with a moderate velocity in a vertical plane, to draw the end of the string through a small ring, or allow it to coil itself round the finger, or round a cylindrical rod held very firmly in a horizontal position. The bullet will then approach the center of motion in a spiral line; and the increase not only of its angular but of its linear velocity, and the rapid diminution of its periodic time when near the center, will express, more clearly than any words, the compensation by which its uniform description of areas is maintained under a constantly diminishing distance. If the motion be reversed, and the thread allowed to uncoil, beginning with a rapid impulse, the velocity will diminish by the same degrees as it before increased. The increasing rapidity of a dancer's *pirouette*, as he draws in his limbs and straightens his whole person, so as to bring every part of his frame as near as possible to the axis of his motion, is another instance where the connection of the observed effect with the central force exerted, though equally real, is much less obvious.

(491.) The second law of Kepler, or that which asserts that the planets describe ellipses about the sun as their focus, involves, as a consequence, the *law* of solar gravitation (so be it allowed to call the force, whatever it be, which urges them towards the sun) as exerted on each individual planet, apart from all connection with the rest.   A straight line, dynamically speaking, is the only path which can be pursued by a body *absolutely free*, and under the action of *no* external force. All *deflection* into a curve is evidence of the exertion of a force ; and the greater the deflection in equal times, the more intense the force.   Deflection from a straight line is only another word for *curvature* of path ; and as a circle is characterized by the uniformity of its curvatures in all its parts — so is every other curve (as an ellipse) characterized by the particular *law* which regulates the increase and diminution of its curvature as we advance along its circumference.   The deflecting force, then, which continually bends a moving body into a curve, may be ascertained, provided its direction, in the first place, and, secondly, the law of curvature of the curve itself, be known.   Both these enter as elements into the expression of the force.   A body may describe, for instance, an ellipse, under a great variety of dispositions of the acting forces: it may glide along it, for example, as a bead upon a polished wire, bent into an elliptic form ; in which case the acting force is always perpendicular to the wire, and the velocity is uniform.   In this case the *force* is directed to *no fixed* center, and there is no equable description of areas at all.   Or it may describe it as we may see done, if we suspend a ball by a *very* long string, and, drawing it a little aside from the perpendicular, throw it *round* with a gentle impulse.   In this case the acting force is directed to the center of the ellipse, about which areas are described equably, and *to* which a force *proportional* to the distance (the decomposed result of terrestrial gravity) perpetually urges it. *   This is at once a very easy experiment, and a very instructive one, and we shall again refer to it. In the case before us, of an ellipse described by the action of

---

* If the suspended body be a vessel full of fine sand, having a small hole at its bottom, the elliptic trace of its orbit will be left in a sand streak on a table placed below it.   This neat illustration is due, to the best of my knowledge, to Mr. Babbage.

a force directed to the *focus*, the steps of the investigation of the law of force are these: 1st, The law of the areas determines the actual *velocity* of the revolving body at every point, or the space really run over by it in a given minute portion of time; 2dly, The law of curvature of the ellipse determines the linear amount of deflection from the tangent *in the direction of the focus*, which corresponds to that space so run over; 3dly, and lastly, The laws of accelerated motion declare that the intensity of the acting force causing such deflection *in its own direction*, is measured by or proportional to the amount of that deflection, and may therefore be calculated in any particular position, or generally expressed by geometrical or algebraic symbols, *as a law* independent of particular positions, when that deflection is so calculated or expressed. We have here the spirit of the process by which Newton has resolved this interesting problem. For its geometrical detail, we must refer to the 3d section of his *Principia*. We know of no artificial mode of imitating this species of elliptic motion; though a rude approximation to it — enough, however, to give a conception of the alternate approach and recess of the revolving body to and from the focus, and the variation of its velocity—may be had by suspending a small steel bead to a fine and very long silk fibre, and setting it to revolve in a small orbit round the pole of a powerful cylindrical magnet, held upright, and vertically under the point of suspension.

(492.) The third law of Kepler, which connects the distances and periods of the planets by a general rule, bears with it, as its theoretical interpretation, this important consequence, viz. that it is one and the same force, modified only by distance from the sun, which retains *all* the planets in their orbits about it. That the attraction of the sun (if such it be) is exerted upon all the bodies of our system indifferently, without regard to the peculiar materials of which they may consist, in the exact proportion of their inertiæ, or quantities of matter; that it is not, therefore, of the nature of the elective attractions of chemistry or of magnetic action, which is powerless on other substances than iron and some one or two more, but is of a more universal character, and extends equally to all the material constituents of our system, and (as we shall hereafter see abundant reason to admit) to those of other systems than our own. This law, important and general as it is, results, as the simplest of corollaries, from the relations established by Newton in the section of the *Principia* referred to (Prop. xv.), from which proposition it results, that if the earth were taken from its actual orbit, and launched anew in space at the place, in the direction, and with the velocity of any of the other planets, it would describe the very same orbit, and in the same period, which that planet actually does, a minute correction of the period only excepted, arising from the difference between the mass of the earth and that of the planet. Small as the planets are compared to the sun, some of them are not, as the earth is, mere atoms in the comparison. The strict wording of Kepler's law, as Newton has proved in his fifty-ninth proposition, is applicable only to the case of planets whose proportion to the central body is absolutely inappretiable. When this is not the case, the periodic time is shortened in the proportion of the square root of the number expressing the sun's mass or inertiæ, to that of the sum of the numbers expressing the masses of the sun and planet; and in general, whatever be the masses of two bodies revolving round each other under the influence of the Newtonian law of gravity, the square of their periodic time will be expressed by a fraction whose numerator is the cube of their mean distance, *i. e.* the greater semi-axis of their elliptic orbit, and whose denominator is the sum of their masses. When one of the masses is incomparably greater than the other, this resolves into Kepler's law; but when this is not the case, the proposition thus generalized stands in lieu of that law. In the system of the sun and planets, however, the numerical correction thus introduced into the results of Kepler's law is too small to be of any importance, the mass of the largest of the planets (Jupiter) being much less than a thousandth part of that of the sun. We shall presently, however, perceive all the importance of this generalization, when we come to speak of the satellites.

(493.) It will first, however, be proper to explain by what process of calculation the expression of a planet's elliptic orbit by its *elements* can be compared with observation, and how we can satisfy ourselves that the numerical data contained in a table of such elements for the whole system does really exhibit a true picture of it, and afford the means of determining its state at every instant of time, by the mere application of Kepler's laws. Now, for each planet, it is necessary for this purpose to know, 1st, the magnitude and form of its ellipse; 2dly, the situation of this ellipse in space, with respect to the ecliptic, and to a fixed line drawn therein; 3dly, the local situation of the planet in its ellipse at some known epoch, and its periodic time or mean angular velocity, or, as it is called, its mean motion.

(494.) The magnitude and form of an ellipse are determined by its greatest length and least breadth, or its two principal axes; but for astronomical uses it is preferable to use the semi-axis major (or half the greatest length), and the excentricity or distance of the focus from the center, which last is usually estimated in parts of the former. Thus, an ellipse, whose length is 10 and breadth 8 parts of any scale, has for its major semi-axis 5, and for its excentricity 3 such parts; but when estimated in parts of the semi-axis, regarded as a unit, the excentricity is expressed by the fraction $\frac{3}{5}$.

(495.) The ecliptic is the plane to which an inhabitant of the earth most naturally refers the rest of the solar system, as a sort of ground-plane; and the axis of its orbit might be taken for a line of departure in that plane or origin of angular reckoning. Were the axis *fixed*, this would be the best possible origin of longitudes; but as it has a motion (though an excessively slow one), there is, in fact, no advantage in reckoning from the axis more than from the line of the equinoxes, and astronomers therefore prefer the latter, taking account of its variation by the effect of precession, and restoring it, by calculation at every instant, to a fixed position. Now, to determine the 'situation of the ellipse described by a planet with respect to this plane, three *elements* require to be known: — 1st, the *inclination* of the plane of the planet's orbit to the plane of the ecliptic; 2dly, the line in which these two planes intersect each other, which of necessity passes through the sun, and whose position with respect to the line of the equinoxes is therefore given by stating its longitude. This line is called the line of the nodes. When the planet is in this line, in the act of passing from the south to the north side of the ecliptic, it is *in its ascending node*, and its longitude at that moment is the element called the *longitude of the node*. These two data determine the situation of *the plane* of the orbit; and there only remains, for the complete determination of the situation of the planet's ellipse, to know how it is placed *in* that plane, which (since its focus is necessarily in the sun) is ascertained by stating the *longitude of its perihelion*, or the place which the extremity of the axis nearest the sun occupies, when orthographically projected on the ecliptic.

(496.) The dimensions and situation of the planet's orbit thus determined, it only remains, for a complete acquaintance with its history, to determine the circumstances of its motion in the orbit so precisely fixed. Now, for this purpose, all that is needed is to know the moment of time when it is either at the perihelion, or at any other precisely determined point of its orbit, and its whole period; for these being known, the law of the areas determines the place at every other instant. This moment is called (when the perihelion is the point chosen) the *perihelion passage*, or, when some point of the orbit is fixed upon, without special reference to the perihelion, the *epoch*.

(497.) Thus, then, we have *seven* particulars or elements, which must be numerically stated, before we can reduce to calculation the state of the system at any given moment. But, these known, it is easy to ascertain the apparent positions of each planet, as it would be seen from the sun, or is seen from the earth at any time. The former is called the *heliocentric*, the latter the *geocentric*, place of the planet.

(498.) To commence with the heliocentric places. Let S represent the sun; P A N the orbit of the planet, being an

ellipse, having the S in its focus, and A for its perihelion;
and let $p\,a\,$N Υ represent the projection of the orbit on the

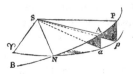

plane of the ecliptic, intersecting the
line of equinoxes S Υ in Υ, which,
therefore, is the origin of longitudes.
Then will S N be the line of nodes;
and if we suppose B to lie on the
south, and A on the north side of the ecliptic, and the direction of the planet's motion to be from B to A, N will be
the ascending node, and the angle Υ S N the *longitude of
the node*.   In like manner, if P be the place of the planet at
any time, and if it and the perihelion A be projected on the
ecliptic, upon the points p, a, the angles Υ S p, Υ S a, will
be the respective heliocentric longitudes of the planet and of
the perihelion, the former of which is to be determined, and
the latter is one of the given elements.   Lastly, the angle
p S P is the heliocentric latitude of the planet, which is also
required to be known.

(499.)   Now, the time being given, and also the moment of
the planet's passing the perihelion, the interval, or the time
of describing the portion A P of the orbit, is given, and the
periodical time, and the whole area of the ellipse being
known, the law of proportionality of areas to the times of
their description gives the magnitude of the area A S P.
From this it is a problem of pure geometry to determine the
corresponding *angle* A S P, which is called the planet's *true
anomaly*.   This problem is of the kind called transcendental,
and has been resolved by a great variety of processes, some
more, some less intricate.   It offers, however, no peculiar
difficulty, and is practically resolved with great facility by
the help of tables constructed for the purpose, adapted to
the case of each particular planet. *

---

* It will readily be understood, that, except in the case of uniform circular
motion, an equable description of *areas* about any center is incompatible with
an equable description of *angles*.   The object of the problem in the text is to
pass from the *area*, supposed known, to the *angle*, supposed unknown : in other
words, to derive the true amount of angular motion from the perihelion, or the
true anomaly from what is technically called the mean anomaly, that is, the mean
angular motion which would have been performed had the motion *in angle* been
uniform instead of the motion *in area*.   It happens fortunately, that this is the

(500.) The true anomaly thus obtained, the planet's angular distance from the node, or the angle N S P, is to be found. Now, the longitudes of the perihelion and node being respectively ♈ a and ♈ N, which are given, their difference a N is also given, and the angle N of the spherical right-angled triangle A N a, being the inclination of the plane of the orbit to the ecliptic, is known. Hence we calculate the arc N A, or the angle N S A, which, added to A S P, gives the angle N S P required. And from this, regarded as the measure of the arc N P, forming the hypothenuse of the right angled spherical triangle P N p, whose angle N, as before, is known, it is easy to obtain the other two sides, N p and P p. The latter, being the measure of the angle p S P, expresses the planet's heliocentric latitude; the former measures the angle N S p, or the planet's distance in longitude from its node, which, added to the known angle ♈ S N, the longitude of the node, gives the heliocentric longitude. This process, however circuitous it may appear, when once well understood may be gone through numerically by the aid of the usual logarithmic and trigonometrical tables, in little more time than it will have taken the reader to peruse its description.

(501.) The geocentric differs from the heliocentric place of a planet by reason of that parallactic change of apparent situation which arises from the earth's motion in its orbit. Were the planets' distances as vast as those of the stars, the earth's orbitual motion would be insensible when viewed from them, and they would always appear to us to hold the same relative situations among the fixed stars as if viewed from the sun, *i. e.* they would then be seen in their *heliocentric* places. The difference, then, between the heliocentric and geocentric places of a planet is, in fact, the same thing with its *parallax,* arising from the earth's removal from the centre

---

simplest of all problems of the transcendental kind, and can be resolved, in the most difficult case, by the rule of " false position," or trial and error, in a very few minutes. Nay, it may even be resolved instantly on inspection by a simple and easily constructed piece of mechanism, of which the reader may see a description in the Cambridge Philosophical Transactions, vol. iv. p. 425., by the author of this work.

of the system and its annual motion. It follows from this, that the first step towards a knowledge of its amount, and the consequent determination of the apparent place of each planet, as referred from the earth to the sphere of the fixed stars, must be to ascertain the proportion of its linear distances from the earth and from the sun, as compared with the earth's distance from the sun, and the angular positions of all three with respect to each other.

(502.) Suppose, therefore, S to represent the sun, E the earth, and P the planet; S ϒ the line of equinoxes, ϒ E the earth's orbit, and P p a perpendicular let fall from the planet on the ecliptic. Then will the angle S P E (according to the general notion of parallax conveyed in art. 69) re-

present the parallax of the planet arising from the change of station from S to E; E P will be the apparent direction of the planet seen from E; and if S Q be drawn parallel to E p, the angle ϒ S Q will be the geocentric longitude of the planet, while ϒ S E represents the heliocentric longitude of the earth, ϒ S p that of the planet. The former of these, S E ϒ, is given by the solar tables; the latter, ϒ S p, is found by the process above described (art. 500). Moreover, S P is the radius vector of the planet's orbit, and S E that of the earth's, both of which are determined from the known dimensions of their respective ellipses, and the places of the bodies in them at the assigned time. Lastly, the angle P S p is the planet's heliocentric latitude.

(503.) Our object, then, is, from all these data, to determine the angle ϒ S Q, and P E p, which is the geocentric latitude. The process, then, will stand as follows: — 1st, In the triangle S P p, right-angled at p, given S P, and the angle P S p (the planet's radius vector and heliocentric latitude), find S p and P p; 2dly, In the triangle S E p, given S p (just found), S E (the earth's radius vector), and the angle E S p (the difference of heliocentric longitudes of the earth and planet), find the angle S p E, and the side E p. The former being equal to the alternate angle p S Q, is the parallactic removal of the planet in longitude, which, added to Υ S p, gives its geocentric longitude. The latter, E p (which is called the *curtate distance* of the planet from the earth), gives at once the geocentric latitude, by means of the right-angled triangle P E p, of which E p and P p are known sides, and the angle P E p is the geocentric latitude sought.

(504.) The calculations required for these purposes are nothing but the most ordinary processes of plane trigonometry; and, though somewhat tedious, are neither intricate nor difficult. When executed, however, they afford us the means of comparing the places of the planets actually observed with the elliptic theory, with the utmost exactness, and thus putting it to the severest trial; and it is upon the testimony of such computations, so brought into comparison with observed facts, that we declare that theory to be a true representation of nature.

(505.) The planets Mercury, Venus, Mars, Jupiter, and Saturn, have been known from the earliest ages in which astronomy has been cultivated. Uranus was discovered by Sir W. Herschel in 1781, March 13th, in the course of a review of the heavens, in which every star visible in a telescope of a certain power was brought under close examination, when the new planet was immediately detected by its disc, under a high magnifying power. It has since been ascertained to have been observed on many previous occasions, with telescopes of insufficient power to show its disc, and even entered in catalogues as a star; and some of the observations which have been so recorded have been used to improve and extend our knowledge of its orbit. The discovery of the ultra-zodiacal planets dates from the first day of 1801, when Ceres was discovered by Piazzi, at Palermo; a discovery speedily followed by those of Juno by professor Harding, of Göttingen, in 1804; and of Pallas and Vesta, by Dr. Olbers, of Bremen, in 1802 and 1807 respectively. It is extremely remarkable that this important addition to our system had been in some sort surmised as a thing not unlikely, on the ground that the interval between the orbit of Mercury and the other planetary orbits, go on doubling as we recede from the sun, or nearly so. Thus, the interval between the orbits of the Earth and Mercury is nearly twice that between those of Venus and Mercury; that between the orbits of Mars and Mercury nearly twice that between the Earth and Mercury; and so on. The interval between the orbits of Jupiter and Mercury, however, is much too great, and would form an exception to this law, which is, however, again resumed in the case of the three planets next in order of remoteness, Jupiter, Saturn, and Uranus. It was therefore thrown out, by the late professor Bode, of Berlin,* as a possible surmise, that a planet not then yet discovered might exist between Mars and Jupiter; and it may easily be imagined what was the astonishment of astronomers on finding not one only, but four planets, differing greatly in all the other elements of their orbits, but agreeing very nearly, both *inter se,* and with the above stated empirical law, in respect of their mean distances from the sun. No account, *à priori* or from theory, was to be given of this singular progression, which is not, like Kepler's laws, strictly exact in numerical verification: but the circumstances we have just mentioned tended to create a strong belief that it was something beyond a mere accidental coincidence, and bore reference to the essential structure of the planetary system. It was even conjectured that the ultra-zodiacal planets are fragments of some greater planet which formerly circulated in that interval, but which has been blown to atoms by an explosion; an idea countenanced by the exceeding minuteness of these bodies which present discs; and it was argued that in that case innumerable more such fragments must exist and might come to be hereafter discovered. Whatever may be thought of such a speculation as a physical hypothesis, this conclusion has been verified to a considerable extent as a matter of fact by subsequent discovery, the result of a careful and minute

---

* The empirical law itself, as we have above stated it, is ascribed by Voiron, not to Bode (who would appear, however, at all events, to have first drawn attention to this interpretation of its interruption,) but to professor Titius of Wittemberg. (*Voiron, Supplement to Bailly.*)

examination and mapping down of the smaller stars in and near the zodiac, undertaken with that express object. Zodiacal charts of this kind, the product of the zeal and industry of many astronomers, have been constructed, in which every star down to the ninth or tenth magnitude is inserted, and these stars being compared with the actual stars of the heavens, the intrusion of any stranger within their limits cannot fail to be noticed when the comparison is systematically conducted. The discovery of Astræa, and that of Hebe by professor Hencke, date respectively from December 8th, 1845, and July 1st, 1847 ; those of Iris and Flora, by Mr. Hind, from August 13th and October 18th, 1847; and that of the new planet, discovered by Mr. Graham, from April 25th, 1848.

(506.) The discovery of Neptune marks in a signal manner the maturity of astronomical science. The proof, or at least the urgent presumption of the existence of such a planet, as a means of accounting (by its attraction) for certain small irregularities observed in the motions of Uranus, was afforded almost simultaneously by the independent researches of two geometers, Messrs. Adams of Cambridge and Leverrier of Paris, who were enabled, *from theory alone,* to calculate whereabouts it ought to appear in the heavens, *if visible ;* the places thus independently calculated agreeing surprisingly. *Within a single degree* of the place assigned by M. Leverrier's calculations, and by him communicated to Dr. Galle of the Royal Observatory at Berlin, it was actually found by that astronomer on the very first night after the receipt of that communication, on turning a telescope on the spot, and comparing the stars in its immediate neighbourhood with those previously laid down in one of the zodiacal charts already alluded to.* This remarkable verification of an indication so extraordinary took place on the 23d of September, 1846.†

---

* Constructed by Dr. Bremiker, of Berlin.   On reading the history of this noble discovery, we are ready to exclaim with Schiller —

" Mit dem Genius steht die Natur in ewigem Bunde,
Was der Eine verspricht leistet die Andre gewiss."

† Professor Challis, of the Cambridge Observatory, directing the Northum-

(507.) The mean distance of Neptune from the sun, however, so far from falling in with the supposed law of planetary distances above mentioned, offers a decided case of discordance. The interval between its orbit and that of Mercury, instead of being nearly double the interval between those of Uranus and Mercury, does not, in fact, exceed the latter interval by much more than half its amount. This remarkable exception may serve to make us cautious in the too ready admission of empirical laws of this nature to the rank of fundamental truths, though, as in the present instance, they may prove useful auxiliaries, and serve as stepping stones, affording a temporary footing in the path to great discoveries. The force of this remark will be more apparent when we come to explain more particularly the nature of the theoretical views which led to the discovery of Neptune itself.

(508.) We shall devote the rest of this chapter to an account of the physical peculiarities and probable condition of the several planets, so far as the former are known by observation, or the latter rest on probable grounds of conjecture. In this, three features principally strike us as necessarily productive of extraordinary diversity in the provisions by which, if they be, like our earth, inhabited, animal life must be supported. These are, first, the difference in their respective supplies of light and heat from the sun; secondly, the difference in the intensities of the gravitating forces which must subsist at their surfaces, or the different ratios which, on their several globes, the *inertiæ* of bodies must bear to their *weights;* and, thirdly, the difference in the nature of the materials of which, from what we know of their mean density, we have every reason to believe they consist. The intensity of solar radiation is nearly seven times greater on Mercury than on the Earth, and on Uranus 330 times less;

berland telescope of that Institution to the place assigned by Mr. Adams's calculations and its vicinity, on the 4th and 12th of August 1846, saw the planet on both those days, and noted its place (among those of other stars) for re-observation. He, however, postponed the *comparison* of the places observed, and, not possessing Dr. Bremiker's chart (which would have at once indicated the presence of an unmapped star), remained in ignorance of the planet's existence as a visible object till its announcement as such by Dr. Galle.

the proportion between the two extremes being that of upwards of 2000 to 1. Let any one figure to himself the condition of our globe, were the sun to be septupled, to say nothing of the greater ratio! or were it diminished to a seventh, or to a 300th of its actual power! Again, the intensity of gravity, or its efficacy in counteracting muscular power and repressing animal activity, on Jupiter, is nearly two and a half times that on the Earth, on Mars not more than one-half, on the Moon one-sixth, and on the smaller planets probably not more than one-twentieth; giving a scale of which the extremes are in the proportion of sixty to one. Lastly, the density of Saturn hardly exceeds one-eighth of the mean density of the Earth, so that it must consist of materials not much heavier than cork. Now, under the various combinations of elements so important to life as these, what immense diversity must we not admit in the conditions of that great problem, the maintenance of animal and intellectual existence and happiness, which seems, so far as we can judge by what we see around us in our own planet, and by the way in which every corner of it is crowded with living beings, to form an unceasing and worthy object for the exercise of the Benevolence and Wisdom which preside over all!

(509.) Quitting, however, the region of mere speculation, we will now show what information the telescope affords us of the actual condition of the several planets within its reach. Of Mercury we can see little more than that it is round, and exhibits phases. It is too small, and too much lost in the constant neighbourhood of the Sun, to allow us to make out more of its nature. The real diameter of Mercury is about 3200 miles: its apparent diameter varies from 5″ to 12″. Nor does Venus offer any remarkable peculiarities: although its real diameter is 7800 miles, and although it occasionally attains the considerable apparent diameter of 61″, which is larger than that of any other planet, it is yet the most difficult of them all to define with telescopes. The intense lustre of its illuminated part dazzles the sight, and exaggerates every imperfection of the telescope; yet we see clearly that its surface is not mottled over with permanent spots like the Moon ; we notice in it neither mountains nor shadows, but a uniform brightness, in which sometimes we may indeed fancy, or perhaps more than fancy, brighter or obscurer portions, but can seldom or never rest fully satisfied of the fact. It is from some observations of this kind that both Venus and Mercury have been concluded to revolve on their axes in about the same time as the Earth, though in the case of Venus, Bianchini, and other more recent observers have contended for a period of twenty-four times that length. The most natural conclusion, from the very rare appearance and want of permanence in the spots, is, that we do not see, as in the Moon, the real surface of these planets, but only their atmospheres, much loaded with clouds, and which may serve to mitigate the otherwise intense glare of their sunshine.

(510.) The case is very different with Mars. In this planet we frequently discern, with perfect distinctness, the outlines of what may be continents and seas. (See Plate I. *fig.* 1., which represents Mars in its gibbous state, as seen on the 16th of August, 1830, in the 20-feet reflector at Slough.) Of these, the former are distinguished by that ruddy colour which characterizes the light of this planet (which always appears red and fiery), and indicates, no doubt, an ochrey tinge in the general soil, like what the red sandstone districts on the Earth may possibly offer to the inhabitants of Mars, only more decided. Contrasted with this (by a general law in optics), the seas, as we may call them, appear greenish.* These spots, however, are not always to be seen equally distinct, but, *when seen*, they offer the appearance of forms considerably definite and highly characteristic,† brought successively into view by the rotation of the planet, from the assiduous observation of which it has even been found practicable to construct a rude chart of the

---

* I have noticed the phænomena described in the text on many occasions, but never more distinct than on the occasion when the drawing was made from which the figure in Plate I. is engraved. — *Author.*

† The reader will find many of these forms represented in Schumacher's *Astronomische Nachrichten*, No. 191, 434, and in the chart in No. 349. by Messrs. Beer and Mädler.

surface of the planet. The variety in the spots may arise
from the planet not being destitute of atmosphere and clouds;
and what adds greatly to the probability of this is the appearance of brilliant white spots at its poles, — one of which
appears in our figure, — which have been conjectured, with
some probability, to be snow; as they disappear when they
have been long exposed to the sun, and are greatest when
just emerging from the long night of their polar winter, the
snow line then extending to about six degrees (reckoned on
a meridian of the planet) from the pole. By watching the
spots during a whole night, and on successive nights, it is
found that Mars has a rotation on an axis inclined about
30° 18′ to the ecliptic, and in a period of $24^h 37^m 23^s$* in the
same direction as the Earth's, or from west to east. The
greatest and least apparent diameters of Mars are 4″ and
18″, and its real diameter about 4100 miles.

(511.) We now come to a much more magnificent planet,
Jupiter, the largest of them all, being in diameter no less
than 87,000 miles, and in bulk exceeding that of the Earth
nearly 1300 times. It is, moreover, dignified by the attendance of four *moons, satellites,* or *secondary planets,* as they are
called, which constantly accompany and revolve about it, as
the Moon does round the Earth, and in the same direction,
forming with their principal, or *primary,* a beautiful miniature
system, entirely analogous to that greater one of which their
central body is itself a member, obeying the same laws, and
exemplifying, in the most striking and instructive manner,
the prevalence of the gravitating power as the ruling principle of their motions: of these, however, we shall speak more
at large in the next chapter.

(512.) The disc of Jupiter is always observed to be
crossed in one certain direction by dark bands or belts presenting the appearance, in Plate III. *fig.* 2., which represents
this planet as seen on the 23d of September, 1832, in the
20-feet reflector at Slough. These belts are, however, by no
means alike at all times; they vary in breadth and in situa-

* Beer and Mädler. *Astr. Nachr.* 349.

tion on the disc (though never in their general direction).
They have even been seen broken up and distributed over
the whole face of the planet; but this phænomenon is extremely rare. Branches running out from them, and subdivisions, as represented in the figure, as well as evident dark
spots, are by no means uncommon; and from these, attentively watched, it is concluded that this planet revolves in
the surprisingly short period of $9^h 55^m 50^s$ (sid. time), on an
axis perpendicular to the direction of the belts. Now, it is
very remarkable, and forms a most satisfactory comment on
the reasoning by which the spheroidal figure of the Earth has
been deduced from its diurnal rotation, that the outline of
Jupiter's disc is evidently not circular, but elliptic, being
considerably flattened in the direction of its axis of rotation.
This appearance is no optical illusion, but is authenticated by
micrometrical measures, which assign 107 to 100 for the
proportion of the equatorial and polar diameters. And to
confirm, in the strongest manner, the truth of those principles
on which our former conclusions have been founded, and
fully to authorize their extension to this remote system, it
appears, on calculation, that this is really the degree of ob-
lateness which corresponds, on those principles, to the dimensions of Jupiter, and to the time of his rotation.

(513.) The parallelism of the belts to the equator of Jupiter,
their occasional variations, and the appearances of spots seen
upon them, render it extremely probable that they subsist
in the atmosphere of the planet, forming tracts of comparatively clear sky, determined by currents analogous to our
trade-winds, but of a much more steady and decided character, as might indeed be expected from the immense velocity
of its rotation. That it is the comparatively darker body of
the planet which appears in the belts is evident from this, —
that they do not come up in all their strength to the edge of
the disc, but fade away gradually before they reach it. (See
Plate III. *fig.* 2.) The apparent diameter of Jupiter varies
from 30″ to 46″.

(514.) A still more wonderful, and, as it may be termed,
elaborately artificial mechanism, is displayed in Saturn, the

next in order of remoteness to Jupiter, to which it is not much inferior in magnitude, being about 79,000 miles in diameter, nearly 1000 times exceeding the earth in bulk, and subtending an apparent angular diameter at the earth, of about 18″ at its mean distance. This stupendous globe, besides being attended by no less than seven satellites, or moons, is surrounded with two broad, flat, extremely thin rings, concentric with the planet and with each other; both lying in one plane, and separated by a very narrow interval from each other throughout their whole circumference, as they are from the planet by a much wider. The dimensions of this extraordinary appendage are as follow * : —

	″	Miles.
Exterior diameter of exterior ring - - -	40·095 =	176,418
Interior ditto - - - - - -	35·289 =	155,272
Exterior diameter of interior ring - - -	34·475 =	151,690
Interior ditto - - - - - - -	26·668 =	117,339
Equatorial diameter of the body - - -	17·991 =	79,160
Interval between the planet and interior ring	4·339 =	19,090
Interval of the rings - - - - -	0·408 =	1,791
Thickness of the rings not exceeding - - -	=	250

The figure (*fig.* 3. Plate III.) represents Saturn surrounded by its rings, and having its body striped with dark belts, somewhat similar, but broader and less strongly marked than those of Jupiter, and owing, doubtless, to a similar cause. † That the ring is a solid opake substance is shown by its throwing its shadow on the body of the planet, on the side nearest the sun, and on the other side receiving that of the body, as shown in the figure. From the parallelism of the belts with the plane of the ring, it may be conjectured that the axis of rotation of the planet is perpendicular to that plane; and this conjecture is confirmed by the occasional appearance of extensive dusky spots on its surface, which

* These dimensions are calculated from Prof. Struve's micrometric measures, Mem. Art. Soc. iii. 301., with the exception of the thickness of the ring, which is concluded from its total disappearance in 1833, in a telescope which would certainly have shown, as a visible object, a line of light one-twentieth of a second in breadth. The interval of the rings here stated is possibly somewhat too small.

† The equatorial bright belt is generally well seen. The subdivision of the dark one by two narrow bright bands is seldom so distinct as represented in the plate.

when watched, like the spots on Mars or Jupiter, indicate a rotation in 10^h 29^m 17^s about an axis so situated.

(515.) The axis of rotation, like that of the earth, preserves its parallelism to itself during the motion of the planet in its orbit; and the same is also the case with the ring, whose plane is constantly inclined at the same, or very nearly the same, angle to that of the orbit, and, therefore, to the ecliptic, viz. 28° 11′; and intersects the latter plane in a line, which makes at present* an angle with the line of equinoxes of 167° 31′. So that the *nodes of the ring* lie in 167° 31′ and 347° 31′ of longitude. Whenever, then, the planet happens to be situated in one or other of these longitudes, as at C, the plane of the ring passes through the sun, which then illuminates only the edge of it. And if the earth at that moment be in F, it will see the ring edgeways, the planet being in opposition, and therefore most favourably situated (*cæteris paribus*) for observation. Under these circumstances the ring, if seen at all, can only appear as a very narrow straight line of light projecting on either side of the body as a prolongation of its diameter. In fact, it is quite invisible in any but telescopes of extraordinary power.†
This remarkable phenomenon takes place at intervals of fifteen years nearly (being a semi-period of Saturn in its orbit). One disappearance at least must take place whenever Saturn passes either node of its orbit; but three must frequently happen, and two are possible. To show this, suppose S to be the sun, A B C D part of Saturn's orbit situated so as to include the node of the ring (at C); E F G H the Earth's orbit; S C the line of the node; E B, G D parallel to S C touching the earth's orbit in E G; and let the direction of motion of both bodies be that indicated by the arrow. Then since the ring preserves its parallelism, its plane can nowhere intersect the earth's orbit, and therefore no disappearance can take place, unless the planet be between B and D: and,

---

* According to Beessl, the longitude of the node of the ring increases be 46″462· per annum. In 1800 it was 166° 53′ 8″·9.

† Its disappearance was *complete* when observed with a reflector eighteen inches in aperture and twenty feet in focal length on the 29th of April, 1833, by the author.

on the other hand, a disappearance is possible (if the earth
be rightly situated) during the whole time of the description

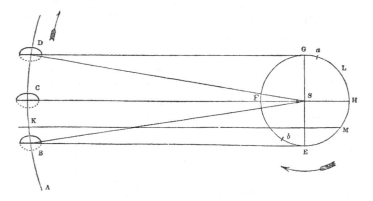

of the arc B D. Now, since S B or S D, the distance of
Saturn from the Sun, is to S E or S G, that of the Earth, as
9·54 to 1, the angle C S D or C S B = 6° 1′, and the whole
angle B S D = 12° 2′, which is described by Saturn (on an
average) in 359·46 days, wanting only 5·8 days of a complete
year.   The Earth then describes very nearly an entire revolution within the limits of time when a disappearance is possible ; and since, in either half of its orbit E F G or G H F,
it may equally encounter the plane of the ring, one such
encounter at least is unavoidable within the time specified.

(516.) Let G a be the arc of the Earth's orbit described
from G in 5·8 days. Then if, at the moment of Saturn's
arrival at B, the Earth be at a, it will encounter the plane
of the ring advancing parallel to itself and to B E to meet
it, somewhere in the quadrant H E, as at M, after which
it will be behind that plane (with reference to the direction
of Saturn's motion) through all the arc M E F G up to G,
where it will again overtake it at the very moment of the
planet quitting the arc B D.   In this state of things there
will be two disappearances.   If, when Saturn is at B, the
Earth be anywhere in the arc a H E, it is equally evident
that it will *meet* and pass through the advancing plane of the
ring somewhere in the quadrant H E, that it will again

overtake and pass through it somewhere in the semicircle E F G, and again meet it in some point of the quadrant G H, so that three disappearances will take place. So, also, if the Earth be at E when Saturn is at B, the motion of the Earth being at that instant directly towards B, the plane of the ring will for a short time leave it behind; but the ground so lost being rapidly regained as the earth's motion becomes oblique to the line of junction, it will soon overtake and pass through the plane in the early part of the quadrant E F, and passing on through G before Saturn arrives at D, will meet the plane again in the quadrant G H. The same will continue up to a certain point *b*, at which, if the earth be initially situated, there will be but two disappearances — the plane of the ring there overtaking the earth for an instant, and being immediately again left behind by it, to be again encountered by it in G H. Finally, if the initial place of the earth (when Saturn is at B) be in the arc *b* F *a*, there will be but one passage through the plane of the ring, viz., in the semicircle G H E, the earth being in advance of that plane throughout the whole of *b* G.

(517.) The appearances will moreover be varied according as the Earth passes from the enlightened to the unenlightened side of the ring, or *vice versâ*. If C be the ascending node of the ring, and if the under side of the paper be supposed south and the upper north of the ecliptic, then, when the Earth *meets* the plane of the ring in the quadrant H E, it passes from the bright to the dark side : where it *overtakes* it in the quadrant E F, the contrary. *Vice versâ*, when it overtakes it in F G, the transition is from the bright to the dark side, and the contrary where it meets it in G H. On the other hand, when the earth is overtaken by the ring-plane in the interval E *b*, the change is from the bright to the dark side. When the dark side is exposed to sight, the aspect of the planet is very singular. It appears as a bright round disc, with its belts, &c., but crossed equatorially by a narrow and perfectly black line. This can never of course happen when the planet is more than 6° 1′ from the node of the ring. Generally, the northern side is enlightened and visible when

the heliocentric longitude of Saturn is between 172° 32′ and
341° 30′, and the southern when between 353° 32′ and 161°
30′. The greatest opening of the ring occurs when the
planet is situated at 90° distance from the node of the ring,
or in longitudes 77° 31′ and 257° 31′, and at these points
the longer diameter of its apparent ellipse is almost exactly
double the shorter.

(518.) It will naturally be asked how so stupendous an
arch, if composed of solid and ponderous materials, can be
sustained without collapsing and falling in upon the planet?
The answer to this is to be found in a swift rotation of the
ring in its own plane, which observation has detected, owing to
some portion of the ring being a little less bright than others,
and assigned its period at 10^h 32^m 15^s, which, from what we
know of its dimensions, and of the force of gravity in the
Saturnian system, is very nearly the periodic time of a satellite revolving at the same distance as the middle of its
breadth. It is the centrifugal force, then, arising from this
rotation, which sustains it; and, although no observation nice
enough to exhibit a difference of periods between the outer and
inner rings have hitherto been made, it is more than probable
that such a difference does subsist as to place each independently of the other in a similar state of equilibrium.

(519.) Although the rings are, as we have said, very
nearly concentric with the body of Saturn, yet micrometrical measurements of extreme delicacy have demonstrated that
the coincidence is not mathematically exact, but that the
center of gravity of the rings oscillates round that of the
body describing a very minute orbit, probably under laws
of much complexity. Trifling as this remark may appear,
it is of the utmost importance to the stability of the system
of the rings. Supposing them mathematically perfect in their
circular form, and exactly concentric with the planet, it is
demonstrable that they would form a system in a state *unstable equilibrium*, which the slightest external power would
subvert — not by causing a rupture in the substance of the
rings — but by precipitating them, *unbroken*, on the surface of
the planet. For the attraction of such a ring or rings on a

point or sphere excentrically within them, is not the same in all directions, but tends to draw the point or sphere towards the nearest part of the ring, or away from the center. Hence, supposing the body to become, from any cause, ever so little excentric to the ring, the tendency of their mutual gravity is not to correct but to increase this excentricity, and to bring the nearest parts of them together. Now, external powers, capable of producing such excentricity, exist in the attractions of the satellites, as will be shown in Chap. XII.; and in order that the system may be *stable*, and possess within itself a power of resisting the first inroads of such a tendency, while yet nascent and feeble, and opposing them by an opposite or maintaining power, it has been shown that it is sufficient to admit the rings to be *loaded* in some part of their circumference, either by some minute inequality of thickness, or by some portions being denser than others. Such a load would give to the whole ring to which it was attached somewhat of the character of a heavy and sluggish satellite maintaining itself in an orbit with a certain energy sufficient to overcome minute causes of disturbance, and establish an average bearing on its center. But even without supposing the existence of any such load, — of which, after all, we have no proof, — and granting, in its full extent, the general instability of the equilibrium, we think we perceive, in the rapid periodicity of all the causes of disturbance, a sufficient guarantee of its preservation. However homely be the illustration, we can conceive nothing more apt, in every way, to give a general conception of this maintenance of equilibrium under a constant tendency to subversion, than the mode in which a practised hand will sustain a long pole in a perpendicular position resting on the finger by a continual and almost imperceptible variation of the point of support. Be that, however, as it may, the observed oscillation of the centers of the rings about that of the planet is in itself the evidence of a perpetual contest between conservative and destructive powers — both extremely feeble, but so antagonizing one another as to prevent the latter from ever acquiring an uncontrollable ascendancy, and rushing to a catastrophe.

(520.) This is also the place to observe, that as the smallest difference of velocity between the body and the rings must infallibly precipitate the latter on the former, never more to separate, (for they would, once in contact, have attained a position of *stable equilibrium*, and be held together ever after by an immense force;) it follows, either that their motions in their common orbit round the sun must have been adjusted to each other by an external power, with the minutest precision, or that the rings must have been formed about the planet while subject to their common orbitual motion, and under the full and free influence of all the acting forces.

(521.) Several astronomers have suspected, and even consider themselves to have certainly observed, the rings of Saturn to be occasionally, at least, streaked with numerous dark lines parallel to the decided black interval which separates the two rings, and which being permanent, and being seen equally and in the same part of the breadth on both sides of the ring, cannot be doubted to be a real separation. As it is equally certain, however, that the ring of Saturn has been admirably well seen by others, with telescopes no way inferior, without giving rise to any suspicion of such a subdivision, the *permanence* of such streaks must at least be considered as undemonstrated, and the phænomenon remanded to the careful attention of observers.*

(522.) The rings of Saturn must present a magnificent spectacle from those regions of the planet which lie above their enlightened sides, as vast arches spanning the sky from horizon to horizon, and holding an almost invariable situation among the stars. On the other hand, in the regions beneath the dark side, a solar eclipse of fifteen years in duration, under their shadow, must afford (to our ideas) an inhospitable asylum to animated beings, ill compensated by the faint light

---

* The passage of Saturn across any considerable star would afford an admirable opportunity of testing the reality of such fissures, as it would flash in succession through them. The opportunity of watching for such occultations — when Saturn traverses the Milky-Way, for instance — should not be neglected.

of the satellites. But we shall do wrong to judge of the fitness or unfitness of their condition from what we see around us, when, perhaps, the very combinations which convey to our minds only images of horror, may be in reality theatres of the most striking and glorious displays of beneficent contrivance.

(523.) Of Uranus we see nothing but a small round uniformly illuminated disc, without rings, belts, or discernible spots. Its apparent diameter is about 4″, from which it never varies much, owing to the smallness of our orbit in comparison of its own. Its real diameter is about 35,000 miles, and its bulk 82 times that of the earth. It is attended by satellites — four at least, probably five or six — whose orbits (as will be seen in the next chapter) offer remarkable peculiarities.

(524.) The discovery of Neptune is so recent, and its situation in the ecliptic at present so little favourable for seeing it with perfect distinctness, that nothing very positive can be stated as to its physical appearance. To two observers it has afforded strong suspicion of being surrounded with a ring very highly inclined. And from the observations of Mr. Lassell, M. Otto Struve, and Mr. Bond, it appears to be attended certainly by one, and very probably by two satellites — though the existence of the second can hardly yet be considered as quite demonstrated.

(525.) If the immense distance of Neptune precludes all hope of coming at much knowledge of its physical state, the minuteness of the ultra-zodiacal planets is no less a bar to any enquiry into theirs. One of them, Pallas, has been said to have somewhat of a nebulous or hazy appearance, indicative of an extensive and vaporous atmosphere, little repressed and condensed by the inadequate gravity of so small a mass. It is probable, however, that the appearance in question has originated in some imperfection in the telescope employed or other temporary causes of illusion. In Vesta and Pallas only have sensible discs been hitherto observed, and those only with very high magnifying powers. Vesta was once seen by Schrœter with the naked eye. No doubt the most remarkable of their peculiarities must lie in this condition of their state.   A man placed on one of them would spring with ease 60 feet high, and sustain no greater shock in his descent than he does on the earth from leaping a yard.   On such planets giants might exist; and those enormous animals, which on earth require the buoyant power of water to counteract their weight, might there be denizens of the land. But of such speculations there is no end.

(526.) We shall close this chapter with an illustration calculated to convey to the minds of our readers a general impression of the relative magnitudes and distances of the parts of our system.   Choose any well levelled field or bowling-green.   On it place a globe, two feet in diameter; this will represent the Sun; Mercury will be represented by a grain of mustard seed, on the circumference of a circle 164 feet in diameter for its orbit; Venus a pea, on a circle 284 feet in diameter ; the Earth also a pea, on a circle of 430 feet; Mars a rather large pin's head, on a circle of 654 feet; Juno, Ceres, Vesta, and Pallas, grains of sand, in orbits of from 1000 to 1200 feet; Jupiter a moderate-sized orange, in a circle nearly half a mile across, Saturn a small orange, on a circle of four-fifths of a mile; Uranus a full-sized cherry, or small plum, upon the circumference of a circle more than a mile and a half, and Neptune a good-sized plum on a circle about two miles and a half in diameter.   As to getting correct notions on this subject by drawing circles on paper, or, still worse, from those very childish toys called orreries, it is out of the question.   To imitate the motions of the planets, in the above-mentioned orbits, Mercury must describe its own diameter in 41 seconds ; Venus in 4^m 14^s; the Earth, in 7 minutes; Mars, in 4^m 48^s; Jupiter, 2^h 56^m; Saturn, in 3^h 13^m; Uranus, in 2^h 16^m; and Neptune in 3^h 30^m.

CHAPTER X.

OF THE SATELLITES.

OF THE MOON, AS A SATELLITE OF THE EARTH. — GENERAL PROX-
IMITY OF SATELLITES TO THEIR PRIMARIES, AND CONSEQUENT
SUBORDINATION OF THEIR MOTIONS. — MASSES OF THE PRIMARIES
CONCLUDED FROM THE PERIODS OF THEIR SATELLITES. — MAIN-
TENANCE OF KEPLER'S LAWS IN THE SECONDARY SYSTEMS. — OF
JUPITER'S SATELLITES. — THEIR ECLIPSES, ETC. — VELOCITY OF
LIGHT DISCOVERED BY THEIR MEANS. — SATELLITES OF SATURN
— OF URANUS — OF NEPTUNE.

(527.) In the annual circuit of the earth about the sun, it is
constantly attended by its satellite, the moon, which revolves
round it, or rather both round their common center of
gravity; while this center, strictly speaking, and not either of
the two bodies thus connected, moves in an elliptic orbit,
undisturbed by their mutual action, just as the center of
gravity of a large and small stone tied together and flung
into the air describes a parabola as if it were a real material
substance under the earth's attraction, while the stones
circulate round it or round each other, as we choose to
conceive the matter.

(528.) If we trace, therefore, the *real* curve actually described by either the moon's or the earth's centers, in virtue
of this compound motion, it will appear to be, not an exact
ellipse, but an undulated curve, like that represented in the
figure to article 324., only that the number of undulations in
a whole revolution is but 13, and their actual deviation from
the general ellipse, which serves them as a central line, is
comparatively very much smaller — so much so, indeed, that
every part of the curve described by either the earth or moon is
concave towards the sun. The excursions of the earth on
either side of the ellipse, indeed, are so very small as to be
hardly appretiable. In fact, the center of gravity of the

earth and moon lies always *within* the surface of the earth, so that the monthly orbit described by the earth's center about the common center of gravity is comprehended within a space less than the size of the earth itself.   The effect *is*, nevertheless, sensible, in producing an apparent monthly displacement of the sun in longitude, of a parallactic kind, which is called the *menstrual equation;* whose greatest amount is, however, less than the sun's horizontal parallax, or than 8·6″.

(529.) The moon, as we have seen, is about 60 radii of the earth distant from the center of the latter. Its proximity, therefore, to its center of attraction, thus estimated, is much greater than that of the planets to the sun; of which Mercury, the nearest, is 84, and Uranus 2026 solar radii from *its* center.   It is owing to this proximity that the moon remains attached to the earth as a satellite. Were it much farther, the feebleness of its gravity towards the earth would be inadequate to produce that alternate acceleration and retardation in its motion about the sun, which divests it of the character of an independent planet, and keeps its movements subordinate to those of the earth.   The one would outrun, or be left behind the other, in their revolutions round the sun (by reason of Kepler's third law), according to the relative dimensions of their heliocentric orbits, after which the whole influence of the earth would be confined to producing some considerable periodical disturbance in the moon's motion, as it passed or was passed by it in each synodical revolution.

(530.) At the distance at which the moon really is from us, its gravity towards the earth is actually less than towards the sun.   That this is the case, appears sufficiently from what we have already stated, that the moon's *real* path, even when between the earth and sun, is *concave towards the latter.* But it will appear still more clearly if, from the known periodic times * in which the earth completes its annual and

---

* R and r radii of two orbits (supposed circular), P and p the periodic times; then the arcs in question (A and a) are to each other as $\frac{R}{P}$ to $\frac{r}{p}$; and since the versed sines are as the squares of the arcs directly and the radii inversely, these are to each other as $\frac{R}{P^2}$ to $\frac{r}{p^2}$; and in this ratio are the forces acting on the revolving bodies in either case.

the moon its monthly orbit, and from the dimensions of those orbits, we calculate the amount of deflection, in either, from their tangents, in equal very minute portions of time, as one second. These are the versed sines of the arcs described in that time in the two orbits, and these are the measures of the acting forces which produce those deflections. If we execute the numerical calculation in the case before us, we shall find 2·233 : 1 for the proportion in which the intensity of the force which retains the earth in its orbit round the sun actually exceeds that by which the moon is retained in *its* orbit about the earth.

(531.) Now the sun is 399 times more remote from the earth than the moon is. And, as gravity increases as the squares of the distances decrease, it must follow that at *equal* distances, the intensity of solar would exceed that of terrestrial gravity in the above proportion, augmented in the further ratio of the square of 400 to 1; that is, in the proportion of 355000 to 1; and therefore, if we grant that the intensity of the gravitating energy is commensurate with the mass or inertia of the attracting body, we are compelled to admit the mass of the earth to be no more than $\frac{1}{355000}$ of that of the sun.

(532.) The argument is, in fact, nothing more than a recapitulation of what has been adduced in Chap VIII. (art. 448.) But it is here re-introduced, in order to show how the mass of a planet which is attended by one or more satellites can be as it were weighed against the sun, provided we have learned from observation the dimensions of the orbits described by the planet about the sun, and by the satellites about the planet, and also the periods in which these orbits are respectively described. It is by this method that the masses of Jupiter, Saturn, Uranus, and Neptune have been ascertained. (See Synoptic Table.)

(533.) Jupiter, as already stated, is attended by four satellites, Saturn by seven; Uranus, certainly by four, and perhaps by six; and Neptune by two or more. These, with their respective *primaries* (as the central planets are called), form in each case miniature systems entirely analogous, in the general laws by which their motions are governed, to the great system in which the sun acts the part of the primary, and the planets of its satellites. In each of these systems the laws of Kepler are obeyed, in the sense, that is to say, in which they are obeyed in the planetary system — approximately, and without prejudice to the effects of mutual perturbation, of extraneous interference, if any, and of that small but not imperceptible correction which arises from the elliptic form of the central body. Their orbits are circles or ellipses of very moderate excentricity, the primary occupying one focus. About this they describe areas very nearly proportional to the times; and the squares of the periodical times of all the satellites belonging to each planet are in proportion to each other as the cubes of their distances. The tables at the end of the volume exhibit a synoptic view of the distances and periods in these several systems, so far as they are at present known; and to all of them it will be observed that the same remark respecting their proximity to their primaries holds good, as in the case of the moon, with a similar reason for such close connection.

(534.) Of these systems, however, the only one which has been studied with attention to all its details, is that of Jupiter; partly on account of the conspicuous brilliancy of its four attendants, which are large enough to offer visible and measurable discs in telescopes of great power; but more for the sake of their eclipses, which, as they happen very frequently, and are easily observed, afford signals of considerable use for the determination of terrestrial longitudes (art. 266.). This method, indeed, until thrown into the back ground by the greater facility and exactness now attainable by lunar observations (art. 267.), was the best, or rather the only one, which could be relied on for great distances and long intervals.

(535.) The satellites of Jupiter revolve from west to east (following the analogy of the planets and moon,) in planes very nearly, although not exactly, coincident with that of the equator of the planet, or parallel to its belts. This latter plane is inclined 3° 5′ 30″ to the orbit of the planet, and is therefore but little different from the plane of the ecliptic. Accordingly, we see their orbits projected very nearly into straight lines, in which they appear to oscillate to and fro, sometimes passing before Jupiter, and casting shadows on his disc, (which are very visible in good telescopes, like small round ink spots, the circular form of which is very evident,) and sometimes disappearing behind the body, or being eclipsed in its shadow at a distance from it. It is by these eclipses that we are furnished with accurate data for the construction of tables of the satellites' motions, as well as with signals for determining differences of longitude.

(536.) The eclipses of the satellites, in their general conception, are perfectly analogous to those of the moon, but in their detail they differ in several particulars. Owing to the much greater distance of Jupiter from the sun, and its greater magnitude, the cone of its shadow or umbra (art. 420.) is greatly more elongated, and of far greater dimension, than that of the earth. The satellites are, moreover, much less in proportion to their primary, their orbits less inclined to *its* ecliptic, and (comparatively to the diameter of the planet) of smaller dimensions, than is the case with the moon. Owing to these causes, the three interior satellites of Jupiter pass through the shadow, and are totally eclipsed, every revolution; and the fourth, though, from the greater inclination of its orbit, it sometimes escapes eclipse, and *may* occasionally graze as it were the border of the shadow, and suffer partial eclipse, yet does so comparatively seldom, and, ordinarily speaking, its eclipses happen, like those of the rest, each revolution.

(537.) These eclipses, moreover, are not seen, as is the case with those of the moon, from the center of their motion, but from a remote station, and one whose situation with respect to the line of shadow is variable. This, of course, makes no difference in the *times* of the eclipses, but a very great one in their visibility, and in their apparent situations with respect to the planet at the moments of their entering and quitting the shadow.

(538.) Suppose S to be the sun, E the earth in its orbit E F G K, J Jupiter, and *a b* the orbit of one of its satellites.

The cone of the shadow, then, will have its vertex at X, a
point far beyond the orbits of all the satellites; and the

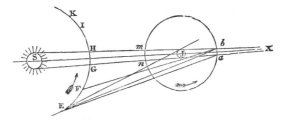

penumbra, owing to the great distance of the sun, and the
consequent smallness of the angle (about 6′ only) its disc
subtends at Jupiter, will hardly extend, within the limits of
the satellites' orbits, to any perceptible distance beyond the
shadow, — for which reason it is not represented in the figure.
A satellite revolving ftom west to east (in the direction of
the arrows) will be eclipsed when it enters the shadow at *a*,
but not suddenly, because, like the moon, it has a considerable
diameter seen from the planet; so that the time elapsing
from the first perceptible loss of light to its total extinction
will be that which it occupies in describing about Jupiter
an angle equal to its apparent diameter as seen from the
center of the planet, or rather somewhat more, by reason of
the penumbra; and the same remark applies to its emergence
at *b*.   Now, owing to the difference of telescopes and of eyes,
it is not possible to assign the *precise* moment of incipient
obscuration, or of total extinction at *a*, nor that of the first
glimpse of light falling on the satellite at *b*, or the complete
recovery of its light.   The observation of an eclipse, then,
in which only the immersion, or only the emersion, is seen,
is incomplete, and inadequate to afford any precise information, theoretical or practical.   But, if both the immersion
and emersion can be observed *with the same telescope, and by
the same person*, the interval of the times will give the duration,
and their mean the exact middle of the eclipse, when the
satellite is in the line S J X, *i.e.* the true moment of its
opposition to the sun.   Such observations, and such only, are
of use for determining the periods and other particulars of

the motions of the satellites, and for affording data of any material use for the calculation of terrestrial longitudes. The intervals of the eclipses, it will be observed, give the *synodic* periods of the satellites' revolutions; from which their sidereal periods must be concluded by the method in art. 418.

(539.) It is evident, from a mere inspection of our figure, that the eclipses take place to the west of the planet, when the earth is situated to the west of the line S J, *i. e.* before the opposition of Jupiter; and to the east, when in the other half of its orbit, or after the opposition. When the earth approaches the opposition, the visual line becomes more and more nearly coincident with the direction of the shadow, and the apparent place where the eclipses happen will be continually nearer and nearer to the body of the planet. When the earth comes to F, a point determined by drawing *b* F to touch the body of the planet, the *emersions* will cease to be visible, and will thenceforth, up to the time of the opposition, happen *behind* the disc of the planet. Similarly, from the opposition till the time when the earth arrives at I, a point determined by drawing *a* I tangent to the eastern limb of Jupiter, the *immersions* will be concealed from our view. When the earth arrives at G (or H) the immersion (or emersion) will happen at the very edge of the visible disc, and when between G and H (a very small space), the satellites will *pass uneclipsed behind the limb* of the planet.

(540.) Both the satellites and their shadows are frequently observed to *transit* or pass across the disc of the planet. When a satellite comes to *m*, *its* shadow will be thrown on Jupiter, and will appear to move across it as a black spot till the satellite comes to *n*. But the satellite itself will not appear to enter on the disc till it comes up to the line drawn from E to the eastern edge of the disc, and will not leave it till it attains a similar line drawn to the western edge. It appears then that the shadow will *precede* the satellite in its progress over the disc *before* the opposition of Jupiter, and *vice versâ*. In these transits of the satellites, which, with very powerful telescopes, may be observed with great precision, it frequently happens that the satellite itself is discernible *on* the disc as a bright spot if projected on a dark belt; but occasionally also as a dark spot of smaller dimensions than the shadow.   This curious fact (observed by Schroeter and Harding) has led to a conclusion that certain of the satellites have occasionally *on their* own bodies, or in their atmospheres, obscure spots of great extent.   We say of great extent ; for the satellites of Jupiter, small as they appear to us, are really bodies of considerable size, as the following comparative table will show : * —

	Mean apparent diameter as seen from the Earth.	Mean apparent diameter as seen from Jupiter.		Diameter in miles.	Mass.†
Jupiter	38″·327			87000	1·0000000
1st satellite	1·017	33′	11″	2508	0·0000173
2d ——	0·911	17	35	2068	0·0000232
3d ——	1·483	18	0	3377	0·0000885
4th ——	1·273	8	46	2890	0·0000427

From which it follows, that the first satellite appears to a spectator on Jupiter, as large as our moon to us ; the second and third nearly equal to each other, and of somewhat more than half the apparent diameter of the first, and the fourth about one quarter of that diameter.   So seen, they will frequently, of course, eclipse one another, and cause eclipses of the sun (the latter visible, however, only over a very small portion of the planet), and their motions and aspects with respect to each other must offer a perpetual variety and singular and pleasing interest to the inhabitants of their primary.

(541.) Besides the eclipses and the transits of the satellites across the disc, they may also disappear to us when not eclipsed, by passing *behind* the body of the planet.   Thus, when the earth is at E, the immersion of the satellite will be seen at *a*, and its emersion at *b*, both to the west of the planet, after which the satellite, still continuing its course in the direction *ab*, will pass behind the body, and again emerge on the opposite side, after an interval of occultation greater or less according to the distance of the satellite.   This interval

---

* Struve, Mem. Art. Soc. iii. 301.   † Laplace, Mec. Cel. liv. viii. § 27.

(on account of the great distance of the earth compared with the radii of the orbits of the satellites) varies but little in the case of each satellite, being nearly equal to the time which the satellite requires to describe an arc of its orbit, equal to the angular diameter of Jupiter as seen from its center, which time, for the several satellites, is as follows: viz., for the first, $2^h\ 20^m$; for the second, $2^h\ 56^m$; for the third, $3^h\ 43^m$; and for the fourth, $4^h\ 56^m$; the corresponding diameters of the planet as seen from these respective satellites being, 19° 49'; 12° 25'; 7° 47'; and 4° 25'.* Before the opposition of Jupiter, these occultations of the satellites happen *after* the eclipses: after the opposition (when, for instance, the earth is in the situation K), the occultations take place before the eclipses. It is to be observed that owing to the proximity of the orbits of the first and second satellites to the planet, *both* the immersion and emersion of either of them can never be observed in any single eclipse, the immersion being concealed by the body, if the planet be past its opposition, the emersion if not yet arrived at it. So also of the occultation. The commencement of the occultation, or the passage of the satellite behind the disc, takes place while obscured by the shadow, before opposition, and its re-emergence after. All these particulars will be easily apparent on mere inspection of the figure (art. 536.). It is only during the short time that the earth is in the arc G H (*i.e.* between the sun and Jupiter, that the cone of the shadow converging (while that of the visual rays diverges) behind the planet, permits their occultations to be completely observed both at ingress and egress, unobscured, the eclipses being then invisible.

(542.) An extremely singular relation subsists between the mean angular velocities or *mean motions* (as they are termed) of the three first satellites of Jupiter. If the mean angular velocity of the first satellite be added to twice that of the third, the sum will equal three times that of the second. From this relation it follows, that if from the mean longitude

---

* These data are taken approximately from Mr. Woolhouse's paper in the supplement to the Nautical Almanack for 1835.

of the first added to twice that of the third, be subducted three times that of the second, the remainder will always be the same, or constant, and observation informs us that this constant is 180°, or two right angles; so that, the situations of any two of them being given, that of the third may be found. It has been attempted to account for this remarkable fact, on the theory of gravity by their mutual action; and Laplace has demonstrated, that if this relation were at any one epoch approximately true, the mutual attractions of the satellites would, in process of time, render it exactly so. One curious consequence is, that these three sastellites cannot be all eclipsed at once; for, in consequence of the last-mentioned relation, when the second and third lie in the *same* direction from the center, the first must lie on the *opposite ;* and therefore, when at such a conjuncture the first is eclipsed, the other two must lie between the sun and planet, throwing its shadow on the disc, and *vice versâ.*

(543.) Although, however, for the above mentioned reason, the satellites cannot be all *eclipsed* at once, yet it may happen, and occasionally does so, that all are either eclipsed, occulted, or projected on the body, in which case they are, generally speaking, equally invisible, since it requires an excellent telescope to discern a satellite on the body, except in peculiar circumstances. Instances of the actual observations of Jupiter thus denuded of its usual attendance and offering the appearance of a solitary disc, though rare, have been more than once recorded. The first occasion in which this was noticed was by Molyneux, on November 2d, (old style) 1681.* A similar observation is recorded by Sir W. Herschel as made by him on May 23d, 1802. The phænomenon has also been observed by Mr. Wallis, on April 15th, 1826; (in which case the deprivation continued two whole hours;) and lastly by Mr. H. Griesbach, on September 27th, 1843.

(544.) The discovery of Jupiter's satellites, one of the first fruits of the invention of the telescope, and of Galileo's early and happy idea of directing its new-found powers to the examination of the heavens, forms one of the most

* Molyneux, Optics, p. 271.

memorable epochs in the history of astronomy.   The first astronomical solution of the great problem of " *the longitude*" — practically the most important for the interests of mankind which has ever been brought under the dominion of strict scientific principles, dates immediately from their discovery. The final and conclusive establishment of the Copernican system of astronomy may also be considered as referable to the discovery and study of this exquisite miniature system, in which the laws of the planetary motions, as ascertained by Kepler, and especially that which connects their periods and distances, were speedily traced, and found to be satisfactorily maintained.   And (as if to accumulate historical interest on this point) it is to the observation of their eclipses that we owe the grand discovery of the aberration of light, and the consequent determination of the enormous velocity of that wonderful element.   This we must explain now at large.

(545.) The earth's orbit being concentric with that of Jupiter and interior to it (see *fig.* art. 536.), their mutual distance is continually varying, the variation extending from the *sum* to the *difference* of the radii of the two orbits; and the difference of the greater and least distances being equal to a diameter of the earth's orbit.   Now, it was observed by Roemer, (a Danish astronomer, in 1675,) on comparing together observations of eclipses of the satellites during many successive years, that the eclipses at and about the opposition of Jupiter (or its nearest point to the earth) took place *too soon* — sooner, that is, than, by calculation from an average, he expected them; whereas those which happened when the earth was in the part of its orbit most remote from Jupiter were always *too late*.   Connecting the observed error in their computed times with the variation of distance, he concluded, that, to make the calculation on an average period correspond with fact, an allowance in respect of time behoved to be made proportional to the excess or defect of Jupiter's distance from the earth above or below its average amount, and such that a difference of distance of one diameter of the earth's orbit should correspond to 16^m $26^s \cdot 6$ of time allowed. Speculating on the probable physical cause, he was naturally led to think of a gradual instead of an instantaneous propagation of light. This explained every particular of the observed phenomenon, but the velocity required (192000 miles per second) was so great as to startle many, and, at all events, to require confirmation. This has been afforded since, and of the most unequivocal kind, by Bradley's discovery of the aberration of light (art. 329.). The velocity of light deduced from this last phænomenon differs by less than one eightieth of its amount from that calculated from the eclipses, and even this difference will no doubt be destroyed by nicer and more rigorously reduced observations.

(546.) The orbits of Jupiter's satellites are but little excentric, those of the two interior, indeed, have no perceptible excentricity. Their mutual action produces in them perturbations analagous to those of the planets about the sun, and which have been diligently investigated by Laplace and others. By assiduous observation it has been ascertained that they are subject to marked fluctuations in respect of brightness, and that these fluctuations happen periodically, according to their position with respect to the sun. From this it has been concluded, apparently with reason, that they turn on their axes, like our moon, in periods equal to their respective sidereal revolutions about their primary.

(547.) The satellites of Saturn have been much less studied than those of Jupiter, being far more difficult to observe. The most distant has its orbit materially inclined (no less than 12° 14′)* to the plane of the ring, with which the orbits of all the rest nearly coincide. Nor is this the only circumstance which separates it by a marked difference of character from the system of the six interior ones, and renders it in some sort an anomalous member of the Saturnian system. Its distance from the planet's center exceeds in the proportion of nearly three to one that of the most distant of all the rest, being no less than 64 times the radius of the globe of Saturn, a distance from the primary to which our own moon (at 60 radii) offers the only parallel. Its variation of light also in different parts of its orbit is very much greater

* Lalande, Astron. Art. 3075.

than in the case of any other secondary planet. Dominic Cassini indeed (its first discoverer, A D. 1671) found it to disappear for nearly half its revolution, when to the east of Saturn, and though the more powerful telescopes now in use enable us to follow it round the whole of its circuit, its diminution of light is so great in the eastern half of its orbit as to render it somewhat difficult to perceive. From this circumstance (viz. from the defalcation of light occurring constantly on the same side of Saturn *as seen from the earth,* the visual ray from which is never very oblique to the direction in which the sun's light falls on it) it is presumed with much certainty that this satellite revolves on its axis in the exact time of rotation about the primary; as we know to be the case with the moon, and as there is considerable ground for believing to be so with all secondaries.

(548.) The next satellite in order proceeding inwards (the first in order of discovery *) is by far the largest and most conspicuous of all, and is probably not much inferior to Mars in size. It is the only one of the number whose theory and perturbations have been at all enquired into † further than to verify Kepler's law of the periodic times, which holds good, *mutatis mutandis,* and under the requisite reservations, in this, as in the system of Jupiter. The three next satellites still proceeding inwards ‡ are very minute and require pretty powerful telescopes to see them; while the two interior satellites which just skirt the edge of the ring § can only be seen with telescopes of extraordinary power and perfection, and under the most favourable atmospheric circumstances. At the epoch of their discovery they were seen to thread, like beads, the almost infinitely thin fibre of light to which the ring then seen edgeways was reduced, and for a short time to advance off it at either end, speedily to return, and hastening to their habitual concealment behind the body. ‖

* By Huyghens, March 25. 1655.
† By Bessel, *Astr. Nachr.* Nos. 193. 214.
‡ Discovered by Dominic Cassini in 1672 and 1684.
§ Discovered by Sir William Herschel in 1789.
‖ Considerable confusion prevails in the nomenclature of the Saturnian system, owing to the order of discovery not coinciding with that of distances. Astronomers have not yet agreed whether to call the two interior satellites the

(549.) Owing to the obliquity of the ring and of the orbits of the satellites to Saturn's ecliptic, there are no eclipses, occultations, or transits of these bodies or their shadows across the disc of their primary (the interior ones excepted), until near the time when the ring is seen edgewise, and when they do take place, their observation is attended with too much difficulty to be of any practical use, like the eclipses of Jupiter's satellites, for the determination of longitudes, for which reason they have been hitherto little attended to by astronomers.

(550.) A remarkable relation subsists between the periodic times of the two interior satellites of Saturn, and those of the two next in order of distance; viz. that the period of the third (Tethys) is double that of the first (Mimas), and that of the fourth (Dione) double that of the second (Enceladus). The coincidence is exact in either case to about one 800th part of the larger period.

(551.) The satellites of Uranus require very powerful and perfect telescopes for their observation. Two are, however, much more conspicuous than the rest, and their periods and distances from the planet have been ascertained with tolerable certainty. They are the second and fourth of those set down in the synoptic table. Of the remaining four, whose existence, though announced with considerable confidence by their original discoverer, could hardly be regarded as

6th and 7th (reckoning inward) and the older ones the 1st, 2d, 3d, 4th, and 5th, reckoning outward; or to commence with the innermost and reckon outwards, from 1 to 7. This confusion has been attempted to be obviated by a mythological nomenclature, in consonance with that at length completely established for the primary planets. Taking the names of the Titanian divinities, the following pentameters afford an easy artificial memory, commencing with the most distant.

> Iapetus, Titan; Rhea, Dione, Tethys; (pron. Tĕthys)
> Enceladus, Mimas——————

It is worth remarking that Simon Marius, who disputed the priority of the discovery of Jupiter's satellites with Galileo, proposed for them mythological names, viz: — Io, Europa, Ganymede, and Callisto. The revival of these names would savour of a preference of Marius's claim, which, even if an absolute priority were conceded (which it is not), would still leave Galileo's general claim to the use of the telescope as a means of astronomical discovery intact. But in the case of Jupiter's satellites there exists no confusion to rectify. They are constantly referred to by their numerical designations in every almanack.

fully demonstrated, two only have been hitherto reobserved; viz. the first of our table, interior to the two larger ones, by the independent observations of Mr. Lassell*, and M. Otto Struve†, and the fourth, intermediate between the larger ones, by the former of these astronomers. The remaining two, if future observation should satisfactorily establish their real existence, will probably be found to revolve in orbits exterior to all these.

(552.) The orbits of these satellites offer remarkable, and, indeed, quite unexpected and unexampled peculiarities. Contrary to the unbroken analogy of the whole planetary system —whether of primaries or secondaries—the planes of their orbits *are nearly perpendicular to the ecliptic,* being inclined no less than 78° 58' to that plane, and in these orbits their motions *are retrograde;* that is to say, their positions, when projected on the ecliptic, instead of advancing *from west to east* round the center of their primary, as is the case with every other planet and satellite, move in the opposite direction. Their orbits are nearly or quite circular, and they do not appear to have any sensible, or, at least, any rapid motion of nodes, or to have undergone any material change of inclination, in the course, at least, of half a revolution of their primary round the sun. When the earth is in the plane of their orbits or nearly so, their apparent paths are straight lines or very elongated ellipses, in which case they become invisible, their feeble light being effaced by the superior light of the planet, long before they come up to its disc, so that the observation of any eclipses or occultations they may undergo is quite out of the question with our present telescopes.

(553.) If the observation of the satellites of Uranus be difficult, those of Neptune, owing to the immense distance of that planet, may be readily imagined to offer still greater difficulties. Of the existence of one, discovered by Mr. Lassel‡, there can remain no doubt, it having also been observed by other astronomers, both in Europe and America. Accord-

---

* September 14th to November 9th, 1847.
† October 8th to December 10th, 1847.
‡ On July 8th, 1847.

ing to M. Otto Struve* its orbit is inclined to the ecliptic at the considerable angle of 35°; but whether, as in the case of the satellites of Uranus, the direction of its motion be retrograde, it is not possible to say, until it shall have been longer observed.

* Astron. Nachr. No. 629., from his own observations, September 11th to December 20th, 1847.

CHAPTER XI.

OF COMETS.

(554.) THE extraordinary aspect of comets, their rapid and
seemingly irregular motions, the unexpected manner in which
they often burst upon us, and the imposing magnitudes which
they occasionally assume, have in all ages rendered them
objects of astonishment, not unmixed with superstitious dread
to the uninstructed, and an enigma to those most conversant
with the wonders of creation and the operations of natural
causes.   Even now, that we have ceased to regard their
movements as irregular, or as governed by other laws than
those which retain the planets in their orbits, their intimate
nature, and the offices they perform in the economy of our
system, are as much unknown as ever.   No distinct and
satisfactory account has yet been rendered of those immensely
voluminous appendages which they bear about with them, and
which are known by the name of their tails, (though impro-
perly, since they often precede them in their motions,) any
more than of several other singularities which they present.

(555.) The number of comets which have been astronomically observed, or of which notices have been recorded

in history, is very great, amounting to several hundreds* ;
and when we consider that in the earlier ages of astronomy,
and indeed in more recent times, before the invention of the
telescope, only large and conspicuous ones were noticed; and
that, since due attention has been paid to the subject,
scarcely a year has passed without the observation of one or
two of these bodies, and that sometimes two and even three
have appeared at once ; it will be easily supposed that their
actual number must be at least many thousands. Multitudes,
indeed, must escape all observation, by reason of their paths
traversing only that part of the heavens which is above the
horizon in the daytime. Comets so circumstanced can only
become visible by the rare coincidence of a total eclipse of the
sun, — a coincidence which happened, as related by Seneca,
sixty-two years before Christ, when a large comet was actually observed very near the sun. Several, however, stand
on record as having been bright enough to be seen with the
naked eye in the daytime, even at noon and in bright sunshine. Such were the comets of 1402, 1532, and 1843, and
that of 43 B. C. which appeared during the games celebrated
by Augustus in honour of Venus shortly after the death of
Cæsar, and which the flattery of poets declared to be the
soul of that hero taking its place among the divinities.

(556.) That feelings of awe and astonishment should be
excited by the sudden and unexpected appearance of a great
comet, is no way surprising ; being, in fact, according to the
accounts we have of such events, one of the most imposing
of all natural phænomena. Comets consist for the most part
of a large and more or less splendid, but ill defined nebulous
mass of light, called the head, which is usually much brighter
towards its center, and offers the appearance of a vivid *nucleus*,

* See catalogues in the Almagest of Riccioli; Pingré's Cometographie ;
Delambre's Astron. vol. iii. ; Astronomische Abhandlungen, No. 1. (which
contains the elements of all the orbits of comets which have been computed to
the time of its publication, 1823); also a catalogue, by the Rev. T. J. Hussey.
Lond. & Ed. Phil. Mag. vol. ii. No. 9. *et seq.* In a list cited by Lalande from
the 1st vol. of the Tables de Berlin, 700 comets are enumerated. See also
notices of the Astronomical Socie+y and Astron. Nachr. passim. A great many
of the more ancient comets are recorded in the Chinese Annals, and in some
cases with sufficient precision to allow of the calculation of rudely approximate
orbits from their motions so described.

like a star or planet. From the head, and in a direction
opposite to that in which the sun is situated from the comet
appear to diverge two streams of light, which grow broader
and more diffused at a distance from the head, and which
most commonly close in and unite at a little distance behind it, but sometimes continue distinct for a great part of
their course; producing an effect like that of the trains left
by some bright meteors, or like the diverging fire of a sky-
rocket (only without sparks or perceptible motion). This is
the tail. This magnificent appendage attains occasionally
an immense apparent length. Aristotle relates of the tail
of the comet of 371 B. C., that it occupied a third of the
hemisphere, or 60°; that of A. D. 1618 is stated to have been
attended by a train no less than 104° in length. The
comet of 1680, the most celebrated of modern times, and
on many accounts the most remarkable of all, with a
head not exceeding in brightness a star of the second magnitude, covered with its tail an extent of more than 70° of
the heavens, or, as some accounts state, 90°; that of the
comet of 1769 extended 97°, and that of the last *great* comet
(1843) was estimated at about 65° when longest. The
figure (*fig.* 2., Plate II.) is a representation of the comet of
1819 — by no means one of the most considerable, but which
was, however, very conspicuous to the naked eye.

(557.) The tail is, however, by no means an invariable
appendage of comets. Many of the brightest have been
observed to have short and feeble tails, and a few great
comets have been entirely without them. Those of 1585
and 1763 offered no vestige of a tail; and Cassini describes
the comets of 1665 and 1682 as being as round * and as well
defined as Jupiter. On the other hand, instances are not
wanting of comets furnished with many tails or streams of
diverging light. That of 1744 had no less than six, spread

---

* This description however applies to the " disc " of the head of these comets
as seen in a telescope. Cassini's expressions are, " aussi rond, aussi net, et aussi
clair que Jupiter," (where it is to be observed that the latter epithet must by
no means be translated *bright*). To understand this passage fully, the reader
must refer to the description given further on, of the " disc " of Halley's comet,
after its perihelion passage in 1835-6.

out like an immense fan, extending to a distance of nearly 30° in length. The small comet of 1823 had two, making an angle of about 160°, the brighter turned as usual from the sun, the fainter towards it, or nearly so. The tails of comets, too, are often somewhat curved, bending, in general, towards the region which the comet has left, as if moving somewhat more slowly, or as if resisted in their course.

(558.) The smaller comets, such as are visible only in telescopes, or with difficulty by the naked eye, and which are by far the most numerous, offer very frequently no appearance of a tail, and appear only as round or somewhat oval vaporous masses, more dense towards the center, where, however, they appear to have no distinct nucleus, or any thing which seems entitled to be considered as a solid body. Stars of the smallest magnitudes remain distinctly visible, though covered by what appears to be the densest portion of their substance; although the same stars would be completely obliterated by a moderate fog, extending only a few yards from the surface of the earth. And since it is an observed fact, that even those larger comets which have presented the appearance of a nucleus have yet exhibited *no phases*, though we cannot doubt that they shine by the reflected solar light, it follows that even these can only be regarded as great masses of thin vapour, susceptible of being penetrated through their whole substance by the sunbeams, and reflecting them alike from their interior parts and from their surfaces. Nor will any one regard this explanation as forced, or feel disposed to resort to a phosphorescent quality in the comet itself, to account for the phænomena in question, when we consider (what will be hereafter shown) the enormous magnitude of the space thus illuminated, and the extremely small *mass* which there is ground to attribute to these bodies. It will then be evident that the most unsubstantial clouds which float in the highest regions of our atmosphere, and seem at sunset to be drenched in light, and to glow throughout their whole depth as if in actual ignition, without any shadow or dark side, must be looked upon as dense and massive bodies compared with the filmy and all but spiritual texture of a comet. Accordingly, whenever powerful telescopes have been turned on these bodies, they have not failed to dispel the illusion which attributes *solidity* to that more condensed part of the head, which appears to the naked eye as a nucleus; though it is true that in some, a very minute stellar point *has* been seen, indicating the existence of a solid body.

(559.) It is in all probability to the feeble coercion of the elastic power of their gaseous parts, by the gravitation of so small a central mass, that we must attribute this extraordinary developement of the atmospheres of comets. If the earth, retaining its present size, were reduced, by any internal change (as by hollowing out its central parts) to one thousandth part of its actual mass, its coercive power over the atmosphere would be diminished in the same proportion, and in consequence the latter would expand to a thousand times its actual bulk; and indeed much more, owing to the still farther diminution of gravity, by the recess of the upper parts from the center.* An atmosphere, however, free to expand equally in all directions, would envelope the nucleus spherically, so that it becomes necessary to admit the action of other causes to account for its enormous extension in the direction of the tail, — a subject to which we shall presently take occasion to recur.

(560.) That the luminous part of a comet is something in the nature of a smoke, fog, or cloud, suspended in a transparent atmosphere, is evident from a fact which has been often noticed, viz. — that the portion of the tail where it comes up, and surrounds the head, is yet separate from it by an interval less luminous, as if sustained and kept off from contact by a transparent stratum, as we often see one layer of clouds over another with a considerable clear space between. These, and most of the other facts observed in

---

* Newton has calculated (Princ. III. p. 512.) that a globe of air of ordinary density at the earth's surface, of one inch in diameter, if reduced to the density due to the altitude above the surface of one radius of the earth, would occupy a sphere exceeding in radius the orbit of Saturn. The tail of a great comet then, for aught we can tell, may consist of only a very few pounds or even ounces of matter.

the history of comets, appear to indicate that the structure
of a comet, as seen in section in the direction of its length,
must be that of a hollow envelope, of a parabolic form,
enclosing near its vertex the nucleus and head, something as
represented in the annexed figure. This would account for
the apparent division of the tail into two principal lateral

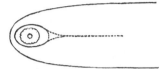

branches, the envelope being oblique to the line of sight at
its borders, and therefore a greater depth of illuminated
matter being there exposed to the eye. In all probability,
however, they admit great varieties of structure, and among
them may very possibly be bodies of widely different physical
constitution, and there is no doubt that one and the same
comet at different epochs undergoes great changes, both in
the disposition of its materials and in their physical state.

(561.) We come now to speak of the motions of comets.
These are apparently most irregular and capricious. Sometimes they remain in sight only for a few days, at others for
many months; some move with extreme slowness, others
with extraordinary velocity; while not unfrequently, the
two extremes of apparent speed are exhibited by the same
comet in different parts of its course. The comet of 1472
described an arc of the heavens of 40° of a great circle* in
a single day. Some pursue a direct, some a retrograde, and
others a tortuous and very irregular course; nor do they
confine themselves, like the planets, within any certain region
of the heavens, but traverse indifferently every part. Their
variations in apparent size, during the time they continue
visible, are no less remarkable than those of their velocity;
sometimes they make their first appearance as faint and slow
moving objects, with little or no tail; but by degrees accelerate, enlarge, and throw out from them this appendage,

---

* 120° in extent in the former editions. But this was the arc described *in
longitude*, and the comet at the time referred to had great north latitude.

which increases in length and brightness till (as always happens in such cases) they approach the sun, and are lost in his beams. After a time they again emerge, on the other side, receding from the sun with a velocity at first rapid, but gradually decaying. It is for the most part after thus passing the sun, that they shine forth in all their splendour, and that their tails acquire their greatest length and developement; thus indicating plainly the action of the sun's rays as the exciting cause of that extraordinary emanation. As they continue to recede from the sun, their motion diminishes and the tail dies away, or is absorbed into the head, which itself grows continually feebler, and is at length altogether lost sight of, in by far the greater number of cases never to be seen more.

(562.) Without the clue furnished by the theory of gravitation, the enigma of these seemingly irregular and capricious movements might have remained for ever unresolved. But Newton, having demonstrated the possibility of any conic section whatever being described about the sun, by a body revolving under the dominion of that law, immediately perceived the applicability of the general proposition to the case of cometary orbits; and the great comet of 1680, one of the most remarkable on record, both for the immense length of its tail and for the excessive closeness of its approach to the sun (within one sixth of the diameter of that luminary), afforded him an excellent opportunity for the trial of his theory. The success of the attempt was complete. He ascertained that this comet described about the sun as its focus an elliptic orbit of so great an excentricity as to be undistinguishable from a parabola, (which is the extreme, or limiting form of the ellipse when the axis becomes infinite,) and that in this orbit the areas described about the sun were, as in the planetary ellipses, proportional to the times. The representation of the apparent motions of this comet by such an orbit, throughout its whole observed course, was found to be as satisfactory as those of the motions of the planets in their nearly circular paths. From that time it became a received truth, that the motions of comets are regulated by the same general laws as those of the planets—the difference of the cases consisting only in the extravagant elongation of their ellipses, and in the absence of any limit to the inclinations of their planes to that of the ecliptic—or any general coincidence in the direction of their motions from west to east, rather than from east to west, like what is observed among the planets.

(563.) It is a problem of pure geometry, from the general laws of elliptic or parabolic motion, to find the situation and dimensions of the ellipse or parabola which shall represent the motion of any given comet. In general, three complete observations of its right ascension and declination, with the times at which they were made, suffice for the solution of this problem, (which is, however, by no means an easy one,) and for the determination of the elements of the orbit. These consist, *mutatis mutandis*, of the same data as are required for the computation of the motion of a planet ; (that is to say, the longitude of the perihelion, that of the ascending node, the inclination to the ecliptic, the semiaxis, excentricity, and time of perihelion passage, as also whether the motion is direct or retrograde ;) and, once determined, it becomes very easy to compare them with the whole observed course of the comet, by a process exactly similar to that of art. 502., and thus at once to ascertain their correctness, and to put to the severest trial the truth of those general laws on which all such calculations are founded.

(564.) For the most part, it is found that the motions of comets may be sufficiently well represented by parabolic orbits,—that is to say, ellipses whose axes are of infinite length, or, at least, so very long that no appretiable error in the calculation of their motions, during all the time they continue visible, would be incurred by supposing them actually infinite. The parabola is that conic section which is the limit between the ellipse on the one hand, which returns into itself, and the hyperbola on the other, which runs out to infinity. A comet, therefore, which should describe an elliptic path, however long its axis, must *have* visited the sun before, and must again return (unless disturbed) in some determinate period, — but should its orbit be of the hyperbolic character, when once it had passed its perihelion, it could never more return within the sphere of our observation, but must run off to visit other systems, or be lost in the immensity of space. A very few comets have been ascertained to move in hyperbolas *, but many more in ellipses. These latter, in so far as their orbits can remain unaltered by the attractions of the planets, must be regarded as permanent members of our system.

(565.) We must now say a few words on the actual dimensions of comets. The calculation of the diameters of their heads, and the lengths and breadths of their tails, offers not the slightest difficulty when once the elements of their orbits are known, for by these we know their real distances from the earth at any time, and the true direction of the tail, which we see only foreshortened. Now calculations instituted on these principles lead to the surprising fact, that the comets are by far the most voluminous bodies in our system. The following are the dimensions of some of those which have been made the subjects of such enquiry.

(566.) The tail of the great comet of 1680, immediately after its perihelion passage, was found by Newton to have been no less than 20000000 of leagues in length, and to have occupied only two days in its emission from the comet's body! a decisive proof this of its being darted forth by some active force, the origin of which, to judge from the direction of the tail, must be sought in the sun itself. Its greatest length amounted to 41000000 leagues, a length much exceeding the whole interval between the sun and earth. The tail of the comet of 1769 extended 16000000 leagues, and that of the great comet of 1811, 36000000. The portion of the head of this last, comprised within the transparent atmospheric envelope which separated it from the tail, was 180000 leagues in diameter. It is hardly conceivable, that matter once projected to such enormous distances should ever be collected

---

* For example, that of 1723, calculated by Burckhardt; that of 1771, by both Burckhardt and Encke; and the second comet of 1818, by Rosenberg and Schwabe.

again by the feeble attraction of such a body as a comet — a consideration which accounts for the surmised progressive diminution of the tails of such as have been frequently observed.

(567.) The most remarkable of those comets which have been ascertained to move in elliptic orbits is that of Halley, so called from the celebrated Edmund Halley, who, on calculating its elements from its perihelion passage in 1682, when it appeared in great splendour, with a tail 30° in length, was led to conclude its identity with the great comets of 1531 and 1607, whose elements he had also ascertained. The intervals of these successive apparitions being 75 and 76 years, Halley was encouraged to *predict* its reappearance about the year 1759. So remarkable a prediction could not fail to attract the attention of all astronomers, and, as the time approached, it became extremely interesting to know whether the attractions of the larger planets might not materially interfere with its orbital motion. The computation of their influence from the Newtonian law of gravity, a most difficult and intricate piece of calculation, was undertaken and accomplished by Clairaut, who found that the action of Saturn would retard its return by 100 days, and that of Jupiter by no less than 518, making in all 618 days, by which the expected return would happen later than on the supposition of its retaining an unaltered period, — and that, in short, the time of the expected perihelion passage would take place within a month, one way or other, of the middle of April, 1759. — It actually happened on the 12th of March in that year. Its next return was calculated by several eminent geometers *, and fixed successively for the 4th, the 7th, the 11th, and the 26th of November, 1835; the two latter determinations appearing entitled to the higher degree of confidence, owing partly to the more complete discussion bestowed on the observations of 1682 and 1759, and partly to the continually improving state of our knowledge of the methods of estimating the disturbing effect

* Damoiseau, Pontecoulant, Rosenberger, and Lehmann.

of the several planets.   The last of these predictions, that of M. Lehmann, was published on the 25th of July.   On the 5th of August the comet first became visible in the clear atmosphere of Rome as an exceedingly faint telescopic nebula, within a degree of its place as predicted by M. Rosenberger for that day.   On or about the 20th of August it became generally visible, and, pursuing very nearly its calculated path among the stars, passed its perihelion on the 16th of November; after which, its course carrying it south, it ceased to be visible in Europe, though it continued to be conspicuously so in the southern hemisphere throughout February, March, and April, 1836, disappearing finally on the 5th of May.

(568.) Although the appearance of this celebrated comet at its last apparition was not such as might be reasonably considered likely to excite lively sensations of terror, even in superstitious ages, yet, having been an object of the most diligent attention in all parts of the world to astronomers, furnished with telescopes very far surpassing in power those which had been applied to it at its former appearance in 1759, and indeed to any of the greater comets on record, the opportunity thus afforded of studying its physical structure, and the extraordinary phænomena which it presented when so examined have rendered this a memorable epoch in cometic history.   Its first appearance, while yet very remote from the sun, was that of a small round or somewhat oval nebula, quite destitute of tail, and having a minute point of more concentrated light excentrically situated within it.   It was not before the 2d of October that the tail began to be developed, and thenceforward increased pretty rapidly, being already 4° or 5° long on the 5th.   It attained its greatest apparent length (about 20°) on the 15th of October.   From that time, though not yet arrived at its perihelion, it decreased with such rapidity, that already on the 29th it was only 3°, and on November the 5th $2\frac{1}{2}$° in length.   There is every reason to believe that before the perihelion, the tail had altogether disappeared, as, though it continued to be observed at Pulkowa up to the very day of its perihelion passage, no mention whatever is made of any tail being then seen.

(569.) By far the most striking phænomena, however, observed in this part of its career, were those which, commencing simultaneously with the growth of the tail, connected themselves evidently with the production of that appendage and its projection from the head. On the 2d of October (the very day of the first observed commencement of the tail) the nucleus, which had been faint and small, was observed suddenly to have become much brighter, and to be in the act of throwing out a jet or stream of light from its anterior part, or that turned *towards* the sun. This ejection after ceasing awhile was resumed, and with much greater apparent violence, on the 8th, and continued, with occasional intermittences, so long as the tail itself continued visible. Both the form of this luminous ejection, and the direction in which it issued from the nucleus, meanwhile underwent singular and capricious alterations, the different phases succeeding each other with such rapidity that on no two successive nights were the appearances alike. At one time the emitted jet was single, and confined within narrow limits of divergence from the nucleus. At others it presented a fan-shaped or swallow-tailed form, analogous to that of a gas-flame issuing from a flattened orifice : while at others again two, three, or even more jets were darted forth in different directions.* (See figures *a*, *b*, *c*, *d*, plate I. fig 4., which represent, highly magnified, the appearances of the nucleus with its jets of light, on the 8th, 9th, 10th, and 12th of October, and in which the direction of the anterior portion of the head, or that fronting the sun, is supposed alike in all, viz. towards the upper part of the engraving. In these representations the head itself is omitted, the scale of the figures not permitting its introduction: *e* represents the nucleus and head as seen October 9th on a less scale.) The direction of the principal jet was observed meanwhile to oscillate to and fro on either side of

* See the exquisite lithographic representations of these phenomena by Bessel. Astron. Nachr. No. 302., and the fine series by Schwabe in No. 297. of that collection, as also the magnificent drawings of Struve, from which our figures *a*, *b*, *c*, *d*, are copied.

a line directed to the sun in the manner of a compass-needle when thrown into vibration and oscillating about a mean position, the change of direction being conspicuous even from hour to hour. These jets, though very bright at their point of emanation from the nucleus, faded rapidly away, and became diffused as they expanded into the coma, at the same time curving backwards as streams of steam or smoke would do, if thrown out from narrow orifices, more or less obliquely in opposition to a powerful wind, against which they were unable to make way, and, ultimately yielding to its force, so as to be drifted back and confounded in a vaporous train, following the general direction of the current.*

(570.) Reflecting on these phænomena, and carefully considering the evidence afforded by the numerous and elaborately executed drawings which have been placed on record by observers, it seems impossible to avoid the following conclusions. 1st. That the matter of the nucleus of a comet is powerfully excited and dilated into a vaporous state by the action of the sun's rays, escaping in streams and jets at those points of its surface which oppose the least resistance, and in all probability throwing that surface or the nucleus itself into irregular motions by its reaction in the act of so escaping, and thus altering its direction.

2dly. That this process chiefly takes place in that portion of the nucleus which is turned towards the sun; the vapour escaping chiefly in that direction.

3dly. That when so emitted, it is prevented from proceeding in the direction originally impressed upon it, by some force directed *from* the sun, drifting it back and carrying it out to vast distances behind the nucleus, forming the tail or so much of the tail as can be considered as consisting of material substance.

4thly. That this force, whatever its nature, acts unequally on the materials of the comet, the greater portion remaining unvaporized, and a considerable part of the vapour actually

---

* On this point Schwabe's and Bessel's drawings are very express and unequivocal. Struve's attention seems to have been more especially directed to the scrutiny of the nucleus.

produced, remaining in its neighbourhood, forming the head and coma.

5thly. That the force thus acting on the materials of the tail cannot possibly be identical with the ordinary gravitation of matter, being centrifugal or repulsive, as respects the sun, and of an energy very far exceeding the gravitating force towards that luminary. This will be evident if we consider the enormous velocity with which the matter of the tail is carried backwards, in opposition both to the motion which it had as part of the nucleus, and to that which it acquired in the act of its emission, both which motions have to be destroyed in the first instance, before any movement in the contrary direction can be impressed.

6thly. That unless the matter of the tail thus repelled from the sun be retained by a peculiar and highly energetic attraction to the nucleus, differing from and exceptional to the ordinary power of gravitation, it must leave the nucleus altogether; being in effect carried far beyond the coercive power of so feeble a gravitating force as would correspond to the minute mass of the nucleus; and it is therefore very conceivable that a comet may lose, at every approach to the sun, a portion of that peculiar matter, whatever it be, on which the production of its tail depends, the remainder being of course less excitable by the solar action, and more impassive to his rays, and therefore, *pro tanto*, more nearly approximating to the nature of the planetary bodies.

(571.) After the perihelion passage, the comet was lost sight of for upwards of two months, and at its reappearance (on the 24th of January 1836) presented itself under quite a different aspect, having in the interval evidently undergone some great physical change which had operated an entire transformation in its appearance. It no longer presented any vestige of tail, but appeared to the naked eye as a hazy star of about the fourth or fifth magnitude, and in powerful telescopes as a small, round, well defined disc, rather more than 2′ in diameter, surrounded with a nebulous chevelure or coma of much greater extent. Within the disc, and somewhat excentrically situated, a minute but bright nucleus appeared,

A A

from which extended towards the posterior edge of the disc (or that remote from the sun) a short vivid luminous ray. (See fig. 4. of pl. I.) As the comet receded from the sun, the coma speedily disappeared, as if absorbed into the disc, which, on the other hand, increased continually in dimensions, and that with such rapidity, that in the week elapsed from January 25th to Febuary 1st, (calculating from micrometrical measures, and from the known distance of the comet from the earth on those days) the actual volume or *real solid content* of the illuminated space had dilated in the ratio of upwards of 40 to 1. And so it continued to swell out with undiminished rapidity, until from this cause alone it ceased to be visible, the illumination becoming fainter as the magnitude increased; till at length the outline became undistinguishable from simple want of light to trace it. While this increase of dimension proceeded, the form of the disc passed, by gradual and successive additions to its length in the direction opposite to the sun, to that of a paraboloid, as represented in *g* fig. 4. plate I., the anterior curved portion preserving its planetary sharpness, but the base being faint and ill-defined. It is evident that had this process continued with sufficient light to render the result visible, a tail would have been ultimately reproduced; but the increase of dimension being accompanied with diminution of brightness, a short, imperfect, and as it were rudimentary tail only was formed, visible as such for a few nights to the naked eye, or in a low magnifying telescope, and that only when the comet itself had begun to fade away by reason of its increasing distance.

(572.) While the parabolic envelope was thus continually dilating and growing fainter, the nucleus underwent little change, but the ray proceeding from it increased in length and comparative brightness, preserving all the time its direction along the axis of the paraboloid, and offering none of those irregular and capricious phænomena which characterized the jets of light emitted anteriorly, previous to the perihelion. If the office of those jets was to feed the tail, the converse office of conducting back its successively condensing matter to the nucleus would seem to be that of the ray now in question. By degrees this also faded, and the last appearance presented by the comet was that which it offered at its first appearance in August; viz. that of a small round nebula with a bright point in or near the center.

(573.) Besides the comet of Halley, several other of the great comets recorded in history have been surmised with more or less probability to return periodically, and therefore to move in elongated ellipses around the sun. Such is the great comet of 1680, whose period is estimated at 575 years, and which is considered, with the highest appearance of probability, to be identical with a magnificent comet observed at Constantinople and in Palestine, and referred by contemporary historians, both European and Chinese, to the year A. D. 1105; with that of A. D. 575, which was seen at noonday close to the sun; with the comet of 43 B. C., already spoken of as having appeared after the death of Cæsar, and which was also observed in the day-time; and finally with two other comets, mention of which occurs in the Sibylline Oracles, and in a passage of Homer, and which are referred, as well as the obscurity of chronology and the indications themselves will allow, to the years 618 and 1194 B. C. It is to the assumed near approach of this comet to the earth about the time of the Deluge, that Whiston ascribed that overwhelming tide wave to whose agency his wild fancy ascribed that great catastrophe — a speculation, it is needless to remark, purely visionary.

(574.) Another great comet, whose return in the year actually current (1848) has been considered by more than one eminent authority in this department of astronomy * highly probable, is that of 1556, to the terror of whose aspect some historians have attributed the abdication of the Emperor Charles V. This comet is supposed to be identical with that of 1264, mentioned by many historians as a great comet, and observed also in China, — the conclusion in this case resting upon the coincidence of elements calculated on the observations, such as they are, which have been recorded. On the subject of this coincidence Mr. Hind has recently entered

* Pingré. Cometographie, i. 411. Lalande. Astr. 3185.

into many elaborate calculations, the result of which is strongly in favour of the supposed identity. This probability is farther increased by the fact of a comet with a tail of 40° and a head bright enough to be visible after sunrise having appeared in A. D. 975; and of two others having been recorded by the Chinese annalists in A. D. 395 and 104. It is true that if these be the same, the mean period would be somewhat short of 292 years. But the effect of planetary perturbation might reconcile even greater differences, and though up to the time of our writing no such comet has yet been observed, at least another year must elapse before its return can be pronounced hopeless.

(575.) In 1661, 1532, 1402, 1145, 891, and 243 great comets appeared — that of 1402 being bright enough to be seen at noon day. A period of 129 years would conciliate all these appearances, and should have brought back the comet in 1789 or 1790 (other circumstances agreeing). That no such comet was observed about that time is no proof that it did not return, since, owing to the situation of its orbit, had the perihelion passage taken place in July it might have escaped observation. Mechain, indeed, from an elaborate discussion of the observations of 1532 and 1661, came to the conclusion that these comets were not the same; but the elements assigned by Olbers to the earlier of them, differ so widely from those of Mechain for the same comet on the one hand, and agree so well with those of the last named astronomer for the other[*], that we are perhaps justified in regarding the question as not yet set at rest.

(576.) We come now, however, to a class of comets of short period, respecting whose return there is no doubt, inasmuch as two at least of them have been indentified as having performed successive revolutions round the sun; have had their return predicted already several times; and have on each occasion scrupulously kept to their appointments. The first of these is the comet of Encke, so called from Professor Encke of Berlin, who first ascertained its periodical

---

[*] See Schumacher's Catal. Astron. Abhandl. i.

return. It revolves in an ellipse of great excentricity (though not comparable to that of Halley's), the plane of which is inclined at an angle of about 13° 22' to the plane of the ecliptic, and in the short period of 1211 days, or about $3\frac{1}{3}$ years. This remarkable discovery was made on the occasion of its fourth recorded appearance, in 1819. From the ellipse then calculated by Encke, its return in 1822 was predicted by him, and observed at Paramatta, in New South Wales, by M. Rümker, being invisible in Europe: since which it has been re-predicted and re-observed in all the principal observatories, both in the northern and southern hemispheres, as a phenomenon of regular occurrence.

(577.) On comparing the intervals between the successive perihelion passages of this comet, after allowing in the most careful and exact manner for all the disturbances due to the actions of the planets, a very singular fact has come to light, viz. that the periods are continually diminishing, or, in other words, the mean distance from the sun, or the major axis of the ellipse, dwindling by slow and regular degrees at the rate of about $0^{d\cdot}11$ per revolution. This is evidently the effect which would be produced by a resistance experienced by the comet from a very rare ethereal medium pervading the regions in which it moves; for such resistance, by diminishing its actual velocity, would diminish also its contrifugal force, and thus give the sun more power over it to draw it nearer. Accordingly this is the solution proposed by Encke, and at present generally received. It will, therefore, probably fall ultimately into the sun, should it not first be dissipated altogether, — a thing no way improbable, when the lightness of its materials is considered.

(578.) By measuring the apparent magnitude of this comet at different distances from the sun, and thence, from a knowledge of its actual distance from the earth at the time, concluding its real volume, it has been ascertained to contract in bulk as it approaches to, and to expand as it recedes from, that luminary. M. Valz, who was the first to notice this fact, accounts for it by supposing it to undergo a real compression or condensation of volume arising from the pressure of an æthereal medium which he conceives to grow more dense in the sun's neighbourhood. But such an hypothesis is evidently inadmissible, since it would require us to assume the exterior of the comet to be in the nature of a skin or bag impervious to the compressing medium. The phenomenon is analogous to the increase of dimension above described as observed in the comet of Halley when in the act of receding from the sun, and is doubtless referable to a similar cause, viz. the alternate conversion of evaporable matter into the states of visible cloud and invisible gas by the alternating action of cold and heat. This comet has no tail, but offers to the view only a small ill-defined nucleus, excentrically situated within a more or less elongated oval mass of vapours, being nearest to that vertex which is towards the sun.

(579.) Another comet of short period is that of *Biela*, so called from M. Biela, of Josephstadt, who first arrived at this interesting conclusion on the occasion of its appearance in 1826. It is considered to be identical with comets which appeared in 1772, 1805, &c., and describes its very excentric ellipse about the sun in 2410 days or about $6\frac{3}{4}$ years; and in a plane inclined 12° 34' to the ecliptic. It appeared again according to the prediction in 1832, and in 1846. Its orbit, by a remarkable coincidence, very nearly intersects that of the earth; and had the latter, at the time of its passage in 1832, been a month in advance of its actual place, it would have passed through the comet, — a singular rencontre, perhaps not unattended with danger. [*]

---

[*] Should calculation establish the fact of a resistance experienced also by this comet, the subject of periodical comets will assume an extraordinary degree of interest. It cannot be doubted that many more will be discovered, and by their resistance questions will come to be decided, such as the following :—What is the law of density of the resisting medium which surrounds the sun? Is it at rest or in motion? If the latter, in what direction does it move? Circularly round the sun, or traversing space? If circularly, in what plane? It is obvious that a circular or vorticose motion of the ether would *accelerate some comets and retard others*, according as their revolution was, relative to such motion, direct or retrograde. Supposing the neighbourhood of the sun to be filled with a material fluid, it is not conceivable that the circulation of the planets in it for ages should not have impressed upon it some degree of rotation in their own direction. And this may preserve them from the extreme effects of accumulated resistance. — *Author.*

(580.) This comet is small and hardly visible to the naked eye, even when brightest. Nevertheless, as if to make up for its seeming insignificance by the interest attaching to it in a physical point of view, it exhibited at its last appearance in 1846, a phænomenon which struck every astronomer with amazement, as a thing without previous example in the history of our system.* It was actually seen to separate itself into two distinct comets, which, after thus parting company, continued to journey along amicably through an arc of upwards of 70° of their apparent orbit, keeping all the while within the same field of view of the telescope pointed towards them. The first indication of something unusual being about to take place, might be, perhaps, referred to the 19th of December 1845, when the comet appeared pear-shaped, the nebulosity being unduly elongated in the north following direction.† But on the 13th of January, at Washington in America, and on the 15th and subsequently in every part of Europe, it was distinctly seen to have become double; a very small and faint cometic body, having a nucleus of its own, being observed appended to it, at a distance of about 2' (in arc) from its center, and in a direction forming an angle of about 328° with the meridian, running northwards from the principal or original comet (see art. 204). From this time the separation of the two comets went on progressively, though slowly. On the 30th of January, the apparent distance of the nucleus had increased to 3', on the 7th of February to 4', and on the 13th to 5', and so on, until on the 5th of March the two comets were separated by an interval of 9' 19", the

---

* Perhaps not quite so. To say nothing of a singular surmise of Kepler, that two great comets *seen at once* in 1618, might be a single comet separated into two, the following passage of Hevelius cited by M. Littrow (Nachr. 564.) does really seem to refer to some phænomenon bearing at least a certain analogy to it. " In ipso disco," he says (Cometographia, p. 326.) *quatuor vel quinque* corpuscula quædam sive nucleos reliquo corpore aliquanto densiores ostendebat.

† According to Mr. Hind's observation. But there can be little doubt that by a mistake of the most common occurrence, when no measure of the position is taken, north following is an error of entry or printing for north preceding (n f for n p). In fact, an elongation from north following to south preceding would agree with the regular direction of the tail and would occasion no remark.

A A 4

apparent direction of the line of junction all the while varying but little with respect to the parallel.*

(581.) During this separation, very remarkable changes were observed to be going on both in the original comet and its companion. Both had nuclei, both had short tails, parallel in direction, and nearly perpendicular to the line of junction, but whereas at its first observation on January 13th, the new comet was extremely small and faint in comparison with the old, the difference both in point of light and apparent magnitude diminished. On the 10th of February, they were nearly equal, although the day before the moonlight had effaced the new one, leaving the other bright enough to be well observed. On the 14th and 16th, however, the new comet had gained a decided superiority of light over the old, presenting at the same time a sharp and starlike nucleus compared by Lieut. Maury to a diamond spark. But this state of things was not to continue. Already, on the 18th, the old comet had regained its superiority, being nearly twice as bright as its companion, and offering an unusually bright and starlike nucleus. From this period the new companion began to fade away, but continued visible up to the 15th of March. On the 24th the comet was again single, and on the 22d of April both had disappeared.

(582.) While this singular interchange of light was going forwards, indications of some sort of communication between the comets were exhibited. The new or companion comet, besides its tail, extending in a direction parallel to that of the other, threw out a faint arc of light which extended as a kind of bridge from the one to the other; and after the restoration of the original comet to its former preeminence, it, on its part, threw forth additional rays, so as to present (on the 22d and 23d February) the appearance of a comet with three faint tails forming angles of about 120° with each other, one of which extended towards its companion.†

---

* By far the greater portion of this increase of apparent distance was due to the comet's increased proximity to the earth. The real increase reduced to a distance = 1 of the comet was at the rate of about 3″ per diem.

† These last mentioned particulars, rest on the testimony of Lieutenant Maury of Washington, who had the advantage of using a nine-inch object glass

(583.) Professor Plantamour, director of the observatory of Geneva, having investigated the orbits of both these comets as separate and independent bodies, from the extensive and careful series of observations made upon them, has arrived at the conclusion that the increase of distance between the two nuclei, *at least during the interval from February* 10*th to March* 22*d,* was simply apparent, being due to the variation of distance from the earth, and to the angle under which their line of junction presented itself to the visual ray; the real distance during all that interval (neglecting small fractions) having been on an average about thirty-nine times the semi-diameter of the earth, or less than two-thirds the distance of the moon from its center. From this it would appear, that already, at this distance, the two bodies had ceased to exercise any perceptible amount of perturbative gravitation on each other; as, indeed, from the probable minuteness of cometary masses we might reasonably expect. Calculating upon the elements assigned by him[*], we find $16^{d\cdot}4$ for the interval of their next perihelion passages. And it will be, therefore, necessary at their next reappearance, to look out for each comet as a separate and independent body, computing its place from these elements as if the other had no existence. Nevertheless, as it is still perfectly possible that some link of connection may subsist between them, (if indeed, by some unknown process the companion has not been actually reabsorbed,) it will not be advisable to rely on this calculation to the neglect of a most vigilant search throughout the whole neighbourhood of the more conspicuous

---

of Munich manufacture. It does not appear that any *large* telescope was turned upon it in Europe on the dates in question.

[*]		Original Comet.		Companion.	
Perihelion passage, 1846, Feb.		11·00476 -		11·07111 Geneva m. t.	
Semiaxis major	- -	- 0·5471002 -		0·5451271	
Perihelion distance	-	- 9·9327011 -		9·9326965	
Angle of excentricity or					
whose sine = e	-	- 49° 12′ 2″·5 -		49° 6′ 14″·4	
Inclination	- -	- 12 34 53 ·3 -		12 34 14 ·3	
Node Ω -	- -	- 245 54 38 ·8 -		245 56 1 ·7	
Perihelion	- -	- 109 2 20 ·1 -		109 2 39 ·6	

Mean equinox of 1846, ·0.

one, lest the opportunity should be lost of pursuing to its conclusion the history of this strange occurrence.

(584.) A third comet of short period has still more recently been added to our list by M. Faye, of the observatory of Paris, who detected it on the 22d of November 1843. A very few observations sufficed to show that no parabola would satisfy the conditions of its motion, and that to represent them completely, it was necessary to assign to it an elliptic orbit of very moderate excentricity. The calculations of M. Nicolai, subsequently revised and slightly corrected by M. Leverrier, have shown that an almost perfect representation of its motions during the whole period of its visibility would be afforded by assuming it to revolve in a period of $2717^{d}\cdot68$ (or somewhat less than $7\frac{1}{2}$ years) in an ellipse whose excentricity is 0·55596, and inclination to the ecliptic 11° 22′ 31″; and taking this for a basis of further calculation, and by means of these data and the other elements of the orbit estimating the effect of planetary perturbation during the revolution now in progress, he has fixed its next return to the perihelion for the 3d of April 1851, with a probable error one way or other not exceeding one or two days.

(585.) The effect of planetary perturbation on the motion of comets has been more than once alluded to in what has been above said. Without going minutely into this part of the subject, which will be better understood after the perusal of a subsequent chapter, it must be obvious, that as the orbits of comets are very excentric, and inclined in all sorts of angles to the ecliptic, they must in many instances, if not actually intersect, at least pass very near to the orbits of some of the planets. We have already seen, for instance, that the orbit of Biela's comet so nearly intersects that of the earth, that an actual collision is not impossible, and indeed (supposing neither orbit variable) must in all likelihood happen in the lapse of some millions of years. Neither are instances wanting of comets having actually approached the earth within comparatively short distances, as that of 1770, which on the 1st of July of that year was within little more than seven times the moon's distance. The same comet in 1767 passed Jupiter at a distance only one 58th of the radius of that planet's orbit, and it has been rendered extremely probable that it is to the disturbance its former orbit underwent during that appulse that we owe its appearance within our own range of vision. This exceedingly remarkable comet was found by Lexell to describe an elliptic orbit with an excentricity of 0·7858, with a periodic time of about five years and a half, and in a plane only 1° 34′ inclined to the ecliptic, having passed its perihelion on the 13th of August 1770. Its return of course was eagerly expected, but in vain, for the comet has never been seen since. Its observation on its first return in 1776 was rendered impossible by the relative situations of the perihelion and of the earth at the time, and before another revolution could be accomplished (as has since been ascertained), viz: about the 23d of August 1779, by a singular coincidence it again approached Jupiter within one 491st part of its distance from the sun, being nearer to that planet by one-fifth than its fourth satellite. No wonder, therefore, that the planet's attraction (which at that distance would exceed that of the sun in the proportion of at least 200 to 1) should completely alter the orbit and deflect it into a curve, not one of whose elements would have the least resemblance to those of the ellipse of Lexell. It is worthy of notice that by this rencontre with the system of Jupiter's satellites, none of *their* motions suffered any perceptible derangement, — a sufficient proof of the smallness of its mass. Jupiter indeed, seems, by some strange fatality, to be constantly in the way of comets, and to serve as a perpetual stumbling-block to them.

(586.) On the 22nd of August, 1844, Signor De Vico, director of the observatory of the Collegio Romano, discovered a comet, the motions of which, a very few observations sufficed to shew, deviated remarkably from a parabolic orbit. It passed its perihelion on the 2nd of September, and continued to be observed until the 7th of December. Elliptic elements of this comet, agreeing remarkably well with each other, were accordingly calculated by several astronomers; from which it appears that the period of revolution is about 1990 days, or $5\frac{1}{2}$ (5·4357) years, which (supposing its orbit

undisturbed in the interim) would bring it back to the perihelion on or about the 13th of January, 1850. As the
assemblage and comparison of these elements thus computed
independently, will serve better, perhaps, than any other
example, to afford the student an idea of the degree of arithmetical certainty capable of being attained in this branch of
astronomy, difficult and complex as the calculations themselves are, and liable to error as *individual* observations of a
body so ill-defined as the smaller comets are for the most
part; we shall present them in a tabular form, as on the next
page: the elements being as usual; the time of perihelion
passage, longitude of the perihelion, that of the ascending
node, the inclination to the ecliptic, semiaxis and excentricity
of the orbit, and the periodic time.

This comet, when brightest, was visible to the naked eye,
and had a small tail. It is especially interesting to astronomers from the circumstance of its having been rendered exceedingly probable by the researches of M. Leverrier, that it
is identical with one which appeared in 1678 with some of its
elements considerably changed by perturbation. This comet
is further remarkable, from having been concluded by
Messrs. Laugier and Mauvais, to be identical with the comet
of 1585 observed by Tycho Brahe, and possibly also with
those of 1743, 1766, and 1819.

(587.) Elliptic elements have in like manner been assigned
to the comet discovered by M. Brorsen, on the 26th of
February, 1846, which, like that last mentioned, speedily
after its discovery began to show evident symptoms of
deviation from a parabola. These elements, with the names
of their respective calculators, are as follow. The dates are
for February 1847, Greenwich time.

Computed by		Brunnow.	Hind.	Van Willingen and De Haan.
Perihelion passage	-	25^d· 37794	25^d· 33109	25^d· 02227
Long. of Perihelion	-	116° 28′ 34″·0	116° 28′ 17″·8	116° 23′ 52″·9
Long. of Ω	- -	102 39 36· 5	102 45 20· 9	103 31 25· 7
Inclination	- -	30 55 6· 5	30 49 3· 6	30 30 30· 2
Semiaxis -	- -	3·15021	3·12292	2·87052
Excentricity	- -	0·79363	0·79771	0·77313
Period (days)	- -	2042	2016	1776

ELEMENTS OF THE PERIODICAL COMET OF DE VICO.

Computed by	Nicolai.	Hind.	Goldschmidt.	Faye.	Schubert.	Brunnow.
Perihelion passage	1844, Sep. 2ᵈ·47594	1844, Sep. 2ᵈ·50412	1844, Sep. 2ᵈ·45430	1844, Sep. 2ᵈ·48745	1844, Sep. 2ᵈ·56348	1844, Sep. 2ᵈ·47402
Longitude of perihelion	342° 31′ 5″·5	342° 32′ 40″·1	342° 29′ 44″·9	342° 31′ 15″·5	342° 34′ 31″·5	342° 30′ 49″·6
Longitude of Ω	63 48 48·9	63 52 24·1	63 48 55·2	63 49 30·6	63 54 40·8	63 49 0·1
Inclination	2 54 45·8	2 54 27·1	2 55 1·9	2 54 45·0	2 52 51·8	2 54 50·3
Semiaxis	3·09853	3·08582	3·11111	3·09946	3·02612	3·10295
Excentricity	0·61716	0·61566	0·61861	0·61726	0·60866	0·61788
Period (days)	1992	1980	2400	1993	1923	1996

This comet is faint, and presents nothing remarkable in its appearance. Its chief interest arises from the great similiarity of its *parabolic* elements to those of the comet of 1532, the place of the perihelion and node, and the inclination of the orbit, being almost identical.

(588.) Elliptic elements have also been calculated by M. D'Arrest, for a comet discovered by M. Peters, on the 26th of June 1846, which go to assign it a place among the comets of short period, viz. $5804^{d}\cdot3$, or very nearly 16 years. The excentricity of the orbit is $0\cdot75672$, its semiaxis $6\cdot32066$, and the inclination of its plane to that of the ecliptic $31°\ 2'\ 14''$. This comet passed its perihelion on the 1st of June 1846.

(589.) By far the most remarkable comet, however, which has been seen during the present century, is that which appeared in the spring of 1843, and whose tail became visible in the twilight of the 17th of March in England as a great beam of nebulous light, extending from a point above the western horizon, through the stars of Eridanus and Lepus, under the belt of Orion. This situation was low and unfavourable; and it was not till the 19th that the head was seen, and then only as a faint and ill-defined nebula, very rapidly fading on subsequent nights. In more southern latitudes, however, not only the tail was seen, as a magnificent train of light extending 50° or 60° in length; but the head and nucleus appeared with extraordinary splendour, exciting in every country where it was seen the greatest astonishment and admiration. Indeed, all descriptions agree in representing it as a stupendous spectacle, such as in superstitious ages would not fail to have carried terror into every bosom. In tropical latitudes in the northern hemisphere, the tail appeared on the 3d of March, and in Van Diemen's Land, so early as the 1st, the comet having passed its perihelion on the 27th of February. Already on the 3d the head was so far disengaged from the immediate vicinity of the sun, as to appear for a short time above the horizon after sunset. On this day when viewed through a 46-inch achromatic telescope it presented a planetary disc, from which rays emerged in the direction of the tail. The tail was double, consisting of two principal lateral streamers, making a very small angle with each other, and divided by a comparatively dark line, of the estimated length of 25°, prolonged however on the north side by a divergent streamer, making an angle of 5° or 6° with the general direction of the axis, and traceable as far as 65° from the head. A similar though fainter lateral prolongation appeared on the south side. A fine drawing of it of this date by C. P. Smyth, Esq. of the Royal Observatory, C. G. H., represents it as highly symmetrical, and gives the idea of a vivid cone of light, with a dark axis, and nearly rectilinear sides, inclosed in a fainter cone, the sides of which curve slightly outwards. The light of the nucleus at this period is compared to that of a star of the first or second magnitude; and on the 11th, of the third; from which time it degraded in light so rapidly, that on the 19th it was invisible to the naked eye, the tail all the while continuing brilliantly visible, though much more so at a distance from the nucleus, with which, indeed, its connexion was not then obvious to the unassisted sight — a singular feature in the history of this body. The tail, subsequent to the 3d, was generally speaking a single straight or slightly curved broad band of light, but on the 11th it is recorded by Mr. Clerihew, who observed it at Calcutta, to have shot forth a lateral tail nearly twice as long as the regular one but fainter, and making an angle of about 18° with its direction on the southern side. The projection of this ray (which was not seen either before or after the day in question) to so enormous a length, (nearly 100°) in a single day conveys an impression of the intensity of the forces acting to produce such a velocity of material transfer through space, such as no other natural phænomenon is capable of exciting. It is clear that *if we have to deal here with matter, such as we conceive it,* viz. *possessing inertia — at all,* it must be under the dominion of forces incomparably more energetic than gravitation.

(590.) There is abundant evidence of the comet in question having been seen in full daylight, and in the sun's immediate vicinity. It was so seen on the 28th of February, the day after its perihelion passage, by every person on board the H. E. I. C. S. Owen Glendower, then off the Cape, as a short dagger-like object close to the sun a little before sunset. On the same day at 3ʰ 6ᵐ P. M., and consequently in full sunshine, the distance of the nucleus from the sun was actually measured with a sextant by Mr. Clarke of Portland, United States, the distance center from center being then only 3° 50′ 43″. He describes it in the following terms : " The nucleus and also every part of the tail were as well defined as the moon on a clear day. The nucleus and tail bore the same appearance, and resembled a perfectly pure white cloud without any variation, except a slight change near the head, just sufficient to distinguish the nucleus from the tail at that point." The denseness of the nucleus was so considerable, that Mr. Clarke had no doubt it might have been visible upon the sun's disc, had it passed between that and the observer. The length of the visible tail resulting from these measures was 59′ or not far from double the apparent diameter of the sun ; and as we shall presently see that on the day in question the distance from the earth of the sun and comet must have been very nearly equal, this gives us about 1700000 miles for the linear dimensions of this the densest portion of that appendage, making no allowance for the foreshortening, which at that time was very considerable.

(591.) The elements of this comet are among the most remarkable of any recorded. They have been calculated by several eminent astronomers, among whose results we shall specify only those which agree best ; the earlier attempts to compute its path having been rendered uncertain by the difficulty attending exact observations of it in the first part of its visible career. The following are those which seem entitled to most confidence : —

	Encke.	Plantamour.	Knorre.	Nicolai.	Peters.
Perihel pass., 1843 Feb., mean time at Greenwich.	27ᵈ·45096	27ᵈ·42935	27ᵈ·39638	27ᵈ·43023	27ᵈ·41319
Long. of perihel.	279° 2′ 30″	278° 18′ 3″	278° 28′ 25″	278° 36′ 33″	279° 59′ 7″
Long of ☊	4 15 5	0 51 4	1 48 3	1 37 55	3 55 17
Inclination	35 12 38	35 8 56	35 35 29	35 36 29	35 15 42
Perihel. dist.	0·00522	0·00581	0·00579	0·00558	0·00428
Motion	Retrograde.	Retrograde.	Retrograde.	Retrograde.	Retrograde.

(592.) What renders these elements so remarkable is the smallness of the perihelion distance. Of all comets which have been recorded this has made the nearest approach to the sun. The sun's radius being the sine of his apparent semi-diameter (16′ 1″ ·5) to a radius equal to the earth's mean distance $= 1$, is represented on that scale by 0·00466, which falls short of 0·00534, the perihelion distance found by taking a mean of all the foregoing results, by only 0·00067, or about one seventh of its whole magnitude. The comet, therefore, approached the *luminous* surface of the sun within about a seventh part of the sun's radius! It is worth while to consider what is implied in such a fact. In the first place, the intensity both of the light and radiant heat of the sun at different distances from that luminary increase proportionally to the spherical area of the portion of the visible hemisphere covered by the sun's disc. This disc, in the case of the earth, at its mean distance has an angular diameter of 32′ 3″. At our comet in perihelio the apparent angular diameter of the sun was no less than 121° 32′. The ratio of the spherical surfaces thus occupied (as appears from spherical geometry) is that of the squares of the sines of the fourth parts of these angles to each other, or that of 1 : 47042. And in this proportion are to each other the amounts of light and heat thrown by the sun on an equal area of exposed surface on our earth and at the comet in equal instants of time. Let any one imagine the effect of so fierce a glare as that of 47000 suns such as we experience the warmth of, on the materials of which the earth's surface is composed. To form some practical idea of it we may compare it with what is recorded of Parker's great lens, whose diameter was $32\frac{1}{2}$ inches and focal length six feet eight inches. The effect of this, supposing all the light and heat transmitted, and the focal concentration perfect, (both conditions very imperfectly satisfied,) would be to enlarge the sun's effective angular diameter to 23° 26′, which, compared on the same principle with a sun of 32′ in diameter, would give a multiplier of only 1915 instead of 47000. The heat to which the comet was subjected therefore surpassed that in the focus of the lens in question, on the lowest calculation, in the proportion of $24\frac{1}{2}$ to 1. Yet that lens melted carnelian, agate, and rock crystal!

(593.) To this extremity of heat however the comet was exposed but for a short time. Its actual velocity in perihelio was no less than 366 miles per second, and the whole of that segment of its orbit above (*i.e.* north of) the plane of the ecliptic, and in which, as will appear from a consideration of the elements, the perihelion was situated, was described in little more than two hours; such being the whole duration of the time from the ascending to the descending node, or in which the comet had north latitude. Arrived at the descending node, its distance from the sun would be already doubled, and the radiation reduced to one fourth of its maximum amount. The comet of 1680, whose perihelion distance was 0·0062, and which therefore approached the sun's surface within one third part of his radius (more than double the distance of the comet now in question) was computed by Newton to have been subjected to an intensity of heat 2000 times that of red-hot iron,—a term of comparison indeed of a very vague description, and which modern thermotics do not recognize as affording a legitimate measure of radiant heat.*

(594.) Although some of the observations of this comet were vague and inaccurate, yet there seem good grounds for believing that its whole course cannot be reconciled with a parabolic orbit, and that it really describes an ellipse. Previous to any calculation, it was remarked that in the year 1668 the tail of an immense comet was seen in Lisbon, at Bologna, in Brazil, and elsewhere, occupying nearly the same situation among the stars, and at the same season of the year, viz. on the 5th of March and the following days. Its brightness was such that its reflected trace was easily

---

* A transit of this comet over the sun's disc must probably have taken place shortly after its passage through its descending node. It is greatly to be regretted that so interesting a phænomenon should have passed unobserved. Whether it be possible that some offset of its tail, darted off so late as the 7th of March, when the comet was already far south of the ecliptic, should have crossed that plane and been seen near the Pleiades, may be doubted. Certain it is, that on the evening of that day, a decidedly cometic ray *was* seen in the immediate neighbourhood of those stars by Mr. Nasmyth. (Ast. Soc. Notices, vol. v. p. 270.)

distinguished on the sea. The head, when it at length came in sight, was comparatively faint and scarce discernible. No precise observations were made of this comet, but the singular coincidence of situation, season of the year, and physical resemblance, excited a strong suspicion of the identity of the two bodies, implying a period of 175 years *within a day or two more or less*. This suspicion has been converted almost into a certainty by a careful examination of what is recorded of the older comet. Locating on a celestial chart the situation of the head, concluded from the direction and appearance of the tail, when only that was seen, and its visible place, when mentioned, according to the descriptions given, it has been found practicable to derive a rough orbit from the course thus laid down: and this agrees in all its features so well with that of the modern comet as nearly to remove all doubt on the subject. Comets, moreover, are recorded to have been seen in A. D. 268, 442–3, 791, 968, 1143, 1317, 1494, which may have been returns of this, since the period above-mentioned would bring round its appearance to the years 268, 443, 618, 793, 968, 1143, 1318, and 1493, and a certain latitude must always be allowed for unknown perturbations.

(595.) But this is not the only comet on record whose identity with the comet of '43 has been maintained. In 1689 a comet bearing a considerable resemblance to it was observed from the 8th to the 23d of December, and from the few and rudely observed places recorded, its elements had been calculated by Pingré, one of the most diligent enquirers into this part of astronomy. * From these it appears that the perihelion distance of that comet was very remarkably small, and a sufficient though indeed rough coincidence in the places of the perihelion and node tended to corroborate the suspicion. But the inclination (69°) assigned to it by Pingré appeared conclusive against it. On recomputing the elements, however, from his data, Professor Pierce has assigned to that comet an inclination widely differing from Pingré's, viz.

---

* Author of the "Cométographie," a work indispensable to all who would study this interesting department of the science.

30° 4′ *, and quite within reasonable limits of resemblance. But how does this agree with the longer period of 175 years before assigned? To reconcile this we must suppose that these 175 years comprise at least eight returns of the comet, and that in effect a mean period of 21ʸ·875 must be allowed for its return. Now it is worth remarking that this period calculated backwards from 1843·156 will bring us upon a series of years remarkable for the appearance of great comets, many of which, as well as the imperfect descriptions we have of their appearance and situation in the heavens, offer at least no obvious contradiction to the supposition of their identity with this. Besides those already mentioned as indicated by the period of 175 years, we may specify as probable or possible intermediate returns, those of the comets of 1733?†, 1689 above-mentioned, 1559?, 1537‡, 1515§, 1471, 1426, 1405-6, 1383, 1361, 1340‖, 1296, 1274, 1230¶, 1208, 1098, 1056, 1034, 1012**, 990?††, 925?, 858??, 684‡‡, 552, 530§§, 421, 245 or 247‖‖, 180¶¶, 158. Should this view of the subject be the true one, we may expect its return about

---

* United States Gazette, May 29. 1843. Considering that all the observations lie *near* the descending node of the orbit, the proximity of the comet at that time to the sun, and the loose nature of the recorded observations, no doubt almost any given inclination *might* be deduced from them. The true test in such cases is not to ascend from the old incorrect data to elements, but to descend from known and certain elements to the older data, and ascertain whether the recorded phænomena can be represented by them (perturbations included) within fair limits of interpretation. Such is the course pursued by Clausen.

† P. Passage 1733·781. The great southern comet of May 17th seems too early in the year.

‡ P. P. 1536·906. In January 1537, a comet was seen in Pisces.

§ P. P. 1515·031. A comet *predicted* the death of Ferdinand the Catholic He died Jan. 23. 1515.

‖ P. P. 1340·031. Evidently a southern comet, and a very probable appearance.

¶ P. P. 1230·656, was perhaps a return of Halley's.

** P. P. 1011·906. In 1012, a very great comet in the southern part of the heavens. "Son éclat blessait les yeux." (Pingré Cométographie, from whom all these recorded appearances are taken.)

†† P. P. 990·031. "Comète fort épouvantable," *some year* between 989 and 998.

‡‡ P. P. 683·781. In 684, appeared two or three comets. Dates begin to be obscure.

§§ Two distinct comets (one probably the comet of Cæsar and 1680) appeared in 530 and 531, the former observed in China, the latter in Europe.

‖‖ P. P. 246·281; both southern comets of the Chinese annals. The year of one or other may be wrong.

¶¶ P. P. 180·656. Nov 6. A.D. 180. A southern comet of the Chinese annals.

the end of 1864 or beginning of 1865, in which event it will be observable in the Southern Hemisphere both before and after its perihelion passage.*

(596.) M. Clausen, from the assemblage of all the observations of this comet known to him, has calculated elliptic elements which give the extraordinarily short period of 6·38 years. And in effect it has been suggested that a still further subdivision of the period of 21·875 into three of 7·292 years would reconcile this with other remarkable comets. This seems going too far, but at all events the possibility of representing its motions by so short an ellipse will easily reconcile us to the admission of a period of 21 years. That it should only be visible in certain apparitions, and not in others, is sufficiently explained by the situation of its orbit.

(597.) We have been somewhat diffuse on the subject of this comet, for the sake of showing the degree and kind of interest which attaches to cometic astronomy in the present state of the science. In fact, there is no branch of astronomy more replete with interest, and we may add more eagerly pursued at present, inasmuch as the hold which exact calculation gives us on it may be regarded as completely established; so that whatever may be concluded as to the motions of any comet which shall henceforward come to be observed, will be concluded on sure grounds and with numerical precision; while the improvements which have been introduced into the calculation of cometary perturbation, and the daily increasing familiarity of numerous astronomers with computations of this nature, enable us to trace their past and future history with a certainty, which at the commencement of the present century could hardly have been looked upon as attainable. Every comet newly discovered is at once subjected to the ordeal of a most rigorous enquiry. Its elements, roughly calculated within a few days of its appearance, are gradually approximated to as observations accumulate, by a multitude of ardent and expert computists. On the least indication of a deviation from a parabolic orbit, its elliptic

* Clausen, Astron. Nachr. No. 485.

elements become a subject of universal and lively interest and
discussion. Old records are ransacked, and old observations
reduced, with all the advantage of improved data and
methods, so as to rescue from oblivion the orbits of ancient
comets which present any similarity to that of the new visitor.
The disturbances undergone in the interval by the action of
the planets are investigated, and the past, thus brought into
unbroken connexion with the present, is made to afford substantial ground for prediction of the future. A great impulse
meanwhile has been given of late years to the discovery of
comets by the establishment in 1840*, by his late Majesty
the King of Denmark, of a prize medal to be awarded for
every such discovery, to the first observer, (the influence of
which may be most unequivocally traced in the great number
of these bodies which every successive year sees added to our
list,) and by the circulation of notices, by special letter †, of
every such discovery (accompanied, when possible, by an
ephemeris), to all observers who have shown that they take
an interest in the enquiry, so as to ensure the full and complete observation of the new comet so long as it remains
within the reach of our telescopes.

(598.) It is by no means merely as a subject of antiquarian
interest, or on account of the brilliant spectacle which comets
occasionally afford, that astronomers attach a high degree of
importance to all that regards them. Apart even from the
singularity and mystery which appertains to their physical
constitution, they have become, through the medium of exact
calculation, unexpected instruments of enquiry into points
connected with the planetary system itself, of no small importance. We have seen that the movements of the comet
of Encke, thus minutely and perseveringly traced by the
eminent astronomer whose name is used to distinguish it, has
afforded ground for believing in the presence of a resisting
medium filling the whole of our system. Similar enquiries,
prosecuted in the cases of other periodical comets, will extend,
confirm, or modify our conclusions on this head. The per-

* See the announcement of this institution in Astron. Nachr. No. 400.
† By Prof. Schumacher, Director of the Royal Observatory of Altona.

turbations, too, which comets experience in passing near any of the planets, may afford, and have afforded, information as to the magnitude of the disturbing masses, which could not well be otherwise obtained.   Thus the approach of this comet to the planet Mercury in 1838 afforded an estimation of the mass of that planet the more precious, by reason of the great uncertainty under which all previous determinations of that element laboured.   Its approach to the same planet in the present year (1848) will be still nearer.   On the 22d of November their mutual distance will be only fifteen times the moon's distance from the earth.

(599.) It is, however, in a physical point of view that these bodies offer the greatest stimulus to our curiosity.   There is, beyond question, some profound secret and mystery of nature concerned in the phænomenon of their tails.   Perhaps it is not too much to hope that future observation, borrowing every aid from rational speculation, grounded on the progress of physical science generally, (especially those branches of it which relate to the æthcrial or imponderable elements), may ere long enable us to penetrate this mystery, and to declare whether it is really *matter* in the ordinary acceptation of the term which is projected from their heads with such extravagant velocity, and if not impelled, at least *directed* in its course by a reference to the sun, as its point of avoidance. In no respect is the question as to the materiality of the tail more forcibly pressed on us for consideration, than in that of the enormous sweep which it makes round the sun in perihelio, in the manner of a straight and rigid rod, in defiance of the law of gravitation, nay, even of the received laws of motion, extending (as we have seen in the comets of 1680 and 1843) from near the sun's surface to the earth's orbit, yet whirled round unbroken ; in the latter case through an angle of 180° in little more than two hours.   It seems utterly incredible that in such a case it is one and the same material object which is thus brandished.   If there could be conceived such a thing as a *negative shadow*, a momentary impression made upon the luminiferous æther behind the comet, this would represent in some degree the conception such a phænomenon

irresistibly calls up. But this is not all. Even such an extraordinary excitement of the æther, conceive it as we will,
will afford no account of the projection of lateral streamers;
of the effusion of light from the nucleus of a comet towards
the sun; and its subsequent *re*jection; of the irregular and
capricious mode in which that effusion has been seen to take
place; none, of the clear indications of alternate evaporation
and condensation going on in the immense regions of space
occupied by the tail and coma, — none, in short, of innumerable other facts which link themselves with almost equally
irresistible cogency to our ordinary notions of matter and
force.

(600.) The great number of comets which appear to move
in parabolic orbits, or orbits at least undistinguishable from
parabolas during their description of that comparatively small
part within the range of their visibility to us, has given rise
to an impression that they are bodies extraneous to our
system, wandering through space, and merely yielding a
local and temporary obedience to its laws during their sojourn.
What truth there may be in this view, we may never have
satisfactory grounds for deciding. On such an hypothesis,
our elliptic comets owe their permanent denizenship within
the sphere of the sun's predominant attraction to the action
of one or other of the planets near which they may have
passed, in such a manner as to diminish their velocity, and
render it compatible with elliptic motion. * A similar cause
acting the other way, might with equal probability, give rise
to a hyperbolic motion. But whereas in the former case, the
comet would remain in the system, and might make an indefinite number of revolutions, in the latter it would return no
more. This may possibly be the cause of the exceedingly
rare occurrence of a hyperbolic comet as compared with
elliptic ones.

(601.) All the planets without exception, and almost all
the satellites, move in one direction round the sun. Retrograde comets, however, are of very common occurrence, which

* The velocity in an ellipse is always less than in a parabola, at equal distances from the sun; in an hyperbola always greater.

certainly would go to assign them an exterior or at least an
independent origin. Laplace, from a consideration of all the
cometary orbits known in the earlier part of the present century, concluded, that the mean or average situation of the
planes of all the cometary orbits, with respect to the ecliptic,
was so nearly that of perpendicularity, as to afford no presumption of any cause biassing their directions in this respect.
Yet we think it worth noticing that among the comets which
are as yet known to describe elliptic orbits, not one whose
inclination is under 17° is retrograde; and that out of thirty-
six comets which have had elliptic elements assigned to them,
whether of great or small excentricities, and without any
limit of inclination, only five are retrograde, and of these,
only two, viz. Halley's and the great comet of 1843, can be
regarded as satisfactorily made out. Finally, of the 125
comets whose elements are given in the collection of Schumacher and Olbers, up to 1823, the number of retrograde
comets under 10° of inclination is only 2 out of 9, and under
20°, 7 out of 23. A plane of motion therefore, nearly coincident with the ecliptic, and a periodical return, are circumstances eminently favourable to direct revolution in the cometary as they are decisive among the planetary orbits.

PART II.

OF THE LUNAR AND PLANETARY PERTURBATIONS.

" Magnus ab integro sæclorum nascitur ordo." — VIRG. *Pollio.*

CHAPTER XII.

SUBJECT PROPOUNDED. — PROBLEM OF THREE BODIES. — SUPER-
POSITION OF SMALL MOTIONS. — ESTIMATION OF THE DISTURBING
FORCE. — ITS GEOMETRICAL REPRESENTATION. — NUMERICAL ES-
TIMATION IN PARTICULAR CASES. — RESOLUTION INTO RECT-
ANGULAR COMPONENTS. — RADIAL, TRANSVERSAL, AND ORTHO-
GONAL DISTURBING FORCES. — NORMAL AND TANGENTIAL. — THEIR
CHARACTERISTIC EFFECTS. —EFFECTS OF THE ORTHOGONAL FORCE.
— MOTION OF THE NODES. — CONDITIONS OF THEIR ADVANCE
AND RECESS. — CASES OF AN EXTERIOR PLANET DISTURBED BY
AN INTERIOR.— THE REVERSE CASE. — IN EVERY CASE THE NODE
OF THE DISTURBED ORBIT RECEDES ON THE PLANE OF THE
DISTURBING ON AN AVERAGE. — COMBINED EFFECT OF MANY SUCH
DISTURBANCES. — MOTION OF THE MOON'S NODES. — CHANGE OF
INCLINATION. — CONDITIONS OF ITS INCREASE AND DIMINUTION.
— AVERAGE EFFECT IN A WHOLE REVOLUTION. — COMPENSATION
IN A COMPLETE REVOLUTION OF THE NODES. — LAGRANGE'S
THEOREM OF THE STABILITY OF THE INCLINATIONS OF THE PLA-
NETARY ORBITS. — CHANGE OF OBLIQUITY OF THE ECLIPTIC. —
PRECESSION OF THE EQUINOXES EXPLAINED. — NUTATION.— PRIN-
CIPLE OF FORCED VIBRATIONS.

(602.) IN the progress of this work, we have more than once
called the reader's attention to the existence of inequalities
in the lunar and planetary motions not included in the
expression of Kepler's laws, but in some sort supplementary
to them, and of an order so far subordinate to those leading
features of the celestial movements, as to require, for their
detection, nicer observations, and longer-continued comparison
between facts and theories, than suffice for the establishment
and verification of the elliptic theory.   These inequalities

are known, in physical astronomy, by the name of *perturbations*. They arise, in the case of the primary planets, from the mutual gravitations of these planets towards each other, which derange their elliptic motions round the sun; and in that of the secondaries, partly from the mutual gravitation of the secondaries of the same system similarly deranging their elliptic motions round their common primary, and partly from the unequal attraction of the sun and planets on them and on their primary. These perturbations, although small, and, in most instances, insensible in short intervals of time, yet, when accumulated, as some of them may become, in the lapse of ages, alter very greatly the original elliptic relations, so as to render the same elements of the planetary orbits, which at one epoch represented perfectly well their movements, inadequate and unsatisfactory after long intervals of time.

(603.) When Newton first reasoned his way from the broad features of the celestial motions, up to the law of universal gravitation, as affecting all matter, and rendering every particle in the universe subject to the influence of every other, he was not unaware of the modifications which this generalization would induce upon the results of a more partial and limited application of the same law to the revolutions of the planets about the sun, and the satellites about their primaries, as their *only* centers of attraction. So far from it, his extraordinary sagacity enabled him to perceive very distinctly how several of the most important of the lunar inequalities take their origin, in this more general way of conceiving the agency of the attractive power, especially the retrograde motion of the nodes, and the direct revolution of the apsides of her orbit. And if he did not extend his investigations to the mutual perturbations of the planets, it was not for want of perceiving that such perturbations *must* exist, and *might* go the length of producing great derangements from the actual state of the system, but was owing to the then undeveloped state of the practical part of astronomy, which had not yet attained the precision requisite to make such an attempt inviting, or indeed feasible. What

Newton left undone, however, his successors have accomplished; and, at this day, it is hardly too much to assert that there is not a single perturbation, great or small, which observation has become precise enough clearly to detect and place in evidence which has not been traced up to its origin in the mutual gravitation of the parts of our system, and minutely accounted for, in its numerical amount and value, by strict calculation on Newton's principles.

(604.) Calculations of this nature require a very high analysis for their successful performance, such as is far beyond the scope and object of this work to attempt exhibiting. The reader who would master them must prepare himself for the undertaking by an extensive course of preparatory study, and must ascend by steps which we must not here even digress to point out. It will be our object, in this chapter, however, to give some general insight into the nature and manner of operation of the acting forces, and to point out what are the circumstances which, in some cases, give them a high degree of efficiency — a sort of *purchase* on the balance of the system; while, in others, with no less amount of intensity, their effective agency in producing extensive and lasting changes is compensated or rendered abortive; as well as to explain the nature of those admirable results respecting the stability of our system, to which the researches of geometers have conducted them; and which, under the form of mathematical theorems of great simplicity and elegance, involve the history of the past and future state of the planetary orbits during ages, of which, contemplating the subject in this point of view, we neither perceive the beginning nor the end.

(605.) Were there no other bodies in the universe but the sun and one planet, the latter would describe an exact ellipse about the former (or both round their common center of gravity), and continue to perform its revolutions in one and the same orbit for ever; but the moment we add to our combination a third body, the attraction of this will draw both the former bodies out of their mutual orbits, and, by acting on them unequally, will disturb their relation to each other, and put an end to the rigorous and mathematical exactness of their elliptic motions, not only about a fixed point in space, but about one another. From this way of propounding the subject, we see that it is not the whole attraction of the newly-introduced body which produces perturbation, but *the difference* of its attractions on the two originally present.

(606.) Compared to the sun, all the planets are of extreme minuteness; the mass of Jupiter, the greatest of them all, being not more than about one 1100th part that of the sun. Their attractions on each other, therefore, are all very feeble, compared with the presiding central power, and the effects of their disturbing forces are proportionally minute. In the case of the secondaries, the chief agent by which their motions are deranged is the sun itself, whose mass is indeed great, but whose disturbing influence is immensely diminished by their near proximity to their primaries, compared to their distances from the sun, which renders the *difference* of attractions on both extremely small, compared to the whole amount. In this case the greatest part of the sun's attraction, viz. that which is common to both, is exerted to retain both primary and secondary in their common orbit about itself, and prevent their parting company. Only the small overplus of force on one as compared with the other acts as a disturbing power. The mean value of this overplus, in the case of the moon disturbed by the sun, is calculated by Newton to amount to no higher a fraction than $\frac{1}{638000}$ of gravity at the earth's surface, or $\frac{1}{179}$ of the principal force which retains the moon in its orbit.

(607.) From this extreme minuteness of the intensities of the disturbing, compared to the principal forces, and the consequent smallness of their *momentary* effects, it happens that we can estimate each of these effects separately, as if the others did not take place, without fear of inducing error in our conclusions beyond the limits necessarily incident to a first approximation. It is a principle in mechanics, immediately flowing from the primary relations between forces and the motions they produce, that when a number of very minute forces act at once on a system, their joint effect is the sum or aggregate of their separate effects, at least within such limits, that the original relation of the parts of the system shall not have been materially changed by their action. Such effects supervening on the greater movements due to the action of the primary forces may be compared to the small riplings caused by a thousand varying breezes on the broad and regular swell of a deep and rolling ocean, which run on as if the surface were a plane, and cross in all directions without interfering, each as if the other had no existence. It is only when their effects become accumulated in lapse of time, so as to alter the primary relations or data of the system, that it becomes necessary to have especial regard to the changes correspondingly introduced into the estimation of their momentary efficiency, by which the *rate* of the subsequent changes is affected, and periods or cycles of immense length take their origin. From this consideration arise some of the most curious theories of physical astronomy.

(608.) Hence it is evident, that in estimating the disturbing influence of several bodies forming a system, in which one has a remarkable preponderance over all the rest, we need not embarrass ourselves with combinations of the disturbing powers one among another, unless where immensely long periods are concerned; such as consist of many hundreds of revolutions of the bodies in question about their common center. So that, in effect, so far as we propose to go into its consideration, the problem of the investigation of the perturbations of a system, however numerous, constituted as ours is, reduces itself to that of a system of three bodies: a predominant central body, a disturbing, and a disturbed; the two latter of which may exchange denominations, according as the motions of the one or the other are the subject of enquiry.

(609.) Both the intensity and direction of the disturbing force are continually varying, according to the relative situation of the disturbing and disturbed body with respect to the sun. If the attraction of the disturbing body M, on the central body S, and the disturbed body P, (by which designations, for brevity, we shall hereafter indicate them,) were equal, and acted in parallel lines, whatever might otherwise be its law of variation, there would be no deviation caused in the elliptic motion of P about S, or of each about the other. The case would be strictly that of art. 454.; the attraction of M, so circumstanced, being at every moment exactly analogous in its effects to terrestrial gravity, which acts in parallel lines, and is equally *intense* on all bodies, great and small. But this is not the case of nature. Whatever is stated in the subsequent article to that last cited, of the disturbing effect of the sun and moon, is, *mutatis mutandis*, applicable to every case of perturbation; and it must be now our business to enter, somewhat more in detail, into the general heads of the subject there merely hinted at.

(610.) To obtain clear ideas of the manner in which the disturbing force produces its various effects, we must ascertain at any given moment, and in any relative situations of the three bodies, its direction and intensity as compared with the gravitation of P towards S, in virtue of which latter force alone P would describe an ellipse about S regarded as fixed, or rather P and S about their common center of gravity in virtue of their mutual gravitation to each other. In the treatment of the problem of three bodies, it is convenient, and tends to clearness of apprehension, to regard one of them as fixed, and refer the motions of the others to it as to a relative center. In the case of two planets disturbing each other's motions, the sun is naturally chosen as this fixed center; but in that of satellites disturbing each other, or disturbed by the sun, the center of their primary is taken as their point of reference, and the sun itself is regarded in the light of a very distant and massive satellite revolving about the primary in a *relative* orbit, equal and similar to that which the primary describes *absolutely* round the sun. Thus the generality of our language is preserved, and when, referring to any particular central body, we speak of an exterior and an interior planet, we include the cases in which the former is the sun and the latter a satellite; as, for example, in the Lunar theory. It is a principle in dynamics, that the relative motions of a system of bodies *inter se* are no way altered by impressing on all of them a common motion or motions, or a common force or forces accelerating or retarding them all equally in common directions, *i. e.* in parallel lines. Suppose, therefore, we apply to all the three bodies, S, P, and M, alike, forces equal to those with which M and P attract S, but in opposite directions. Then will the relative motions both of M and P about S be unaltered; but S, being now urged by equal and opposite forces to and from both M and P, will remain at rest. Let us now consider how either of the other bodies, as P, stands affected by these newly introduced forces, in addition to those which before acted on it. It is clear that now P will be simultaneously acted on by *four* forces; firstly, the attraction *of* S in the direction P S; secondly, an additional force, in the same direction, equal to its attraction *on* S; thirdly, the attraction *of* M in the direction P M; and fourthly, a force parallel to M S, and equal to M's attraction *on* S. Of these, the two first, following the same law of the inverse square of the distance S P, may be regarded as one force, precisely as if the sum of the masses of S and P were collected in S; and in virtue of their joint action, P will describe an ellipse about S, except in so far as that elliptic motion is disturbed by the other two forces. Thus we see that in this view of the subject the *relative* disturbing force acting on P is no longer the mere single attraction of M, but a force resulting from the composition of that attraction with M's attraction on S transferred to P in a contrary direction.

(611.) Let C P A be part of the relative orbit of the disturbed, and M B of the disturbing body, their planes intersecting in the line of nodes S A B, and having to each other the inclination expressed by the spherical angle P Aa. In M P, produced if required, take M N : M S :: M S^2 : M P^2. Then, if S M* be taken to represent, in quantity and direction, the accelerative attraction of M on S, M S will represent

---

* The reader will be careful to observe the order of the letters, where forces are represented by lines. M S represents a force acting from M towards S, S M from S towards M.

in quantity and direction the new force applied to P, parallel to that line, and N M will represent on the same scale the accelerative attraction of M on P. Consequently, the disturbing force acting on P will be the resultant of two forces applied at P, represented respectively by N M and M S,

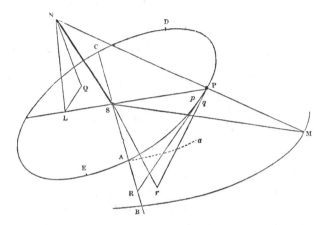

which by the laws of dynamics are equivalent to a single force represented in *quantity* and *direction* by N S, *but having* P *for its point of application.*

(612.) The *line* M S, is easily calculated by trigonometry, when the relative situations and real distances of the bodies are known; and the *force* expressed by that line is directly comparable with the attractive forces of S on P by the following proportions, in which M, S, represent the masses of those bodies which are supposed to be known, and to which, at equal distances, their attractions are proportional : —

Disturbing force : M's attraction on S :: N S : S M;

M's attraction on S : S's attraction on M :: M : S;

S's attraction on M : S's attraction on P :: $S P^2$: $S M^2$:

by compounding which proportions we collect as follows : —

Disturbing force : S's attraction on P :: $M . N S . S P^2$: $S . S M^3$.

A few numerical examples are subjoined, exhibiting the results of this calculation in particular cases, chosen so as to exemplify its application under very various circumstances,

throughout the planetary system. In each case the numbers
set down express the proportion in which the central force
retaining the disturbed body in its elliptic orbit exceeds the
disturbing force, to the nearest whole number. The calculation is made for three positions of the disturbing body — viz.
at its greatest, its least, and its mean distance from the disturbed.

Disturbing Body.	Disturbed Body.	Ratio at the greatest Distance : 1.	Ratio at the mean Distance. : 1.	Ratio at the least Distance. : 1.
The Sun -	The Moon -	90	179	89
Jupiter - -	Saturn - -	354	312	128
Jupiter - -	The Earth -	95683	147575	53268
Venus -	The Earth -	255208	210245	26833
Neptune	Uranus - -	57420	56592	5519
Mercury -	Neptune -	526	526	526
Jupiter -	Ceres -	6433	6937	1033
Saturn - -	Jupiter - -	20248	21579	3065

(613.) If the orbit of the disturbing body be circular, S M
is invariable. In this case, N S will continue to represent the
disturbing force *on the same invariable scale*, whatever may
be the configuration of the three bodies with respect to each
other. If the orbit of M be but little elliptic, the same will
be nearly the case. In what follows throughout this chapter,
except where the contrary is expressly mentioned, we shall
neglect the excentricity of the disturbing orbit.

(614.) If P be nearer to M than S is, M N is greater than
M P, and N lies in M P prolonged, and therefore on the
opposite side of the plane of P's orbit from that on which M
is situated. The force N S therefore urges P towards that
plane, and towards a point X, situated between S and M, in
the line S M. If the distance M P be equal to M S as when
P is situated, suppose, at D or E, M N is also equal to M P
or M S, so that N coincides with P, and therefore X with S,
the disturbing forces being in these cases directed *towards*
the central body. But if M P be greater than M S, M N is
less than M P, and N lies between M and P, or on the same
side of the plane of P's orbit that M is situated on. The
force N S, therefore, applied at P, urges P towards the contrary side of that plane towards a point in the line M S pro-

duced, so that X now shifts to the farther side of S.   In all cases, the disturbing force is wholly effective in the plane M P S, in which the three bodies lie.

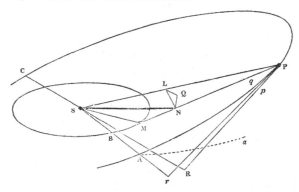

It is very important for the student to fix distinctly and bear constantly in his mind these relations of the disturbing agency considered as a *single unresolved force*, since their re- collection will preserve him from many mistakes in con- ceiving the mutual actions of the planets, &c. on each other. For example, in the figures here referred to, that of Art. 611. corresponds to the case of a nearer disturbed by a more distant body, as the earth by Jupiter, or the moon by the Sun; and that of the present article to the converse case: as, for instance, of Mars disturbed by the earth.   Now, in this latter class of cases, whenever M P is greater than M S, or S P greater than 2 S M, N lies on the same side of the plane of P's orbit with M, so that N S, the disturbing force, contrary to what might at first be supposed, always urges the disturbed planet out of the plane of its orbit towards the opposite side to that on which the disturbing planet lies.   It will tend greatly to give clearness and definiteness to his ideas on the subject, if he will trace out on various sup- positions as to the relative magnitude of the disturbing and disturbed orbits (supposed to lie in one plane) the form of the oval about M considered as a fixed point, in which the point N lies when P makes a complete revolution round S.

(615.) Although it is necessary for obtaining in the first

instance a clear conception of the action of the disturbing force, to consider it in this way as a single force having a definite direction in space and a determinate intensity, yet as that direction is continually varying with the position of N S, both with respect to the radii S P, S M, the distance P M, and the direction of P's motion, it would be impossible, by so considering it, to attain clear views of its dynamical effect after any considerable lapse of time, and it therefore becomes necessary to resolve it into other equivalent forces acting in such directions as shall admit of distinct and separate consideration. Now this may be done in several different modes. First, we may resolve it into three forces acting in fixed directions in space rectangular to one another, and by estimating its effect in each of these three directions separately, conclude the total or joint effect. This is the mode of procedure which affords the readiest and most advantageous handle to the problem of perturbations when taken up in all its generality, and is accordingly that resorted to by geometers of the modern school in all their profound researches on the subject. Another mode consists in resolving it also into three rectangular components, not, however, in fixed directions, but in variable ones, viz. in the directions of the lines N Q, Q L, and L S, of which L S is in the direction of the radius vector S P, Q L in a direction perpendicular to it, and in the plane in which S P and a tangent to P's orbit as P both lie ; and lastly, N Q in a direction perpendicular to the plane in which P is at the instant moving about S. The first of these resolved portions we may term the *radial* component of the disturbing force, or simply the radial disturbing force; the second the *transversal*; and the third the *orthogonal*.* When the disturbed orbit is one of small excentricity, the transversal component acts nearly in the direction of the tangent to P's orbit at P, and is therefore confounded with that resolved component which we shall presently describe (art. 618.) under the name of the *tangential*

---

* This is a term coined for the occasion. The want of some appellation for this component of the disturbing force is often felt.

force. This is the mode of resolving the disturbing force followed by Newton and his immediate successors.

(616.) The immediate actions of these components of the disturbing force are evidently independent of each other, being rectangular in their directions; and they affect the movement of the disturbed body in modes perfectly distinct and characteristic. Thus, the radial component, being directed to or from the central body, has no tendency to disturb either the plane of P's orbit, or the equable description of areas by P about S, since the law of areas proportional to the times is not a character of the force of gravity only, but holds good equally, whatever be the force which retains a body in an orbit, *provided only* its direction is always towards a fixed center.* Inasmuch, however, as its law of variation is not conformable to the simple law of gravity, it alters the elliptic form of P's orbit, by directly affecting both its curvature and velocity at every point. In virtue, therefore, of the action of this disturbing force, the orbit deviates from the elliptic form by the approach or recess of P to or from S, so that the effect of the perturbations produced by this part of the disturbing force falls wholly on the radius vector of the disturbed orbit.

(617.) The transversal disturbing force represented by Q L, on the other hand, has no direct action to draw P to or from S. Its whole efficiency is directed to accelerate or retard P's motion in a direction at right angles to S P. Now the area momentarily described by P about S, is, *cæteris paribus*, directly as the velocity of P in a direction perpendicular to S P. Whatever force, therefore, increases this transverse velocity of P, accelerates the description of areas, and *vice versâ*. With the area A S P is directly connected, by the nature of the ellipse, the angle A S P described or to be described by P from a fixed line in the plane of the orbit, so that any change in the rate of description of areas ultimately resolves itself into a change in the amount of angular motion about S, and gives rise to a departure from the elliptic laws. Hence arise what are called in the perturbational theory

* Newton, i. 1.

c c 3

equations (*i. e.* changes or fluctuations to and fro about an average quantity) of the mean motion of the disturbed body.

(618.) There is yet another mode of resolving the disturbing force into rectangular components, which, though not so well adapted to the computation of results, in reducing to numerical calculation the motions of the disturbed body, is fitted to afford a clearer insight into the nature of the modifications which the form, magnitude, and situation of its orbit undergo in virtue of its action, and which we shall therefore employ in preference.   It consists in estimating the components of the disturbing force, which lie in the plane of the orbit, not in the direction we have termed radial and transversal, i. e. in that of the radius vector P S and perpendicular to it, but in the direction of a tangent to the orbit at P, and in that of a normal to the curve, and at right angles to the tangent, for which reason these components may be called the *tangential* and *normal* disturbing forces.   When the orbit of the disturbed body is circular, or nearly so, this mode of resolution coincides with or differs but little from the former, but, when the ellipticity is considerable, these directions may deviate from the radial and transversal directions to any extent.   As in the Newtonian mode of resolution, the effect of the one component falls wholly upon the approach and recess of the body P to the central body S, and of the other wholly on the rate of description of areas by P round S, so in this which we are now considering, the direct effect of the one component (the normal) falls wholly on the curvature of the orbit at the point of its action, increasing that curvature when the normal force acts inwards, or towards the concavity of the orbit, and diminishing it when in the opposite direction; while, on the other hand, the tangential component is directly effective on the velocity of the disturbed body, increasing or diminishing it according as its direction conspires with or opposes its motion.   It is evident enough that where the object is to trace simply the changes produced by the disturbing force, in *angle* and *distance* from the central body, the former mode of resolution must have the advantage in perspicuity of view and applicability to calculation. It is less obvious, but will abundantly appear in the sequel that the latter offers peculiar advantages in exhibiting to the eye and the reason the momentary influence of the disturbing force on the *elements* of the orbit itself.

(619.) Neither of the last mentioned pairs of resolved portions of the disturbing force tends to draw P out of the plane of its orbit P S A. But the remaining or orthogonal portion N Q acts directly and solely to produce that effect. In consequence, under the influence of this force, P must quit that plane, and (the same cause continuing in action) must describe a *curve of double curvature* as it is called, no two consecutive portions of which lie in the same plane passing through S. The effect of this is to produce a continual variation in those elements of the orbit of P on which *the situation of its plane* in space depends; *i. e.* on its inclination to a fixed plane, and the position in such a plane of the node or line of its intersection therewith. As this, among all the various effects of perturbation, is that which is at once the most simple in its conception, and the easiest to follow into its remoter consequences, we shall begin with its explanation.

(620.) Suppose that up to P (Art. 611. 614.) the body were describing an undisturbed orbit C P. Then at P it would be moving in the direction of a tangent P R to the ellipse P A, which prolonged will intersect the plane of M's orbit somewhere in the line of nodes, as at R. Now, at P, let the disturbing force parallel to N Q act momentarily on P; then P will be deflected in the direction of that force, and instead of the arc P *p*, which it would have described in the next instant if undisturbed, will describe the arc P *q* lying in the state of things represented in Art. 611. below, and in Art. 614. above, P *p* with reference to the plane P S A. Thus, by this action of the disturbing force, the plane of P's orbit will have shifted its position in space from P S *p* (an elementary portion of the old orbit) to P S *q*, one of the new. Now the line of nodes S A B in the former is determined by prolonging

P p into the tangent P R, intersecting the plane M S B in R, and joining S R. And in like manner, if we prolong P q into the tangent P r, meeting the same plane in r, and join S r, this will be the new line of nodes. Thus we see that, under the circumstances expressed in the former figure, the momentary action of the orthogonal disturbing force will have caused the line of nodes to *retrograde* upon the plane of the orbit of the disturbing body, and under those represented in the latter to advance. And it is evident that the action of the other resolved portions of the disturbing force will not in the least interfere with this result, for neither of them tends either to carry P out of its former plane of motion, or to prevent its quitting it. Their influence would merely go to transfer the points of intersection of the tangents P p or P q from R or r to R′ or r′, points nearer to or farther from S than R r, but in the same lines.

(621.) Supposing, now, M to lie to the left instead of the right side of the line of nodes in fig. 1., P retaining its situation, and M P being less than M S, so that X shall still lie between M and S. In this situation of things (or *configuration*, as it is termed of the three bodies with respect to each other), N will lie *below* the plane A S P, and the disturbing force will tend to raise the body P above the plane, the resolved orthogonal portion N Q in this case acting upwards. The disturbed arc P q will therefore lie above P p, and when prolonged to meet the plane M S B, will intersect it in a point *in advance* of R; so that in this configuration the node will advance upon the plane of the orbit of M, provided always that the latter orbit remains fixed, or, at least, does not itself shift its position in such a direction as to defeat this result.

(622.) Generally speaking, the node of the disturbed orbit will recede *upon any plane which we may consider as fixed*, whenever the action of the orthogonal disturbing force tends to bring the disturbed body nearer to that plane; and *vice versâ*. This will be evident on a mere inspection of the annexed figure, in which C A represents a semicircle of the projection of the fixed plane as seen from S on the sphere of the heavens, and C P A that of the plane of P's undisturbed orbit, the motion of P being in the direction of the arrow, from C the ascending, to A the descending node. It is at

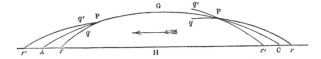

once seen, by prolonging P q, P q' into arcs of great circles, P r, P r', (forwards or backwards, as the case may be) to meet C A, that the node will have retrograded through the arc A r, or C r, whenever P q lies between C P A and C A, or when the perturbing force carries P towards the fixed plane, but will have advanced through A r' or C r' when P q' lies above C P A, or when the disturbing impulse has lifted P above its old orbit or away from the fixed plane, *and this without any reference to whether the undisturbed orbital motion of* P *at the moment is carrying it towards the plane* C A *or from it,* as in the two cases represented in the figure.

(623.) Let us now consider the mutual disturbance of two bodies M and P, in the various configurations in which they may be presented to each other and to their common central body. And first, let us take the case, as the simplest, where the disturbed orbit is exterior to that of the disturbing body (as in fig. art. 614.), and the distance between the orbits greater than the semiaxis of the smaller. First, let both planets lie on the same side of the line of nodes. Then (as in art. 620.) the direction of the whole disturbing force, and therefore also that of its orthogonal component, will be towards the opposite side of the plane of P's orbit from that on which M lies. Its effect therefore will be, to draw P out of its plane in a direction *from* the plane of M's orbit, so that in this state of things the node will advance on the latter plane, however P and M may be situated in these semicircumferences of their respective orbits. Suppose, next, M transferred to the opposite side of the line of nodes, then will the direction of its action on P, with respect to the plane of P's orbit, be reversed, and P in quitting that plane will now approach to instead of receding from the plane of M's orbit, so that its node will now recede on that plane.

(624.) Thus, while M and P revolve about S, and in the course of many revolutions of each are presented to each other and to S in all possible configurations, the node of P's orbit will always advance on A's when both bodies are on the same side of the line of nodes, and recede when on the opposite. They will therefore, on an average, advance and recede during equal times (supposing the orbits nearly circular). And, therefore, if their advance were at each instant of its duration equally rapid with their recess at each corresponding instant during that phase of the movement, they

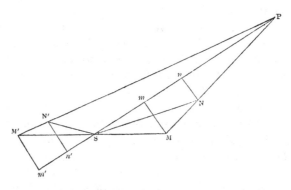

would merely oscillate to and fro about a mean position, without any permanent motion in either direction. But this is not the case. The rapidity of their recess in every position favourable to recess is greater than that of their advance in the corresponding opposite position. To show this, let us consider any two configurations in which M's phases are diametrically opposite, so that the triangles P S M, P S M', shall lie in one plane, having any inclination to P's orbit, according to the situation of P. Produce P S, and draw M m, M'm' perpendicular to it, which will therefore be equal. Take M N : M S :: M S² : M P², and M' N' : M' S :: M' S² : M' P²: then, if the orbits be nearly circles, and therefore M S = M' S, N'M' will be less than M N; and therefore (since P M' is greater than P M) P N' : P M' in a greater ratio than P N : P M; and consequently, by similar triangles, drawing N n, N' n' perpendicular to PS, N' n' : M' m' in a greater ratio than Nn : Mm, and therefore N'n' is greater than Nn. Now the plane P M M' intersects P's orbit in P S, and being inclined to that orbit at the same angle through its whole extent, if from n and n' perpendiculars be conceived let fall on that orbit, these will be to each other in the proportion of N n, N' n' ; and therefore the perpendicular from n' will be greater than that from n. Now since by art. 611. N' S and N S represent in quantity and direction the total disturbing forces of M' and M on P respectively, therefore these perpendiculars express (art. 615.) the orthogonal disturbing forces, the former of which tends (as above shown) to make the nodes recede, and the latter to advance ; and therefore the preponderance in every such pair of situations of M is in favour of a retrograde motion.

(625.) Let us next consider the case where the distance between the orbits is less than the semiaxis of the interior, or in which the least distance of M from P is less than M S.

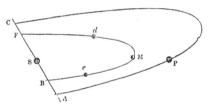

Take any situation of P with respect to the line of nodes A C. Then two points d and e, distant by less than 120°, can be taken on the orbit of M equidistant from P with S. Suppose M to occupy successively every possible situation in its orbit, P retaining its place ; — then, if it were not for the existence of the arc $d e$, in which the relations of art. 624. are reversed, it would appear by the reasoning of that article that the motion of the node is direct when M occupies any part of the semiorbit F M B, and retrograde when it is in the opposite, but that the retrograde motion on the whole would predominate. Much more then will it predominate when there exists an arc $d\,\mathrm{M}e$ within which if M be placed, its action will produce a retrograde instead of a direct motion.

(626.) This supposes that the arc *de* lies wholly in the semicircle F*d*B.　But suppose it to lie, as in the annexed figure, partly within and partly without that circle.　The *greater* part *d*B necessarily lies within it, and not only so,

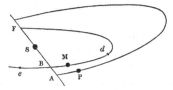

but within that portion, the point of M's orbit nearest to P, in which, therefore, the retrograding force has its maximum, is situated.　Although, therefore, in the por-

tion B *e*, it is true, the retrograde tendency otherwise general over the whole of *that* semicircle (Art. 624.) will be reversed, yet the effect of this will be much more than counterbalanced by the more energetic and more prolonged retrograde action over *d*B ; and, therefore, in this case also, on the average of every possible situation of M, the motion of the node will be retrograde.

(627.) Let us lastly consider an interior planet disturbed by an exterior.　Take M D and M E (fig. of art. 611.) each equal to M, S.　Then first, when P is between D and the node A, being nearer than S to M, the disturbing force acts towards M's orbit on the side on which M lies, and the node recedes.　It also recedes when (M retaining the same situation) P is in any part of the arc E C from E to the other node, because in that situation the direction of the disturbing force, it is true, is reversed, but that portion of P's orbit being also reversely situated with respect to the plane of M's, P is still urged towards the latter plane, but on the side opposite to M. Thus, (M holding its place) whenever P is anywhere in D A or E C, the node recedes.　On the other hand, it advances whenever P is between A and E or between C and D, because, in these arcs, only one of the two determining elements (viz. the direction of the disturbing force with respect to the plane of P's orbit ; and the situation of the one plane with respect to the other as to above and below) has undergone reversal.　Now first, whenever M is anywhere but in the line of nodes, the sum of the arcs D A and E C exceeds a semicircle, and *that* the more, the nearer M is to a position at right angles to the line of nodes. Secondly, the arcs favourable to the recess of the node comprehend those situations in which the orthogonal disturbing force is most powerful, and *vice versâ*. This is evident, because as P approaches D or E, this component decreases, and vanishes at those points (612.). The movement of the node itself also vanishes when P comes to the node, for although in this position the disturbing orthogonal force neither vanishes nor changes its direction, yet, since at the instant of P's passing the node (A) the recess of the node is changed into an advance, it must necessarily at that point be stationary.* Owing to both these causes, therefore, (that the node recedes during a longer time than it advances, and that a more energetic force acting in its recess causes it to recede more rapidly,) the retrograde motion will preponderate on the whole in each complete synodic revolution of P. And it is evident that the reasoning of this and the foregoing articles, is no way vitiated by a moderate amount of excentricity in either orbit.

(628.) It is therefore a general proposition, that on the average of each complete synodic revolution, the node of every disturbed planet recedes upon the orbit of the disturbing one, or in other words, that in every pair of orbits, the

---

* It would seem, at first sight, as if a change *per saltum* took place here, but the continuity of the node's motion will be apparent from an inspection of the annexed figure, where *b a d* is a portion of P's disturbed path near the node A,

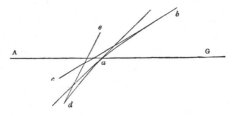

concave towards the plane G A. The momentary place of the moving node is determined by the intersection of the tangent *b e* with A G, which as *b* passes through *a* to *d*, recedes from A to *a*, rests there for an instant, and then advances again.

node of each recedes upon the other, and of course upon any intermediate plane which we may regard as fixed. On a plane not intermediate between them, however, the node of one orbit will advance, and that of the other will recede. Suppose for instance, C A C to be a plane intermediate between P P

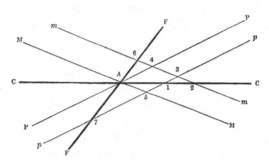

and M M the two orbits. If *p p* and *m m* be the new positions of the orbits, the node of P on M will have receded from A to 5, that of M on P from A to 4, that of P and M on C C respectively from A to 1 and from A to 2. But if F A F be a plane not intermediate, the node of M on that plane has receded from A to 6, but that of P will have *advanced* from A to 7. If the fixed plane have not a common intersection with those of both orbits, it is equally easy to see that the node of the disturbed orbit may either recede on both that plane and the disturbing orbit or advance on the one and recede on the other, according to the relative situation of the planes.

(629.) This is the case with the planetary orbits. They do not all intersect each other in a common node. Although perfectly true, therefore, that the node of any one planet would recede on the orbit of any and each other by the individual action of that other, yet, when all act together, recess on one plane may be equivalent to advance on another, so that the motion of the node of any one orbit on a given plane, arising from their joint action, taking into account the different situations of all the planes, becomes a curiously complicated phænomenon whose law cannot be very easily expressed in words, though reducible to strict numerical statement, being, in fact, a mere geometrical result of what is above shown.

(630.) The nodes of all the planetary orbits on the *true* ecliptic, as a matter of fact, are retrograde, though they are not all so on a fixed plane, such as we may conceive to exist in the planetary system, and to be a plane of reference unaffected by their mutual disturbances. It is, however, to the ecliptic, that we are under the necessity of referring their movements from our station in the system; and if we would transfer our ideas to a fixed plane, it becomes necessary to take account of the variation of the ecliptic itself, produced by the joint action of all the planets.

(631.) Owing to the smallness of the masses of the planets, and their great distances from each other, the revolutions of their nodes are excessively slow, being in every case less than a single degree per century, and in most cases not amounting to half that quantity. It is otherwise with the moon, and that owing to two distinct reasons. First, that the disturbing force itself arising from the sun's action, (as appears from the table given in art. 612.) bears a much larger proportion to the earth's central attraction on the moon than in the case of any planet disturbed by any other. And secondly, because the synodic revolution of the moon, within which the average is struck, (and always on the side of recess) is only $29\frac{1}{2}$ days, a period much shorter than that of any of the planets, and vastly so than that of several among them. All this is agreeable to what has already been stated (art. 407, 408.) respecting the motion of the moon's nodes, and it is hardly necessary to mention that, when calculated, as it has been, *à priori* from an exact estimation of all the acting forces, the result is found to coincide with perfect precision with that immediately derived from observation, so that not a doubt can subsist as to this being the real process by which so remarkable an effect is produced.

(632.) So far as the physical condition of each planet is concerned, it is evident that the position of their nodes can be of little importance. It is otherwise with the mutual inclinations of their orbits with respect to each other, and to the equator of each. A variation in the position of the ecliptic, for instance, by which its pole should shift its distance from the pole of the equator, would disturb our seasons. Should the plane of the earth's orbit, for instance, ever be so changed as to bring the ecliptic to coincide with the equator, we should have perpetual spring over all the world; and, on the other hand, should it coincide with a meridian, the extremes of summer and winter would become intolerable. The enquiry, then, of the variations of inclination of the planetary orbits *inter se,* is one of much higher practical interest than those of their nodes.

(633.) Referring to the figures of art. 610. *et seq.,* it is evident that the plane SPq, in which the disturbed body moves during an instant of time from its quitting P, is differently inclined to the orbit of M, or to a fixed plane, from the original or undisturbed plane P S *p*. The difference of absolute position of these two planes in space is the angle made between the planes P S R and P S *r*, and is therefore calculable by spherical trigonometry, when the angle R S *r* or the momentary recess of the node is known, and also the inclination of the planes of the orbits to each other. We perceive, then, that between the momentary change of inclination, and the momentary recess of the node there exists an intimate relation, and that the research of the one is in fact bound up in that of the other. This may be, perhaps, made clearer, by considering the orbit of P to be not merely an imaginary line, but an actual circle or elliptic hoop of some rigid material, without inertia, on which, as on a wire, the body P may slide as a bead. It is evident that the position of this hoop will be determined at any instant, by its inclination to the ground plane to which it is referred, and by the place of its intersection therewith, or node. It will also be determined by the momentary direction of P's motion, which (having no inertia) it must obey; and any change by which P should, in the next instant, alter its orbit, would be equivalent to a shifting, bodily, of the whole hoop, changing at once its inclination and nodes.

(634.) One immediate conclusion from what has been

pointed out above, is that where the orbits, as in the case of
the planetary system and the moon, are slightly inclined to
one another, the momentary variations of the inclination are
of an order much inferior in magnitude to those in the place
of the node.   This is evident on a mere inspection of our
figure, the angle R P r being, by reason of the small inclination of the planes S P R and R S r, necessarily much
smaller than the angle R S r.   In proportion as the planes
of the orbits are brought to coincidence, a very trifling
angular movement of P p about P S as an axis will make a
great variation in the situation of the point r, where its
prolongation intersects the ground plane.

(635.) Referring to the figure of art. 622., we perceive
that although the motion of the node is retrograde whenever
the momentary disturbed arc P Q lies between the planes
C A and C G A of the two orbits, and *vice versâ*, indifferently
whether P be in the act of receding from the plane C A, as
in the quadrant C G, or of approaching to it, as in G A, yet
the same identity as to the character of the change does not
subsist in respect of the inclination.   The inclination of the
disturbed orbit (*i. e.* of its momentary element) P q or Pq',
is measured by the spherical angle P r H or P r' H.   Now
in the quadrant C G, P r H is less, and P r' H greater than
P C H; but in G A, the converse.   Hence this rule : —
1st., If the disturbing force urge P towards the plane of
M's orbit, and the undisturbed motion of P carry it also
towards that plane ; and 2dly, if the disturbing force urge
P from that plane, while P's undisturbed motion also carries
it from it, in either case the inclination momentarily increases ; but if, 3dly, the disturbing force act to, and P's
motion carry it from — or if the force act from, and the
motion carry it to, that plane, the inclination momentarily
diminishes.   Or (including all the cases under one alternative)
if the action of the disturbing force and the undisturbed
motion of P with reference to the plane of M's orbit be of
the same character, the inclination increases; if of contrary
characters, it diminishes.

(636.) To pass from the momentary changes which take

D D

place in the relations of nature to the accumulated effects
produced in considerable lapses of time by the continued
action of the same causes, under circumstances varied by
these very effects, is the business of the integral calculus.
Without going into any calculations, however, it will be
easy for us to demonstrate from the principles above laid
down, the leading features of this part of the planetary
theory, viz. the periodic nature of the change of the inclinations of two orbits to each other, the re-establishment of
their original values, and the consequent oscillation of each
plane about a certain mean position.   As in explaining the
motion of the nodes, we will commence, as the simplest case,
with that of an exterior planet disturbed by an interior one
at less than half its distance from the central body.   Let
A C A′ be the great circle of the heavens into which M's
orbit seen from S is projected, extended into a straight line,
and A *g* C *h* A′ the corresponding projection of the orbit
of P so seen.   Let M occupy some fixed situation, suppose
in the semicircle A C, and let P describe a complete revolution from A through *g* C *h* to A′.   Then while it is between
A and *g* or in its first quadrant, its motion is *from* the plane
of M's orbit, and at the same time the orthogonal force acts

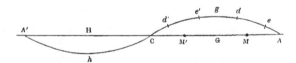

from that plane : the inclination, therefore, (art. 635.) increases.   In the second quadrant the motion of P is *to*, but
the force continues to act from, the plane, and the inclination
again decreases.   A similar alternation takes place in its
course through the quadrants C *h* and *h* A.   Thus the plane
of P's orbit oscillates to and fro about its mean position twice
in each revolution of P.   During this process if M held a
fixed position at G, the forces being symmetrically alike on
either side, the extent of these oscillations would be exactly
equal, and the inclination at the end of one revolution of P

would revert precisely to its original value. But if M be elsewhere, this will not be the case, and in a single revolution of P, only a partial compensation will be operated, and an overplus on the side, suppose of diminution, will remain outstanding. But when M comes to M', a point equidistant from G on the other side, this effect will be precisely reversed (supposing the orbits circular). On the average of both situations, therefore, the effect will be the same as if M were divided into two equal portions, one placed at M and the other at M', which will annihilate the preponderance in question and effect a perfect restoration. And on an average of all possible situations of M, the effect will in like manner be the same as if its mass were distributed over the whole circumference of its orbit, forming a ring, each portion of which will exactly destroy the effect of that similarly situated on the opposite side of the line of nodes.

(637.) The reasoning is precisely similar for the more complicated cases of arts. (625.) and (627.). Suppose that owing either to the proximity of the two orbits, (in the case of an exterior disturbed planet) or to the disturbed orbit being interior to the disturbing one, there were a larger or less portion, $d\,e$, of P's orbit in which these relations were reversed. Let M be the position of M' corresponding to $d\,e$, then taking $G\,M' = G\,M$, there will be a similar portion $d'\,e'$ bearing precisely the same reversed relation to M', and therefore, the actions of M' M, will equally neutralize each other in this as in the former state of things.

(638.) To operate a complete and rigorous compensation, however, it is necessary that M should be presented to P in every possible configuration, not only with respect to P itself, but to the line of nodes, to the position of which line the whole reasoning bears reference. In the case of the moon for example, the disturbed body (the moon) revolves in $27^{d}\cdot322$, the disturbing (the sun) in $365^{d}\cdot256$, and the line of nodes in $6793^{d}\cdot391$, numbers in proportion to each other about as 1 to 13 and 249 respectively. Now in 13 revolutions of P, and one of M, if the node remained fixed, P would have been presented to M so nearly in every configuration as to operate an almost exact compensation. But in 1 revolution of M, or 13 of P, the node itself has shifted $\frac{15}{249}$ or about $\frac{1}{19}$ of a revolution, in a direction opposite to the revolutions of M and P, so that although P has been brought back to the same configuration with respect to M, both are $\frac{1}{19}$ of a revolution in advance of the same configuration as respects the node. The compensation, therefore, will not be exact, and to make it so, this process must be gone through 19 times, at the end of which both the bodies will be restored to the same relative position, not only with respect to each other, but to the node. The fractional parts of entire revolutions, which in this explanation have been neglected, are evidently no farther influential than as rendering the compensation thus operated in a revolution of the node slightly inexact, and thus giving rise to a compound period of greater duration, at the end of which a compensation almost mathematically rigorous, will have been effected.

(639.) It is clear then, that if the orbits be circles, the lapse of a very moderate number of revolutions of the bodies will very nearly, and that of a revolution of the node almost exactly, bring about a perfect restoration of the inclinations. If, however, we suppose the orbits excentric, it is no less evident, owing to the want of symmetry in the distribution of the forces, that a perfect compensation will not be effected either in one or in any number of revolutions of P and M, independent of the motion of the node itself, as there will always be some configuration more favourable to either an increase of inclination than its opposite is unfavourable. Thus will arise a change of inclination which, were the nodes and apsides of the orbits fixed, would be always progressive in one direction until the planes were brought to coincidence. But, 1st, half a revolution of the nodes would of itself reverse the direction of this progression by making the position in question favour the opposite movement of inclination; and, 2dly, the planetary apsides are themselves in motion with unequal velocities, and thus the configuration whose influence destroys the balance, is, itself, always shifting its place on the orbits. The variations of inclination dependent on the excentricities are therefore, like those independent of them, periodical, and being, moreover, of an order more minute (by reason of the smallness of the excentricities) than the latter, it is evident that the total variation of the planetary inclinations must fluctuate within very narrow limits. Geometers have accordingly demonstrated by an accurate analysis of all the circumstances, and an exact estimation of the acting forces, that such is the case; and this is what is meant by asserting the stability of the planetary system as to the mutual inclinations of its orbits. By the researches of Lagrange (of whose analytical conduct it is impossible here to give any idea), the following elegant theorem has been demonstrated: —

"*If the mass of every planet be multiplied by the square root of the major axis of its orbit, and the product by the square of the tangent of its inclination to a fixed plane, the sum of all these products will be constantly the same under the influence of their mutual attraction.*" If the present situation of the plane of the ecliptic be taken for that fixed plane (the ecliptic itself being variable like the other orbits), it is found that this sum is actually very small: it must, therefore, always remain so. This remarkable theorem alone, then, would guarantee the stability of the orbits of the greater planets; but from what has above been shown of the tendency of each planet to work out a compensation on every other, it is evident that the minor ones are not excluded from this beneficial arrangement.

(640.) Meanwhile, there is no doubt that the plane of the ecliptic does actually vary by the actions of the planets. The amount of this variation is about 48″ per century, and has long been recognized by astronomers, by an increase of the latitudes of all the stars in certain situations, and their diminution in the opposite regions. Its effect is to bring the ecliptic by so much per annum nearer to coincidence with the equator; but from what we have above seen, this diminution of the obliquity of the ecliptic will not go on beyond certain very moderate limits, after which (although in an immense period of ages, being a compound cycle resulting from the joint action of all the planets,) it will again increase, and thus oscillate backward and forward about a mean position, the extent of its deviation to one side and the other being less than 1° 21′.

(641.) One effect of this variation of the plane of the ecliptic,— that which causes its nodes on a fixed plane to change,— is mixed up with the precession of the equinoxes, and undistinguishable from it, except in theory. This last-mentioned phænomenon is, however, due to another cause, analagous, it is true, in a general point of view, to those above considered, but singularly modified by the circumstances under which it is produced. We shall endeavour to render these modifications intelligible, as far as they can be made so without the intervention of analytical formulæ.

(642.) The precession of the equinoxes, as we have shown in art. 312., consists in a continual retrogradation of the node of the earth's equator on the ecliptic ; and is, therefore, obviously an effect so far analogous to the general phæ-nomenon of the retrogradation of the nodes of the orbits on each other. The immense distance of the planets, however, compared with the size of the earth, and the smallness of their masses compared to that of the sun, puts *their* action out of the question in the enquiry of its cause, and we must, therefore, look to the massive though distant sun, and to our near though minute neighbour, the moon, for its explanation. This will, accordingly, be found in their dis-turbing action on the redundant matter accumulated on the equator of the earth, by which its figure is rendered spheroidal, combined with the earth's rotation on its axis. It is to the sagacity of Newton that we owe the discovery of this singular mode of action.

(643.) Suppose in our figure (art. 611.) that instead of one body, P, revolving round S, there were a succession of particles not coherent, but forming a kind of fluid ring, free to change its form by any force applied. Then, while this ring revolved round S in its own plane, under the disturbing influence of the distant body M, (which now represents the moon or the sun, as P does one of the particles of the earth's equator,) two things would happen : 1st, its figure would be bent out of a plane into an undulated form, those parts of it within the arcs D A and E C being rendered more inclined to the plane of M's orbit, and those within the arcs A E, C D, less so than they would otherwise be; 2dly, the nodes of this ring, regarded as a whole, without respect to its change of figure, would retreat upon that plane.

(644.) But suppose this ring, instead of consisting of discrete molecules free to move independently, to be rigid and incapable of such flexure, like the *hoop* we have supposed in art. 633., but having inertia, then it is evident that the effort of those parts of it which tend to become more inclined will act through the medium of the ring itself (as a mechanical engine or lever) to counteract the effort of those which have *at the same instant* a contrary tendency. In so far only, then, as there exists an excess on the one or the other side will the inclination change, an average being struck at every moment of the ring's motion; just as was shown to happen in the view we have taken of the inclinations, in every complete revolution of a single disturbed body, under the influence of a fixed disturbing one.

(645.) Meanwhile, however, the nodes of the rigid ring will retrograde, the general or average tendency of the nodes of every molecule being to do so. Here, as in the other case, a struggle will take place by the counteracting efforts of the molecules contrarily disposed, propagated through the solid substance of the ring; and thus at every instant of time, an average will be struck, which being identical in its nature with that effected in the complete revolution of a single disturbed body, will, in every case, be in favour of a recess of the node, save only when the disturbing body, be it sun or moon, is situated in the plane of the earth's equator.

(646.) This reasoning is evidently independent of any consideration of the cause which maintains the rotation of the ring; whether the particles be small satellites retained in circular orbits under the equilibrated action of attractive and centrifugal forces, or whether they be small masses conceived as attached to a set of imaginary spokes, as of a wheel, centering in S, and free only to shift their planes by a motion of those spokes perpendicular to the plane of the wheel. This makes no difference in the *general* effect; though the different velocities of rotation, which may be impressed on such a system, may and will have a very great influence both on the absolute and relative magnitudes of the two effects in question — the motion of the nodes and change of inclination. This will be easily understood, if we suppose the ring *without* a rotatory motion, in which extreme case it is obvious that so long as M remained fixed there would take place no recess of nodes at all, but only a tendency of the ring to tilt its plane round a diameter perpendicular to the position of M, bringing it towards the line S M.

(647.) The motion of such a ring, then, as we have been considering, would imitate, so far as the recess of the nodes goes, the precession of the equinoxes, only that its nodes would retrograde far more rapidly than the observed precession, which is excessively slow. But now conceive this ring to be loaded with a spherical mass enormously heavier than itself, placed concentrically within it, and cohering firmly to it, but indifferent, or very nearly so, to any such cause of motion; and suppose, moreover, that instead of one such ring there are a vast multitude heaped together around the equator of such a globe, so as to form an elliptical protuberance, enveloping it like a shell on all sides, but whose mass, taken together, should form but a very minute fraction of the whole spheroid. We have now before us a tolerable representation of the case of nature *; and it is evident that the rings, having to drag round with them in their nodal revolution this great inert

* That a perfect sphere would be so inert and indifferent as to a revolution of the nodes of its equator under the influence of a distant attracting body appears from this, — that the direction of the resultant attraction of such a body, or of that single force which, opposed, would neutralize and destroy its whole action, is necessarily in a line passing through the center of the sphere, and, therefore, can have no tendency to turn the sphere one way or other. It may be objected by the reader, that the whole sphere may be conceived as consisting of rings parallel to its equator, of every possible diameter, and that, therefore, its nodes should retrograde even without a protuberant equator. The inference is incorrect, but our limits will not allow us to go into an exposition of the fallacy. We should, however, caution him, generally, that no dynamical subject is open to more mistakes of this kind, which nothing but the closest attention, in every varied point of view, will detect.

mass, will have their velocity of retrogradation proportionally diminished. Thus, then, it is easy to conceive how a motion similar to the precession of the equinoxes, and, like it, characterized by extreme slowness, will arise from the causes in action.

(648.) Now a recess of the node of the earth's equator, upon a given plane, corresponds to a conical motion of its axis round a perpendicular to that plane. But in the case before us, that plane is not the ecliptic, but the moon's orbit for the time being; and it may be asked how we are to reconcile this with what is stated in art. 317. respecting the nature of the motion in question. To this we reply, that the nodes of the lunar orbit, being in a state of continual and rapid retrogradation, while its inclination is preserved nearly invariable, the point in the sphere of the heavens round which the pole of the earth's equator revolves (with that extreme slowness characteristic of the precession) is itself in a state of continual circulation round the pole of the ecliptic, with that much more rapid motion which belongs to the lunar node. A glance at the annexed figure will explain this better than words. P is the

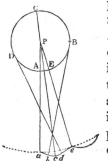

pole of the ecliptic, A the pole of the moon's orbit, moving round the small circle A B C D in 19 years; a the pole of the earth's equator, which at each moment of its progress has a *direction* perpendicular to the varying position of the line A a, and a *velocity* depending on the varying intensity of the acting causes during the period of the nodes. This velocity however being extremely small, when A comes to B, C, D, E, the line A a will have taken up the positions B b, C c, D d, E e, and the earth's pole a will thus, in one tropical revolution of the node, have arrived at e, having described not an exactly circular arc $a\,e$, but a single undulation of a wave-shape or epicycloidal curve, $a\,b\,c\,d\,e$, with a velocity alternately greater and less than its mean motion, and this will be repeated in every succeeding revolution of the node.

(649.) Now this is precisely the kind of motion which, as we have seen in art. 325., the pole of the earth's equator really has round the pole of the ecliptic, in consequence of the joint effects of precession and nutation, which are thus uranographically represented. If we superadd to the effect of lunar precession that of the solar, which alone would cause the pole to describe a circle uniformly about P, this will only affect the undulations of our waved curve, by extending them in length, but will produce no effect on the depth of the waves, or the excursions of the earth's axis to and from the pole of the ecliptic. Thus we see that the two phenomena of nutation and precession are intimately connected, or rather both of them essential constituent parts of one and the same phenomenon. It is hardly necessary to state that a rigorous analysis of this great problem, by an exact estimation of all the acting forces and summation of their dynamical effects, leads to the precise value of the co-efficients of precession and nutation, which observation assigns to them. The solar and lunar portions of the precession of the equinoxes, that is to say, those portions which are uniform, are to each other in the proportion of about 2 to 5.

(650.) In the nutation of the earth's axis we have an example (the first of its kind which has occurred to us), of a periodical movement in one part of the system, giving rise to a motion having the same precise period in another. The motion of the moon's nodes is here, we see, represented, though under a very different form, yet in the same exact periodic time, by a movement of a peculiar oscillatory kind impressed on the solid mass of the earth. We must not let the opportunity pass of generalizing the principle involved in this result, as it is one which we shall find again and again exemplified in every part of physical astronomy, nay, in every department of natural science. It may be stated as " the principle of forced oscillations, or of forced vibrations, " and thus generally announced : —

If one part of any system connected either by material ties, or by the mutual attractions of its members, be continually maintained by any cause, whether inherent in the constitution of the system or external to it, in a state of regular periodic motion, that motion will be propagated throughout the whole systems and will give rise, in every member of it, and in every part of each member, to periodic movements executed in equal period, with that to which they owe their origin, though not necessarily synchronous with them in their maxima and minima. *

The system may be favourably or unfavourably constituted for such a transfer of periodic movements, or favourably in some of its parts and unfavourably in others; and accordingly as it is the one or the other, the *derivative* oscillation (as it may be termed) will be imperceptible in one case, of appreciable magnitude in another, and even more perceptible in its visible effects than the original cause in a third; of this last kind we have an instance in the moon's acceleration, to be hereafter noticed.

(651.) It so happens that our situation on the earth, and the delicacy which our observations have attained, enable us to make it as it were an instrument to *feel* these forced vibrations, — these derivative motions, communicated from various quarters, especially from our near neighbour, the moon, much in the same way as we detect, by the trembling of a board beneath us, the secret transfer of motion by which the sound of an organ pipe is dispersed through the air, and carried down into the earth. Accordingly, the monthly revolution of the moon, and the annual motion of the sun, produce, each of them, small *nutations* in the earth's axis, whose periods are respectively half a month and half a year, each of which, in this view of the subject, is to be regarded as one portion of a period consisting of two equal and similar parts. But the most remarkable instance, by far, of this propagation of periods, and one of high importance to mankind, is that of the tides, which are forced oscillations, excited by the rotation of the earth in an ocean disturbed from its figure by the varying attractions of the sun and moon, each

---

* See a demonstration of this theorem for the forced vibrations of systems connected by material ties of imperfect elasticity, in my treatise on Sound, Encyc. Metrop. art. 323. The demonstration is easily extended and generalized to take in other systems.

revolving in its own orbit, and propagating its own period into the joint phenomenon. The explanation of the tides, however, belongs more properly to that part of the general subject of perturbations which treats of the action of the radial component of the disturbing force, and is therefore postponed to a subsequent chapter.

CHAPTER XIII.

THEORY OF THE AXES, PERIHELIA, AND EXCEN-
TRICITIES.

VARIATION OF ELEMENTS IN GENERAL. — DISTINCTION BETWEEN
PERIODIC AND SECULAR VARIATIONS. — GEOMETRICAL EXPRESSION
OF TANGENTIAL AND NORMAL FORCES.—VARIATION OF THE MAJOR
AXIS PRODUCED ONLY BY THE TANGENTIAL FORCE.—LAGRANGE'S
THEOREM OF THE CONSERVATION OF THE MEAN DISTANCES AND
PERIODS. — THEORY OF THE PERIHELIA AND EXCENTRICITIES.—
— GEOMETRICAL REPRESENTATION OF THEIR MOMENTARY VARIA-
TIONS. — ESTIMATION OF THE DISTURBING FORCES IN NEARLY
CIRCULAR ORBITS. — APPLICATION TO THE CASE OF THE MOON. —
THEORY OF THE LUNAR APSIDES AND EXCENTRICITY. — EXPE-
RIMENTAL ILLUSTRATION. — APPLICATION OF THE FOREGOING
PRINCIPLES TO THE PLANETARY THEORY. — COMPENSATION IN
ORBITS VERY NEARLY CIRCULAR. — EFFECTS OF ELLIPTICITY. —
GENERAL RESULTS. — LAGRANGE'S THEOREM OF THE STABILITY
OF THE EXCENTRICITIES.

(652.) IN the foregoing chapter we have sufficiently explained the action of the orthogonal component of the disturbing force, and traced it to its results in a continual
displacement of the plane of the disturbed orbit, in virtue of
which the nodes of that plane alternately advance and recede
upon the plane of the disturbing body's orbit, with a general
preponderance on the side of advance, so as after the lapse
of a long period to cause the nodes to make a complete revolution and come round to their former situation. At the
same time the inclination of the plane of the disturbed motion continually changes, alternately increasing and diminishing; the increase and diminution however compensating each
other, nearly in single revolutions of the disturbed and disturbing bodies, more exactly in many, and with perfect
accuracy in long periods, such as those of a complete revolution of the nodes and apsides. In the present and follow-

ing chapters we shall endeavour to trace the effects of the other components of the disturbing force, — those which act in the plane (for the time being) of the disturbed orbit, and which tend to derange the elliptic form of the orbit, and the laws of elliptic motion in that plane.   The small inclination, generally speaking, of the orbits of the planets and satellites to each other, permits us to separate these effects in theory one from the other, and thereby greatly to simplify their consideration. Accordingly, in what follows, we shall throughout neglect the mutual inclination of the orbits of the disturbed and disturbing bodies, and regard all the forces as acting and all the motions as performed in one plane.

(653.) In considering the changes induced by the mutual action of two bodies in different aspects with respect to each other on the magnitudes and forms of their orbits and in their positions therein, it will be proper in the first instance to explain the conventions under which geometers and astronomers have alike agreed to use the language and laws of the elliptic system, and to continue to apply them to disturbed orbits, although those orbits so disturbed are no longer, in mathematical strictness, ellipses, or any known curves. This they do, partly on account of the convenience of conception and calculation which attaches to this system, but much more for this reason, — that it is found, and may be demonstrated from the dynamical relations of the case, that the departure of each planet from its ellipse, as determined at any epoch, is capable of being truly represented, by supposing the ellipse itself to be slowly variable, to change its magnitude and excentricity, and to shift its position and the plane in which it lies according to certain laws, while the planet all the time continues to move in this ellipse, just as it would do if the ellipse remained invariable and the disturbing forces had no existence. By this way of considering the subject, the whole effect of the disturbing forces is regarded as thrown upon the orbit, while the relations of the planet to that orbit remain unchanged. This course of procedure, indeed, is the most natural, and is in some sort forced upon us by the extreme slowness with which the variations of the elements, at least where the planets only are concerned, develope themselves. For instance, the fraction expressing the excentricity of the earth's orbit changes no more than 0·00004 in its amount in a *century*; and the place of its perihelion, as referred to the sphere of the heavens, by only 19′ 39″ in the same time. For several years, therefore, it would be next to impossible to distinguish between an ellipse so varied and one that had not varied at all; and in a single revolution, the difference between the original ellipse and the curve really represented by the varying one, is so excessively minute, that, if accurately drawn on a table, six feet in diameter, the nicest examination with microscopes, continued along the whole outlines of the two curves, would hardly detect any perceptible interval between them. Not to call a motion so minutely conforming itself to an elliptic curve, *elliptic*, would be affectation, even granting the existence of trivial departures alternately on one side or on the other; though, on the other hand, to neglect a variation, which continues to accumulate from age to age, till it forces itself on our notice, would be wilful blindness.

(654.) Geometers, then, have agreed in each single revolution, or for any moderate interval of time, to regard the motion of each planet as elliptic, and performed according to Kepler's laws, with a reserve in favour of those very small and transient fluctuations which take place within that time, but at the same time to regard all the *elements* of each ellipse as in a continual, though extremely slow, state of change; and, in tracing the effects of perturbation on the system, they take account principally, or entirely, of this change of the elements, as that upon which any material change in the great features of the system will ultimately depend.

(655.) And here we encounter the distinction between what are termed secular variations, and such as are rapidly periodic, and are compensated in short intervals. In our exposition of the variation of the inclination of a disturbed orbit (art. 636.), for instance, we showed that, in each single revolution of the disturbed body, the plane of its motion underwent fluctuations to and fro in its inclination to that of the disturbing body, which nearly compensated each other; leaving, however, a portion outstanding, which again is nearly compensated by the revolution of the disturbing body, yet still leaving outstanding and uncompensated a minute portion of the change which requires a whole revolution of the node to compensate and bring it back to an average or mean value. Now, the two first compensations which are operated by the planets going through the succession of configurations with each other, and therefore in comparatively short periods, are called periodic variations; and the deviations thus compensated are called *inequalities depending on configurations ;* while the last, which is operated by a period of the node (one of the *elements*), has nothing to do with the configurations of the individual planets, requires a very long period of time for its consummation, and is, therefore, distinguished from the former by the term *secular* variation.

(656.) It is true, that, to afford an exact representation of the motions of a disturbed body, whether planet or satellite, both periodical and secular variations, with their corresponding inequalities, require to be expressed; and, indeed, the former even more than the latter; seeing that the secular inequalities are, in fact, nothing but what remains after the mutual destruction of a much larger amount (as it very often is) of periodical. But these are in their nature transient and temporary: they disappear in short periods, and leave no trace. The planet is temporarily drawn from its orbit (its slowly varying orbit), but forthwith returns to it, to deviate presently as much the other way, while the varied orbit accommodates and adjusts itself to the average of these excursions on either side of it ; and thus continues to present, for a succession of indefinite ages, a kind of medium picture of all that the planet has been doing in their lapse, in which the expression and character is preserved; but the individual features are merged and lost. These periodic inequalities, however, are, as we have observed, by no means neglected, but it is more convenient to take account of them by a separate process, independent of the secular variations of the elements.

(657.) In order to avoid complication, while endeavouring

to give the reader an insight into both kinds of variations, we shall henceforward conceive all the orbits to lie in one plane, and confine our attention to the case of two only, that of the disturbed and disturbing body, a view of the subject which (as we have seen) comprehends the case of the moon disturbed by the sun, since any one of the bodies may

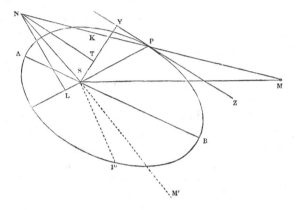

be regarded as fixed at pleasure, provided we conceive all its motions transferred in a contrary direction to each of the others.   Let therefore A P B be the undisturbed elliptic orbit of a planet P; M a disturbing body, join M P, and supposing M K=M S take M N : M K :: M K² : M P². Then if S N be joined, N S will represent the disturbing force of M or P, on the same scale that S M represents M's attraction on S.   Suppose Z P Y a tangent at P, S Y perpendicular to it, and N T, N L perpendicular respectively to S Y and P S produced.   Then will N T represent the tangential, T S the normal, N L the transversal, and L S the radial components of the disturbing force.   In circular orbits or orbits only slightly elliptic, the directions P S L and S Y are nearly coincident, and the former pair of forces will differ but slightly from the latter.   We shall here, however, take the general case, and proceed to investigate in an elliptic orbit of any degree of excentricity the momentary changes produced by the action of the disturbing force in those elements

E E

on which the magnitude, situation, and form of the orbit depends (*i.e.* the length and position of the major axis and the excentricity), in the same way as in the last chapter we determined the momentary changes of the inclination and node similarly produced by the orthogonal force.

(658.) We shall begin with the momentary variation in the *length* of the axis, an element of the first importance, as on it depends (art. 487) the periodic time and mean angular motion of the planet, as well as the average supply of light and heat it receives in a given time from the sun, any permanent or constantly progressive change in which would alter most materially the conditions of existence of living beings on its surface. Now it is a property of elliptic motion performed under the influence of gravity, and in conformity with Kepler's laws, that if the velocity with which a planet moves at any point of its orbit be given, and also the distance of that point from the sun, the major axis of the orbit is thereby also given. It is no matter in what *direction* the planet may be moving at that moment. This will influence the excentricity and the position of its ellipse, but not its length. This property of elliptic motion has been demonstrated by Newton, and is one of the most obvious and elementary conclusions from his theory. Let us now consider a planet describing an indefinitely small arc of its orbit about the sun, under the joint influence of its attraction, and the disturbing power of another planet. This arc will have some certain curvature and direction, and, therefore, may be considered as an arc of a certain ellipse described about the sun as a focus, for this plain reason, — that whatever be the curvature and direction of the arc in question, an ellipse may always be assigned, whose focus shall be in the sun, and which shall coincide with it throughout the whole interval (supposed indefinitely small) between its extreme points. This is a matter of pure geometry. It does not follow, however, that the ellipse thus instantaneously determined will have the same elements as that similarly determined from the arc described in either the previous or the subsequent instant. If the disturbing force did not exist, this would be the case ; but, by its action, a variation of the element from instant to instant is produced, and the ellipse so determined is in a continual state of change. Now when the planet has reached the end of the small arc under consideration, the question whether it will in the next instant describe an arc of an ellipse having the same or a varied axis will depend, not on the new *direction* impressed upon it by the acting forces, — for the axis, as we have seen, is independent of that direction, — not on its change of distance from the sun, while describing the former arc, — for the elements of that arc are accommodated to it, so that one and the same axis must belong to its beginning and its end. The question, in short, whether in the next arc it shall take up a new major axis or go on with the old one will depend solely on this — whether its *velocity* has or has not undergone a change by the action of the *disturbing* force. For the central force residing in the focus can impress on it no such change of velocity as to be incompatible with the permanence of its ellipse, seeing that it is by the action of that force that the velocity is maintained in that due proportion to the distance which elliptic motion, as such, requires.

(659.) Thus we see that the momentary variation of the major axis depends on nothing but the momentary deviation from the law of elliptic velocity produced by the disturbing force, without the least regard to the direction in which that extraneous velocity is impressed, or the distance from the sun at which the planet may be situated, at the moment of its impression. Nay, we may even go farther, for, as this holds good at every instant of its motion, it will follow, that after the lapse of any time, however great, the total amount of change which the axis may have undergone will be determined only by the total deviation produced by the action of the disturbing force in the velocity of the disturbed body from that which it would have had in its undisturbed ellipse, at the same distance from the center, and that therefore the total amount of change produced in the axis in any lapse of time may be estimated, if we know at every instant the efficacy of the disturbing force to alter the velocity of the body's motion, and *that* without any regard to the alterations which the action of that force may have produced in the other elements of the motion in the same time.

(660.) Now it is not the whole disturbing force which is effective in changing P's velocity, but only its tangential component. The normal component tends merely to alter the curvature of the orbit or to deflect it into conformity with a circle of curvature of greater or lesser radius, as the case may be, and in no way to alter the velocity. Hence it appears that the *variation of the length of the axis is due entirely to the tangential force, and is quite independent on the normal.* Now it is easily shown that as the velocity increases, the axis increases (the distance remaining unaltered *) though not in the same exact proportion. Hence it follows that if the tangential disturbing force conspires with the motion of P, its momentary action increases the axis of the disturbed orbit, whatever be the situation of P in its orbit, and *vice versâ.*

(661.) Let A S B (fig. art. 657.) be the major axis of the ellipse A P B, and on the opposite side of A B take two points P′ and M′, similarly situated with respect to the axis with P and M on their side. Then if at P′ and M′ bodies equal to P and M be placed, the forces exerted by M′ on P′ and S will be equal to those exerted by M on P and S, and therefore the tangential disturbing force of M′ on P′ exerted in the direction P′ Z′ (suppose) will equal that exerted by M on P in the direction P Z. P′ therefore (supposing it to revolve in the same direction round S as P) will be retarded (or accelerated, as the case may be) by *precisely* the same force by which P is accelerated (or retarded), so that the variation in the axis of the respective orbits of P and P′ will be equal in amount, but contrary in character. Suppose now M's orbit to be circular. Then (*if the periodic times of* M *and* P *be not commensurate, so that a moderate number of revolutions may bring them back to the same precise relative positions*) it will necessarily happen, that in the course of a very great

---

* If a be the semiaxis, r the radius vector, and v the velocity of P in any point of an ellipse, a is given by the relation $v^2 = \dfrac{2}{r} - \dfrac{1}{a}$, the units of velocity and force being properly assumed.

number of revolutions of both bodies, P will have been presented to M on one side of the axis, at some one moment, in the same manner as at some other moment on the other. Whatever variation may have been effected in its axis in the one situation will have been reversed in that symmetrically opposite, and the ultimate result, on a general average of an infinite number of revolutions, will be a complete and exact compensation of the variations in one direction by those in the direction opposite.

(662.) Suppose, next, P's orbit to be circular. If now M's orbit were so also, it is evident that in one complete synodic revolution, an exact restoration of the axis to its original length would take place, because the tangential forces would be symmetrically equal and opposite during each alternate quarter revolution. But let M, during a synodic revolution, have *receded* somewhat from S, then will its disturbing power have become gradually weaker, so that, in a synodic revolution the tangential force in each quadrant, though reversed in direction being inferior in power, an exact compensation will not have been effected, but there will be left an outstanding uncompensated portion, the excess of the stronger over the feebler effects. But now suppose M to approach by the same gradations as it before receded. It is clear that this result will be reversed; since the uncompensated stronger actions will all lie in the opposite direction. Now suppose M's orbit to be elliptic. Then during its recess from S or in the half revolution from its perihelion to its aphelion, a continual uncompensated variation will go on accumulating in one direction. But from what has been said, it is clear that this will be destroyed, during M's approach to S in the other half of its orbit, so that here again, on the average of a multitude of revolutions during which P *has been* presented to M *in every situation for every distance of* M *from* S, the restoration will be effected.

(663.) If neither P's nor M's orbit be circular, and if moreover the directions of their axes be different, this reasoning, drawn from the symmetry of their relations to each other, does not apply, and it becomes necessary to take a more general

view of the matter.   Among the fundamental relations of
dynamics, relations which presuppose no particular law of
force like that of gravitation, but which express in general
terms the results of the action of *force* on *matter* during *time,*
to produce or change *velocity,* is one usually cited as the
" Principle of the conservation of the *vis viva,*" which applies
directly to the case before us.   This principle (or rather this
theorem) declares that if a body subjected at every instant of
its motion to the action of forces directed to fixed centers (no
matter how numerous), and having their intensity dependent
only on the distances from their respective centers of action,
travel from one point of space to another, the velocity which
it has on its arrival at the latter point will differ from that
which it had on setting out from the former, by a quantity
depending only on the different relative situations of these two
points in space, without the least reference to the form of the
curve in which it may have moved in passing from one point
to the other, whether that curve have been described freely
under the simple influence of the central forces, or the body
have been compelled to glide upon it, as a bead upon a smooth
wire.   Among the forces thus acting may be included any
constant forces, acting in parallel directions, which may be
regarded as directed to fixed centers infinitely distant.   It
follows from this theorem, that, if the body return to the point
P from which it set out, its velocity of arrival will be the same
with that of its departure ; a conclusion which (for the purpose
we have in view) sets us free from the necessity of entering
into any consideration of the laws of the disturbing force,
the change which its action may have induced in the form of
the orbit of P, or the successive steps by which velocity generated at one point of its intermediate path is destroyed at
another, by the reversed action of the tangential force.   Now
to apply this theorem to the case in question, let M be supposed to retain a fixed position during one whole revolution
of P.   P then is acted on, during that revolution, by three
forces : 1st. by the central attraction of S directed always to
S ; 2nd. by that to M, always directed to M ; 3rd. by a force
equal to M's attraction on S ; but in the direction M S, which

therefore is a constant force, acting always in parallel directions. On completing its revolution, then, P's velocity, and therefore the major axis of its orbit, will be found unaltered, at least neglecting that excessively minute difference which will result from the non-arrival after a revolution at the *exact* point of its departure by reason of the perturbations in the orbit produced in the interim by the disturbing force, which for the present we may neglect.

(664.) Now suppose M to revolve, and it will appear, by a reasoning precisely similar to that of art. 662., that whatever uncompensated variation of the velocity arises in successive revolutions of P during M's recess from S will be destroyed by contrary uncompensated variations arising during its approach. Or, more simply and generally thus: whatever M's situation may be, for every place which P can have, there must exist some other place of P (as P'), in which the action of M shall be precisely reversed. Now *if the periods be incommensurable*, in an indefinite number of revolutions of both bodies, for every possible combination of situations (M, P) there will occur, *at some time or other*, the combination (M, P') which neutralizes the effect of the other, when carried to the general account; so that ultimately, and when very long periods of time are embraced, a complete compensation will be found to be worked out.

(665.) This supposes, however, that in such long periods the orbit of M is not so altered as to render the occurrence of the compensating situation (M, P') impossible. This would be the case if M's orbit were to dilate or contract indefinitely by a variation in *its* axis. But the same reasoning which applies to P, applies also to M. P retaining a *fixed* situation, M's velocity, and therefore the axis of its orbit, would be exactly restored at the end of a revolution of M; so that for every position P M there exists a compensating position P M'. Thus M's orbit is maintained of the same magnitude, and the possibility of the occurrence of the compensating situation (M, P') is secured.

(666.) To demonstrate as a rigorous mathematical truth the complete and absolute ultimate compensation of the variations in question, it would be requisite to show that the minute outstanding changes due to the non-arrivals of P and M at the same *exact* points at the end of each revolution, cannot accumulate in the course of infinite ages in one direction. Now it will appear in the subsequent part of this chapter, that the effect of perturbation on the excentricities and apsides of the orbits is to cause the former to undergo only periodical variations, and the latter to revolve and take up in succession every possible situation. Hence in the course of infinite ages, the points of arrival of P and M at fixed lines of direction, S P, S M, in successive revolutions, though at one time they will approach S, at another will recede from it, fluctuating to and fro about mean points from which they never greatly depart. And if the arrival of either of them at P, at a point *nearer* S, at the end of a complete revolution, cause an *excess* of velocity, its arrival at a more distant point will cause a deficiency, and thus, as the fluctuations of distance to and fro ultimately balance each other, so will also the excesses and defects of velocity, though in periods of enormous length, being no less than that of a complete revolution of P's apsides for the one cause of inequality, and of a complete restoration of its excentricity for the other.

(667.) The dynamical proposition on which this reasoning is based is general, and applies equally well to cases wherein the forces act in one plane, or are directed to centers anywhere situated in space. Hence, if we take into consideration the inclination of P's orbit to that of M, the same reasoning will apply. Only that in this case, upon a complete revolution of P, the variation of inclination and the motion of the nodes of P's orbit will prevent its returning to a point in the exact *plane* of its original orbit, as that of the excentricity and perihelion prevent its arrival at the same exact *distance* from S. But since it has been shown that the inclination fluctuates round a mean state from which it never departs much, and since the node revolves and makes a complete circuit, it is obvious that in a complete period of the latter the points of arrival of P at the same longitude will deviate as often and by the same quantities above as below its original point of departure from exact coincidence; and, therefore, that on the average of an infinite number of revolutions, the effect of this cause of non-compensation will also be destroyed.

(668.) It is evident, also, that the dynamical proposition in question being general, and applying equally to any *number* of fixed centers, as well as to any distribution of them in space, the conclusion would be precisely the same whatever be the number of disturbing bodies, only that the periods of compensation would become more intricately involved. We are, therefore, conducted to this most remarkable and important conclusion, viz. that the major axes of the planetary (and lunar) orbits, and, consequently, also their mean motions and periodic times, are subject to none but periodical changes; that the length of the year, for example, in the lapse of infinite ages, has no preponderating tendency either to increase or diminution, — that the planets will neither recede indefinitely from the sun, nor fall into it, but continue, so far as their mutual perturbations at least are concerned, to revolve for ever in orbits of very nearly the same dimensions as at present.

(669.) This theorem (the *Magna Charta* of our system), the discovery of which is due to Lagrange, is justly regarded as the most important, as a single result, of any which have hitherto rewarded the researches of mathematicians in this application of their science; and it is especially worthy of remark, and follows evidently from the view here taken of it, that it would not be true but for the influence of the perturbing forces on other elements of the orbit, viz. the perihelion and excentricity, and the inclination and nodes; since we have seen that the revolution of the apsides and nodes, and the periodical increase and diminution of the excentricities and inclinations, are both essential towards operating that final and complete compensation which gives it a character of mathematical exactness. We have here an instance of a perturbation of one kind operating on a perturbation of another to annihilate an effect which would otherwise accumulate to the destruction of the system. It must, however, be borne in mind, that it is the smallness of the excentricities of the more influential planets, which gives this theorem its *practical* importance, and distinguishes it from a mere barren speculative result. Within the limits of ultimate restoration, it is this alone which keeps the periodical fluctuations of the axis to and fro about a mean value within moderate and reasonable limits. Although the earth might not fall into the sun, or recede from it beyond the present limits of our system, any considerable increase or diminution of its mean distance, to the extent, for instance, of a tenth of its actual amount, would not fail to subvert the conditions on which the existence of the present race of animated beings depends. Constituted as our system is, however, changes to anything like this extent are utterly precluded. The greatest departure from the mean value of the axis of any planetary orbit yet recognized by theory or observation (that of the orbit of Saturn disturbed by Jupiter), does not amount to a thousandth part of its length.* The effects of these fluctuations, however, are very sensible, and manifest themselves in alternate accelerations and retardations in the angular motions of the disturbed about the central body, which cause it alternately to outrun and to lag behind its *elliptic* place in its orbit, giving rise to what are called equations in its motion, some of the chief instances of which will be hereafter specified when we come to trace more particularly in detail the effects of the tangential force in various configurations of the disturbed and disturbing bodies, and to explain the consequences of a near approach to commensurability in their periodic times. An exact commensurability in this respect, such, for instance, as would bring both planets round to the same configuration in two or three revolutions of one of them, would appear at first sight to destroy one of the essential elements of our demonstration. But even supposing such an exact adjustment to subsist at any epoch, it could

---

* Greater deviations will probably be found to exist in the orbits of the small extra-tropical planets. But these are too insignificant members of our system to need special notice in a work of this nature.

not remain permanent, since by a remarkable property of perturbations of this class, which geometers have demonstrated, but the reasons of which we cannot stop to explain, any change produced on the axis of the disturbed planet's orbit is necessarily accompanied by a change *in the contrary direction* in that of the disturbing, so that the periods would recede from commensurability by the mere effect of their mutual action.   Cases are not wanting in the planetary system of a certain approach to commensurability, and in one very remarkable case (that of Uranus and Neptune) of a considerably near one, not near enough, however, in the smallest degree to affect the validity of the argument, but only to give rise to inequalities of very long periods, of which more presently.*

(670.) The variation of the length of the axis of the disturbed orbit is due solely to the action of the tangential disturbing force.   It is otherwise with that of its excentricity and of the position of its axis, or, which is the same thing, the longitude of its perihelion.   Both the normal and tangential components of the disturbing force affect these elements. We shall, however, consider separately the influence of each,

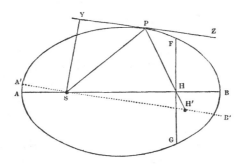

and, commencing, as the simplest case, with that of the tangential force ; — let P be the place of the disturbed planet in its elliptic orbit A P B, whose axis at the moment is A S B and focus S.   Suppose Y P Z to be a tangent to this orbit at

---

* 41 revolutions of Neptune are nearly equal to 81 of Uranus, giving rise to an inequality, having 6805 years for its period.

P. Then, if we suppose A B = 2 a, the other focus of the ellipse, H, will be found by making the angle Z P H = Y P S or Y P H = 180° — Y P Z, or S P H = 180° — 2 Y P S, and taking P H = 2 a — S P. This is evident from the nature of the ellipse, in which lines drawn from any point to the two foci make equal angles with the tangent, and have their sum equal to the major axis. Suppose, now, the tangential force to act on P and to increase its velocity. It will therefore increase the axis, so that the new value assumed by a (viz. a') will be greater than a. But the tangential force does not alter the angle of tangency, so that to find the new position (H') of the upper focus, we must measure off along *the same line* P H, a distance P H' (= 2 a' — P H) greater than P H. Do this then, and join S H and produce it. Then will A' B' be the new position of the axis, and ½ S H' the new excentricity. Hence we conclude, 1st, that the new position of the perihelion A will deviate from the old one A towards the same side of the axis A B on which P is when the tangential force acts to increase the velocity, whether P be moving from perihelion to aphelion, or the contrary. 2dly, That on the same supposition as to the action of the tangential force, the excentricity increases when P is between the perihelion and the perpendicular to the axis F H G drawn through the upper focus, and diminishes when between the aphelion and the same perpendicular. 3dly, That for a given change of velocity, *i. e.* for a given value of the tangential force, the momentary variation in the place of the perihelion is a maximum when P is at F or G, from which situation of P to the perihelion or aphelion, it decreases to nothing, the perihelion being stationary when P is at A or B. 4thly, That the variation of the excentricity due to this cause is complementary in its law of increase and decrease to that of the perihelion, being a maximum for a given tangential force when P is at A or B, and vanishing when at G or F. And lastly, that where the tangential force acts to diminish the velocity, all these results are reversed. If the orbit be very nearly circular* the points F, G, will be

* So nearly that the cube of the excentricity may be neglected.

so situated that, although not at opposite extremities of a *diameter*, the *times* of describing A F, F M, M G, and G A will be all equal, and each of course one quarter of the whole periodic time of P.

(671.) Let us now consider the effects of the normal component of the disturbing force upon the same elements.   The direct effect of this force is to increase or diminish the curvature of the orbit at the point P of its action, without producing any change on the velocity, so that the length of the

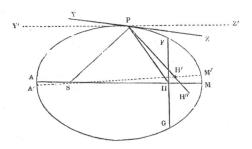

axis remains unaltered by its action.   Now, an increase of curvature at P is synonymous with a decrease in the angle of tangency S P Y when P is approaching towards S, and with an increase in that angle when receding from S.   Suppose the former case, and while P approaches S (or is moving from aphelion to perihelion), let the normal force act *inwards* or towards the concavity of the ellipse.   Then will the tangent P Y by the action of that force have taken up the position P Y'.   To find the corresponding position H' taken up by the focus of the orbit so disturbed, we must make the angle S P H′ = 180° — 2 S P Y', or, which comes to the same, draw P H′ on the side of P H opposite to S, making the angle H P H′ = twice the angle of deflection Y P Y' and in P H take P H′ = P H.   Joining, then, S H' and producing it, A′ S′ H′ M′ will be the new position of the axis, A′ the new perihelion, and ½ S H′ the new excentricity.   Hence we conclude, 1st, that the normal force acting *inwards,* and P moving *towards* the perihelion, the new direction S A′ of the perihelion is in advance (with reference to the direction of P's revolution) of the old — or the apsides advance — when P is anywhere situated between F and A (since when at F the point H′ falls upon H M between H and M). When P is at G the apsides are stationary, but when P is anywhere between M and G the apsides retrograde, H′ in this case lying on the opposite side of the axis. 2dly, That the same directions of the normal force and of P's motion being supposed, the excentricity increases while P moves through the whole semiellipse from aphelion to perihelion — the rate of its increase being a maximum when P is at F, and nothing at the aphelion and perihelion. 3dly, That these effects are reversed in the opposite half of the orbit, A G M, in which P passes from perihelion to aphelion or recedes from S. 4thly, That they are also reversed by a reversal of the direction of the normal force, outwards, in place of inwards. 5thly, That here also the variations of the excentricity and perihelion are complementary to each other; the one variation being most rapid when the other vanishes, and *vice versâ*. 6thly, And lastly, that the changes in the situation of the focus H produced by the actions of the tangential and normal components of the disturbing force are at right angles to each other in every situation of P, and therefore where the tangential force is most efficacious (in proportion to its intensity) in varying either the one or the other of the elements in question, the normal is least so, and *vice versâ*.

(672.) To determine the momentary effect of the whole disturbing force then, we have only to resolve it into its tangential and normal components, and estimating by these principles separately the effects of either constituent on both elements, add or subtract the results according as they conspire or oppose each other. Or we may at once make the angle H P H″ equal to twice the angle of deflection produced by the normal force, and lay off P H″ = P H + twice the variation of *a* produced in the same moment of time by the tangential force, and H″ will be the new focus. The momentary velocity generated by the tangential force is calculable from a knowledge of that force by the ordinary principles of dynamics; and from this, the variation of the axis is easily derived.* The momentary velocity generated by the normal force in its own direction is in like manner calculable from a knowledge of that force, and dividing this by the linear velocity of P at that instant, we deduce the angular velocity of the tangent about P, or the momentary variation of the angle of tangency S P Y, corresponding.

(673.) The following *résumé* of these several results in a tabular form includes every variety of case according as P is approaching to or receding from S; as it is situated in the arc F A G of its orbit *about the perihelion* or in the remoter arc G M F *about the aphelion*, as the tangential force *accelerates* or *retards* the disturbed body, or as the normal acts *inwards* or outwards with reference to the concavity of the orbit.

EFFECTS OF THE TANGENTIAL DISTURBING FORCE.

Direction of P's motion.	Situation of P in orbit.	Action of Tangential Force.	Effect on Elements.
Approaching S.	Anywhere.	Accelerating P.	Apsides recede.
Ditto.	Ditto.	Retarding P.	advance.
Receding from S.	Ditto.	Accelerating P.	advance.
Ditto.	Ditto.	Retarding P.	recede.
Indifferent.	About Aphelion.	Accelerating P.	Excentr. decreases.
Ditto.	Ditto.	Retarding P.	increases.
Ditto.	About Perihelion.	Accelerating P.	increases.
Ditto.	Ditto.	Retarding P.	decreases.

EFFECTS OF THE NORMAL DISTURBING FORCE.

Direction of P's motion.	Situation of P in orbit.	Action of Normal Force.	Effect on Elements.
Indifferent.	About Aphelion.	Inwards.	Apsides recede.
Ditto.	Ditto.	Outwards.	advance.
Ditto.	About Perihelion.	Inwards.	advance.
Ditto.	Ditto.	Outwards.	recede.
Approaching S.	Anywhere.	Inwards.	Excentr. increases.
Ditto.	Ditto.	Outwards.	decreases.
Receding from S.	Ditto.	Inwards.	decreases.
Ditto.	Ditto.	Outwards.	increases.

* $\frac{1}{a} = \frac{2}{r} - v^2$, and $\frac{1}{a'} = \frac{2}{r} - v'^2$ ∴ $\frac{1}{a'} - \frac{1}{a} = v^2 - v'^2 = (v + v') (v - v')$ or when infinitesimal variations only are considered $\frac{a' - a}{a^2} = 2v (v' - v)$ or $a' - a = 2a^2 v (v' - v)$ from which it appears that the variation of the axis arising from a given variation of velocity is independent of r, or is the same at whatever distance from S the change takes place, and that *cæteris paribus* it is *greater* for a given change of velocity (or for a given tangential force) in *the direct ratio of the velocity itself.*

(674.) From the momentary changes in the elements of the disturbed orbit corresponding to successive situations of P and M, to conclude the total amount of change produced in any given time is the business of the integral calculus, and lies far beyond the scope of the present work. Without its aid, however, and by general considerations of the periodical recurrence of configurations of the same character, we have been able to demonstrate many of the most interesting conclusions to which geometers have been conducted, examples of which have already been given in the reasoning by which the permanence of the axes, the periodicity of the inclinations, and the revolutions of the nodes of the planetary orbits have been demonstrated. We shall now proceed to apply similar considerations to the motion of the apsides, and the variations of the excentricities. To this end we must first trace the changes induced on the disturbing forces themselves, with the varying positions of the bodies, and here as in treating of the inclinations we shall suppose, unless the contrary is expressly indicated, both orbits to be very nearly circular, without which limitation the complication of the subject would become too embarrassing for the reader to follow, and defeat the end of explanation.

(675.) On this supposition the directions of S P and S Y, the perpendicular on the tangent at P, may be regarded as coincident, and the normal and radial disturbing forces become nearly identical in quantity, also the tangential and transversal, by the near coincidence of the points T and L (fig. art. 687.).

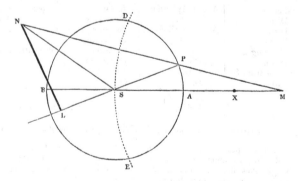

So far then as the *intensity* of the forces is concerned, it will make very little difference in which way the forces are resolved, nor will it at all materially affect our conclusions as to the effects of the normal and tangential forces, if in estimating their *quantitative* values, we take advantage of the simplification introduced into their numerical expression by the neglect of the angle P S Y, *i. e.* by the substitution for them of the radial and transversal components. The *character* of these effects depends (art. 670, 671.) on the *direction* in which the forces act, which we shall suppose normal and tangential as before, and it is only on the estimation of their quantitative effects that the error induced by the neglect of this angle can fall. In the lunar orbit this angle never exceeds 3° 10′, and its influence on the quantitative estimation of the acting forces may therefore be safely neglected in a first approximation. Now M N being found by the proportion $M P^2 : M S^2 ::$ M S : M N, N P ($= M N - M P$) is also known, and therefore N L = N P . sin N P S = N P . sin (A S P + S M P) and L S = P L - P S = N P . cos N P S - P S = N P . cos (A S P + S M P) - S P become known, which express respectively the tangential and normal forces on the same scale that S M represents M's attraction on S.* Suppose P to revolve in the direction E A D B. Then, by drawing the figure in various situations of P throughout the whole circle, the reader will easily satisfy himself —1st. That the tangential force accelerates P, as it moves from E towards A, and from D towards B, but retards it as it passes from A to D, and from B to E. 2nd. That the tangential force vanishes at the four points A, D, E, B, and attains a maximum at some intermediate points. 3rdly. That the normal force is directed outwards at the

---

* M S = R ; S P = r ; M P = f ; A S P = θ; A M P = M ; M N = $\dfrac{R^3}{f^2}$; N P

= $\dfrac{R^3 - f^3}{f^2}$ = (R − f) $\left(1 + \dfrac{R}{f} + \dfrac{R^2}{f^2}\right)$; whence we have N L = (R − f) . sin (θ + M)

$\cdot \left(1 + \dfrac{R}{f} + \dfrac{R^2}{f^2}\right)$; L S = (R − f) . cos (θ + M) . $\left(1 + \dfrac{R}{f} + \dfrac{R^2}{f^2}\right) - r$. When R and f, owing to the great distance of M, are nearly equal, we have R − f = P V, $\dfrac{R}{f}$ = 1 nearly, and the angle M may be neglected ; so that we have N P = 3 P V.

syzigies A, B, and inwards at the points D, E, at which points respectively its outward and inward intensities attain their maxima. Lastly, that this force vanishes at points intermediate between A D, D B, B E, and E A, which points, when M is considerably remote, are situated nearer to the quadrature than the syzygies.

(676.) In the lunar theory, to which we shall now proceed to apply these principles, both the geometrical representation and the algebraic expression of the disturbing forces admit of great simplification. Owing to the great distance of the sun M, at whose center the radius of the moon's orbit never subtends an angle of more than about 8', N P may be regarded as parallel to A B. And D S E becomes a straight line coincident with the line of quadratures, so that V P becomes the cosine of A S P to radius S P, and $N L = N P \cdot \sin A S P$; $L P = N P \cdot \cos A S P$. Moreover, in this case (see the note on the last article) $N P = 3 P V = 3 S P \cdot \cos A S P$; and consequently $N L = 3 S P \cdot \cos A S P \cdot \sin A S P = \frac{3}{2} S P \cdot \sin 2 A S P$, and $L P = S P (3 \cdot \cos A S P^2 - 1) = \frac{1}{2} S P (1 + 3 \cdot \cos 2 A S P)$ which vanishes when $\cos A S P^2 = \frac{1}{3}$, or at 64° 14' from the syzygy. Suppose through every point of P's orbit there be drawn $S Q = 3 S P \cdot \cos A S P^2$, then will Q trace out a certain looped oval, as in the figure, cutting the orbit in four points 64° 14' from A and B respectively, and P Q will always represent in quantity and direction the normal force acting at P.

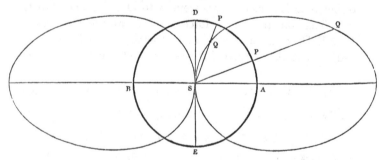

(677.) It is important to remark here, because upon this the whole lunar theory and especially that of the motion of

the apsides hinges, that all the acting disturbing forces, at
equal angles of elongation A S P of the moon from the sun,
are *cæteris paribus* proportional to S P, the moon's distance
from the earth, and are therefore greater when the moon is
near its apogee than when near its perigee; the extreme
proportion being that of about 28 : 25. This premised, let
us first consider the effect of the normal force in displacing
the lunar apsides. This we shall best be enabled to do by
examining separately those cases in which the effects are most
strongly contrasted; viz. when the major axis of the moon's
orbit is directed towards the sun, and when at right angles
to that direction. First, then, let the line of apsides be

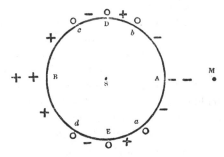

directed to the sun as in the annexed figure, where A is the
perigee, and take the arcs A *a*, A *b*, B *c*, B *d* each=64° 14'.
Then while P is between *a* and *b* the normal force acting outwards, and the moon being near its perigee, by art. 671. the
apsides will recede, but when between *c* and *d*, the force there
acting outwards, but the moon being near its apogee, they
will advance. The rapidity of these movements will be respectively at its maxima at A and B, not only because the
disturbing forces are then most intense, but also because
(see art. 671.) they act most advantageously at those points
to displace the axis. Proceeding from A and B towards the
neutral points *a b c d* the rapidity of their recess and advance
diminishes, and is nothing (or the apsides are stationary)
when P is at either of these points. From *b* to D, or rather
to a point some little beyond D (art. 671.) the force acts
inwards, and the moon is still near perigee, so that in this arc

of the orbit the apsides advance.   But the rate of advance
is feeble, because in the early part of that arc the normal
force is small, and as P approaches D and the force gains
power, it acts disadvantageously to move the axis, its effect
vanishing altogether when it arrives beyond D at the extremity of the perpendicular to the upper focus of the lunar
ellipse.   Thence up to c this feeble advance is reversed and
converted into a recess, the force still acting inwards, but
the moon now being near its apogee.   And so also for
the arcs d E, E a.   In the figure these changes are indicated
by + + for rapid advance, — — for rapid recess, + and —
for feeble advance and recess, and 0 for the stationary points.
Now if the forces were equal on the sides of + and — it is
evident that there would be an exact counterbalance of
advance and recess on the average of a whole revolution.
But this is not the case.   The force in apogee is greater than
that in perigee in the proportion of 28 : 25, while in the
quadratures about D and E they are equal.   Therefore,
while the feeble movements + and — in the neighbourhood of
these points destroy each other almost exactly, there will
necessarily remain a considerable balance in favour of advance,
in this situation of the line of apsides.

(678.) Next, suppose the apogee to lie at A, and the perigee at B.   In this case it is evident that, so far as the direction of the motions of the apsides is concerned, all the
conclusions of the foregoing reasoning will be reversed by the
substitution of the word perigee for apogee, and *vice versâ;*
and all the signs in the figure referred to will be changed.
But now the most powerful forces act on the side of A, that
is to say, still on the side of advance, this condition also being
reversed.   In either situation of the orbit, then, the apsides
advance.

(679.) (Case 3.) Suppose, now, the major axis to have the
situation D E, and the perigee to be on the side of D.   Here,
in the arc b c of P's motion the normal force acts inwards,
and the moon is near perigee, consequently the apsides
advance, but with a moderate rapidity, the maximum of the
inward normal force being only half that of the outward.

In the arcs A b and c B the moon is still near perigee, and
the force acts outwards, but though powerfully towards A
and B, yet at a constantly increasing disadvantage (art. 671.)
Therefore in these arcs the apsides recede, but moderately.
In a A and B d (being towards apogee) they again advance,
still with a moderate velocity. Lastly, throughout the arc
$d\,a$, being about apogee with an inward force, they recede.
Here as before, if the perigee and apogee forces were equal,
the advance and recess would counterbalance; but as in fact
the apogee forces preponderate, there will be a balance on
the entire revolution in favour of recess. The same reasoning
of course holds good if the perigee be towards E. But now,
between these cases and those in the foregoing articles, there
is this difference, viz. that in this the dominant effect results
from the inward action of the normal force in quadratures,
while in the others it results from its outward, and doubly
powerful action in syzygies. The recess of the apsides in
their quadratures arising from the action of the normal force
will therefore be less than their advance in their syzygies;
and not only on this account, but also because of the much
less extent of the arcs $b\,c$ and $d\,a$ on which the balance is
mainly struck in this case, than of $a\,b$ and $c\,d$, the corresponding most influential arcs in the other.

(680.) In intermediate situations of the line of apsides, the
effect will be intermediate, and there will of course be a situation of them in which on an average of a whole revolution,
they are stationary. This situation it is easy to see will be
nearer to the line of quadratures than of syzygies, and the
preponderance of advance will be maintained over a much
more considerable arc than that of recess, among the possible
situations which they can hold. On every account, therefore
the action of the normal force causes the lunar apsides to
progress *in a complete revolution of* M or in a synodical year,
during which the motion of the sun round the earth (as we
consider the earth at rest) brings the line of syzygies into all
situations with respect to that of apsides.

(681.) Let us next consider the action of the tangential
force. And as before (Case 1.), supposing the perigee of the

moon at A, and the direction of her revolution to be A D B E, the tangential force *retards* her motion through the quadrant A D, in which she *recedes* from S, therefore by art. 370. the apsides *recede.*   Through D B the force *accelerates,* while the moon still *recedes,* therefore they *advance.*   Through B E the force retards, and the moon approaches, therefore they continue to advance, and finally throughout the quadrant E A the force accelerates and the moon approaches, therefore they recede.   In virtue therefore of this force, the apsides recede, during the description of the arc E A D, and advance during D B E, but the force being in this case as in that of the normal force more powerful at apogee, the latter will preponderate, and the apsides will advance on an average of a whole revolution.

(682.) (Case 2.) The perigee being towards B, we have to substitute in the foregoing reasoning approach to S, for recess from it, and *vice versâ,* the accelerations and retardations remaining as before.   Therefore the results, as far as direction is concerned, will be reversed in each quadrant, the apsides advance during E A D and recede during D B E. But the situation of the apogee being also reversed, the predominance remains on the side of E A D, that is, of advance.

(683.) (Case 3.) Apsides in quadratures, perigee near D.— Over quadrant A D, approach and retardation, therefore advance of apsides.   Over D B recess and acceleration, therefore again advance; over B E recess and retardation with recess of apsides, and lastly over E A approach and acceleration, producing their continued recess.   Total result: advance during the half revolution A D B, and recess during B E A, the acting forces being more powerful in the latter, whence of course a preponderant recess.   The same result when the perigee is at E.

(684.) So far the analogy of reasoning between the action of the tangential and normal forces is perfect.   But from this point they diverge.   It is not here as before.   The recess of the apsides in quadratures does not now arise from the predominance of feeble over feebler forces, while that in syzygies results from that of powerful over powerful ones.   The maxi-

mum accelerating action of the tangential force is equal to
its maximum retarding, while the inward action of the normal
at its maximum is only half the maximum of its outward.
Neither is there that difference in the extent of the arcs over
which the balance is struck in this, as in the other case, the
action of the tangential force being inward and outward
alternately over equal arcs, each a complete quadrant.
Whereas, therefore, in tracing the action of the normal force,
we found reason to conclude it much more effective to produce
progress of the apsides in their syzygy, than in their quadrature
situations, we can draw no such conclusion in that of the
tangential forces: there being, as regards *that* force, a *complete
symmetry* in the four *quadrants*, while in regard of the normal
force the symmetry is only a *half-symmetry* having relation
to two *semicircles*.

(685.) Taking the average of many revolutions of the sun
about the earth, in which it shall present itself in every possible variety of situations to the line of apsides, we see that the
effect of the normal force is to produce a rapid advance in the
syzygy of the apsides, and a less rapid recess in their quadrature, and on the whole, therefore, a moderately rapid general
advance, while that of the tangential is to produce an equally
rapid advance in syzygy, and recess in quadrature. Directly,
therefore, the tangential force would appear to have no ultimate influence in causing either increase or diminution in the
mean motion of the apsides resulting from the action of the
normal force. It does so, however, indirectly, conspiring in
that respect with, and greatly increasing, an indirect action
of the normal force in a manner which we shall now proceed
to explain.

(686.) The sun moving uniformly, or nearly so, in the
same direction as P, the line of apsides when in or near the
syzygy, in advancing follows the sun, and therefore remains
materially longer in the neighbourhood of syzygy than if it
rested. On the other hand, when the apsides are in quadrature
they recede, and moving therefore contrary to the sun's motion,
remain a shorter time in that neighbourhood, than if they
rested. Thus the advance, already preponderant, is made to

preponderate more by its longer continuance, and the recess, already deficient, is rendered still more so by the shortening of its duration.* Whatever cause, then, increases directly the rapidity of both advance and recess, *though it may do both equally,* aids in this indirect process, and it is thus that the tangential force becomes effective through the medium of the progress already produced, in doing and aiding the normal force to do that which alone it would be unable to effect. Thus we have perturbation exaggerating perturbation, and thus we see what is meant by geometers, when they declare that a considerable part of the motion of the lunar apsides is due to the square of the disturbing force, or, in other words, arises out of a second approximation in which the influence of the first in altering the data of the problem is taken into account.

(687.) The curious and complicated effect of perturbation, described in the last article, has given more trouble to geometers than any other part of the lunar theory. Newton himself had succeeded in tracing that part of the motion of the apogee which is due to the direct action of the radial force; but finding the amount only half what observation assigns, he appears to have abandoned the subject in despair. Nor, when resumed by his successors, did the inquiry, for a very long period, assume a more promising aspect. On the contrary, Newton's result appeared to be even minutely verified, and the elaborate investigations which were lavished upon the subject without success began to excite strong doubts whether this feature of the lunar motions could be explained at all by the Newtonian law of gravitation. The doubt was removed, however, almost in the instant of its origin, by the same geometer, Clairaut, who first gave it currency, and who gloriously repaired the error of his momentary hesitation, by demonstrating the exact coincidence between theory and observation, when the effect of the tangential force is properly taken into the account. The lunar apogee circulates, in $3232^{d}\cdot575343$, or about $9\frac{1}{2}$ years.

* Newton, Princ. i. 66. Cor. 8.

(688.) Let us now proceed to investigate the influence of the disturbing forces so resolved on the excentricity of the lunar orbit, and the foregoing articles having sufficiently familiarized the reader with our mode of following out the changes in different situations of the orbit, we shall take at once a more general situation, and suppose the line of apsides in any position with respect to the sun, such as Z Y, the perigee being at Z, a point between the lower syzygy and the quadrature next following it, the direction of P's motion as all along supposed being A D B E. Then (commencing with the normal force) the momentary change of excentricity

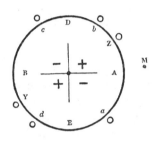

will vanish at a, b, c, d, by the vanishing of that force, and at Z and Y by the effect of situation in the orbit annulling its action (art. 671.). In the arcs Z b and Y d therefore the change of excentricity will be small, the acting force nowhere attaining either a great magnitude or an advantageous situation within their limits. And the force within these two arcs having the same character as to inward and outward, but being oppositely influential by reason of the approach of P to S in one of them and its recess in the other, it is evident that, so far as these arcs are concerned, a very near compensation of effects will take place, and though the apogeal arc Y d will be somewhat more influential, this will tell for little upon the average of a revolution.

(689.) The arcs b D c and d E a are each much less than a quadrant in extent, and the force acting inwards throughout them (which at its maximum in D and E is only half the outward force at A, B) degrades very rapidly in intensity towards either syzygy (see art. 676.). Hence whether Z be between b c or b A, the effects of the force in these arcs will not produce very extensive changes on the excentricity, and the changes which it does produce will (for the reason already given) be opposed to each other. Although, then, the arc $a\,d$ be farther from perigee than $b\,c$, and therefore the force in it is greater, yet the predominance of effect here will not be very marked, and will moreover be partially neutralized by the small predominance of an opposite character in Y d over Z b.   On the other hand, the arcs a Z, c Y are both larger in extent than either of the others, and the seats of action of forces doubly powerful.   Their influence, therefore, will be of most importance, and their preponderance one over the other, (being opposite in their tendencies,) will decide the question whether on an average of the revolution, the   excentricity   shall increase or diminish.   It is clear that the decision must be in favour of c Y, the apogeal arc, and, since in this the force is *outwards* and the moon *receding* from the earth, an *increase* of the excentricity will arise from its influence.   A similar reasoning will, evidently, lead to the same   conclusion   were   the   apogee   and   perigee   to   change places, for the directions of P's motion as to approach and recess to S will be indeed reversed, but at the same time the dominant forces will have changed sides, and the arc a A Z will now give the character to the result.   But when Z lies between A and E, as the reader may easily satisfy himself, the case will be altogether different, and the reverse conclusion will obtain.   Hence the changes of excentricity emergent on the average of single revolutions from the action of the normal force   will   be   as   represented   by   the   signs $+$ and $-$ in the figure above annexed.

(690.) Let us next consider the effect of the tangential force.   This retards P in the quadrants A D, B E, and

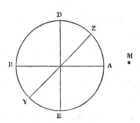

accelerates it in the alternate ones.   In the whole quadrant A D, therefore, the effect is of one   character,   the   perigee being less than 90° from every point in it, and in the whole quadrant B E it is of the opposite,   the   apogee   being so situated (art. 670.)   Moreover, in the middle of each quadrant, the tangential force is at its maximum. Now, in the other quadrants, E A and D B, the change from perigeal to apogeal vicinity takes place, and the tangential force, however powerful, has its effect annulled by situation (art. 670.), and this happens more or less nearly about the points where the force is a maximum. These quadrants, then, are far less influential on the total result, so that the character of that result will be decided by the predominance of one or other of the former quadrants, and will lean to that which has the apogee in it. Now in the quadrant B E the force *retards* the moon and the moon is in apogee. Therefore the excentricity increases. In this situation therefore of the apogee, *such* is the average result of a complete revolution of the moon. Here again also if the perigee and apogee change places, so will also the character of all the partial influences, arc for arc. But the quadrant A D will now preponderate instead of D E, so that under this double reversal of conditions the result will be identical. Lastly, if the line of apsides be in A E, B D, it may be shown in like manner that the excentricity will diminish on the average of a revolution.

(691.) Thus it appears, that in varying the excentricity, precisely as in moving the line of apsides, the direct effect of the tangential force conspires with that of the normal, and tends to increase the extent of the deviations to and fro on either side of a mean value which the varying situation of the sun with respect to the line of apsides gives rise to, having for their period of restoration a synodical revolution of the sun and apse. Supposing the sun and apsis to start together, the sun of course will outrun the apsis (whose period is nine years), and in the lapse of about $(\frac{1}{4} + \frac{1}{32})$ part of a year will have gained on it 90°, during all which interval the apse will have been in the quadrant A E of our figure, and the excentricity continually decreasing. The decrease will then cease, but the excentricity itself will be a minimum, the sun being now at right angles to the line of apsides. Thence it will increase to a maximum when the sun has gained another 90°, and again attained the line of apsides, and so on alternately. The actual effect on the numerical value of the lunar excentricity is very considerable, the greatest and least excentricities being in the ratio of 3 to 2.*

(692.) The motion of the apsides of the lunar orbit may be illustrated by a very pretty mechanical experiment, which is otherwise instructive in giving an idea of the mode in which orbitual motion is carried on under the action of central forces variable according to the situation of the revolving body. Let a leaden weight be suspended by a brass or iron wire to a hook in the under side of a firm beam, so as to allow of its free motion on all sides of the vertical, and so that when in a state of rest it shall just clear the floor of the room, or a table placed ten or twelve feet beneath the hook. The point of support should be well secured from wagging to and fro by the oscillation of the weight, which should be sufficient to keep the wire as tightly stretched as it will bear, with the certainty of not breaking. Now, let a very small motion be communicated to the weight, not by merely withdrawing it from the vertical and letting it fall, but by giving it a slight impulse sideways. It will be seen to describe a regular ellipse about the point of rest as its center. If the weight be heavy, and carry attached to it a pencil, whose point lies exactly in the direction of the string, the ellipse may be transferred to paper lightly stretched and gently pressed against it. In these circumstances, the situation of the major and minor axes of the ellipse will remain for a long time very nearly the same, though the resistance of the air and the stiffness of the wire will gradually diminish its dimensions and excentricity. But if the impulse communicated to the weight be considerable, so as to carry it out to a great angle (15° or 20° from the vertical), this permanence of situation of the ellipse will no longer subsist. Its axis will be seen to shift its position at every revolution of the weight, advancing in the same direction with the weight's motion, by an uniform and regular progression, which at length will entirely reverse its situation, bringing the direction of the longest excursions to coincide with that

---

* Airy. Gravitation, p. 106.

in which the shortest were previously made; and so on, round the whole circle; and, in a word, imitating to the eye, very completely, the motion of the apsides of the moon's orbit.

(693.) Now, if we inquire into the cause of this progression of the apsides, it will not be difficult of detection. When a weight is suspended by a wire, and drawn aside from the vertical, it is urged to the lowest point (or rather in a direction at every instant perpendicular to the wire) by a force which varies as the sine of the deviation of the wire from the perpendicular. Now, the sines of very small arcs are nearly in the proportion of the arcs themselves; and the more nearly, as the arcs are smaller. If, therefore, the deviations from the vertical be so small that we may neglect the curvature of the spherical surface in which the weight moves, and regard the curve described as coincident with its projection on a horizontal plane, it will be then moving under the same circumstances as if it were a revolving body attracted to a center by a force varying directly as the distance; and, in this case, the curve described would be an ellipse, having its centre of attraction not in the focus, but in the center *, and the apsides of this ellipse would remain fixed. But if the excursions of the weight from the vertical be considerable, the force urging it towards the center will deviate in its law from the simple ratio of the distances; being as the *sine*, while the distances are as the *arc*. Now the sine, though it continues to increase as the arc increases, yet does not increase so fast. So soon as the arc has any sensible extent, the sine begins to fall somewhat short of the magnitude which an exact numerical proportionality would require; and therefore the force urging the weight towards its center or point of rest at great distances falls, in like proportion, somewhat short of that which would keep the body in its precise elliptic orbit. It will no longer, therefore, have, at those greater distances, the same command over the weight, *in proportion to its speed*, which would enable it to

* Newton, Princip. i. 47.

deflect it from its rectilinear tangential course into an ellipse.
The true path which it describes will be *less curved in the
remoter parts* than is consistent with the elliptic figure, as in
the annexed cut; and, therefore, it will not so soon have its
motion brought to be again at right angles to the radius.
It will require a longer continued action of the central force
to do this; and before it is accomplished, more than a quadrant

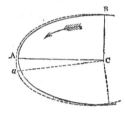

of its revolution must be passed over in
angular motion round the center.   But
this is only stating at length, and in a
more circuitous manner, that fact which
is more briefly and summarily expressed
by saying that *the apsides of its orbit
are progressive.* Nothing beyond a familiar illustration is of course intended in
what is above said.   The case is not an exact parallel with
that of the lunar orbit, the disturbing force being simply
radial, whereas in the lunar orbit a transversal force is also concerned, and even were it otherwise, only a confused and in-
distinct view of apsidal motion can be obtained from this kind
of consideration of the curvature of the disturbed path.   If we
would obtain a clear one, the two foci of the instantaneous
ellipse must be found from the laws of elliptic motion performed under the influence of a force directly as the distance,
and the radial disturbing force being decomposed into its
tangential and normal components, the momentary influence
of either in altering their positions and consequently the
directions and lengths of the axis of the ellipse must be
ascertained.   The student will find it neither a difficult nor
an uninstructive exercise to work out the case from these
principles, which we cannot afford the space to do.

(694.) The theory of the motion of the planetary apsides
and the variation of their excentricities is in one point of
view much more simple, but in another much more complicated
than that of the lunar.   It is simpler, because owing to the
exceeding minuteness of the changes operated in the course
of a single revolution, the angular position of the bodies with
respect to the line of apsides is very little altered by the

motion of the apsides themselves. The line of apsides neither follows up the motion of the disturbing body in its state of advance, nor *vice versâ*, in any degree capable of pro-longing materially their advancing or shortening materially their receding phase. Hence no second approximation of the kind explained in (art. 686.), by which the motion of the lunar apsides is so powerfully modified as to be actually doubled in amount, is at all required in the planetary theory. On the other hand, the latter theory is rendered more com-plicated than the former, at least in the cases of planets whose periodic times are to each other in a ratio much less than 13 to 1, by the consideration that the disturbing body shifts its position with respect to the line of apsides by a much greater angular quantity in a revolution of the disturbed body than in the case of the moon. In that case we were at liberty to suppose (for the sake of explanation), without any very egregious error, that the sun held nearly a fixed position during a single lunation. But in the case of planets whose times of revolution are in a much lower ratio this cannot be permitted. In the case of Jupiter disturbed by Saturn for example, in one sidereal revolution of Jupiter, Saturn has advanced in its orbit with respect to the line of apsides of Jupiter by more than 140°, a change of direction which entirely alters the conditions under which the disturbing forces act. And in the case of an exterior disturbed by an interior planet, the situation of the latter with respect to the line of the apsides varies even more rapidly than the situation of the exterior or disturbed planet with respect to the central body. To such cases then the reasoning which we have applied to the lunar perturbations becomes totally inappli-cable; and when we take into consideration also the excen-tricity of the orbit of the disturbing body, which in the most important cases is exceedingly influential, the subject becomes far too complicated for verbal explanation, and can only be successfully followed out with the help of algebraic expression and the application of the integral calculus. To Mercury, Venus, and the earth indeed, as disturbed by Jupiter, and planets superior to Jupiter, this objection to the reasoning in question does not apply; and in each of these cases therefore we are entitled to conclude that the apsides are kept in a state of progression by the action of all the larger planets of our system. Under certain conditions of distance, excentricity, and relative situation of the axes of the orbits of the disturbed and disturbing planets, it is perfectly possible that the reverse may happen, an instance of which is afforded by Venus, whose apsides recede under the combined action of the earth and Mercury more rapidly than they advance under the joint actions of all the other planets. Nay, it is even possible under certain conditions that the line of apsides of the disturbed planet, instead of revolving always in one direction, may librate to and fro within assignable limits, and in a definite and regularly recurring period of time.

695.) Under any conditions, however, as to these particulars, the view we have above taken of the subject enables us to assign at every instant, and in every configuration of the two planets, the momentary effect of each upon the perihelion and excentricity of the other. In the simplest case, that in which the two orbits are so nearly circular, that the relative situation of their perihelia shall produce no appreciable difference in the intensities of the disturbing forces, it is very easy to show that whatever temporary oscillations to and fro in the positions of the line of apsides, and whatever temporary increase and diminution in the excentricity of either planet may take place, the final effect on the average of a great multitude of revolutions, presenting them to each other in all possible configurations, must be *nil*, for both elements.

(696.) To show this, all that is necessary is to cast our eyes on the synoptic table in art. 673. If M, the disturbing body, be supposed to be successively placed in two diametrically opposite situations in its orbit, the aphelion of P will stand related to M in one of these situations precisely as its perihelion in the other. Now the orbits being so nearly circles as supposed, the distribution of the disturbing forces, whether normal or tangential, is symmetrical relative to their common diameter passing through M, or to the line of syzygies. Hence it follows that the half of P's orbit " about perihelion " (art. 673.) will stand related to all the acting forces in the one situation of M, precisely as the half " about aphelion " does in the other: and also, that the half of the orbit in which P " approaches S," stands related to them in the one situation precisely as the half in which it " recedes from S " in the other. Whether as regards, therefore, the normal or tangential force, the conditions of advance or recess of apsides, and of increase or diminution of excentricities, are reversed in the two supposed cases. Hence it appears that whatever situation be assigned to M, and whatever influence it may exert on P in that situation, that influence will be annihilated in situations of M and P, diametrically opposite to those supposed, and thus, on a general average, the effect on both apsides and excentricities is reduced to nothing.

(697.) If the orbits, however, be excentric, the symmetry above insisted on in the distribution of the forces does not exist. But, in the first place, it is evident that if the excentricities be moderate, (as in the planetary orbits,) by far the larger part of the effects of the disturbing forces destroys itself in the manner described in the last article, and that it is only a residual portion, viz. that which arises from the greater proximity of the orbits at one place than at another, which can tend to produce permanent or secular effects. The precise estimation of these effects is too complicated an affair for us to enter upon; but we may at least give some idea of the process by which they are produced, and the order in which they arise. In so doing, it is necessary to distinguish between the effects of the normal and tangential forces. The effects of the former are greatest at the point of conjunction of the planets, because the normal force itself is there always at its maximum; and although, where the conjunction takes place at 90° from the line of apsides, its effect to move the apsides is nullified by situation, and when *in* that line its effect on the excentricities is similarly nullified, yet, in the situations rectangular to these, it acts to its greatest advantage. On the other hand, the tangential force vanishes at conjunction,

G G

whatever be the place of conjunction with respect to the line
of apsides, and where it is at its maximum its effect is still
liable to be annulled by situation. Thus it appears that
the normal force is most influential, and mainly determines
the character of the general effect. It is, therefore, at conjunction that the most influential effect is produced, and
therefore, on the long average, those conjunctions which
happen about the place where the orbits are nearest will
determine the general character of the effect. Now, the
nearest points of approach of two ellipses which have a
common focus may be very variously situated with respect
to the perihelion of either. It may be at the perihelion or
the aphelion of the disturbed orbit, or in any intermediate
position. Suppose it to be at the perihelion. Then, if the
disturbed orbit be *interior* to the disturbing, the force acts
outwards, and therefore the apsides recede : if exterior, the
force acts inwards, and they advance. In neither case does
the excentricity change. If the conjunction take place at
the aphelion of the disturbed orbit, the effects will be reversed : if intermediate, the apsides will be less, and the
excentricity more affected.

(698.) Supposing only two planets, this process would go
on till the apsides and excentricities had so far changed as to
alter the point of nearest approach of the orbits so as either
to accelerate or retard and perhaps reverse the motion of the
apsides, and give to the variation of the excentricity a corresponding periodical character. But there are many planets
all disturbing one another. And this gives rise to variations
in the points of nearest approach of all the orbits taken two
and two together, of a very complex nature.

(699.) It cannot fail to have been remarked, by any one
who has followed attentively the above reasonings, that
a close analogy subsists between two sets of relations; viz.
that between the inclinations and nodes on the one hand, and
between the excentricity and apsides on the other. In fact,
the strict geometrical theories of the two cases present a
close analogy, and lead to final results of the very same
nature. What the variation of excentricity is to the motion

of the perihelion, the change of inclination is to the motion of the node.   In either case, the period of the one is also the period of the other; and while the perihelia describe considerable angles by an oscillatory motion to and fro, or circulate in immense periods of time round the entire circle, the excentricities increase and decrease by comparatively small changes, and are at length restored to their original magnitudes.   In the lunar orbit, as the rapid rotation of the nodes prevents the change of inclination from accumulating to any material amount, so the still more rapid revolution of its apogee effects a speedy compensation in the fluctuations of its excentricity, and never suffers them to go to any material extent; while the same causes, by presenting *in quick succession* the lunar orbit in every possible situation to all the disturbing forces, whether of the sun, the planets, or the protuberant matter at the earth's equator, prevent any secular accumulation of small changes, by which, in the lapse of ages, its ellipticity might be materially increased or diminished.   Accordingly, observation shows the *mean* excentricity of the moon's orbit to be the same now as in the earliest ages of astronomy.

(700.) The movements of the perihelia, and variations of excentricity of the planetary orbits, are interlaced and complicated together in the same manner and nearly by the same laws as the variations of their nodes and inclinations.   Each acts upon every other, and every such mutual action generates its own peculiar period of circulation or compensation; and every such period, in pursuance of the principle of art. 650., is thence propagated throughout the system.   Thus arise cycles upon cycles, of whose compound duration some notion may be formed, when we consider what is the length of one such period in the case of the two principal planets — Jupiter and Saturn.   Neglecting the action of the rest, the effect of their mutual attraction would be to produce a secular variation in the excentricity of Saturn's orbit, from $0 \cdot 08409$, its *maximum*, to $0 \cdot 01345$, its *minimum* value: while that of Jupiter would vary between the narrow limits, $0 \cdot 06036$ and $0 \cdot 02606$: the greatest excentricity of Jupiter corresponding

to the least of Saturn, and *vice versâ*. The period in which
these changes are gone through, would be 70414 years.
After this example, it will be easily conceived that many
millions of years will require to elapse before a complete
fulfilment of the joint cycle which shall restore the whole
system to its original state as far as the excentricities of its
orbits are concerned.

(701.) The place of the perihelion of a planet's orbit is of
little consequence to its well-being; but its excentricity is
most important, as upon this (the axes of the orbits being
permanent) depends the mean temperature of its surface, and
the extreme variations to which its seasons may be liable.
For it may be easily shown that the *mean annual amount* of
light and heat received by a planet from the sun is, *cæteris
paribus*, as the minor axis of the ellipse described by it.   Any
variation, therefore, in the excentricity, by changing the
minor axis will alter the *mean* temperature of the surface.
How such a change will also influence the extremes of temperature appears from art. 368.   Now it may naturally be
inquired whether (in the vast cycle above spoken of, in which,
at some period or other, conspiring changes may accumulate
on the orbit of one planet from several quarters,) it may not
happen that the excentricity of any one planet — as the earth
— may become exorbitantly great, so as to subvert those
relations which render it habitable to man, or to give rise
to great changes, at least, in the physical comfort of his state.
To this the researches of geometers have enabled us to answer
in the negative.   A relation has been demonstrated by
Lagrange between the masses, axes of the orbits, and excentricities of each planet, similar to what we have already
stated with respect to their inclinations, viz. *that if the mass
of each planet be multiplied by the square root of the axis of its
orbit, and the product by the square of its excentricity, the sum
of all such products throughout the system is invariable ;* and
as, in point of fact, this sum is extremely small, so it will
always remain.   Now, since the axes of the orbits are liable
to no secular changes, this is equivalent to saying that no
one orbit shall increase its excentricity, unless at the expense

of a common fund, the whole amount of which is, and must for ever remain, extremely minute.*

* There is nothing in this relation, however, taken *per se*, to secure the smaller planets — Mercury, Mars, Juno, Ceres, &c. — from a catastrophe, could they accumulate on themselves, or any one of them, the whole amount of this *excentricity fund*. But that can never be : Jupiter and Saturn will always retain the lion's share of it. A similar remark applies to the *inclination fund* of art. 639. These *funds*, be it observed, can never get into debt. Every term of them is essentially positive.

CHAPTER XIV.

(702.) To calculate the actual place of a planet or the moon, in longitude and latitude at any assigned time, it is not enough to know the changes produced by perturbation in the elements of its orbit, still less to know the *secular* changes so produced, which are only the outstanding or uncompensated portions of much greater changes induced in short periods of configuration. We must be enabled to estimate the actual effect on its longitude of those periodical accelerations and retardations in the rate of its mean angular motion, and on its latitude of those deviations above and below the mean plane of its orbit, which result from the continued action of the perturbative forces, not as compensated in long periods, but as in the act of their generation and destruction

in short ones.   In this chapter we purpose to give an account
of some of the most prominent of the *equations* or inequalities
thence arising, several of which are of high historical interest,
as having become known by observation previous to the
discovery of their theoretical causes, and as having, by their
successive explanations from the. theory of gravitation, removed what were in some instances regarded as formidable
objections against that theory, and afforded in all most
satisfactory and triumphant verifications of it.

(703.) We shall begin with those which compensate themselves in a synodic revolution of the disturbed and disturbing
body, and which are independent of any permanent excentricity of either orbit, going through their changes and
effecting their compensations in orbits slightly elliptic, almost
precisely as if they were circular.   These inequalities result,
in fact, from a circulation of the true upper focus of the
disturbed ellipse about its mean place in a curve whose
form and magnitude the principles laid down in the last
chapter enable us to assign in any proposed case.   If the
disturbed orbit be circular, this mean place coincides with its
centre : if elliptic, with the situation of its upper focus, as
determined from the principles laid down in the last chapter.

(704.) To understand the nature of this circulation, we
must consider the joint action of the two elements of the
disturbing force.   Suppose H to be the place of the upper

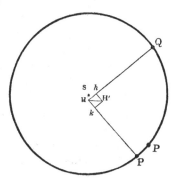

focus, corresponding to any situation P of the disturbed body,

and let P P' be an infinitesimal element of its orbit, described in an instant of time.   Then supposing no disturbing
force to act, P P' will be a portion of an ellipse, having H for
its focus, equally whether the point P or P' be regarded.
But now let the disturbing forces act during the instant of
describing P P'.   Then the focus H will shift its position to
H' to find which point we must recollect, 1st. What is demonstrated in art. (671.), viz. that the effect of the normal
force is to vary the position of the line P' H so as to make
the angle H P H' equal to double the variation of the angle
of tangency due to the action of that force, without altering
the distance P H : so that in virtue of the normal force alone,
H would move to a point h, along the line H Q, drawn from
H to a point Q, 90° in advance of P, (because S H being
exceedingly small, the angle P H Q may be taken as a right
angle when P S Q is so,) H *approaching* Q if the normal
force act *outwards*, but receding from Q if inwards.   And
similarly the effect of the tangential force (art. 670.) is to
vary the position of H in the direction H P or P H, according
as the force retards or accelerates P's motion.   To find H'
then from H draw H P, H Q, to P and to a point of P's
orbit 90° in advance of P.   On H Q take H h, the motion
of the focus due to the normal force, and on H P take H k the
motion due to the tangential force; complete the parallelogram
H H', and its diagonal H H' will be the element of the true
path of H in virtue of the joint action of both forces.

(705.) The most conspicuous case in the planetary system
to which the above reasoning is applicable, is that of the
moon disturbed by the sun.   The inequality thus arising
is known by the name of the moon's variation, and was discovered so early as about the year 975 by the Arabian
astronomer Aboul Wefa.*   Its magnitude (or the extent of
fluctuation to and fro in the moon's longitude which it produces) is considerable, being no less than 1° 4', and it is
otherwise interesting as being the first inequality produced
by perturbation, which Newton succeeded in explaining by

* Sedillot, Nouvelles Recherches pour servir à l'Histoire de l'Astronomie chez
les Arabes.

the theory of gravity. A good general idea of its nature may be formed by considering the direct action of the disturbing forces on the moon, supposed to move in a circular orbit. In such an orbit undisturbed, the velocity would be uniform; but the tangential force acting to accelerate her motion through the quadrants *preceding* her conjunction and opposition, and to retard it through the alternate quadrants, it is evident that the velocity will have two maxima and two minima, the former at the syzygies, the latter at the quadratures. Hence at the syzygies,the velocity will exceed that which corresponds to a circular orbit, and at quadratures will fall short of it. The true orbit will therefore be less curved or more flattened than a circle in syzygies, and more curved (*i. e.* protuberant beyond a circle) in quadratures. This would be the case even were the normal force not to act. But the action of that force increases the effect in question, for at the syzygies, and as far as 64° 14′ on either side of them, it acts outwards, or in counteraction of the earth's attraction, and thereby prevents the orbit from being so much curved as it otherwise would be; while at quadratures, and for 25° 46′ on either side of them, it acts inwards, aiding the earth's attraction, and rendering that portion of the orbit more curved than it otherwise would be. Thus the joint action of both forces distorts the orbit from a circle into a flattened or elliptic form, having the longer axis in quadratures, and the shorter in syzygies; and in this orbit the moon moves faster than with her mean velocity at syzygy (*i. e.* where she is nearest the earth) and slower at quadratures where farthest. Her *angular* motion about the earth is therefore for both reasons greater in the former than in the latter situation. Hence at syzygy her true longitude seen from the earth will be in the act of gaining on her mean, — in quadratures of losing, and at some intermediate points (not very remote from the octants) will neither be gaining nor losing. But at these points, *having been gaining or losing* through the whole previous 90°, the *amount of gain or loss* will have attained its maximum. Consequently at the octants the true longitude will deviate most from the mean in excess and defect, and the inequality in question is therefore *nil* at syzygies and quadratures, and attains its maxima in advance or retardation at the octants, which is agreeable to observation.

(706.) Let us, however, now see what account can be rendered of this inequality by the simultaneous variations of the axis and excentricity as above explained. The tangential force, as will be recollected, is *nil* at syzygies and quadratures, and a maximum at the octants, accelerative in the quadrants E A and D B, and retarding in A D and B E. In the two former then the axis is in process of lengthening; in the two latter, shortening. On the other hand the normal force vanishes at (*a, b, d, e*) 64° 14' from the syzygies. It acts outwards over *e* A *a*, *b* B *d*, and inwards over *a* D *b* and *d* E *e*. In virtue of the tangential force, then, the point H moves towards P when P is in A D, B E, and from it when in D B, E A, the motion being *nil* when at A, B, D, E, and most rapid when at the octant D, at which points, therefore, (so far as this force is concerned,) the focus H would have its mean situation. And in virtue of the normal focus, the motion of H in the direction H Q will be at its maximum of rapidity towards Q at A, or B, from Q at D or

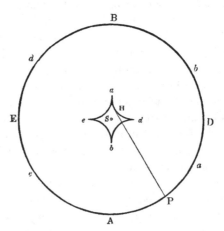

E, and *nil* at *a, b, d, e*. It will assist us in following out these indications to obtain a notion of the form of the curve

really described by H, if we trace separately the paths which
H would pursue in virtue of either motion separately, since
its true motion will necessarily result from the superposition
of these partial motions, because at every instant they are
at right angles to each other, and therefore cannot interfere.
First, then, it is evident, from what we have said of the
tangential force, that when P is at A, H is for an instant at
rest, but that as P removes from A towards D, H continually
approaches P along their line of junction H P, which is,
therefore, at each instant a tangent to the path of H.   When
P is in the octant, H is at its mean distance from P (equal to
P S), and is then in the act of approaching P most rapidly.
From thence to the quadrature D the movement of H towards P decreases in rapidity till the quadrature is attained,
when H rests for an instant, and then begins to reverse its
motion, and travel *from* P at the same rate of progress as
before *towards* it.   Thus it is clear that, in virtue of the
tangential force alone, H would describe a four-cusped
curve *a, d, b, e,* its direction of motion round S in this curve
being opposite to that of P, so that A and *a,* D and *d,* B and
b, E and *e,* shall be corresponding points.

(707.) Next as regards the normal force.   When the

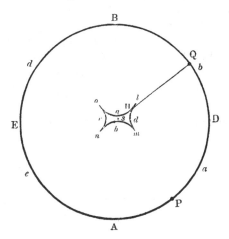

moon is at A the motion of H is towards D, and is at its

maximum of rapidity, but slackens as P proceeds towards D
and as Q proceeds towards B.   To the curve described, H Q
will be always a tangent, and since at the neutral point of
the normal force (or when P is 64° 14′ from A, and Q 64°
14′ from D), the motion of H is for an instant *nil* and is then
reversed, the curve will have a cusp at *l* corresponding, and
H will then begin to travel along the arc *l m*, while P describes the corresponding arc from neutral point to neutral
point through D.   Arrived at the neutral point between D
and B, the motion of H along Q H will be again arrested
and reversed, giving rise to another cusp at *m*, and so on.
Thus, in virtue of the normal force acting alone, the path of
H would be the four-cusped, elongated curve *l m n o*, described with a motion round S the reverse of P's, and having
a, d, b, e for points corresponding to A, B, D, E, places
of P.

(708.) Nothing is now easier than to superpose these motions.   Supposing H_1, H_2 to be the points in either curve corresponding to P, we have nothing to do but to set from off S, S*h*
equal and parallel to S H_1 in the one curve and from *h, h* H
equal and parallel to S H_2 in the other.   Let this be done
for every corresponding point in the two curves, and there
results an oval curve *a d b e*, having for its semiaxes S*a* = Sa_1
+ Sa_2; and S*d* = Sd_1 + Sd_2.   And this will be the true path
of the upper focus, the points *a, d, b, e*, corresponding to
A, D, B, E, places of P.   And from this it follows, 1st,
that at A, B, the syzygies, the moon is in perigee in her momentary ellipse, the lower focus being nearer than the upper.
2dly, That in quadratures D, E, the moon is in apogee in her
then momentary ellipse, the upper focus being then nearer

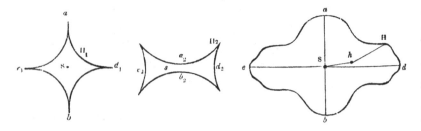

than the lower. 3dly, That H revolves in the oval $a\,d\,b\,e$ the contrary way to P in its orbit, making a complete revolution from syzygy to syzygy in one synodic revolution of the moon.

(709.) Taking 1 for the moon's mean distance from the earth, suppose we represent Sa_1 or Sd_1 (for they are equal) by $2a$, Sa_2 by $2b$, and Sd_2 by $2c$, then will the semiaxes of the oval $a\,d\,b\,e$, Sa and Sd be respectively $2a+2b$ and $2a+2c$, so that the excentricities of the momentary ellipses at A and D will be respectively $a+b$ and $a+c$. The total amount of the effect of the tangential force on the *axis*, in passing from syzygy to quadrature, will evidently be equal to the length of the curvilinear arc $a_1\,d_1$ (*fig.* art. 708.), which is necessarily less than $Sa_1 + Sd_1$ or $4a$. Therefore the total effect on the *semiaxis* or distance of the moon is less than $2a$, and the excess and defect of the greatest and least values of this distance thus varied above and below the mean value $S\,A = 1$ (which call α) will be less than a. The moon then is moving at A *in* the *perigee* of an ellipse whose semiaxis is $1 + \alpha$ and excentricity $a + b$, so that its actual distance from the earth there is $1 + \alpha - a - b$, which (because α is less than a) is less than $1 - b$. Again, at D it is moving *in apogee* of an ellipse whose semiaxis is $1 - \alpha$ and excentricity $a + c$, so that its distance then from the earth is $1 - \alpha + a + c$, which (a being greater than α) is greater than $1 + c$, the latter distance exceeding the former by $2a - 2\alpha + b + c$.

(710.) Let us next consider the corresponding changes induced upon the angular velocity. Now it is a law of elliptic motion that at different points of differentellipses, each differing very little from a circle, the angular velocities are to each other as the square roots of the semiaxes directly, and as the squares of the distances inversely. In this case the semiaxes at A and D are to each other as $1 + \alpha$ to $1 - \alpha$, or as $1 : 1 - 2\alpha$, so that their square roots are to each other as $1 : 1 - \alpha$. Again, the distances being to each other as $1 + \alpha - a - b : 1 - \alpha + a + c$, the inverse ratio of their squares (since α, a, b, c, are all very small quantities) is that of $1 - 2\alpha + 2a + 2c : 1 + 2\alpha - 2a - 2b$, or as $1 : 1 - 4\alpha - 4a - 2b - 2c$. The angular velocities then are to each other in a ratio compounded of these two proportions, that is in the ratio of

$$1 : 1 + 3\alpha - 4a - 2b - 2c,$$

which is evidently that of a greater to a less quantity. It is obvious also, from the constitution of the second term of this ratio, that the normal force is far more influential in producing this result than the tangential.

(711.) In the foregoing reasoning the sun has been regarded as fixed. Let us now suppose it in motion (in a circular orbit), then it is evident that at equal angles of *elongation* (of P from M seen from S), equal disturbing forces, both tangential and normal, will act : only the syzygies and quadratures, as well as the neutral points of the normal force, instead of being points fixed in longitude on the orbit of the moon, will advance on that orbit with a uniform angular motion equal to the angular motion of the sun. The cuspidated curves $a_1 \, d_1 \, b_1 \, e_1$ and $a_2 \, d_2 \, b_2 \, e_2$, *fig.* art. 708., will, therefore, no longer be re-entering curves ; but each will have its cusps *screwed round* as it were in the direction of the sun's motion, so as to increase the angles between them in the ratio of the synodical to the sidereal revolution of the moon (art. 418.). And if, in like manner, the motions in these two curves, thus separately described by H, be compounded, the resulting curve, though still (loosely speaking) a species of oval, will not return into itself, but will make successive spiroidal convolutions about S, its farthest and nearest points being in the same ratio more than 90° asunder. And to this movement that of the moon herself will conform, describing a species of elliptic spiroid, having its least distances always in the line of syzygies and its greatest in that of quadratures. It is evident also, that, owing to the longer continued action of both forces, *i. e.* owing to the greater arc over which their intensities increase and decrease by equal steps, the branches of each curve between the cusps will be longer, and the cusps themselves will be more remote from S, and in the same degree will the dimensions of the resulting oval be enlarged, and with them the amount of the inequality in the moon's motion which they represent.

(712.) In the above reasoning the sun's distance is supposed so great, that the disturbing forces in the semi-orbit nearer to it shall not sensibly differ from those in the more remote. The sun, however, is actually nearer to the moon in conjunction than in opposition by about one two-hundredth part of its whole distance, and this suffices to give rise to a very sensible inequality (called the *parallactic inequality*) in the lunar motions, amounting to about 2′ in its effect on the moon's longitude, and having for its period one synodical revolution or one lunation. As this inequality, though subordinate in the case of the moon to the great inequality of the variation with which it stands in connexion, becomes a prominent feature in the system of inequalities corresponding to it in the planetary perturbations (by reason of the very great variations of their distances from conjunction to opposition), it will be necessary to indicate what modifications this consideration will introduce into the forms of our focus curves, and of their superposed oval. Recurring then to our figures in art. 706, 707., and supposing the moon to set out from E, and the upper focus, in each curve from *e*, it is evident that the intercuspidal arcs *e a*, *a d*, in the one, and *e p*, *p a l*, *l d*, in the other, being described under the influence of more powerful forces, will be greater than the arcs *d b*, *b e*, and *d m*, *m b n*, *n e* corresponding in the other half revo-

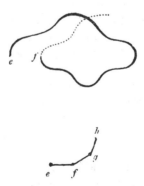

lution. The two extremities of these curves then, the initial and terminal places of *e* in each, will not meet, and the same

conclusion will hold respecting those of the compound oval
in which the focus really revolves, which will, therefore, be
as in the annexed figure. Thus, at the end of a complete
lunation, the focus will have shifted its place from e to f in a
line parallel to the line of quadratures. The next revolution,
and the next, the same thing would happen. Meanwhile,
however, the sun has advanced in its orbit, and the line of
quadratures has changed its situation by an equal angular
motion. In consequence, the next terminal situation (g) of
the forces will not lie in the line ef prolonged, but in a line
parallel to the new situation of the line of quadratures, and
this process continuing, will evidently give rise to a movement of circulation of the point e, round a mean situation in
an annual period; and this, it is evident, is equivalent to
an annual circulation of the central point of the compound
oval itself, in a small orbit about its mean position S. Thus
we see that no permanent and indefinite increase of excentricity can arise from this cause ; which would be the case,
however, but for the annual motion of the sun.

(713.) Inequalities precisely similar in principle to the
variation and parallactic inequality of the moon, though
greatly modified by the different relations of the dimensions
of the orbits, prevail in all cases where planet disturbs planet.
To what extent this modification is carried will be evident,
if we cast our eyes on the examples given in art. 612., where
it will be seen that the disturbing force in conjunction often
exceeds that in opposition in a very high ratio, (being in the
case of Neptune disturbing Uranus more than ten times as
great). The effect will be, that the orbit described by the
center of the compound oval about S, will be much greater
relatively to the dimensions of that oval itself, than in the
case of the moon. Bearing in mind the nature and import
of this modification, we may proceed to enquire, apart from
it, into the number and distribution of the undulations in the
contour of the oval itself arising from the alternations of direction *plus* and *minus* of the disturbing forces in the course of
a synodic revolution. But first it should be mentioned that,
in the case of an exterior disturbed by an interior planet,

the disturbing body's angular motion exceeds that of the disturbed. Hence P, though advancing in its orbit, recedes relatively to the line of syzygies, or, which comes to the same thing, the neutral points of either force overtake it in succession, and each, as it comes up to it, gives rise to a cusp in the corresponding *focus curve*. The angles between the successive cusps will therefore be to the angles between the corresponding neutral points for a fixed position of M, in the same constant ratio of the synodic to the sidereal period of P, which however is now a ratio of less inequality. These angles then will be contracted in amplitude, and, for the same reason as before, the excursions of the focus will be diminished, and the more so the shorter the synodic revolution.

(714.) Since the cusps of either curve recur, in successive synodic revolutions in the same order, and at the same angular distances from each other, and from the line of conjunction, the same will be true of all the corresponding points in the curve resulting from their superposition. In that curve, every cusp, of either constituent, will give rise to a convexity, and every intercuspidal arc to a relative concavity. It is evident then that the compound curve or true path of the focus so resulting, but for the cause above mentioned, would return into itself, whenever the periodic times of the disturbing and disturbed bodies are commensurate, because in that case the synodic period will also be commensurate with either, and the arc of longitude

 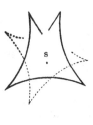

intercepted between the sidereal place of any one conjunction, and that next following it, will be an aliquot part of 360°.

In all other cases it would be a non-reentering, more or less undulating and more or less regular, spiroid, according to the number of cusps in each of the constituent curves (that is to say, according to the number of neutral points or changes of direction from inwards to outwards, or from accelerating to retarding, and *vice versa*, of the normal and tangential forces,) in a complete synodic revolution, and their distribution over the circumference.

(715.) With regard to these changes, it is necessary to distinguish three cases, in which the perturbations of planet by planet are very distinct in character.   1st. When the disturbing planet is exterior.   In this case there are four neutral points of either force.   Those of the tangential force occur at the syzygies, and at the points of the disturbed orbit (which we shall call points of equidistance), equidistant from the sun and the disturbing planet (at which points, as we have shown (art. 614.), the total disturbing force is always directed inwards towards the sun).   Those of the normal force occur at points intermediate between these last mentioned points, and the syzygies, which, if the disturbing planet be *very* distant, hold nearly the situation they do in the lunar theory, *i. e.* considerably nearer the quadratures than the syzygies.   In proportion as the distance of the disturbing planet diminishes, two of these points, viz. those nearest the syzygy, approach to each other, and to the syzygy, and in the extreme case, when the dimensions of the orbits are equal, coincide with it.

(716.) The second case is that in which the disturbing planet is interior to the disturbed, but at a distance from the sun greater than half that of the latter.   In this case there are four neutral points of the tangential force, and only two of the normal.   Those of the tangential force occur at the syzygies, and at the points of equidistance.   The force retards the disturbed body from conjunction to the first such points after conjunction, accelerates it thence to the opposition, thence again retards it to the next point of equidistance, and finally again accelerates it up to the conjunction.   As the disturbing orbit contracts in dimension the points of equidistance approach; their distance from syzygy from 60° (the extreme case) diminishing to nothing, when they coincide with each other, and with the conjunction. In the case of Saturn disturbed by Jupiter, that distance is only 23° 33'. The neutral points of the normal force lie somewhat beyond the quadratures, on the side of the opposition, and do not undergo any very material change of situation with the contraction of the disturbing orbit.

(717.) The third case is that in which the diameter of the disturbing interior orbit is less than half that of the disturbed. In this case there are only two points of evanescence for either force. Those of the tangential force are the syzygies. The disturbed planet is accelerated throughout the whole semi-revolution from conjunction to opposition, and retarded from opposition to conjunction, the maxima of acceleration and retardation occurring not far from quadrature. The neutral points of the normal force are situated nearly as in the last case; that is to say, beyond the quadratures towards the opposition. All these varieties the student will easily trace out by simply drawing the figures, and resolving the forces in a series of cases, beginning with a very large and ending with a very small diameter of the disturbing orbit. It will greatly aid him in impressing on his imagination the general relations of the subject, if he construct, as he proceeds,

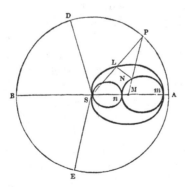

for each case, the elegant and symmetrical ovals in which the points N and L (*fig.* art. 675.) always lie, for a fixed position

of M, and of which the annexed figure expresses the forms
they respectively assume in the third case now under consideration.   The second only differs from this, in having the
common vertex *m* of both ovals outside of the disturbed orbit
A P, while in the case of an exterior disturbing planet the
oval *m* L assumes a four-lobed form; its lobes respectively
touching the oval *m* N in its vertices, and cutting the orbit
A P in the points of equidistance and of tangency, (*i. e.* where
M P S is a right angle) as in this figure.

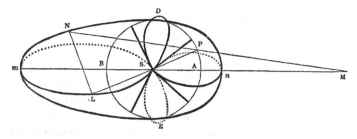

(718.) It would be easy now, bearing these features in mind,
to trace in any proposed case the form of the spiroid curve,
described, as above explained, by the upper focus.   It will
suffice, however, for our present purpose to remark, 1st, That
between every two successive conjunctions of P and M the
same general form, the same subordinate undulations, and the
same terminal displacement of the upper focus are continually
repeated.   2dly, That the motion of the focus in this curve
is retrograde whenever the disturbing planet is exterior, and
that in consequence the apsides of the momentary ellipse also
recede, with a mean velocity such as, but for that displacement, would bring them round at each conjunction to
the same relative situation with respect to the line of syzygies.
3dly, That in consequence of this retrograde movement of the
apse, the disturbed planet, apart from that consideration,
would be twice in perihelio and twice in aphelio in its
momentary ellipse in each synodic revolution, just as in the
case of the moon disturbed by the sun — and that in
consequence of this and of the undulating movement of
the focus H itself, an inequality will arise, analogous,

mutatis mutandis in each case, to the moon's variation, under which term we comprehend (not exactly in conformity to its strict technical meaning in the lunar theory) not only the principal inequality thus arising, but all its subordinate fluctuations.   And on this the parallactic inequality thus violently exaggerated is superposed.

(719.) We come now to the class of inequalities which depend for their existence on an appretiable amount of *permanent* excentricity in the orbit of one or of both the disturbing and disturbed planets, in consequence of which all their conjunctions do not take place at equal distances either from the central body or from each other, and therefore that symmetry in every synodic revolution on which depends the exact restoration of both the axis and excentricity to their original values at the completion of each such revolution no longer subsists.   In passing from conjunction to conjunction, then, there will no longer be effected either a complete restoration of the upper focus to the same relative situation, or of the axis to the same length which they respectively had at the outset.   At the same time it is not less evident that the differences in both respects are only what remain outstanding after the compensation of by far the greater part of the deviations to and fro from a mean state which occur in the course of the revolution; and that they amount to but small fractions of the total excursions of the focus from its first position, or of the increase and decrease in the length of the axis effected by the direct action of the tangential force, — so small, indeed, that, unless owing to peculiar adjustments they be enabled to accumulate again and again at successive conjunctions in the same direction, they would be altogether undeserving of any especial notice in a work of this nature.   Such adjustments, however, would evidently exist if the periodic times of the planets were exactly commensurable; since in that case all the possible conjunctions which could ever happen (the elements not being materially changed) would take place at fixed points in longitude, the intermediate points being never visited by a conjunction.   Now, of the conjunctions thus distributed, their relations to the lines of symmetry in the orbits being all dissimilar, some one must be more influential than the rest on each of the elements (not necessarily the *same* upon *all*). Consequently, in a complete cycle of conjunctions, wherein each has been visited in its turn, the influence of that one on the element to which it stands so especially related will preponderate over the counteracting and compensating influence of the rest, and thus, although in such a cycle as above specified, a further and much more exact compensation will have been effected in its value than in a single revolution; still that compensation will not be complete, but a portion of the effect (be it to increase or to diminish the excentricity or the axis, or to cause the apse to advance or to recede,) will remain outstanding. In the next cycle of the same kind this will be repeated, and the result will be of the same character, and so on, till at length a sensible and ultimately a large amount of change shall have taken place, and in fact until the axis (and with it the mean motion) shall have so altered as to destroy the commensurability of periods, and the apsides have so shifted as to alter the place of the most influential conjunction.

(720.) Now, although it is true that the mean motions of no two planets are exactly commensurate, yet cases are not wanting in which there exists an approach to this adjustment. For instance, in the case of Jupiter and Saturn, a cycle composed of five periods of Jupiter and two of Saturn, although it does not *exactly* bring about the same configuration, does so pretty nearly. Five periods of Jupiter are 21663 days, and two periods of Saturn, 21519 days. The difference is only 146 days, in which Jupiter describes, on an average, 12°, and Saturn about 5°; so that after the lapse of the former interval they will only be 7° from a conjunction in the same parts of their orbits as before. If we calculate the time which will exactly bring about, on the average, three conjunctions of the two planets, we shall find it to be 21760 days, their synodical period being 7253·4 days. In this interval Saturn will have described 8° 6′ in excess of two sidereal revolutions, and Jupiter the same angle in excess of five. Every third conjunction, then, will take place 8° 6′ in advance of the preceding, which is near enough to establish, not, it is true, an identity with, but still a great approach to the case in question.   The excess of action, for several such triple conjunctions (7 or 8) in succession, will lie the same way, and at each of them the elements of P's orbit and its angular motion will be similarly influenced, so as to accumulate the effect upon its longitude; thus giving rise to an irregularity of considerable magnitude and very long period, which is well known to astronomers by the name of the great inequality of Jupiter and Saturn.

(721.) The arc 8° 6' is contained $44\frac{4}{9}$ times in the whole circumference of 360°; and accordingly, if we trace round this particular conjunction, we shall find it will return to the same point of the orbit in so many times 21760 days, or in 2648 years.   But the conjunction we are now considering is only one out of three.   The other two will happen at points of the orbit about 123° and 246° distant, and *these points also will advance* by the same arc of 8° 6' in 21760 days.   Consequently the period of 2648 years will bring them *all* round, and in that interval each of them will pass through that point of the two orbits from which we commenced: hence *a conjunction* (one or other of the three) will happen at that point once in one third of this period, or in 883 years; and this is, therefore, the cycle in which the "great inequality" would undergo its full compensation, did the elements of the orbits continue all that time invariable.   Their variation, however, is considerable in so long an interval; and, owing to this cause, the period itself is prolonged to about 918 years.

(722.) We have selected this inequality as the most remarkable instance of this kind of action on account of its magnitude, the length of its period, and its high historical interest.   It had long been remarked by astronomers, that on comparing together modern with ancient observations of Jupiter and Saturn, the mean motions of these planets did not appear to be uniform.   The period of Saturn, for instance, appeared to have been lengthening throughout the whole of the seventeenth century, and that of Jupiter shortening — that is to say, the one planet was constantly lagging behind, and the

other getting in advance of its calculated place.   On the other
hand, in the eighteenth century, a process precisely the reverse
seemed to be going on.   It is true the whole retardations
and accelerations observed were not very great ; but, as their
influence went on accumulating, they produced, at length,
material differences between the observed and calculated
places of both these planets, which as they could not then be
accounted for by any theory, excited a high degree of attention,
and were even, at one time, too hastily regarded as almost
subversive of the Newtonian doctrine of gravity.   For a long
while this difference baffled every endeavour to account for
it ; till at length Laplace pointed out its cause in the near
commensurability of the mean motions, as above shown, and
succeeded in calculating its period and amount.

(723.) The inequality in question amounts, at its maximum,
to an alternate gain and loss of about 0° 49′ in the longitude
of Saturn, and a corresponding loss and gain of about 0° 21′
in that of Jupiter.   That an acceleration in the one planet
must necessarily be accompanied by a retardation in the other,
might appear at first sight self-evident, if we consider, that
action and reaction being equal, and in contrary directions,
whatever momentum Jupiter communicates to Saturn in the
direction P M, the same momentum must Saturn communicate
to Jupiter in the direction M P.   The one, therefore, it might
seem to be plausibly argued, will be dragged forward,
whenever the other is pulled back in its orbit.   The inference
is correct, *so far as the general and final result goes ;* but the
reasoning by which it would, on the first glance, appear to be
thus summarily established is fallacious, or at least incomplete.
It is perfectly true that whatever momentum Jupiter communicates directly to Saturn, Saturn communicates an equal
momentum to Jupiter in an opposite linear direction.   But it
is not with the absolute motions of the two planets in space
that we are now concerned, but with the relative motion of
each separately, with respect to the sun regarded as at rest.
The *perturbative* forces (the forces which disturb these relative
motions) do not act along the line of. junction of the planets
(art. 614.).   In the reasoning thus objected to, the attraction

of each on the sun has been left out of the account *, and it
remains to be shown that these attractions neutralize and
destroy each other's effects in considerable periods of time,
as bearing upon the result in question.   Suppose then that
we for a moment abandon the point of view, in which we
have hitherto all along considered the subject, and regard the
sun as free to move, and liable to be displaced by the attractions of the two planets.   Then will the movements of all be
performed about the common center of gravity, just as they
would have been about the sun's center regarded as immoveable, the sun all the while circulating in a small orbit (with
a motion compounded of the two elliptic motions it would
have in virtue of their separate attractions) about the same
center.   Now in this case M still disturbs P, and P, M, but
the whole *disturbing* force now acts along their line of junction,
and since it remains true that whatever momentum M generates in P, P will generate the same in M in a contrary
direction; it will also be strictly true that, so far as a disturbance of their elliptic motions *about the common center of
gravity of the system is alone regarded,* whatever disturbance
of velocity is generated in the one, a contrary disturbance of
velocity (only in the inverse ratio of the masses and modified,
though never contradicted, by the directions in which they
are respectively moving), will be generated in the other.
Now when we are considering only inequalities of long period
comprehending many complete revolutions of both planets,
and which arise from changes in the axes of the orbits,
affecting their *mean motions*, it matters not whether we
suppose these motions performed about the common center
of gravity, or about the sun, which never departs from
that center to any material extent (the mass of the sun
being such in comparison with that of the planets, that
that center always lies within his surface).   The *mean* motion

* We are here reading a sort of recantation.   In the edition of 1833 the
remarkable result in question *is* sought to be established by this vicious reasoning.   The mistake is a very natural one, and is so apt to haunt the ideas of
beginners in this department of physics, that it is worth while expressly to warn
them against it.

therefore, regarded as the average angular velocity during a revolution, is the same whether estimated by reference to the sun's center, or to the center of gravity, or, in other words, the relative mean motion referred to the sun is identical with the absolute mean motion referred to the center of gravity.

(724.) This reasoning applies equally to every case of mutual disturbance resulting in a long inequality such as may arise from a slow and long-continued periodical increase and diminution of the axes, and geometers have accordingly demonstrated as a consequence from it, that the proportion in which such inequalities affect the *longitudes* of the two planets concerned, or the maxima of the excesses and defects of their longitudes above and below their elliptic values, thence arising, in each, are to each other in the inverse ratio of their masses multiplied by the square roots of the major axes of their orbits, and this result is confirmed by observation, and will be found verified in the instance immediately in question as nearly as the uncertainty still subsisting as to the masses of the two planets will permit.

(725.) The inequality in question, as has been observed in general, (art. 718.) would be much greater, were it not for the partial compensation which is operated in it in every triple conjunction of the planets. Suppose P Q R to be Saturn's orbit, and *p q r* Jupiter's; and suppose a conjunction to take place at P *p*, on the line S A; a second at 123° distance. on the line S B; a third at 246° distance, on S C ; and the next at 368°, on S D. This last-mentioned conjunction, taking place nearly in the situation of the first, will produce 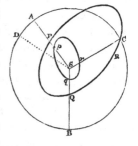 nearly a repetition of the first effect in retarding or accelerating the planets; but the other two, being in the most remote situations possible from the first, will happen under entirely different circumstances as to the position of the perihelia of the orbits. Now, we have seen that a presentation of the one planet to the other in conjunction, in a variety of situations, tends to produce compensation; and, in fact, the greatest possible amount of compensation which can be produced by only three conjunctions is when they are thus equally distributed round the center. Hence we see that it is not the whole amount of perturbation which is thus accumulated in each triple conjunction, but only that small part which is left uncompensated by the intermediate ones. The reader, who possesses already some acquaintance with the subject, will not be at a loss to perceive how this consideration is, in fact, equivalent to that part of the geometrical investigation of this inequality which leads us to seek its expression in terms of the third order, or involving the cubes and products of three dimensions of the excentricities and inclinations; and how the continual accumulation of small quantities, during long periods, corresponds to what geometers intend when they speak of small terms receiving great accessions of magnitude by the introduction of large coefficients in the process of integration.

(726.) Similar considerations apply to every case of approximate commensurability which can take place among the mean motions of any two planets. Such, for instance, is that which obtains between the mean motion of the earth and Venus, — 13 times the period of Venus being very nearly equal to 8 times that of the earth. This gives rise to an extremely near coincidence of every fifth conjunction, in the same parts of each orbit (within $\frac{1}{240}$th part of a circumference), and therefore to a correspondingly extensive accumulation of the resulting uncompensated perturbation. But, on the other hand, the part of the perturbation thus accumulated is only that which remains outstanding after passing the equalizing ordeal of five conjunctions equally distributed round the circle; or, in the language of geometers, is dependent on powers and products of the excentricities and inclinations of the fifth order. It is, therefore, extremely minute, and the whole resulting inequality, according to the elaborate calculations of Mr. Airy, to whom it owes its detection, amounts to no more than a few seconds at its maximum, while its period is no less than 240 years. This example will

serve to show to what minuteness these enquiries have been
carried in the planetary theory.

(727.) That variations of long period arising in the way
above described are necessarily accompanied by similarly
periodical displacements of the upper focus, equivalent in their
effect to periodical fluctuations in the magnitude of the
excentricity, and in the position of the line of apsides, is
evident from what has been already said respecting the motion
of the upper focus under the influence of the disturbing forces.
In the case of circular orbits the *mean* place of H coincides
with S the center of the sun, but if the orbits have any independent ellipticity, this coincidence will no longer exist—and
the *mean* place of the upper focus will come to be inferred
from the average of all the situations which it actually holds
during an entire revolution.   Now the fixity of this point
depends on the equality of each of the branches of the cuspidated curves, and consequent equality of excursion of
the focus in each particular direction, in every successive
situation of the line of conjunction.   But if there be some
one line of conjunction in which these excursions are greater
in any one particular direction than in another, the mean
place of the focus will be displaced, and if this process be repeated, that mean place will continue to deviate more and
more from its original position, and thus will arise a circulation of the *mean* place *of the focus for a revolution* about
another mean situation, the average of all the former mean
places during a complete cycle of conjunctions.   Supposing
S to be the sun, O the situation the upper focus would have,
had these inequalities no existence, and H K the path of the
upper focus, which it pursues about O by reason of them,
then it is evident that in the course of a complete cycle of
the inequality in question, the excentricity will have fluc-
tuated between the extreme limits S J and S I and the direction of the longer axis between the extreme position S H
and S K, and that if we suppose $i\,j\,h\,k$ to be the corresponding
mean places of the focus, $i\,j$ will be the extent of the fluctuation of the mean excentricity, and the angle $h\,s\,k$, that of the
longitude of the perigee.

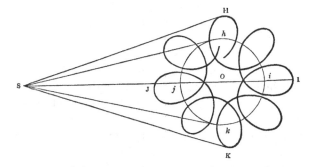

(728.) The periods then in which these fluctuations go through their phases are necessarily equal in duration with that of the inequality in longitude, with which they stand in connection. But it by no means follows that their maxima all coincide. The variation of the axis to which that of the mean motion corresponds, depends on the tangential force only whose maximum is not at conjunction or opposition, but at points remote from either, while the excentricity depends both on the normal and tangential forces, the maximum of the former of which is at the conjunction. That particular conjunction therefore, which is most influential on the axis, is not so on the excentricity, so that it can by no means be concluded that either the maximum value of the axis coincides with the maximum, or the minimum of the excentricity, or with the greatest excursion to or fro of the line of apsides from its mean situation, all that can be safely asserted is, that as either the axis or the excentricity of the one orbit varies, that of the other will vary in the opposite direction.

(729.) The primary elements of the lunar and planetary orbits, which may be regarded as variable, are the longitude of the node, the inclination, the axis, excentricity, longitude of the perihelion, and epoch (art. 496.). In the foregoing articles we have shown in what manner each of the first five of these elements are made to vary, by the direct action of the perturbing forces. It remains to explain in what manner the last comes to be affected by them. And here it is necessary, in the first instance, to remove some degree of obscurity

which may be thought to hang about the sense in which the
term itself is to be understood in speaking of an orbit, every
other element of which is regarded as in a continual state of
variation.   Supposing, then, that we were to reverse the process of calculation described in arts. 499. and 500. by which a
planet's heliocentric longitude in an elliptic orbit is computed
for a given time ; and setting out with a heliocentric longitude
ascertained by observation, all the *other* elements being known,
we were to calculate either what mean longitude the planet
had at a given epochal time, or, which would come to the same
thing, at what moment of time (thenceforward to be assumed
as an epoch) it had a given mean longitude.   It is evident
that by this means the epoch, if not otherwise known, would
become known, whether we consider it as the moment of time
corresponding to a convenient mean longitude, or as the mean
longitude corresponding to a convenient time.   The latter way
of considering it has some advantages in respect of general
convenience, and astronomers are in agreement in employing,
as an element under the title " Epoch of the mean longitude,"
the mean longitude of the planet so computed for a fixed date ;
as, for instance, the commencement of the year 1800, mean
time at a given place.   Supposing now all the elements of the
orbit invariable, if we were to go through this reverse process,
and thus ascertain the epoch (so defined) from any number of
different perfectly correct heliocentric longitudes, it is clear
we should always come to the same result.   One and the
same " epoch " would come out from all the calculations.

(730.) Considering then the "epoch" in this light, as
merely a result of this reversed process of calculation, and not
as the direct result of an observation instituted for the purpose at the precise epochal moment of time, (which would be,
generally speaking, impracticable,) it might be conceived
subject to variation in two distinct ways, viz. dependently
and independently.   Dependently it must vary, as a necessary
consequence of the variation of the other elements ; because,
if setting out from one and the same observed heliocentric
longitude of the planet, we calculate back to the epoch with
two different sets of intermediate elements, the one set con-

sisting of those which it had immediately before its arrival at that longitude, the other that which it takes up immediately after (*i. e.* with an unvaried and a varied system), we cannot (unless by singular accident of mutual counteraction) arrive at the same result; and the difference of the results is evidently the variation of the epoch. On the other hand, however, it cannot vary independently; for since this is the only mode in which the unvaried and varied epochs can become known, and as both result from direct processes of calculation involving only given data, the results can only differ by reason of the difference of those data. Or we may argue thus. The change in the path of the planet, and its place in that path so changed, at any future time (supposing it to undergo no further variation), are entirely owing to the change in its velocity and direction, produced by the disturbing forces at the point of disturbance; now these latter changes (as we have above seen) are *completely* represented by the momentary change in the situation of the upper focus, taken in combination with the momentary variation in the plane of the orbit; and these therefore express the total effect of the disturbing forces. There is, therefore, no direct and specific action on the epoch as an independent variable. It is simply left to accommodate itself to the altered state of things in the mode already indicated.

(731.) Nevertheless, should the effects of perturbation by inducing changes on these other elements affect the mean longitude of the planet in any other way than can be considered as properly taken account of, by the varied periodic time due to a change of axis, such effects must be regarded as incident on the epoch. This is the case with a very curious class of perturbations which we are now to consider, and which have their origin in an alteration of the average distance at which the disturbed body is found at every instant of a complete revolution, distinct from, and not brought about by the variation of the major semi-axis, or *momentary "mean distance"* which is an imaginary magnitude, to be carefully distinguished from the average of the actual distances now contemplated. Perturbations of this class (like the moon's variation, with which they are intimately connected) are independent on the excentricity of the disturbed orbit; for which reason we shall simplify our treatment of this part of the subject, by supposing that orbit to have no permanent excentricity, the upper focus in its successive displacements merely revolving about a mean position coincident with the lower. We shall also suppose M very distant, as in the lunar theory.

(732.) Referring to what is said in arts. 706. and 707., and to the figures accompanying those articles, and considering first the effect of the tangential force, we see that besides the effect of that force in changing the length of the axis, and consequently the periodic time, it causes the upper focus H to describe, in each revolution of P, a four-cusped curve, *a, d, b, e,* about S, all whose intercuspidal arcs are similar and equal. This supposes M fixed, and at an invariable distance,— suppositions which simplify the relations of the subject, and (as we shall afterwards show) do not affect the general nature of the conclusions to be drawn. In virtue, then, of the excentricity thus given rise to, P will be at the perigee of its momentary ellipse at syzygies and in its apogee at quadratures. *Apart, therefore, from the change arising from the variation of axis,* the distance of P from S will be less at syzygies, and greater at quadratures, than in the original circle. But the average of all the distances during a whole revolution will be unaltered; because the distances of *a, d, b, e* from S being equal, and the arcs symmetrical, the approach in and about perigee will be equal to the recess in and about apogee. And, in like manner, the effect of the changes going on in the length of the axis itself, on the average in question, is *nil,* because the alternate increases and decreases of that length balance each other in a complete revolution. Thus we see that the tangential force is excluded from all influence in producing the class of perturbations now under consideration.

(733.) It is otherwise with respect to the normal force. In virtue of the action of that force the upper focus describes, in each revolution of P, the four-cusped curve (*fig.* art. 707.), whose intercuspidal arcs are alternately of very unequal extent, arising, as we have seen, from the longer duration and greater energy of the outward than of the inward action of the disturbing force.   Although, therefore, in perigee at syzygies and in apogee at quadratures, the apogeal recess is much greater than the perigeal approach, inasmuch as S d greatly exceeds S a.   On the average of a whole revolution, then, the recesses will preponderate, and the average distance will therefore be greater in the disturbed than in the undisturbed orbit. And it is manifest that this conclusion is quite independent of any change in the length of the axis, which the normal force has no power to produce.

(734.) But neither does the normal force operate any change of linear velocity in the disturbed body.   When carried out, therefore, by the effect of that force to a greater distance from S, the angular velocity of its motion round S will be diminished: and contrariwise when brought nearer. The average of all the momentary angular motions, therefore, will decrease with the increase in that of the momentary distances; and in a higher ratio, since the angular velocity, under an equable description of areas, is inversely as the square of the distance, and the disturbing force, being (in the case supposed) directed to or from the center, does not disturb that equable description (art. 616.).   Consequently, on the average of a whole revolution, the angular motion is slower, and therefore the time of completing a revolution, and returning to the same longitude, longer than in the undisturbed orbit, and *that* independent of and without any reference to the length of the momentary axis, and the "periodic time" or "mean motion" dependent thereon.   We leave to the reader to follow out (as is easy to do) the same train of reasoning in the cases of planetary perturbation, when M is not very remote, and when it is interior to the disturbed orbit.   In the latter case the preponderant effect changes from a retardation of angular velocity to an acceleration, and the dilatation of the average dimensions of P's orbit to a contraction.

(735.) The above is an accurate analysis, according to strict dynamical principles, of an effect which, speaking roughly, may be assimilated to an alteration of M's gravitation towards S by the mean preponderant amount of the outward and inward action of the normal forces constantly exerted—nearly as would be the case if the mass of the disturbing body were formed into a ring of uniform thickness, concentric with S, and of such diameter as to exert an action on P everywhere equal to such mean preponderant force, and in the same direction as to inwards or outwards. For it is clear that the action of such a ring on P, will be the difference of its attractions on the two points P and S, of which the latter occupies its center, the former is excentric. Now the attraction of a ring on its center is manifestly equal in all directions, and therefore, estimated in any one direction, is zero. On the other hand, on a point P out of its center, if *within* the ring, the resulting attraction will always be *outwards*, towards the nearest point of the ring, or directly from the center.* But if P lie without the ring, the resulting force will act always *inwards*, urging P towards its center. Hence it appears that the mean effect of the radial force of the ring will be different in its direction, according as the orbit of the disturbing body is exterior or interior to that of the disturbed. In the former

---

* As this is a proposition which the equilibrium of Saturn's ring renders not merely speculative or illustrative, it will be well to demonstrate it; which may be done very simply, and without the aid of any calculus. Conceive a spherical shell, and a point within it: every line passing through the point, and terminating both ways in the shell, will, of course, be equally inclined to its surface at either end, being a chord of a spherical surface, and, therefore symmetrically related to all its parts. Now, conceive a small double cone, or pyramid, having its apex at the point, and formed by the conical motion of such a line round the point. Then will the two portions of the spherical shell, which form the bases of both the cones, or pyramids, be similar and equally inclined to their axes. Therefore their areas will be to each other as the squares of their distances from the common apex. Therefore their attractions on it will be equal, because the attraction is as the attracting matter directly, and the square of its distance inversely. Now, these attractions act in opposite directions, and therefore counteract each other. Therefore the point is in equilibrium between them; and as the sâme is true of every such pair of areas into which the spherical shell can be broken up, therefore the point will be in equilibrium *however situated within* such a spherical shell. Now take a ring, and treat it similarly, breaking its circumference up into pairs of elements, the bases of *triangles* formed by lines passing through the attracted point. Here the attracting elements being *lines*, not *surfaces*, are in the *simple* ratio of the distances, not the *duplicate*, as they should be to maintain the equilibrium. Therefore it will not be maintained, but the *nearest* elements will have the superiority, and the point will, on the whole, be urged towards the nearest part of the ring. The same is true of every *linear* ring, and is therefore true of any assemblage of concentric ones forming a flat annulus, like the ring of Saturn.

case it will act in diminution, in the latter in augmentation of the central gravity.

(736.) Regarding, still, only the mean effect, as produced in a great number of revolutions of both bodies, it is evident that such an increase of central force will be accompanied with a diminution of periodic time and distance of a body revolving with a stated velocity, and *vice versâ*. This, as we have shown, is the first and most obvious effect of the radial part of the disturbing force, when exactly analyzed. It alters permanently, and by a certain mean amount, the distances and times of revolution of all the bodies composing the planetary system, from what they would be, did each planet circulate about the sun uninfluenced by the attraction of the rest; the angular motion of the interior bodies of the system being thus rendered less, and those of the exterior greater, than on that supposition. The latter effect, indeed, might be at once concluded from this obvious consideration,—that all the planets revolving interiorly to any orbit may be considered as adding to the general aggregate of the attracting matter within, which is not the less efficient for being distributed over space, and maintained in a state of circulation.

(737.) This effect, however, is one which we have no means of measuring, or even of detecting, otherwise than by calculation. For our knowledge of the periods of the planets is drawn from observations made on them in their actual state, and therefore under the influence of this, which may be regarded as a sort of *constant part* of the perturbative action. Their observed mean motions are therefore affected by the whole amount of its influence ; and we have no means of distinguishing this by observation from the direct effect of the sun's attraction, with which it is blended. Our knowledge, however, of the masses of the planets assures us that it is extremely small; and this, in fact, is all which it is at all important to us to know, in the theory of their motions.

(738.) The action of the sun upon the moon, in like manner, tends, by its mean influence during many successive revolutions of both bodies, to increase permanently the moon's distance and periodic time. But this general average is not established, either in the case of the moon or planets, without a series of surbordinate fluctuations, which we have purposely neglected to take account of in the above reasoning, and which obviously tend, in the average of a great multitude of revolutions, to neutralize each other.   In the lunar theory, however, some of these subordinate fluctuations are very sensible to observation.   The most conspicuous of these is the moon's annual equation ; so called because it consists in an alternate increase and decrease in her longitude, corresponding with the earth's situation in its annual orbit; *i. e.* to its angular distance from the perihelion, and therefore having a year instead of a month, or aliquot part of a month, for its period.   To understand the mode of its production, let us suppose the sun, still holding a fixed position in longitude, to approach gradually nearer to the earth.   Then will all its disturbing forces be gradually increased in a very high ratio compared with the diminution of the distance (being inversely as its cube ; so that its effects of every kind are three times greater in respect of any change of distance, than they would be by the simple law of proportionality).   Hence, it is obvious that the focus H (art. 707.) in the act of describing each intercuspidal arc of the curve *a, d, b, e,* will be continually carried out farther and farther from S ; and the curve, instead of returning into itself at the end of each revolution, will open out into a sort of cuspidated spiral, as in the figure annexed.   Retracing now the reasoning of art. 733. as adapted to this state of things, it will be seen that so long as this dilatation goes on, so long will the difference between M's recess from S in aphelio and its approach in perihelio (which is equal to the difference of consecutive long and short semidiameters of this curve) also continue to increase, and with it the average of the distances of M from S in a whole revolution, and consequently also the time of performing such a revolution. The reverse process will go on as the sun again recedes. Thus it appears that, as the sun approaches the earth, the

mean angular motion of the moon on the average of a whole
revolution will diminish, and the duration of each lunation
will therefore exceed that of the foregoing, and *vice versâ.*

(739.) The moon's orbit being supposed circular, the sun's
orbital motion will have no other effect than to keep the
moon longer under the influence of every gradation of the
disturbing force, than would have been the case had his
situation in longitude remained unaltered (art. 711). The
same effects, therefore, will take place only on an increased
scale in the proportion of the increased time; *i. e.* in the
proportion of the synodic to the sidereal revolution of the
moon. Observation confirms these results, and assigns to the
inequality in question a maximum value of between 10′ and
11′, by which the moon is at one time in advance of, and at
another behind, its mean place, in consequence of this perturbation.

(740.) To this class of inequalities we must refer one of
great importance, and extending over an immense period of
time, known by the name of the *secular acceleration of the
moon's mean motion.* It had been observed by Dr. Halley,
on comparing together the records of the most ancient lunar
eclipses of the Chaldean astronomers with those of modern
times, that the period of the moon's revolution at present is
sensibly shorter than at that remote epoch; and this result
was confirmed by a further comparison of both sets of
observations with those of the Arabian astronomers of the
eighth and ninth centuries. It appeared, from these comparisons, that the rate at which the moon's mean motion
increases is about 11 seconds per century,—a quantity small
in itself, but becoming considerable by its accumulation
during a succession of ages. This remarkable fact, like the
great equation of Jupiter and Saturn, had been long the
subject of toilsome investigation to geometers. Indeed, so
difficult did it appear to render any exact account of, that
while some were on the point of again declaring the theory
of gravity inadequate to its explanation, others were for
rejecting altogether the evidence on which it rested, although
quite as satisfactory as that on which most historical events

are credited. It was in this dilemma that Laplace once more stepped in to rescue physical astronomy from its reproach, by pointing out the real cause of the phænomenon in question, which, when so explained, is one of the most curious and instructive in the whole range of our subject,— one which leads our speculations farther into the past and future, and points to longer vistas in the dim perspective of changes which our system has undergone and is yet to undergo, than any other which observation assisted by theory has developed.

(741.) The year is not an exact number of lunations. It consists of twelve and a fraction. Supposing then the sun and moon to set out from conjunction together; at the twelfth conjunction subsequent the sun will not have returned precisely to the same point of its annual orbit, but will fall somewhat short of it, and at the thirteenth will have overpassed it. Hence in twelve lunations the gain of longitude during the first half year will be somewhat under and in thirteen somewhat over-compensated. In twenty-six it will be nearly twice as much over-compensated, in thirty-nine not quite so nearly three times as much, and so on, until, after a certain number of such multiples of a lunation have elapsed, the sun will be found half a revolution in advance, and in place of receding farther at the expiration of the next, it will have begun to approach. From this time every succeeding cycle will destroy some portion of that over-compensation, until a complete revolution of the sun in excess shall be accomplished. Thus arises a subordinate or rather supplementary inequality, having for its period as many years as is necessary to multiply the deficient arc into a whole revolution, at the end of which time a much more exact compensation will have been operated, and so on. Thus after a moderate number of years an almost perfect compensation will be effected, and if we extend our views to centuries we may consider it as quite so. Such at least would be the case if the solar ellipse were invariable. But that ellipse is kept in a continual but excessively slow state of change by the action of the planets on the earth.

Its axis, it is true, remains unaltered; but its excentricity is, and has been since the earliest ages, diminishing; and this diminution will continue (there is little reason to doubt) till the excentricity is annihilated altogether, and the earth's orbit becomes a perfect circle; after which it will again open out into an ellipse, the excentricity will again increase, attain a certain moderate amount, and then again decrease. The time required for these evolutions, though calculable, has not been calculated, further than to satisfy us that it is not to be reckoned by hundreds or by thousands of years. It is a period, in short, in which the whole history of astronomy and of the human race occupies but as it were a point, during which all its changes are to be regarded as uniform. Now, it is by this variation in the excentricity of the earth's orbit that the secular acceleration of he moon is caused. The compensation above spoken of (even after the lapse of centuries) will now, we see, be only imperfectly effected, owing to this slow shifting of one of the essential data. The steps of restoration are no longer identical with, nor equal to, those of change. The struggle up hill is not maintained on equal terms with the downward tendency. The ground is all the while slowly sliding beneath the feet of the antagonists. During the whole time that the earth's excentricity is diminishing, a preponderance is given to the re-action over the action; and it is not till that diminution shall cease, that the tables will be turned, and the process of ultimate restoration will commence. Meanwhile, a minute, outstanding, and uncompensated effect in favour of acceleration is left at each recurrence, or near recurrence, of the same configurations of the sun, the moon, and the solar perigee. These accumulate, and at length affect the moon's longitude to an extent not to be overlooked.

(742.) The phænomenon, of which we have now given an account, is another and very striking example of the propagation of a periodic change from one part of a system to another. The planets, with one exception, have no direct appretiable action on the lunar motions as referred to the earth. Their masses are too small, and their distances too

great, for their difference of action on the moon and earth
ever to become sensible. Yet their effect on the earth's
orbit is thus, we see, propagated through the sun to that of
the moon; and, what is very remarkable, the transmitted
effect thus indirectly produced on the angle described by the
moon round the earth is more sensible to observation than
that directly produced by them on the angle described by the
earth round the sun.

(743.) Referring to the reasoning of art. 738., we shall
perceive that if, owing to any other cause than its elliptic
motion, the sun's distance from the earth be subject to a
periodical increase and decrease, that variation will give rise
to a lunar inequality of equal period analogous to the annual
equation.   It thus happens that very minute changes impressed on the orbit of the earth, by the direct action of the
planets, (provided their periods, though not properly speaking
secular, be of considerable length,) may make themselves
sensible in the lunar motions.   The longitude of that satellite,
as observed from the earth, is, in fact, singularly sensible to
this kind of reflected action, which illustrates in a striking
manner the principle of forced vibrations laid down in art.
(650.).   The reason of this will be readily apprehended, if
we consider that however trifling the increase of her longitude
which would arise in a single revolution, from a minute and
almost infinitesimal increase of her mean angular velocity,
that increase is not only repeated in each subsequent revolution, but is reinforced during each by a similar fresh accession of angular motion generated in its lapse.   This process goes on so long as the angular motion continues to
increase, and only begins to be reversed when lapse of time,
bringing round a contrary action on the angular motion,
shall have destroyed the excess of velocity previously gained,
and begun to operate a retardation.   In this respect, the
advance gained by the moon on her undisturbed place may
be assimilated, during its increase, to the space described
from rest under the action of a continually accelerating force.
The velocity gained in each instant is not only effective in
carrying the body forward during each subsequent instant,

but new velocities are every instant generated, and go on adding their cumulative effects to those before produced.

(744.) The distance of the earth from the sun, like that of the moon from the earth, may be affected in its average value estimated over long periods embracing many revolutions, in two modes, conformably to the theory above delivered. 1st, it may vary by a variation in the length of the axis major of its orbit, arising from the direct action of some tangential disturbing force on its velocity, and thereby producing a change of mean motion and periodic time in virtue of the Keplerian law of periods, which declares that the periodic times are in the sesquiplicate ratio of the mean distances. Or, 2dly, it may vary by reason of that peculiar action on the average of actual distances during a revolution, which arises from variations of excentricity and perihelion only, and which produces that sort of change in the mean motion which we have characterized as incident on the epoch. The change of mean motion thus arising, has nothing whatever to do with any variation of the major axis. It does not depend on the change of distance by the Keplerian law of *periods*, but by that of *areas*. The altered mean motion is not sub-sesquiplicate to the altered axis of the ellipse, which in fact does not alter at all, but is *sub-duplicate* to the altered *average of distances* in a revolution; a distinction which must be carefully borne in mind by every one who will clearly understand either the subject itself, or the force of Newton's explanation of it in the 6th Corollary of his celebrated 66th Proposition. In whichever mode, however, an alteration in the mean motion is effected, if we accommodate the general sense of our language to the specialties of the case, it remains true that every change in the mean motion is accompanied with a corresponding change in the mean distance.

(745.) Now we have seen, art. (726.), that Venus produces in the earth a perturbation in longitude, of so long a period (240 years), that it cannot well be regarded without violence to ordinary language, otherwise than as an equation of the mean motion. Of course, therefore, it follows that during that half of this long period of time, in which the earth's motion is retarded, the distance between the sun and earth is on the increase, and *vice versâ*. Minute as is the equation in question, and consequent alteration of solar distance, and almost inconceivably minute as is the effect produced on the moon's mean angular velocity in a single lunation, yet the great number of lunations (1484), during which the effect goes on accumulating in one direction, causes the moon at the moment when that accumulation has attained its maximum to be very sensibly in advance of its undisturbed place (viz. by 23″ of longitude), and after 1484 more lunations, as much in arrear. The calculations by which this curious result has been established, formidable from their length and intricacy, are due to the industry, as the discovery of its origin is to the sagacity, of Professor Hansen.

(746.) The action of Venus, just explained, is indirect, being as it were a sort of reflection of its influence on the earth's orbit.   But a very remarkable instance of its influence, in actually perturbing the moon's motions by its direct attraction, has been pointed out, and the inequality due to it computed by the same eminent geometer.*   As the details of his processes have not yet appeared, we can here only explain, in general terms, the principle on which the result in question depends, and the nature of the peculiar adjustment of the mean angular velocities of the earth and Venus which render it effective.   The disturbing forces of Venus on the moon are capable of being represented or expressed (as is indeed generally the case with all the forces concerned in producing planetary disturbance) by the substitution for them of a series of other forces, each having a period or cycle within which it attains a maximum in one direction, decreases to nothing, reverses its action, attains a maximum in the opposite direction, again decreases to nothing, again reverses its action and re-attains its former magnitude, and so on.   These cycles differ for each particular constituent or *term*, as it is called, of the total forces considered as so broken up into partial ones, and generally speaking, every combination which

* Astronomische Nachrichten, No. 597.

can be formed by subtracting a multiple of the mean motion
of one of the bodies concerned from a multiple of that of the
other, and, when there are three bodies disturbing one another,
every such triple combination becomes, under the technical
name of an argument, the cyclical representative of a force
acting in the manner and according to the law described.
Each of these periodically acting forces produces its perturbative effect, according to the law of the superposition of
small motions, as if the others had no existence.   And if it
happen, as in an immense majority of cases it does, that the
cycle of any particular one of these partial forces has no relation to the periodic time of the disturbed body, so as to
bring it to the same, or very nearly the same point of its
orbit, or to any situation favourable to any particular form
of disturbance, over and over again when the force is at its
maximum; that force will, in a few revolutions, neutralize
its own effect, and nothing but fluctuations of brief duration
can result from its action.   The contrary will evidently be
the case, if the cycle of the force coincide so nearly with the
cycle of the moon's anomalistic revolution, as to bring round
the maximum of the force acting in one and the same direction
(whether tangential or normal) either accurately, or very
nearly indeed to some definite point, as, for example, the
apogee of her orbit.   Whatever the effect produced by such
a force on the angular motion of the moon, if it be not
exactly compensated in one cycle of its action, it will go on
accumulating, being repeated over and over again under
circumstances very nearly the same, for many successive
revolutions, until at length, owing to the want of precise
accuracy in the adjustment of that cycle to the anomalistic
period, the maximum of the force (in the same phase of its
action) is brought to coincide with a point in the orbit (as
the perigee), determinative of an opposite effect, and thus, at
length, a compensation will be worked out; in a time, however, so much the longer as the difference between the cycle
of the force and the moon's anomalistic period is less.

(747.) Now, in fact, in the case of Venus disturbing the
moon, there exists a cyclical combination of this kind.   Of

course the disturbing force of Venus on the moon varies with
her distance from the earth, and this distance again depends
on her configuration with respect to the earth and the sun,
taking into account the ellipticity of both their orbits. Among
the combinations which take their rise from this latter consideration, and which, as may easily be supposed, are of great
complexity, there is a term (an exceedingly minute one),
whose argument or cycle is determined by subtracting 16
times the mean motion of the earth from 18 times that of
Venus. The difference is so very nearly the mean motion of
the moon in her anomalistic revolution, that whereas the
latter revolution is completed in 27^d 13^h 18^m $32 \cdot 3^s$, the cycle
of the force is completed in 27^d 13^h 7^m $35 \cdot 6^s$, differing from
the other by no more than 10^m $56 \cdot 7^s$, or about one 3625th
part of a complete period of the moon from apogee to apogee.
During half of this very long interval (that is to say, during
about $136\frac{1}{2}$ years), the perturbations produced by a force of
this character, go on increasing and accumulating, and are
destroyed in another equal interval. Although therefore
excessively minute in their actual effect on the angular
motion, this minuteness is compensated by the number of
repeated acts of accumulation, and by the length of time
during which they continue to act on the longitude. Accordingly M. Hansen has found the total amount of fluctuation to and fro, or the value of the equation of the moon's
longitude, so arising to be $27 \cdot 4''$. It is exceedingly interesting to observe that the two equations considered in
these latter paragraphs, account satisfactorily for the only
remaining material differences between theory and observation in the modern history of this hitherto rebellious satellite.
We have not thought it necessary (indeed it would have
required a treatise on the subject) to go into a special account of the almost innumerable other lunar inequalities
which have been computed and tabulated, and which are
necessary to be taken into account in every computation of
her place from the tables. Many of them are of very much
larger amount than these. We ought not, however, to pass
unnoticed, that the parallactic inequality, already explained

art. (712.), is interesting, as affording a measure of the sun's distance. For this equation originates, as there shown, in the fact that the disturbing forces are not precisely alike in the two halves of the moon's orbit nearest to and most remote from the sun, all their values being greater in the former half. As a knowledge of the relative dimensions of the solar and lunar orbits enables us to calculate à priori, the amount of this inequality, so a knowledge of that amount deduced by the comparison of a great number of observed places of the moon with tables in which every inequality but this should be included, would enable us conversely to ascertain the ratio of the distances in question. Owing to the smallness of the inequality, this is not a very accurate mode of obtaining an element of so much importance in astronomy as the sun's distance, but were it larger (i. e. were the moon's orbit considerably larger than it actually is), this would be, perhaps, the most exact method of any by which it could be concluded.

(748.) The greatest of all the lunar inequalities, produced by perturbation, is that called the *evection*. It arises directly from the variation of the excentricity of her orbit, and from the fluctuation to and fro in the general progress of the line of apsides, caused by the different situation of the sun, with respect to that line (arts. 685. 691.). Owing to these causes the moon is alternately in advance, and in arrear of her elliptic place by about 1° 20′ 30″. This equation was known to the ancients, having been discovered by Ptolemy, by the comparison of a long series of observations handed down to him from the earliest ages of astronomy. The mode in which the effects of these several sources of inequality become grouped together under one principal argument common to them all, belongs, for its explanation, rather to works specially treating of the lunar theory than to a treatise of this kind.

(749.) Some small perturbations are produced in the lunar orbit by the protuberant matter of the earth's equator. The attraction of a sphere is the same as if all its matter were condensed into a point in its center; but that is not the

case with a spheroid. The attraction of such a mass is
neither exactly directed to its center, nor does it exactly
follow the law of the inverse squares of the distances. Hence
will arise a series of perturbations, extremely small in amount,
but still perceptible in the lunar motions ; by which the node
and the apogee will be affected. A more remarkable consequence of this cause, however, is a small nutation of the
lunar orbit, exactly analogous to that which the moon causes
in the plane of the earth's equator, by its action on the same
elliptic protuberance. And, in general, it may be observed,
that in the systems of planets which have satellites, the
elliptic figure of the primary has a tendency to bring the
orbits of the satellites to coincide with its equator, — a
tendency which, though small in the case of the earth, yet in
that of Jupiter, whose ellipticity is very considerable, and of
Saturn especially, where the ellipticity of the body is reinforced by the attraction of the rings, becomes predominant
over every external and internal cause of disturbance, and
produces and maintains an almost exact coincidence of the
planes in question. Such, at least, is the case with the
nearer satellites. The more distant are comparatively less
affected by this cause, the difference of attractions between
a sphere and spheroid diminishing with great rapidity as the
distance increases. Thus, while the orbits of all the interior satellites of Saturn lie almost exactly in the plane of
the ring and equator of the planet, that of the external
satellite, whose distance from Saturn is between sixty and
seventy diameters of the planet, is inclined to that plane considerably. On the other hand, this considerable distance,
while it permits the satellite to retain its actual inclination,
prevents (by parity of reasoning) the ring and equator of the
planet from being perceptibly disturbed by its attraction, or
being subjected to any appretiable movements analogous to
our nutation and precession. If such exist, they must be
much slower than those of the earth; the mass of this satellite being, as far as can be judged by its apparent size, a
much smaller fraction of that of Saturn than the moon is of

the earth; while the solar precession, by reason of the immense distance of the sun, must be quite imperceptible.

(750.) The subject of the tides, though rather belonging to terrestrial physics than properly to astronomy, is yet so directly connected with the theory of the lunar perturbations, that we cannot omit some explanatory notice of it, especially since many persons find a strange difficulty in conceiving the manner in which they are produced. That the sun, or moon, should by its attraction heap up the waters of the ocean under it, seems to them very natural. That it should at the same time heap them up on the opposite side seems, on the contrary, palpably absurd. The error of this class of objectors is of the same kind with that noticed in art. 723., and consists in disregarding the attraction of the disturbing body on the mass of the earth, and looking on it as wholly effective on the superficial water. Were the earth indeed absolutely fixed, held in its place by an external force, and the water left free to move, no doubt the effect of the disturbing power would be to produce a single accumulation vertically under the disturbing body. But it is not by its whole attraction, but by the difference of its attractions on the superficial water at both sides, and on the central mass, that the waters are raised: just as in the theory of the moon, the difference of the sun's attractions on the moon and on the earth (regarded as moveable and as obeying that amount of attraction which is due to its situation) gives rise to a *relative* tendency in the moon to recede from the earth in conjunction and opposition, and to approach it in quadratures. Referring to the figure of art. 675., instead of supposing A D B C to represent the moon's orbit, let it be supposed to represent a section of the (comparatively) thin film of water reposing on the globe of the earth, in a great circle, the plane of which passes through the disturbing body M, which we shall suppose to be the moon. The disturbing force on a particle at P will then (exactly as in the lunar theory) be represented in amount and direction by N S, on the same scale on which S M represents the moon's whole attraction on a particle situated at S. This force, applied at P, will urge it in the direction P X parallel

to N S; and therefore, when compounded with the direct force
of gravity which (neglecting as of no account in this theory

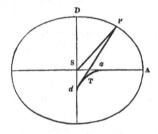

the spheroidal form of the earth)
urges P towards S, will be equivalent to a single force deviating from the direction P S towards X.   Suppose P T to be
the direction of this force, which,
it is easy to see, will be directed
towards a point in D S *produced*, at an extremely small
distance below S, because of the excessive minuteness of the
disturbing force compared to gravity.*   Then if this be done
at every point of the quadrant A D, it will be evident that the
direction P T of the resultant force will be always that of a
tangent to the small cuspidated curve *a d* at T, to which tangent the surface of the ocean at P must everywhere be perpendicular, by reason of that law of hydrostatics which requires the direction of gravity to be everywhere perpendicular
to the surface of a fluid *in equilibrio*.   The form of the curve
D P A, to which the surface of the ocean will tend to conform
itself, so as to place itself everywhere in equilibrio under two
acting forces, will be that which always has P T for its radius
of curvature.   It will therefore be slightly less curved at D,
and more so at A, being in fact no other than an ellipse,
having S for its center, *d a* for its *evolute*, and S A, S D for
its longer and shorter semi-axes respectively; so that the
whole surface (supposing it covered with water) will tend to
assume, as its form of equilibrium, that of an oblongated
ellipsoid, having its longer axis directed towards the disturbing body, and its shorter of course at right angles to that
direction.   The difference of the longer and shorter semiaxes of this ellipsoid due to the moon's attraction would be

* According to Newton's calculation, the maximum disturbing force of the
sun on the water does not exceed one 25736400th part of its gravity.   That of
the moon will therefore be to this fraction as the cube of the sun's distance to
that of the moon's directly, and as the mass of the sun to that of the moon inversely, *i. e.* as $(400)^3 \times 0.012517 : 354936$, which, reduced to numbers, gives,
for the moon's maximum of power to disturb the waters, about one 11400000th
of gravity, or somewhat less than $2\frac{1}{4}$ times the sun's.

about 58 inches: that of the ellipsoid, similarly formed in virtue of the sun, about $2\frac{1}{2}$ times less, or about 23 inches.

(751.) Let us suppose the moon only to act, and to have no orbitual motion; then if the earth also had no diurnal motion, the ellipsoid of equilibrium would be quietly formed, and all would be thenceforward tranquil. There is never time, however, for this spheroid to be fully formed. Before the waters can take their level, the moon has advanced in her orbit, both diurnal and monthly, (for in this theory it will answer the purpose of clearness better, if we suppose the earth's diurnal motion transferred to the sun and moon in the contrary direction,) the vertex of the spheroid has shifted on the earth's surface, and the ocean has to seek a new bearing. The effect is to produce an immensely broad and excessively flat wave (not a circulating *current*), which follows, or endeavours to follow, the apparent motions of the moon, and must, in fact, by the principle of forced vibrations, imitate, by equal though not by *synchronous* periods, all the periodical inequalities of that motion. When the higher or lower parts of this wave strike our coasts, they experience what we call high and low water.

(752.) The sun also produces precisely such a wave, whose vertex tends to follow the apparent motion of the sun in the heavens, and also to imitate its periodic inequalities. This solar wave co-exists with the lunar — is sometimes superposed on it, sometimes transverse to it, so as to partly neutralize it, according to the monthly synodical configuration of the two luminaries. This alternate mutual reinforcement and destruction of the solar and lunar tides cause what are called the spring and neap tides — the former being their sum, the latter their difference. Although the real amount of either tide is, at present, hardly within the reach of exact calculation, yet their proportion at any one place is probably not very remote from that of the ellipticities which would belong to their respective spheroids, could an equilibrium be attained. Now these ellipticities, for the solar and lunar spheroids, are respectively about two and five feet; so that the average spring tide will be to the neap as 7 to 3, or thereabouts.

(753.) Another effect of the combination of the solar and lunar tides is what is called the *priming* and *lagging* of the tides. If the moon alone existed, and moved in the plane of the equator, the tide-day (*i. e.* the interval between two successive arrivals at the same place of the same vertex of the tide-wave) would be the lunar day (art. 143.), formed by the combination of the moon's sidereal period and that of the earth's diurnal motion. Similarly, did the sun alone exist, and move always on the equator, the tide-day would be the mean solar day. The actual tide-day, then, or the interval of the occurrence of two successive *maxima* of their superposed waves, will vary as the separate waves approach to or recede from coincidence; because, when the vertices of two waves do not coincide, their joint height has its maximum at a point intermediate between them. This variation from uniformity in the lengths of successive tide-days is particularly to be remarked about the time of the new and full moon.

(754.) Quite different in its origin is that deviation of the time of high and low water at any port or harbour, from the culmination of the luminaries, or of the theoretical maximum of their superposed spheroids, which is called the " establishment " of that port. If the water were without inertia, and free from obstruction, either owing to the friction of the bed of the sea, the narrowness of channels along which the wave has to travel before reaching the port, their length, &c. &c., the times above distinguished would be identical. But all these causes tend to create a difference, and to make that difference not alike at all ports. The observation of the establishments of harbours is a point of great maritime importance ; nor is it of less consequence, theoretically speaking, to a knowledge of the true distribution of the tide-waves over the globe. In making such observations, care must be taken not to confound the time of " slack water," when the current caused by the tide ceases to flow visibly one way or the other, and that of *high* or *low water*, when the level of the surface ceases to rise or fall. These are totally distinct phænomena, and depend on entirely different causes, though it is true they may sometimes coincide in point of time. They are, it is feared, too often mistaken one for the other by practical men; a circumstance which, whenever it occurs, must produce the greatest confusion in any attempt to reduce the system of the tides to distinct and intelligible laws.

(755.) The declination of the sun and moon materially affects the tides at any particular spot. As the vertex of the tide-wave tends to place itself vertically under the luminary which produces it, when this vertical changes its point of incidence on the surface, the tide-wave must tend to shift accordingly, and thus, by monthly and annual periods, must tend to increase and diminish alternately the principal tides. The period of the moon's nodes is thus introduced into this subject; her excursions in declination in one part of that period being 29°, and in another only 17°, on either side the equator.

(756.) Geometry demonstrates that the efficacy of a luminary in raising tides is inversely proportional to the cube of its distance. The sun and moon, however, by reason of the ellipticity of their orbits, are alternately nearer to and farther from the earth than their mean distances. In consequence of this, the efficacy of the sun will fluctuate between the extremes 19 and 21, taking 20 for its mean value, and that of the moon between 43 and 59. Taking into account this cause of difference, the highest spring tide will be to the lowest neap as $59 + 21$ to $43 - 19$, or as 80 to 24, or 10 to 3. Of all the causes of differences in the height of tides however, local situation is the most influential. In some places the tide-wave, rushing up a narrow channel, is suddenly raised to an extraordinary height. At Annapolis, for instance, in the Bay of Fundy, it is said to rise 120 feet. Even at Bristol the difference of high and low water occasionally amounts to 50 feet.

(757.) It is by means of the perturbations of the planets, as ascertained by observation and compared with theory, that we arrive at a knowledge of the masses of those planets which having no satellites, offer no other hold upon them for this purpose. Every planet produces an amount of perturbation in the motions of every other, proportioned to its mass, and to the degree of advantage or *purchase* which its situation in the system gives it over their movements. The latter is a subject of exact calculation ; the former is unknown, otherwise than by observation of its effects. In the determination, however, of the masses of the planets by this means, theory lends the greatest assistance to observation, by pointing out the combinations most favourable for eliciting this knowledge from the confused mass of superposed inequalities which affect every observed place of a planet ; by pointing out the laws of each inequality in its periodical rise and decay ; and by show-ing how every particular inequality depends for its magnitude on the mass producing it. It is thus that the mass of Jupiter itself (employed by Laplace in his investigations, and inter-woven with all the planetary tables) has been ascertained, by observations of the derangements produced by it in the mo-tions of the ultra-zodiacal planets, to have been insufficiently determined, or rather considerably mistaken, by relying too much on observations of its satellites, made long ago by Pound and others, with inadequate instrumental means. The same conclusion has been arrived at, and nearly the same mass ob-tained, by means of the perturbations produced by Jupiter on Encke's comet. The error was one of great importance ; the mass of Jupiter being by far the most influential element in the planetary system, after that of the sun. It is satis-factory, then, to have ascertained, as Mr. Airy has done, the cause of the error ; to have traced it up to its source, in insufficient micrometric measurements of the greatest elon-gations of the satellites ; and to have found it disappear when measures, taken with more care and with infinitely superior instruments, are substituted for those before employed.

(758.) In the same way that the perturbations of the planets lead us to a knowledge of their masses, as compared with that of sun, so the perturbations of the satellites of Ju-piter have led, and those of Saturn's attendants will no doubt hereafter lead, to a knowledge of the proportion *their* masses bear to their respective primaries. The system of Jupiter's satellites has been elaborately treated by Laplace ; and it is from his theory, compared with innumerable observations of their eclipses, that the masses assigned to them, in art. 540. have been fixed. Few results of theory are more surprising than to see these minute atoms weighed in the same balance, which we have applied to the ponderous mass of the sun, which exceeds the least of them in the enormous proportion of 65000000 to 1.

(759.) The mass of the moon is concluded, 1st, from the proportion of the lunar to the solar tide, as observed at various stations, the effects being separated from each other by a long series of observations of the relative heights of spring and neap tides which, we have seen, (art. 752.) depends on the proportional influence of the two luminaries. 2dly, from the phænomenon of nutation, which, being the result of the moon's attraction alone, affords a means of calculating her mass, independent of any knowledge of the sun's. Both methods agree in assigning to our satellite a mass about one seventy-fifth that of the earth. *

(760.) Not only, however, has a knowledge of the perturbations produced on other bodies of our system enabled us to estimate the mass of a disturbing body already known to exist, and to produce disturbance. It has done much more, and enabled geometers to satisfy themselves of the existence, and even to indicate the situation of a planet previously unknown, with such precision, as to lead to its immediate discovery on the very first occasion of pointing a telescope to the place indicated. We have already (art. 506.) had occasion to mention in general terms this great discovery; but its importance, and its connexion with the subject before us, calls for a more specific notice of the circumstances attending it. When the regular observation of Uranus, consequent on its discovery in 1781, had afforded some certain knowledge of the elements of its orbit, it became possible to calculate backwards into time past, with a view to ascertain whether certain stars of about the same apparent magnitude, observed by Flamsteed, and since reported as *missing,* might not possibly be this planet. No less than six ancient observations of it as a supposed star were thus found to have

* Laplace, Expos. du Syst. du Monde, pp. 285. 300.
KK 3

been recorded by that astronomer, — one in 1690, one in 1712, and four in 1715. On further inquiry, it was also ascertained to have been observed by Bradley in 1753, by Mayer in 1756, and no less than twelve times by Le Monnier, in 1750, 1764, 1768, 1769, and 1771, all the time without the least suspicion of its planetary nature. The observations, however, so made, being all circumstantially registered, and made with instruments the best that their respective dates admitted, were quite available for correcting the elements of the orbit, which, as will be easily understood, is done with so much the greater precision the larger the arc of the ellipse embraced by the extreme observations employed. It was, therefore, reasonably hoped and expected, that, by making use of the data thus afforded, and duly allowing for the perturbations produced since 1690, by Saturn, Jupiter, and the inferior planets, elliptic elements would be obtained, which, taken in conjunction with those perturbations, would represent not only all the observations up to the time of executing the calculations, but also all future observations, in as satisfactory a manner as those of any of the other planets are actually represented. This expectation, however, proved delusive. M. Bouvard, one of the most expert and laborious calculators of whom astronomy has had to boast, and to whose zeal and indefatigable industry we owe the tables of Jupiter and Saturn in actual use, having undertaken the task of constructing similar tables for Uranus, found it impossible to reconcile the ancient observations above mentioned with those made from 1781 to 1820, so as to represent both series by means of the same ellipse and the same system of perturbations. He therefore rejected altogether the ancient series, and grounded his computations solely on the modern, although evidently not without serious misgivings as to the grounds of such a proceeding, and " leaving it to future time to determine whether the difficulty of reconciling the two series arose from inaccuracy in the older observations, or whether it depend on some extraneous and unperceived influence which may have acted on the planet."

(761.) But neither did the tables so calculated continue

to represent, with due precision, observations subsequently made. The " error of the tables" after attaining a certain amount, by which the true longitude of Uranus was in advance of the computed, and which advance was steadily maintained from about the year 1795 to 1822, began, about the latter epoch, rapidly to diminish, till, in 1830–31, the tabular and observed longitudes agreed. But, far from remaining in accordance, the planet, still losing ground, fell, and continued to fall behind its calculated place, and that with such rapidity as to make it evident that the existing tables could no longer be received as representing, with any tolerable precision, the true laws of its motion.

(762.) The reader will easily understand the nature and progression of these discordancies by casting his eye on fig. 1. Plate A, in which the horizontal line or *abscissa* is divided into equal parts, each representing 50° of heliocentric longitude in the motion of Uranus round the sun, and in which the distances between the horizontal lines represent each 100″ of error in longitude. The result of each year's observation of Uranus (or of the mean of all the observations obtained during that year) in longitude, is represented by a black dot placed above or below the point of the *abscissa*, corresponding to the mean of the observed longitudes for the year above, if the observed longitude be in excess of the calculated, below if it fall short of it, and on the line if they agree ; and at a distance from the line corresponding to their difference on the scale above mentioned.* Thus in Flamsteed's earliest observations in 1690, the dot so marked is placed above the line at 65″·9 above the line, the observed longitude being so much greater than the calculated.

(763.) If, neglecting the individual points, we draw a curve (indicated in the figure by a fine unbroken line) through their general course, we shall at once perceive a

---

* The points are laid down from M. Levenier's comparison of the whole series of observations of Uranus, with an ephemeris of his own calculation, founded on a complete and searching revision of the tables of Bouvard, and a rigorous computation of the perturbations caused by all the known planets capable of exercising any influence on it. The differences of longitude are *geocentric*, but for our present purpose it matters not in the least whether we consider the errors in heliocentric or in geocentric longitude.

certain regularity in its undulations. It presents two great
elevations above, and one nearly as great intermediate depression below the medial line or abscissa. And it is evident
that these undulations would be very much reduced, and the
errors in consequence greatly palliated, if each dot were
removed in the vertical direction through a distance and in the
direction indicated by the corresponding point of the curve
A, B, C, D, E, F, G, H, intersecting the abscissa at points 180°
distant, and making equal excursions on either side. Thus
the point a for 1750 being removed upwards or in the direction towards b through a distance equal to cb would be
brought almost to precise coincidence with the point e in
the abscissa. Now, this is a clear indication that a very large
part of the differences in question are due, not to perturbation,
but simply to error in the elements of Uranus which have
been assumed as the basis of calculation. For such excesses
and defects of longitude alternating over arcs of 180° are
precisely what would arise from error in the excentricity, or
in the place of the perihelion, or in both. In ellipses slightly
excentric, the true longitude alternately exceeds and falls
short of the mean during 180° for each deviation, and the
greater the excentricity, the greater these alternate fluctuations to and fro. If then the excentricity of a planet's orbit be
assumed erroneously (suppose too great) the observed longitudes will exhibit a less amount of such fluctuation above
and below the mean than the computed, and the difference of
the two, instead of being, as it ought to be, always *nil*, will
be alternately + and − over arcs of 180°. If then a difference
be observed following such a law, it may arrive from erroneously assumed excentricity, provided always the longitudes
at which they agree (supposed to differ by 180°) be coincident
with those of the perihelion and aphelion; for in elliptic
motion nearly circular, these are the points where the mean
and true longitudes agree, so that any fluctuation of the
nature observed, if this condition be not satisfied, cannot
arise from error of excentricity. Now the longitude of the
perihelion of Uranus in the elements employed by Bouvard is
(neglecting fractions of a degree) 168°, and of the aphelion

348°.   These points then, in our figure, fall at π, and α respectively, that is to say, nearly half way between A C, C E, E G, &c.   It is evident therefore that it is not to error of excentricity that the fluctuation in question is mainly due.

(764.) Let us now consider the effect of an erroneous assumption of the *place* of the perihelion.   Suppose in fig. 2. Plate A, ox to represent the longitude of a planet, and xy the excess of its true above its mean longitude, due to ellipticity.   Then if R be the place of the perihelion, and P, or T, the aphelion in longitude, y will always lie in a certain undulating curve P Q R S T, above * P T, between R and T, and below it between P and R.   Now suppose the place of the perihelion shifted forward to r, or the whole curve shifted bodily forward into the situation $pqrst$, then at the same longitude ox, the excess of the true above the mean longitude will be xy' only ; in other words, this excess will have diminished by the quantity yy' below its former amount.   Take therefore in o N (*fig.* 3. Pl. A) $oy=ox$ and yy' always$=yy'$ in *fig.* 2., and having thus constructed the curve K L M N O, the ordinate yy' will always represent the effect of the supposed change of perihelion.   It is evident (the excentricity being always supposed small), that this curve will consist also of alternate superior and inferior waves of 180° each in amplitude, and the points L, N of its intersection with the axis will occur at longitudes corresponding to X, Y intermediate between the *maxima* Q, q ; and S, s of the original curves, that is to say (if these intervals Qq, Ss, or Rr to which both are equal, be very small) very nearly at 90° from the perihelion and aphelion.   Now this agrees with the conditions of the case in hand, and we are therefore authorised to conclude that the major portion of the errors in question *has* arisen from error in the place of the perihelion of Uranus itself, and not from perturbation, and that to correct this portion, the perihelion must be shifted somewhat forward. As to the amount of this shifting, our only object being explanation, it will not be necessary here to enquire into it. It will suffice that it must be such as shall make the curve

* The curves, figs. 2, 3, are inverted in the engraving.

A B C D E F G as nearly as possible similar, equal, and opposite to the curve traced out by the dots on the other side. And this being done, we may next proceed to lay down a curve of the residual differences between observation and theory in the mode indicated in art. (763.)

(765.) This being done, by laying off at each point of the line of longitudes an ordinate equal to the difference of the ordinates of the two curves in fig. 1. when on opposite, and their sum when on the same side of the abscissa, the result will be as indicated by the dots in fig. 4. And here it is at once seen that a still farther reduction of the differences under consideration would result, if, instead of taking the line A B for the line of longitudes, a line *a b* slightly inclined to it were substituted, in which case the whole of the differences between observation and theory from 1712 to 1800 would be annihilated, or at least so far reduced as hardly to exceed the ordinary errors of observation ; and as respects the observation of 1690, the still outstanding difference of about 35″ would not be more than might be attributed to a not very careful observation at so early an epoch. Now the assumption of such a new line of longitudes as the correct one is in effect equivalent to the admission of a slight amount of error in the periodic time and epoch of Uranus; for it is evident that by reckoning from the inclined instead of the horizontal line, we in effect alter all the apparent outstanding errors by an amount proportional to the time before or after the date at which the two lines intersect (viz. about 1789). As to the direction in which this correction should be made, it is obvious by inspection of the course of the dots, that if we reckon from A B, or any line parallel to it, the observed planet on the long run keeps falling more and more behind the calculated one ; *i. e.* its assigned mean angular velocity by the tables is too great and must be diminished, or its periodic time requires to be increased.

(766.) Let this increase of period be made, and in correspondence with that change let the longitudes be reckoned on *a b*, and the residual differences from that line instead of A B, and we shall have then done all that can be done in the way of reducing and palliating these differences, and that, with such success, that up to the year 1804 it might have been safely asserted that positively no ground whatever existed for suspecting any disturbing influence. But with this epoch an action appears to have commenced, and gone on increasing, producing an acceleration of the motion in longitude, in consequence of which, Uranus continually gains on its elliptic place, and continued to do so till 1822, when it ceased to gain, and the excess of longitude was at its maximum, after which it began rapidly to lose ground, and has continued to do so up to the present time. It is perfectly clear, then, that in this interval some extraneous cause must have come into action which was not so before, or not in sufficient power to manifest itself by any marked effect, and that *that* cause must have ceased to act, or rather begun to reverse its action, in or about the year 1822, the reverse action being even more energetic than the direct.

(767.) Such is the phænomenon in the simplest form we are *now* able to present it. Of the various hypotheses formed to account for it, during the progress of its developement, none seemed to have any degree of rational probability but that of the existence of an exterior, and hitherto undiscovered, planet, disturbing, according to the received laws of planetary disturbance, the motion of Uranus by its attraction, or rather superposing its disturbance on those produced by Jupiter and Saturn, the only two of the old planets which exercise any sensible disturbing action on that planet. Accordingly, this was the explanation which naturally, and almost of necessity, suggested itself to those conversant with the planetary perturbations who considered the subject with any degree of attention. The idea, however, of setting out from the observed anomalous deviations, and employing them as data to ascertain the distance and situation of the unknown body, or, in other words, to resolve the inverse problem of perturbations, "*given the disturbances to find the orbit, and place in that orbit* of the *disturbing planet*," appears to have occurred only to two mathematicians, Mr. Adams in England and M. Leverrier in France, with sufficient distinctness and hopefulness of success to induce them to attempt its solution. Both succeeded, and their solutions, arrived at with perfect independence, and by each in entire ignorance of the other's attempt, were found to agree in a surprising manner when the nature and difficulty of the problem is considered; the calculations of M. Leverrier assigning for the heliocentric longitude of the disturbing planet for the 23d Sept. 1846, 326° 0′, and those of Mr. Adams (brought to the same date) 329° 19′, differing only 3° 19′; the plane of its orbit deviating very slightly, if at all, from that of the ecliptic.

(768.) On the day above mentioned — a day for ever memorable in the annals of astronomy — Dr. Galle, one of the astronomers of the Royal Observatory at Berlin, received a letter from M. Leverrier, announcing to him the result he had arrived at, and requesting him to look for the disturbing planet in or near the place assigned by his calculation. He did so, and *on that very night actually found it.* A star of the eighth magnitude was seen by him and by M. Encke in a situation where no star was marked as existing in Dr. Bremiker's chart, then recently published by the Berlin Academy. The next night it was found to have moved from its place, and was therefore assuredly a planet. Subsequent observations and calculations have fully demonstrated this planet, to which the name of Neptune has been assigned, to be really that body to whose disturbing attraction, according to the Newtonian law of gravity, the observed anomalies in the motion of Uranus were owing. The geocentric longitude determined by Dr. Galle from this observation was 325° 53′, which, converted into heliocentric, gives 326° 52′, differing 0° 52′ from M. Leverrier's place, 2° 27′ from that of Mr. Adams, and only 47′ from a mean of the two calculations.

(769.) It would be quite beyond the scope of this work, and far in advance of the amount of mathematical knowledge we have assumed our readers to possess, to attempt giving more than a superficial idea of the course followed by these geometers in their arduous investigations. Suffice it to say, that it consisted in regarding, as unknown quantities, to be determined, the mass, and all the elements of the unknown planet (supposed to revolve in the same plane and the same direction with Uranus), except its major semiaxis. This was assumed in the first instance (in conformity with " Bode's law," art. (505.), and certainly at the time with a high *primâ facie* probability,) to be double that of Uranus, or 38·364 radii of the Earth's orbit. Without *some* assumption as to the value of this element, owing to the peculiar form of the analytical expression of the perturbations, the analytical investigation would have presented difficulties apparently insuperable.   But besides these, it was also necessary to regard as unknown, or at least as liable to corrections of unknown magnitude of the same order as the perturbations, all the elements of Uranus itself, a circumstance whose necessity will easily be understood, when we consider that the received elements could only be regarded as provisional, and must *certainly* be erroneous, the places from which they were obtained being affected by at least some portions of the very perturbations in question.   This consideration, though indispensable, added vastly both to the complication and the labour of the inquiry.   The axis (and therefore the mean motion) of the one orbit, then, being known very nearly, and that of the other thus hypothetically assumed, it became practicable to express in terms, partly algebraic, partly numerical, the amount of perturbation at any instant, by the aid of general expressions delivered by Laplace in his " *Mécanique Céleste*" and elsewhere.   These, then, together with the corrections due to the altered elements of Uranus itself, being applied to the tabular longitudes, furnished, when compared with those observed, a series of *equations*, in which the elements and mass of Neptune, and the corrections of those of Uranus entered as the *unknown quantities*, and by whose resolution (no slight effort of analytical skill) all their values were at length obtained.   The calculations were then repeated, reducing at the same time the value of the assumed distance of the new planet, the discordances between the given and calculated results indicating it to have been assumed too large when the results were found to agree better, and the solutions to be, in fact, more satisfactory.

Thus, at length, elements were arrived at for the orbit of the unknown planet, as below.

	Leverrier.	Adams.
Epoch of Elements - - -	Jan. 1. 1847.	Oct. 6. 1846.
Mean longitude in Epoch - -	318° 47′ 4	323° 2′
Semiaxis Major - - -	36·1539	37·2474
Excentricity - - - -	0·107610	0·120615
Longitude of Perihelion - - -	284° 45′ 8	299° 11′
Mass (the Sun being 1) - -	0·00010727	0·00015003

The elements of M. Leverrier were obtained from a consideration of the observations up to the year 1845, those of Mr. Adams, only as far as 1840. On subsequently taking into account, however, those of the five years up to 1845, the latter was led to conclude that the semiaxis ought to be still much further diminished, and that a mean distance of 33·33 (being to that of Uranus as 1 : 0·574) would probably satisfy all the observations very nearly.*

(770.) On the actual discovery of the planet, it was, of course, assiduously observed, and it was soon ascertained that a mean distance, even less than Mr. Adams's last presumed value, agreed better with its motion ; and no sooner were elements obtained from direct observation, sufficiently approximate to trace back its path in the heavens for a considerable interval of time, than it was ascertained to have been observed as a star by Lalande on the 8th and 10th of May, 1795, the latter of the two observations, however, having been rejected by him as faulty, by reason of its non-agreement with the former (a consequence of the motion of the planet in the interval). From these observations, combined with those since accumulated, the elements calculated by Prof. Walker, U.S., result as follows : —

Epoch of Elements - - -	Jan. 1. 1847, M. noon, Greenwich.
Mean longitude at Epoch -	328° 32′ 44″ 2
Semiaxis major - - -	30·0367
Excentricity - - -	0·00871946
Longitude of the Perihelion -	47° 12′ 6″·50
Ascending Node - - -	130° 4′ 20″·81
Inclination - - - -	1° 46′ 58″·97
Periodic time - - -	164·6181 tropical year.
Mean annual Motion - - -	2°·18688

* In a letter to the Astronomer Royal, dated Sept. 2. 1846, — i. e. three weeks previous to the optical discovery of the planet.

(771.) The great disagreement between these elements and those assigned either by M. Leverrier or Mr. Adams will not fail to be remarked; and it will naturally be asked how it has come to pass, that elements so widely different from the truth should afford anything like a satisfactory representation of the perturbation in question, and that the true situation of the planet in the heavens should have been so well, and indeed accurately, pointed out by them.   As to the latter point, any one may satisfy himself by half an hour's calculation that both sets of elements do really place the planet, on the day of its discovery, not only in the longitudes assigned in art. 763., *i. e.* extremely near its apparent place, but also at a distance from the Sun very much more approximately correct than the *mean* distances or semiaxes of the respective orbits.   Thus the radius vector of Neptune, calculated from M. Leverrier's elements for the day in question, instead of 36·1539 (the mean distance) comes out almost exactly 33; and indeed, if we consider that the excentricity assigned by those elements gives for the perihelion distance 32·2634, the longitude assigned to the perihelion brings the whole arc of the orbit (more than 83°), described in the interval from 1806 to 1847 to lie within 42° one way or the other of the perihelion, and therefore, during the whole of that interval, the hypothetical planet would be moving within limits of distance from the sun, 32·6 and 33·0.   The following comparative tables of the relative situations of Uranus, the real and hypothetical planet, will exhibit more clearly than any lengthened statement, the near imitation of the motion of the former by the latter within that interval.   The longitudes are heliocentric.*

---

* The calculations are carried only to tenths of degrees, as quite sufficient for the object in view.

A. D.	Uranus.	Neptune.		Leverrier.		Adams.	
	Long.	Long.	Rad. Vec.	Long.	Rad. Vec.	Long.	Rad. Vec.
1805·0	197°·8	235°·9	30·3	241°·2	33·1	246°·5	34·2
1810·0	220·9	247·0	30·3	251·1	32·8	255·9	33·7
1815·0	243·2	258·0	30·3	261·2	32·5	265·5	33·3
1820·0	264·7	268·8	30·2	271·4	32·4	275·4	33·1
21·0	269·0	271·0	30·2	273·5	32·3	277·4	33·0
22·0	273·3	273·2	30·2	275·6	32·3	279·5	33·0
23·0	277·6	275·3	30·2	277·6	32·3	281·5	32·9
24·0	281·8	277·4	30·2	279·7	32·3	283·6	32·9
1825·0	285·8	279·6	30·2	281·8	32·3	285·6	32·8
1830·0	306·1	290·5	30·1	292·1	32·3	296·0	32·8
1835·0	326·0	301·4	30·1	302·5	32·4	306·3	32·8
1840·0	345·7	312·2	30 1	312·6	32·6	316·3	32·9
1845·0	365·3	323·1	30·0	322·6	32·9	326·0	33·1
1847·0	373·3	327·6	30·0	326·5	33 1	329·3	33·2

(772.) From this comparison it will be seen that Uranus arrived at its conjunction with Neptune at or immediately before the commencement of 1822, with the calculated planet of Leverrier at the beginning of the following year 1823, and with that of Adams about the end of 1824. Both the theoretical planets, and especially that of M. Leverrier, therefore, during the whole of the above interval of time, so far as the *directions* of their attractive forces on Uranus are concerned, would act nearly on it as the true planet must have done. As regards the intensity of the relative disturbing forces, if we estimate these by the principles of art. (612.) at the epochs of conjunction, and for the commencement of 1805 and 1845, we find for the respective denominators of the fractions of the sun's attraction on Uranus regarded as unity, which express the *total* disturbing force, N S, in each case, as below :

			1805.	Conjunction.	1845.
Neptune with	Peirce's mass	$\dfrac{1}{19840}$	27540	7508	32390
	Struve's mass	$\dfrac{1}{14496}$	20244	5519	23810
Leverrier's theoretical Planet, mass		$\dfrac{1}{9322}$	20837	5193	19935

The masses here assigned to Neptune are those respectively deduced by Prof. Peirce and M. Struve from observations of the satellite discovered by Mr. Lassell made with the large telescopes of Fraunhofer in the observatories of Cambridge, U.S. and Pulkova respectively. These it will be perceived differ very considerably, as might reasonably be expected in the results of micrometrical measurements of such difficulty, and it is not possible at present to say to which the preference ought to be given. As compared with the mass assigned by M. Struve, an agreement on the whole more satisfactory could not have been looked for within the interval immediately in question.

(773.) Subject then to this uncertainty as to the real mass of Neptune, the theoretical planet of Leverrier must be considered as representing with quite as much fidelity as could possibly be expected in a research of such exceeding delicacy, the particulars of its motion and perturbative action during the forty years elapsed from 1805 to 1845, an interval which (as is obvious from the rapid diminution of the forces on either side of the conjunction indicated by the numbers here set down) comprises all the most influential range of its action. This will, however, be placed in full evidence by the construction of curves representing the normal and tangential forces on the principles laid down (as far as the normal constituent is concerned) in art. (717.), one slight change only being made, which, for the purpose in view, conduces greatly to clearness of conception. The force L s (in the figure of that article) being supposed applied at P in the direction L s, we here construct the curve of the normal force by erecting at P (*fig.* 5. Plate A) P W always perpendicular to the disturbed orbit, A P, at P, measured from P in the same direction that S lies from L, and equal in length to L S. P W then will always represent both the direction and magnitude of the normal force *acting at* P. And in like manner, if we take always P Z on the tangent to the disturbed orbit at P, equal to N L of the former figure, and measured in the same direction from P that L is from N, P Z will represent both in magnitude and direction the tangential force acting at P. Thus will be traced out the two curious ovals represented in our figure of their proper forms and proportions for the case in question. That expressing the normal force is formed of four lobes, having a common point in S, viz., S W *m* X S *a* S *n* S *b* S W, and that expressing the tangential, A Z *c f* B *e d* Y A Z, consisting of four mutually intersecting loops, surrounding and touching the disturbed orbit in four points, A B *c d*. The normal force acts outwards over all that part of the orbit, both in conjunction and opposition, corresponding to the portions of the lobes *m, n,* exterior to the disturbed orbit, and inwards in every other part. The figure sets in a clear light the great disproportion between the energy of this force near the conjunction, and in any other configuration of the planets; its exceedingly rapid degradation as P approaches the point of neutrality (whose situation is 35° 5′ on either side of the conjunction, an arc described synodically by Uranus in $16^{y}.72$); and the comparatively short duration and consequent inefficacy to produce any great amount of perturbation, of the more intense part of its inward action in the small portions of the orbit corresponding to the lobes *a, b,* in which the line representing the inward force exceeds the radius of the circle. It exhibits, too, with no less distinctness, the gradual developement, and rapid degradation and extinction of the tangential force from *its* neutral points, *c, d,* on either side up to the conjunction, where its action is reversed, being accelerative over the arc *d* A, and retardative over A *c,* each of which arcs has an amplitude of 71° 20′, and is described by Uranus synodically in $34^{y}.00$. The insignificance of the tangential force in the configurations remote from conjunction throughout the arc *c* B *d* is also obviously expressed by the small comparative developement of the loops *e, f.*

(774.) Let us now consider how the action of these forces results in the production of that peculiar character of perturbation which is exhibited in our curve, *fig.* 4. Plate A. It is at once evident that the increase of the longitude from 1800 to 1822, the cessation of that increase in 1822, and its conversion into a decrease during the subsequent interval is in complete accordance with the growth, rapid decay, extinction at conjunction, and subsequent reproduction in a reversed sense of the tangential force: so that we cannot hesitate in attributing the greater part of the perturbation expressed by the swell and subsidence of the curve between the years 1800 and 1845, — all that part, indeed, which is symmetrical on either side of 1822 — to the action of the tangential force.

(775.) But it will be asked, — has then the normal force (which, on the plain showing of *fig.* 5., is nearly twice as powerful as the tangential, and which does not reverse its action, like the latter force, at the point of conjunction, but, on the contrary, is there most energetic,) no influence in producing the observed effects?   We answer, very little, within the period to which observation had extended up to 1845.   The effect of the tangential force on the longitude is direct and immediate (art. 660.), that of the normal indirect, consequential, and cumulative with the progress of time (art. 734.).   The effect of the tangential force on *the mean motion* takes place through the medium of the change it produces on the axis, and is transient : the reversed action after conjunction (supposing the orbits circular), exactly destroying all the previous effect, and leaving the mean motion on the whole unaffected.   In the passage through the conjunction, then, the tangential force produces a sudden and powerful acceleration, succeeded by an equally powerful and equally sudden retardation, which done, its action is completed, and no trace remains in the subsequent motion of the planet that it ever existed, for its action on the perihelion and excentricity is in like manner also nullified by its reversal of direction. But with the normal force the case is far otherwise.   Its *immediate* effect on the angular motion is *nil.*   It is not till it has acted long enough to produce a perceptible change in the distance of the disturbed planet from the sun that the angular velocity begins to be sensibly affected, and it is not till its whole outward action has been exerted (*i.e.* over the whole interval from neutral point to neutral point) that its maximum effect in lifting the disturbing planet away from the sun has been produced, and the full amount of diminution in angular velocity it is capable of causing has been developed.

This continues to act in producing a retardation in longitude long after the normal force itself has reversed its action, and from a powerful outward force has become a feeble inward one. A certain portion of this perturbation is incident on the epoch in the mode described in art. (731.) et seq., and permanently disturbs the mean motion from what it would have been, had Neptune no existence. The rest of its effect is compensated in a single synodic revolution, not by the reversal of the action of the force (for that reversed action is far too feeble for this purpose), but by the effect of the permanent alteration produced in the excentricity, which (the axis being unchanged) compensates by increased proximity in one part of the revolution, for increased distance in the other. Sufficient time has not yet elapsed since the conjunction to bring out into full evidence the influence of this force. Still its commencement is quite unequivocally marked in the more rapid descent of our curve *fig.* 4., subsequent to the conjunction than ascent previous to that epoch, which indicates the commencement of a series of undulations in its future course *of an elliptic character*, consequent on the altered excentricity and perihelion (the total and ultimate effect of this constituent of the disturbing force) which will be maintained till within about 20 years from the next conjunction, with the exception, perhaps, of some trifling inequalities about the time of the opposition, similar in character, but far inferior in magnitude to those now under discussion.

(776.) Posterity will hardly credit that, with a full knowledge of all the circumstances attending this great discovery — of the calculations of Leverrier and Adams — of the communication of its predicted place to Dr. Galle — and of the new planet being actually found by him in that place, in the remarkable manner above commemorated; not only have doubts been expressed as to the validity of the calculations of those geometers, and the legitimacy of their conclusions, but these doubts have been carried so far as to lead the objectors to attribute the acknowledged fact of a planet previously unknown occupying that precise place in the heavens

at that precise time, to sheer accident !* What share accident
may have had in the successful issue of the calculations, we
presume the reader, after what has been said, will have little
difficulty in satisfying himself. As regards the time when
the discovery was made, much has also been attributed to
fortunate coincidence. The following considerations will, we
apprehend, completely dissipate this idea, if still lingering in
the mind of any one at all conversant with the subject. The
period of Uranus being 84·0140 years, and that of Neptune
164·6181, their synodic revolution (art. 418.), or the interval
between two successive conjunctions, is 171·58 years. The
late conjunction having taken place about the beginning of
1822; that next preceding must have happened in 1649, or
more than 40 years before the first recorded observation of
Uranus in 1690, to say nothing of its discovery as a planet.
In 1690, then, it must have been effectually out of reach of any
perturbative influence worth considering, and so it remained
during the whole interval from thence to 1800. From that
time the effect of perturbation began to become sensible, about
1805 prominent, and in 1820 had nearly reached its maximum.
At this epoch an alarm was sounded. The maximum was
not attained, — the event, so important to astronomy, was still

---

* These doubts seem to have originated partly in the great disagreement
between the predicted and real elements of Neptune, partly in the near (*possibly*
precise) commensurability of the mean motions of Neptune and Uranus. We
conceive them however to be founded in a total misconception of the nature of
the problem, which was not, from such obviously uncertain indications as the
observed discordances could give, to determine as astronomical quantities the
axis, excentricity and mass of the disturbing planet; but practically to discover
where to look for it: when, if once found, these elements would be far better
ascertained. To do this, *any axis, excentricity, perihelion, and mass, however wide
of the truth*, which would represent, even roughly the amount, but *with tolerable
correctness the direction* of the disturbing force during the very moderate interval when the departures from theory were really considerable, would equally
serve their purposes; and with an excentricity, mass, and perihelion disposable,
it is obvious that any assumption of the axis between the limits 30 and 38, nay,
even with a much wider inferior limit, would serve the purpose. In his attempt
to assign an inferior limit to the axis, and in the value so assigned, M. Leverrier,
it must be admitted, was not successful. Mr. Adams, on the other hand, influenced by no considerations of the kind which appear to have weighed with
his brother geometer, fixed ultimately (as we have seen) on an axis not very
egregiously wrong. Still it were to be wished, for the satisfaction of all parties,
that some one would undertake the problem *de novo*, employing formulæ not
liable to the passage through infinity, which, technically speaking, hampers, or
may be supposed to hamper the *continuous* application of the usual perturbational
formulæ when cases of commensurability occur.

in progress of developement,—when the fact (any thing rather than a striking one) was noticed, and made matter of complaint. But the time for discussing its cause with any prospect of success was not yet come. Every thing turns upon the precise determination of the epoch of the maximum, when the perturbing and perturbed planet were in conjunction, and upon the law of increase and diminution of the perturbation itself on either side of that point. Now it is always difficult to assign the time of the occurrence of a maximum by observations liable to errors bearing a ratio far from inconsiderable to the whole quantity observed. Until the lapse of some years from 1822 it would have been impossible to have fixed that epoch with any certainty, and as respects the law of degradation and total arc of longitude over which the sensible perturbations extend, we are hardly yet arrived at a period when this can be said to be completely determinable from observation alone. In all this we see nothing of accident, unless it be accidental that an event which must have happened between 1781 and 1953, actually happened in 1822; and that we live in an age when astronomy has reached that perfection, and its cultivators exercise that vigilance which neither permit such an event, nor its scientific importance, to pass unnoticed. The blossom had been watched with interest in its developement, and the fruit was gathered in the very moment of maturity.*

* The student who may wish to see the perturbations of Uranus produced by Neptune, as computed from a knowledge of the elements and mass of that planet, such as we now know to be pretty near the truth, will find them stated at length from the calculations of Mr. Walker, (of Washington, U. S.) in the "Proceedings of the American Academy of Arts and Sciences," vol. i. p. 334. et seq. On examining the comparisons of the results of Mr. Walker's formulæ with those of Mr. Adams's theory in p. 342, he will perhaps be surprised at the enormous difference between the actions of Neptune and Mr. Adams's "hypothetical planet" on the longitude of Uranus. This is easily explained. Mr. Adams's perturbations are deviations from Bouvard's orbit of Uranus, as it stood immediately previous to the late conjunction. Mr. Walker's are the deviations from a mean or undisturbed orbit freed from the influence of the long inequality resulting from the near commensurability of the motions.

PART III.

OF SIDEREAL ASTRONOMY.

CHAPTER XV.

OF THE FIXED STARS. — THEIR CLASSIFICATION BY MAGNITUDES. — PHOTOMETRIC SCALE OF MAGNITUDES. — CONVENTIONAL OR VULGAR SCALE. — PHOTOMETRIC COMPARISON OF STARS. — DISTRIBUTION OF STARS OVER THE HEAVENS. — OF THE MILKY WAY OR GALAXY. — ITS SUPPOSED FORM THAT OF A FLAT STRATUM PARTIALLY SUBDIVIDED. — ITS VISIBLE COURSE AMONG THE CONSTELLATIONS. — ITS INTERNAL STRUCTURE. — ITS APPARENTLY INDEFINITE EXTENT IN CERTAIN DIRECTIONS. — OF THE DISTANCE OF THE FIXED STARS. — THEIR ANNUAL PARALLAX. — PARALLACTIC UNIT OF SIDEREAL DISTANCE.—EFFECT OF PARALLAX ANALOGOUS TO THAT OF ABERRATION. — HOW DISTINGUISHED FROM IT. — DETECTION OF PARALLAX BY MERIDIONAL OBSERVATIONS. — HENDERSON'S APPLICATION TO α CENTAURI. — BY DIFFERENTIAL OBSERVATIONS. — DISCOVERIES OF BESSEL AND STRUVE. — LIST OF STARS IN WHICH PARALLAX HAS BEEN DETECTED.— OF THE REAL MAGNITUDES OF THE STARS. — COMPARISON OF THEIR LIGHTS WITH THAT OF THE SUN.

(777.) BESIDES the bodies we have described in the foregoing chapters, the heavens present us with an innumerable multitude of other objects, which are called generally by the name of stars. Though comprehending individuals differing from each other, not merely in brightness, but in many other essential points, they all agree in one attribute, — a high degree of permanence as to apparent relative situation. This has procured them the title of " fixed stars ; " an expression which is to be understood in a comparative and not an absolute sense, it being certain that many, and probable that all, are in a state of motion, although too slow to be perceptible unless by means of very delicate observations, continued during a long series of years.

(778.) Astronomers are in the habit of distinguishing the stars into classes, according to their apparent brightness. These are termed magnitudes. The brightest stars are said to be of the first magnitude; those which fall so far short of the first degree of brightness as to make a strongly marked distinction are classed in the second; and so on down to the sixth or seventh, which comprise the smallest stars visible to the naked eye, in the clearest and darkest night. Beyond these, however, telescopes continue the range of visibility, and magnitudes from the 8th down to the 16th are familiar to those who are in the practice of using powerful instruments; nor does there seem the least reason to assign a limit to this progression; every increase in the dimensions and power of instruments, which successive improvements in optical science have attained, having brought into view multitudes innumerable of objects invisible before; so that, for any thing experience has hitherto taught us, the number of the stars may be really infinite, in the only sense in which we can assign a meaning to the word.

(779.) This classification into magnitudes, however, it must be observed, is entirely arbitrary. Of a multitude of bright objects, differing probably, intrinsically, both in size and in splendour, and arranged at unequal distances from us, one must of necessity appear the brightest, one next below it, and so on. An order of succession (relative, of course, to our local situation among them) *must* exist, and it is a matter of absolute indifference, where, in that infinite progression downwards, from the one brightest to the invisible, we choose to draw our lines of demarcation. All this is a matter of pure convention. Usage, however, has established such a convention; and though it is impossible to determine exactly, or *à priori*, where one magnitude ends and the next begins, and although different observers have differed in their magnitudes, yet, on the whole, astronomers have restricted their first magnitude to about 23 or 24 principal stars; their second to 50 or 60 next inferior; their third to about 200 yet smaller, and so on; the numbers increasing very rapidly as we descend in the scale of brightness, the whole number of stars already registered down to the seventh magnitude, inclusive, amounting to from 12000 to 15000.

(780.) As we do not see the actual disc of a star, but judge only of its brightness by the total impression made upon the eye, the apparent " magnitude " of any star will, it is evident, depend, 1st, on the star's distance from us; 2d, on the absolute magnitude of its illuminated surface; 3d, on the intrinsic brightness of that surface. Now, as we know nothing, or next to nothing, of any of these data, and have every reason for believing that each of them may differ in different individuals, in the proportion of many millions to one, it is clear that we are not to expect much satisfaction in any conclusions we may draw from numerical statements of the number of individuals which have been arranged in our artificial classes antecedent to any general or definite *principle* of arrangement. In fact, astronomers have not yet agreed upon any principle by which the magnitudes may be photometrically classed *à priori*, whether for example a scale of brightnesses decreasing in geometrical progression should be adopted, each term being one half of the preceding, or one third, or any other ratio, or whether it would not be preferable to adopt a scale decreasing as the squares of the terms of an harmonic progression, *i. e.* according to the series 1, $\frac{1}{4}, \frac{1}{9}, \frac{1}{16}, \frac{1}{25}$, &c. The former would be a purely photometric scale, and would have the apparent advantage that the light of a star of any magnitude would bear a fixed proportion to that of the magnitude next above it, an advantage, however, merely apparent, as it is certain, from many optical facts, that the unaided eye forms very different judgments of the proportions existing between bright lights, and those between feeble ones. The latter scale involves a physical idea, that of supposing the scale of magnitudes to correspond to the appearance of a first magnitude standard star, removed successively to twice, three times, &c. its original or standard distance. Such a scale, which would make the nominal magnitude a sort of index to the *presumable* or average distance, on the supposition of an equality among the real lights of the stars, would facilitate the expression of speculative ideas on the constitution of the sidereal heavens. On the other hand, it would at first sight appear to make too small a difference between the lights in the lower magnitudes. For example, on this principle of nomenclature, the light of a star of the seventh magnitude would be thirty-six 49ths of that of one of the sixth, and of the tenth 81 hundredths of the ninth, while between the first and the second the proportion would be that of four to one. So far, however, from this being really objectionable, it falls in well with the general tenor of the optical facts already alluded to, inasmuch as the eye (in the absence of disturbing causes) does actually discriminate with greater precision between the relative intensities of feeble lights than of bright ones, so that the fraction $\frac{36}{49}$ for instance, expresses quite as great a step downwards (physiologically speaking) from the sixth magnitude, as $\frac{1}{4}$ does from the first. As the choice, therefore, so far as we can see, lies between these two scales, in drawing the lines of demarcation between what may be termed the *photometrical magnitudes* of the stars, we have no hesitation in adopting, and recommending others to adopt, the latter system in preference to the former.

(781.) The conventional magnitudes actually in use among astronomers, so far as their usage is consistent with itself, conforms moreover very much more nearly to this than to the geometrical progression. It has been shown * by direct photometric measurement of the light of a considerable number of stars from the first to the fourth magnitude, that if M be the number expressing the magnitude of a star on the above system, and *m* the number expressing the magnitude of the same star in the loose and irregular language at present conventionally or rather provisionally adopted, so far as it can be collected from the conflicting authorities of different observers, the difference between these numbers, or M—*m*, is the same in all the higher parts of the scale, and is less than half a mag-

* See " Results of Observations made at the Cape of Good Hope, &c. &c." p. 371. By the Author.

nitude (0^m. 414).   The standard star assumed as the unit of magnitude in the measurements referred to, is the bright southern star α Centauri, a star somewhat superior to Arcturus in lustre.   If we take the distance of this star for unity, it follows that when removed to the distances 1·414, 2·414, 3·414, &c. its apparent lustre would equal those of average stars of the 1st, 2d, 3d, &c. magnitudes, *as ordinarily reckoned,* respectively.

(782.)   The difference of lustre between stars of two consecutive magnitudes is so considerable as to allow of many intermediate gradations being perfectly well distinguished. Hardly any two stars of the first or of the second magnitude would be judged by an eye practised in such comparisons to be exactly equal in brightness.   Hence, the necessity, if anything like accuracy be aimed at, of subdividing the magnitudes and admitting fractions into our nomenclature of brightness.   When this necessity first began to be felt, a simple bisection of the interval was recognized, and the intermediate degree of brightness was thus designated, viz. 1.2 m, 2.3 m, and so on.   At present it is not unfrequent to find the interval trisected thus : 1 m, 1.2 m, 2.1 m, 2 m, &c. where the expression 1.2 m denotes a magnitude intermediate between the first and second, but nearer 1 than 2 ; while 2.1 m designates a magnitude also intermediate, but nearer 2 than 1.   This may suffice for common parlance, but as this department of astronomy progresses towards exactness, a decimal subdivision will of necessity supersede these rude forms of expression, and the magnitude will be expressed by an integer number followed by a decimal fraction ; as for instance, 2.51 which expresses the magnitude of γ Geminorum on the vulgar or conventional scale of magnitudes, by which we at once perceive that its place is almost exactly half way between the 2d and 3d average magnitudes, and that its light is to that of an average first magnitude star in that scale (of which α Orionis in its usual or normal state * may be taken as a typical specimen) as 1^2 : $(2·51)^2$, and to that of α

* In the interval from 1836 to 1839 this star underwent considerable and remarkable fluctuations of brightness.

Centauri as 1^2: $(2\cdot924)^2$, making its place in the photometric scale (so defined) $2\cdot924$. Lists of stars northern and southern, comprehending those of the vulgar first, second, and third magnitudes, with their magnitudes decimally expressed in both systems, will be found at the end of this work. The light of a star of the sixth magnitude may be roughly stated as about the hundredth part of one of the first. Sirius would make between three and four hundred stars of that magnitude.

(783.) The exact photometrical determination of the comparative intensities of light of the stars is attended with many and great difficulties, arising partly from their differences of colour; partly from the circumstance that no invariable standard of *artificial* light has yet been discovered; partly from the physiological cause above alluded to, by which the eye is incapacitated from judging correctly of the proportion of two lights, and can only decide (and that with not very great precision) as to their equality or inequality; and partly from other physiological causes. The least objectionable method hitherto proposed would appear to be the following. A natural standard of comparison is in the first instance selected, brighter than any of the stars, so as to allow of being equalized with any of them by a *reduction* of its light optically effected, and at the same time either invariable, or at least only *so* variable that its changes can be exactly calculated and reduced to numerical estimation. Such a standard is offered by the planet Jupiter, which, being much brighter than any star, subject to no phases, and variable in light only by the variation of its distance from the sun, and which moreover comes in succession above the horizon at a convenient altitude, simultaneously with all the fixed stars, and, in the absence of the moon, twilight, and other disturbing causes (which fatally affect all observations of this nature), combines all the requisite conditions. Let us suppose, now, that Jupiter being at A and the star to be compared with it at B, a glass prism C, is so placed that the light of the planet deflected by *total internal reflexion* at its base, shall emerge parallel to B E the direction of the star's visual ray. After reflexion, let it be received on a lens D, in whose focus

F, it will form a small bright star-like image capable of being
viewed by an eye placed at E, so far out of the axis of the cone
of diverging rays as to admit of seeing at the same time, and

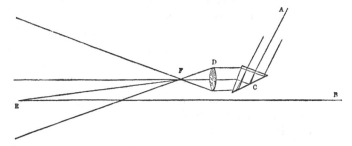

with the same eye, and so comparing, this image with the
star seen directly. By bringing the eye nearer to or further
from the focus F, the apparent brightness of the focal point
will be varied in the inverse ratio of the square of the distance E F, and therefore may be *equalized*, as well as the eye
can judge of such equalities, with the star. If this be done
for two stars several times alternately, and a mean of the
results taken, by measuring E F, their relative brightness
will be obtained: that of Jupiter, the temporary standard of
comparison, being altogether eliminated from the result.

(784.) A moderate number of well selected stars being thus
photometrically determined by repeated and careful measurements, so as to afford an ascertained and graduated scale of
brightness among the stars themselves, the intermediate steps
or grades of magnitude may be filled up, by inserting between
them, according to the judgment of the eye, other stars,
forming an ascending or descending *sequence*, each member
of such a sequence being brighter than that below, and less
bright than that above it; and thus at length, a scale of numerical magnitudes will become established, complete in all its
members, from Sirius, the brightest of the stars, down to the
least visible magnitude.* It were much to be wished that

---

* For the method of combining and treating such sequences, where accumulated in considerable numbers, so as to eliminate from their results the influence
of erroneous judgment, atmospheric circumstances, &c., which often give rise to
contradictory arrangements in the order of stars differing but little in magnitude,

this branch of astronomy, which at present can hardly be said to be advanced beyond its infancy, were perseveringly and systematically cultivated. It is by no means a subject of mere barren curiosity, as will abundantly appear when we come to speak of the phænomena of variable stars, and being moreover, one in which amateurs of the science may easily chalk out for themselves a useful and available path, may naturally be expected to receive large and interesting accessions at their hands.

(785.) If the comparison of the apparent magnitudes of the stars with their numbers leads to no immediately obvious conclusion, it is otherwise when we view them in connection with their local distribution over the heavens. If indeed we confine ourselves to the three or four brightest classes, we shall find them distributed with a considerable approach to impartiality over the sphere : a marked preference however being observable, especially in the southern hemisphere, to a zone or belt, following the direction of a great circle passing through ε Orionis and α Crucis. But if we take in the whole amount visible to the naked eye, we shall perceive a great increase of number as we approach the borders of the Milky Way. And when we come to telescopic magnitudes, we find them crowded beyond imagination, along the extent of that circle, and of the branches which it sends off from it; so that in fact its whole light is composed of nothing but stars of every magnitude, from such as are visible to the naked eye down to the smallest point of light perceptible with the best telescopes.

(786.) These phænomena agree with the supposition that the stars of our firmament, instead of being scattered in all directions indifferently through space, form a stratum of which the thickness is small, in comparison with its length and breadth ; and in which the earth occupies a place somewhere about the middle of its thickness, and near the point where it subdivides into two principal laminæ, inclined at a small angle to each other (art. 302.). For it is certain that, to an eye so

as well as for an account of a series of photometric comparisons (in which however, not Jupiter, but the moon was used as an intermediate standard), see the work above cited, note on p. 353. (Results of Observations, &c.)

situated, the apparent density of the stars, supposing them
pretty equally scattered through the space they occupy, would
be least in a direction of the visual ray (as S A), perpendicular to the lamina, and greatest in that of its breadth, as
S B, S C, S D ; increasing rapidly in passing from one to the
other direction, just as we see a slight haze in the atmosphere
thickening into a decided fog bank near the horizon, by the
rapid increase of the mere length of the visual ray.   Such is
the view of the construction of the starry firmament taken by
Sir William Herschel, whose powerful telescopes first effected
a complete analysis of this wonderful zone, and demonstrated
the fact of its entirely consisting of stars.*   So crowded are

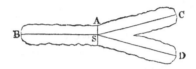

they in some parts of it, that by counting the stars in a single
field of his telescope, he was led to conclude that 50000 had
passed under his review in a zone two degrees in breadth,
during a single hour's observation.   In that part of the milky
way which is situated in 10h 30m R A and between the
147th and 150th degree of N P D, upwards of 5000 stars
have been reckoned to exist in a square degree.   The immense distances at which the remoter regions must be situated
will sufficiently account for the vast predominance of small
magnitudes which are observed in it.

(787.) The course of the Milky Way as traced through the
heavens by the unaided eye, neglecting occasional deviations
and following the line of its greatest brightness as well as its
varying breadth and intensity will permit, conforms nearly
to that of a great circle inclined at an angle of about 63° to
the equinoctial, and cutting that circle in R A 0 h 47 m and

---

* Thomas Wright of Durham (Theory of the Universe, London, 1750) appears so early as 1734 to have entertained the same general view as to the constitution of the Milky Way and starry firmament, founded, quite in the spirit of
just astronomical speculation on a partial resolution of a portion of it with a
" one-foot reflector " (a reflector one foot in focal length). See an account of this
rare work by M. de Morgan in Phil. Mag. Ser. 3. xxxii. p. 241. et seq.

12h 47m, so that its northern and southern poles respectively
are situated in R. A. 12h 47m NPD 63° and R. A. 0h 47m
NPD 117°. Throughout the region where it is so remarkably subdivided (art. 186.), this great circle holds an intermediate situation between the two great streams; with a
nearer approximation however to the brighter and continuous stream, than to the fainter and interrupted one. If we
trace its course in order of right ascension, we find it traversing the constellation Cassiopeia, its brightest part passing
about two degrees to the north of the star δ of that constellation, *i. e.* in about 62° of north declination, or 28° NPD.
Passing thence between γ and ε Cassiopeiæ it sends off a
branch to the south-preceding side, towards α Persei, very
conspicuous as far as that star, prolonged faintly towards
ε of the same constellation, and possibly traceable towards
the Hyades and Pleiades as remote outliers. The main
stream however (which is here very faint), passes on through
Auriga, over the three remarkable stars, ε, ζ, η, of that constellation preceding Capella, called the Hœdi, preceding
Capella, between the feet of Gemini and the horns of the
Bull (where it intersects the ecliptic nearly in the Solstitial
Colure) and thence over the club of Orion to the neck of Monoceros, intersecting the equinoctial in R. A. 6h 54m. Up to
this point, from the offset in Perseus, its light is feeble and
indefinite, but thenceforward it receives a gradual accession
of brightness, and where it passes through the shoulder of
Monoceros and over the head of Canis Major it presents a
broad, moderately bright, very uniform, and to the naked eye,
starless stream up to the point where it enters the prow of the
ship Argo, nearly on the southern tropic.* Here it again
subdivides (about the star m Puppis), sending off a narrow and
winding branch on the preceding side as far as γ Argûs, where
it terminates abruptly. The main stream pursues its southward course to the 123d parallel of NPD, where it diffuses

---

* In reading this description a celestial globe will be a necessary companion.
It may be thought needless to detail the course of the Milky Way verbally, since
it is mapped down on all celestial charts and globes. But in the generality of
them, indeed in all which have come to our knowledge, this is done so very loosely
and incorrectly, as by no means to dispense with a verbal description.

itself broadly and again subdivides, opening out into a wide fan-like expanse, nearly 20° in breadth formed of interlacing branches, all which terminate abruptly, in a line drawn nearly through λ and γ Argûs.

(788.) At this place the continuity of the Milky Way is interrupted by a wide gap, and where it recommences on the opposite side it is by a somewhat similar fan-shaped assemblage of branches which converge upon the bright star η Argûs. Thence it crosses the hind feet of the Centaur, forming a curious and sharply defined semicircular concavity of small radius, and enters the Cross by a very bright neck or isthmus of not more than 3 or 4 degrees in breadth, being the narrowest portion of the Milky Way. After this it immediately expands into a broad and bright mass, enclosing the stars α and β Crucis, and β Centauri, and extending almost up to α of the latter constellation. In the midst of this bright mass, surrounded by it on all sides, and occupying about half its breadth, occurs a singular dark pear-shaped vacancy, so conspicuous and remarkable as to attract the notice of the most superficial gazer, and to have acquired among the early southern navigators the uncouth but expressive appellation of the *coal-sack*. In this vacancy which is about 8° in length, and 5° broad, only one very small star visible to the naked eye occurs, though it is far from devoid of telescopic stars, so that its striking blackness is simply due to the effect of contrast with the brilliant ground with which it is on all sides surrounded. This is the place of nearest approach of the Milky Way to the South Pole. Throughout all this region its brightness is very striking, and when compared with that of its more northern course already traced, conveys strongly the impression of greater proximity, and would almost lead to a belief that our situation as spectators is separated on all sides by a considerable interval from the dense body of stars composing the Galaxy, which in this view of the subject would come to be considered as a flat ring of immense and irregular breadth and thickness, within which we are excentrically situated, nearer to the southern than to the northern part of its circuit.

M M

(789.) At α Centauri, the Milky Way again subdivides *, sending off a great branch of nearly half its breadth, but which thins off rapidly, at an angle of about 20° with its general direction, towards the preceding side, to η and d Lupi, beyond which it loses itself in a narrow and faint streamlet. The main stream passes on increasing in breadth to γ Normæ, where it makes an abrupt elbow and again subdivides into one principal and continuous stream of very irregular breadth and brightness on the following side, and a complicated system of interlaced streaks and masses on the preceding, which covers the tail of Scorpio, and terminates in a vast and faint effusion over the whole extensive region occupied by the preceding leg of Ophiuchus, extending northwards to the parallel of 103° NPD, beyond which it cannot be traced; a wide interval of 14°, free from all appearance of nebulous light, separating it from the great branch on the north side of the equinoctial of which it is usually represented as a continuation.

(790.) Returning to the point of separation of this great branch from the main stream, let us now pursue the course of the latter. Making an abrupt bend to the following side, it passes over the stars ι Aræ, θ and ι Scorpii, and γ Tubi to γ Sagittarii, where it suddenly collects into a vivid oval mass about 6° in length and 4° in breadth, so excessively rich in stars that a very moderate calculation makes their number exceed 100,000. Northward of this mass, this stream crosses the ecliptic in longitude about 276°, and proceeding along the bow of Sagittarius into Antinous has its course rippled by three deep concavities, separated from each other by remarkable protuberances, of which the larger and brighter (situated between Flamstead's stars 3 and 6 Aquilæ) forms the most conspicuous patch in the southern portion of the Milky Way visible in our latitudes.

(791.) Crossing the equinoctial at the 19th hour of right ascension, it next runs in an irregular, patchy, and winding stream through Aquila, Sagitta and Vulpecula up to Cygnus;

* All the maps and globes place this subdivision at β Centauri, but erroneously.

at ε of which constellation its continuity is interrupted, and a
very confused and irregular region commences, marked by a
broad dark vacuity, not unlike the southern "coal-sack," occupying the space between ε, α, and γ Cygni, which serves as a
kind of center for the divergence of three great streams; one,
which we have already traced; a second, the continuation of
the first (across the interval) from α northward, between La-
certa and the head of Cepheus to the point in Cassiopeia whence
we set out, and a third branching off from γ Cygni, very vivid
and conspicuous, running off in a southern direction through
β Cygni, and s Aquilæ almost to the equinoctial, where it
loses itself in a region thinly sprinkled with stars, where in
some maps the modern constellation Taurus Poniatovii is
placed. This is the branch which, if continued across the
equinoctial, might be supposed to unite with the great southern effusion in Ophiuchus already noticed (art. 789.). A
considerable offset, or protuberant appendage, is also thrown
off by the northern stream from the head of Cepheus directly
towards the pole, occupying the greater part of the quartile
formed by α, β, ι, and δ of that constellation.

(792.) We have been somewhat circumstantial in describing the course and principal features of the Via Lactea,
not only because there does not occur any where (so far as
we know) any correct account of it, but chiefly by reason of
its high interest in sidereal astronomy, and that the reader
may perceive how very difficult it must necessarily be to form
any just conception of the real, solid form, as it exists in
space, of an object so complicated, and which we see from a
point of view so unfavourable. The difficulty is of the same
kind which we experience when we set ourselves to conceive
the real shape of an auroral arch or of the clouds, but far
greater in degree, because we know the laws which regulate
the formation of the latter, and limit them to certain conditions of altitude — because their motion presents them to us
in various aspects, but chiefly because we contemplate them
from a station considerably below their general plane, so as to
allow of our mapping out some kind of ground-plan of their
shape. All these aids are wanting when we attempt to map

and model out the Galaxy, and beyond the obvious conclusion that its form must be, generally speaking, *flat*, and of a thickness small in comparison with its area in length and breadth, the laws of perspective afford us little further assistance in the inquiry. Probability may, it is true, here and there enlighten us as to certain features. Thus when we see, as in the coal-sack, a sharply defined oval space free from stars, insulated in the midst of a uniform band of not much more than twice its breadth, it would seem much less probable that a conical or tubular hollow traverses the whole of a starry stratum, continuously extended from the eye outwards, than that a *distant* mass of comparatively moderate thickness should be simply perforated from side to side, or that an oval vacuity should be seen foreshortened in a *distant* foreshortened area, not really exceeding two or three times its own breadth. Neither can we without obvious improbability refuse to admit that the long lateral offsets which at so many places quit the main stream and run out to great distances, are either planes seen edgeways, or the convexities of curved surfaces viewed tangentially, rather than cylindrical or columnar excrescences bristling up obliquely from the general level.   And in the same spirit of probable surmise we may account for the intricate reticulations above described as existing in the region of Scorpio, rather by the accidental crossing of streaks thus originating, at very different distances, or by a cellular structure of the mass, than by real intersections.   Those cirrous clouds which are often seen in windy weather, convey no unapt impression either of the kind of appearance in question, or of the structure it suggests. It is to other indications however, and chiefly to the telescopic examination of its intimate constitution, and to the law of the distribution of stars, not only within its bosom, but generally over the heavens, that we must look for more definite knowledge respecting its true form and extent.

(793.) It is on observations of this latter class, and not on merely speculative or conjectural views, that the generalization in Art. 786., which refers the phænomena of the starry firmament to the system of the Galaxy as their embodying fact,

is brought to depend.   The process of "gauging" the heavens
was devised by Sir W. Herschel for this purpose.   It consisted in simply counting the stars of all magnitudes which
occur in single fields of view, of 15′ in diameter, visible
through a reflecting telescope of 18 inches aperture, and 20
feet focal length, with a magnifying power of 180°: the
points of observation being very numerous and taken indiscriminately in every part of the surface of the sphere
visible in our latitudes.   On a comparison of many hundred
such "gauges" or local enumerations it appears that the
density of star-light (or the number of stars existing on an
average of several such enumerations in any one immediate
neighbourhood) is least in the pole of the *Galactic circle**,
and increases on all sides, with the *Galactic polar distance*
(and that nearly equally in all directions) down to the Milky
Way itself, where it attains its maximum.   The progressive
rate of increase in proceeding from the pole is at first slow,
but becomes more and more rapid as we approach the plane
of that circle according to a law of which the following
numbers, deduced by M. Struve from a careful analysis of
all the gauges in question, will afford a correct idea.

Galactic † North Polar Distance.	Average Number of Stars in a Field 15′ in Diameter.
0°	4·15
15°	4·68
30°	6·52
45°	10·36
60°	17·68
75°	30·30
90°	122·00

From which it appears that the *mean* density of the stars in the
galactic circle exceeds in a ratio of very nearly 30 to 1 that

* From γαλα, γαλακτος, milk ; meaning the great circle spoken of in
Art. 787., to which the course of the Via Lactea most nearly conforms.
Every subject has its technical or conventional terms, by whose use circumlo-
cution is avoided, and ideas rendered definite.   This circle is to sidereal what
the invariable ecliptic is to planetary astronomy — a plane of ultimate reference,
the ground-plane of the sidereal system.
   † Etudes d'Astronomie Stellaire, p. 71.

in its pole, and in a proportion of more than 4 to 1 that in a direction 15° inclined to its plane.

(794.) These numbers fully bear out the statement in Art. 786. and even draw closer the resemblance by which that statement is there illustrated. For the rapidly increasing density of a fog-bank as the visual ray is depressed towards the plane of the horizon is a consequence not only of the mere increase in length of the foggy space traversed, but also of an actual increase of density in the fog itself in its lower strata. Now this very conclusion follows from a comparison *inter se* of the numbers above set down, as M. Struve has clearly shown from a mathematical analysis of the empirical formula, which faithfully represents their law of progression, and of which he states the result in the following table, expressing the densities of the stars at the respective distances, 1, 2, 3, &c., from the galactic plane, taking the mean density of the stars in that plane itself for unity.

Distances from the Galactic Plane.	Density of Stars.	Distances from the Galactic Plane.	Density of Stars.
0·00	1·00000	0·50	0·08646
0·05	0·48568	0·60	0·05510
0·10	0·33288	0·70	0·03079
0·20	0·23895	0·80	0·01414
0·30	0·17980	0·866	0·00532
0·40	0·13021		

The unit of distance, of which the first column of this table expresses fractional parts, is the distance at which such a telescope is capable of rendering just *visible* a star of average magnitude, or, as it is termed, its *space-penetrating* power. As we ascend therefore from the galactic plane into this kind of stellar atmosphere, we perceive that the density of its parallel strata decreases with great rapidity. At an altitude above that plane equal to only one-twentieth of the telescopic limit, it has already diminished to one-half, and at an altitude of 0·866, to hardly more than one-two-hundredth of its amount in that plane. So far as we can perceive there is no flaw in this reasoning, if only it be granted, 1st, that the level planes are continuous, and of equal density throughout; and, 2dly, *that an absolute and definite limit is set to telescopic vision, beyond which, if stars exist, they elude our sight, and are to us as if they existed not :* a postulate whose probability the reader will be in a better condition to estimate, when in possession of some other particulars respecting the constitution of the Galaxy to be described presently.

(795.) A similar course of observation followed out in the southern hemisphere, leads independently to the same conclusion as to the law of the visible distribution of stars over the southern galactic hemisphere, or that half of the celestial surface which has the south galactic pole for its center. A system of gauges, extending over the whole surface of that hemisphere taken with the same telescope, field of view and magnifying power employed in Sir William Herschel's gauges, has afforded the average numbers of stars *per* field of 15' in diameter, within the areas of zones encircling that pole at intervals of 15°, set down in the following table.

Zones of Galactic South Polar Distance.	Average Number of Stars per Field of 15'.
0° to 15°	6·05
15 to 30	6·62
30 to 45	9·08
45 to 60	13·49
60 to 75	26·29
75 to 90	59·06

(796.) These numbers are not directly comparable with those of M. Struve, given in Art. 793. because the latter corresponds to the limiting polar distances, while these are the averages for the included zones.   That eminent astronomer, however, has given a table of the average gauges appropriate to *each degree* of north galactic polar distance*, from which it is easy to calculate averages for the whole extent of each zone. How near a parallel the results of this calculation for the northern hemisphere exhibit with those above stated for the southern, will be seen by the following table.

* Etudes d' Astronomie Stellaire, p. 34.

Zones of Galactic North Polar Distance.	Average Number of Stars per Field of 15' from M. Struve's Table.
0° to 15°	4·32
15 to 30	5·42
30 to 45	8·21
45 to 60	13·61
60 to 75	24·09
75 to 90	53·43

It would appear from this that, with an almost exactly similar law of apparent density in the two hemispheres, the southern were somewhat richer in stars than the northern, which may, and not improbably does arise, from our situation not being precisely in the middle of its thickness, but somewhat nearer to its northern surface.

(797.) When examined with powerful telescopes, the constitution of this wonderful zone is found to be no less various than its aspect to the naked eye is irregular. In some regions the stars of which it is wholly composed are scattered with remarkable uniformity over immense tracts, while in others the irregularity of their distribution is quite as striking, exhibiting a rapid succession of closely clustering rich patches separated by comparatively poor intervals, and indeed in some instances by spaces absolutely dark *and completely void of any star*, even of the smallest telescopic magnitude. In some places not more than 40 or 50 stars on an average occur in a " gauge" field of 15', while in others a similar average gives a result of 400 or 500. Nor is less variety observable in the character of its different regions in respect of the magnitudes of the stars they exhibit, and the proportional numbers of the larger and smaller magnitudes associated together, than in respect of their aggregate numbers. In some, for instance, extremely minute stars, though never altogether wanting, occur in numbers so moderate as to lead us irresistibly to the conclusion that in these regions we see *fairly through* the starry stratum, since it is impossible otherwise (supposing their light not intercepted) that the numbers of the smaller

TELESCOPIC CONSTITUTION OF THE GALAXY.   537

magnitudes should not go on continually increasing ad
infinitum.   In such cases moreover the ground of the heavens,
as seen between the stars, is for the most part perfectly dark,
which again would not be the case, if innumerable multitudes
of stars, too minute to be individually discernible, existed beyond.   In other regions we are presented with the phænomenon of an almost uniform degree of brightness of the
individual stars, accompanied with a very even distribution
of them over the ground of the heavens, both the larger and
smaller magnitudes being strikingly deficient.   In such cases
it is equally impossible not to perceive that we are looking
through a sheet of stars nearly of a size, and of no great thickness compared with the distance which separates them from us.
Were it otherwise we should be driven to suppose the more
distant stars uniformly the larger, so as to compensate by
their greater intrinsic brightness for their greater distance, a
supposition contrary to all probability.   In others again, and
that not unfrequently, we are presented with a double phænomenon of the same kind, viz. a tissue as it were of large stars
spread over another of very small ones, the intermediate magnitudes being wanting.   The conclusion here seems equally
evident that in such cases we look through two sidereal sheets
separated by a starless interval.

(798.) Throughout by far the larger portion of the extent
of the Milky Way in both hemispheres, the general blackness
of the ground of the heavens on which its stars are projected,
and the absence of that innumerable multitude and excessive
crowding of the smallest visible magnitudes, and of glare
produced by the aggregate light of multitudes too small to
affect the eye singly, which the contrary supposition would
appear to necessitate, must, we think, be considered unequivocal indications that its dimensions *in directions where these
conditions obtain,* are not only not infinite, but that the space-
penetrating power of our telescopes suffices fairly to pierce
through and beyond it.   It is but right however to warn our
readers that this conclusion has been controverted, and that
by an authority not lightly to be put aside, on the ground of
certain views taken by Olbers as to a defect of perfect trans-

parency in the celestial spaces, in virtue of which the light of the more distant stars is enfeebled more than in proportion to their distance. The extinction of light thus originating, proceeding in geometrical progression while the distance increases in arithmetical, a limit, it is argued, is placed to the space-penetrating powers of telescopes, far within that which distance alone apart from such obscuration would assign. It would lead us too far aside of the objects of a treatise of this nature to enter upon any discussion of the grounds (partly metaphysical) on which these views rely. It must suffice here to observe that the objection alluded to, if applicable to any, is equally so to every part of the galaxy. We are not at liberty to argue that at one part of its circumference, our view is limited by this sort of cosmical veil which extinguishes the smaller magnitudes, cuts off the nebulous light of distant masses, and closes our view in impenetrable darkness; while at another we are compelled by the clearest evidence telescopes can afford to believe that star-strown vistas *lie open*, exhausting their powers and stretching out beyond their utmost reach, as is proved by that very phænomenon which the existence of such a veil would render impossible, viz. infinite increase of number and diminution of magnitude, terminating in complete irresolvable nebulosity. Such is, in effect, the spectacle afforded by a very large portion of the Milky Way in that interesting region near its point of bifurcation in Scorpio (arts. 789, 792.) where, through the hollows and deep recesses of its complicated structure we behold what has all the appearance of a wide and indefinitely prolonged area strewed over with discontinuous masses and clouds of stars which the telescope at length refuses to analyse.* Whatever other conclusions we may draw, this must any how be regarded as the direction of the greatest linear extension of the ground-plan of the galaxy. And it

---

* It would be doing great injustice to the illustrious astronomer of Pulkova (whose opinion, if we here seem to controvert, it is with the utmost possible deference and respect) not to mention that at the time of his writing the remarkable essay already more than once cited, in which the views in question are delivered, he could not have been aware of the important facts alluded to in the text, the work in which they are described being then unpublished.

would appear to follow, also, as a not less obvious consequence, that in those regions where that zone is clearly resolved into stars well separated and *seen projected on a black ground*, and where by consequence it is *certain* if the foregoing views be correct that we look out beyond them into space, the smallest visible stars appear as such, not by reason of excessive distance, but of a real inferiority of size or brightness.

(799.) When we speak of the comparative remoteness of certain regions of the starry heavens beyond others, and of our own situation in them, the question immediately arises, what is the distance of the nearest fixed star? What is the scale on which our visible firmament is constructed? And what proportion do its dimensions bear to those of our own immediate system? To these questions astronomy has at length been enabled to afford an answer.

(800.) The diameter of the earth has served us for the base of a triangle, in the *trigonometrical survey* of our system (art. 274.), by which to calculate the distance of the sun; but the extreme minuteness of the sun's parallax (art. 357.) renders the calculation from this "ill-conditioned" triangle (art. 275.) so delicate, that nothing but the fortunate combination of favourable circumstances, afforded by the transits of Venus (art. 479.), could render its results even tolerably worthy of reliance. But the earth's diameter is too small a base for direct triangulation to the verge even of our own system (art. 526.), and we are, therefore, obliged to substitute the *annual parallax* for the diurnal, or, which comes to the same thing, to ground our calculation on the relative velocities of the earth and planets in their orbits (art. 486.), when we would push our triangulation to that extent. It might be naturally enough expected, that by this enlargement of our base to the vast diameter of the earth's orbit, the next step in our survey (art. 275.) would be made at a great advantage; — that our change of station, from side to side of it, would produce a considerable and easily measurable amount of annual parallax in the stars, and that by its means we should come to a knowledge of their distance. But, after exhausting every refinement of observation, astronomers were, up to a very late period,

unable to come to any positive and coincident conclusion
upon this head; and the amount of such parallax, even for
the nearest fixed star examined with the requisite attention,
remained mixed up with, and concealed among, the errors
incidental to all astronomical determinations. The nature of
these errors has been explained in the earlier part of this
work, and we need not remind the reader of the difficulties
which must necessarily attend the attempt to disentangle an
element not exceeding a few tenths of a second or at most a
whole second from the host of uncertainties entailed on the
results of observations by them: none of them individually
perhaps of greater magnitude, but embarrassing by their
number and fluctuating amount. Nevertheless, by successive
refinements in instrument making, and by constantly pro
gressive approximation to the exact knowledge of the Uranographical corrections, that assurance had been obtained, even
in the earlier years of the present century, viz. that no star
visible in northern latitudes, to which attention had been
directed, manifested an amount of parallax exceeding a single
second of arc. It is worth while to pause for a moment to
consider what conclusions would follow from the admission of
a parallax to this amount.

(801.) Radius is to the sine of 1″ as 206265 to 1. In this
proportion then *at least* must the distance of the fixed stars
from the sun exceed that of the sun from the earth. Again,
the latter distance, as we have already seen (art. 357.), exceeds
the earth's radius in the proportion of 23984 to 1. Taking
therefore the earth's radius for unity, a parallax of 1″ supposes
a distance of 4947059760 or nearly five thousand millions of
such units: and lastly, to descend to ordinary standards,
since the earth's radius may be taken at 4000 of our miles,
we find 19788239040000 or about twenty billions of miles
for our resulting distance.

(802.) In such numbers the imagination is lost. The only
mode we have of conceiving such intervals at all is by the
time which it would require for light to traverse them.
Light, as we know (art. 545.), travels at the rate of 192000
miles per second, traversing a semidiameter of the earth's

orbit in $8^m 13^s{\cdot}3$. It would, therefore, occupy 206265 times this interval or 3 years and 83 days to traverse the distance in question. Now as this is an inferior limit which it is already ascertained that even the brightest and therefore (in the absence of all other indications) the nearest stars exceed, what are we to allow for the distance of those innumerable stars of the smaller magnitudes which the telescope discloses to us! What for the dimensions of the galaxy in whose remoter regions, as we have seen, the united lustre of myriads of stars is perceptible only in powerful telescopes as a feeble nebulous gleam!

(803.) The space-penetrating power of a telescope or the comparative distance to which a given star would require to be removed in order that it may appear of the same brightness in the telescope as before to the naked eye, may be calculated from the aperture of the telescope compared with that of the pupil of the eye, and from its reflecting or transmitting power, $i.\ e.$ the proportion of the incident light it conveys to the observer's eye. Thus it has been computed that the space-penetrating power of such a reflector as that used in the star-gauges above referred to is expressed by the number 75. A star then of the sixth magnitude removed to 75 times its distance would still be perceptible *as a star* with that instrument, and admitting such a star to have 100th part of the light of a standard star of the first magnitude, it will follow that such a standard star, if removed to 750 times its distance, would excite in the eye, when viewed through the gauging telescope, the same impression as a star of the sixth magnitude does to the naked eye. Among the infinite multitude of such stars in the remoter regions of the galaxy, it is but fair to conclude that innumerable individuals equal in intrinsic brightness to those which immediately surround us must exist. The light of such stars then must have occupied upwards of 2000 years in travelling over the distance which separates them from our own system. It follows then that when we observe the places and note the appearances of such stars, we are only reading their history of two thousand years' anterior date thus wonderfully recorded. We cannot escape this conclusion but by adopting as an alternative an intrinsic inferiority of light in *all* the smaller stars of the galaxy. We shall be better able to estimate the probability of this alternative when we shall have made acquaintance with other sidereal systems whose existence the telescope discloses to us, and whose analogy will satisfy us that the view of the subject here taken is in perfect harmony with the general tenor of astronomical facts.

(804.) Hitherto we have spoken of a parallax of 1″ as a mere limit below which that of any star yet examined assuredly, or at least very probably falls, and it is not without a certain convenience to regard this amount of parallax as a sort of unit of reference, which, connected in the reader's recollection with a parallactic unit of distance from our system of 20 billions of miles, and with a $3\frac{1}{4}$ year's journey of light, may save him the trouble of such calculations, and ourselves the necessity of covering our pages with such enormous numbers, when speaking of stars whose parallax has actually been ascertained with some approach to certainty, either by direct meridian observation or by more refined and delicate methods. These we shall proceed to explain, after first pointing out the theoretical peculiarities which enable us to separate and disentangle its effects from those of the Uranographical corrections, and from other causes of error which being periodical in their nature add greatly to the difficulty of the subject. The effects of precession and proper motion (see art. 852.) which are uniformly progressive from year to year, and that of nutation which runs through its period in nineteen years, it is obvious enough, separate themselves at once by these characters from that of parallax; and, being known with very great precision, and being certainly independent, as regards their causes, of any individual peculiarity in the stars affected by them, whatever small uncertainty may remain respecting the numerical elements which enter into their computation (or in mathematical language their *co-efficients*), can give rise to no embarrassment. With regard to aberration the case is materially different. This correction affects the place of a star by a fluctuation annual in its period, and therefore, so far, agreeing with parallax. It is also very similar in the *law* of its variation at different seasons of the year, parallax having for its apex (see art. 343, 344.) the apparent place of the sun in the ecliptic, and aberration a point in the same great circle 90° behind that place, so that in fact the formulæ of calculation (the co-efficients excepted) are the same for both, substituting only for the sun's longitude in the expression for the one, that longitude diminished by 90° for the other. Moreover, in the absence of *absolute certainty* respecting the nature of the propagation of light, astronomers have hitherto considered it necessary to assume at least as a *possibility* that the velocity of light may be to some slight amount dependent on in-dividual peculiarities in the body emitting it. *

(805.) If we suppose a line drawn from the star to the earth at all seasons of the year, it is evident that this line will sweep over the surface of an exceedingly acute, oblique cone, having for its axis the line joining the sun and star, and for its base the earth's annual orbit, which, for the present purpose, we may suppose circular. The star will therefore appear to describe each year about its mean place regarded as fixed, and in virtue of parallax alone, a minute ellipse, the section of this cone by the surface of the celestial sphere, perpendicular to the visual ray. But there is also another way in which the same fact may be represented. The ap-parent orbit of the star about its mean place as a center, will be precisely that which it would appear to describe, if seen from the sun, supposing it really revolved about that place in a circle exactly equal to the earth's annual orbit, in a plane parallel to the ecliptic. This is evident from the equality and parallelism of the lines and directions concerned. Now the effect of aberration (disregarding the slight variation of the earth's velocity in different parts of its orbit) is precisely similar in law, and differs only in amount, and in its bearing

* In the actual state of astronomy and photology this necessity can hardly be considered as still existing, and it is desirable, therefore, that the practice of astronomers of introducing an unknown correction for the constant of aberration into their "equations of condition" for the determination of parallax, should be disused, since it actually tends to introduce error into the final result.

reference to a direction 90° different in longitude. Suppose,
in order to fix our ideas, the maximum of parallax to be 1″
and that of aberration 20· 5″, and let A B, *a b*, be two circles
imagined to be described separately, as above, by the star
about its mean place S, in virtue of these two causes respectively, S ♈ being a line parallel to that of the line of equinoxes.   Then if in virtue of parallax alone, the star would
be found at *a* in the smaller orbit, it would in virtue of
aberration alone be found at A, in the larger, the angle *a* S A
being a right angle.   Drawing then A C equal and parallel
to S *a*, and joining S C, it will in virtue of both simultaneously be found in C, *i. e.* in the circumference of a circle
whose radius is S C, and at a point in that circle, in advance
of A, the aberrational place, by the angle A S C.   Now since
S A : A C :: 20·5 : 1, we find for the angle A S C 2° 47′ 35″,

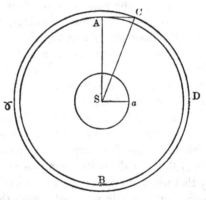

and for the length of the radius S C of the circle representing
the compound motion 20″·524.   The difference (0″·024)
between this and S C, the radius of the aberration circle, is
quite imperceptible, and even supposing a quantity so minute
to be capable of detection by a prolonged series of observations, it would remain a question whether it were produced
by parallax or by a specific difference of aberration from the
general average 20″·5 in the star itself.   It is therefore to
the difference of 2° 48′ between the angular situation of the
displaced star in this hypothetical orbit, *i. e.* in the *arguments*

(as they are called) of the joint correction (Υ S C) and that of aberration alone (Υ S A), that we have to look for the resolution of the problem of parallax. The reader may easily figure to himself the delicacy of an inquiry which turns wholly (even when stripped of all its other difficulties) on the *precise* determination of a quantity of this nature, and of such very moderate magnitude.

(806.) But these other difficulties themselves are of no trifling order. All astronomical instruments are affected by differences of temperature. Not only do the materials of which they are composed expand and contract, but the masonry and solid piers on which they are erected, nay even the very soil on which these are founded, participate in the general change from summer warmth to winter cold. Hence arise slow oscillatory movements of exceedingly minute amount, which levels and plumblines afford but very inadequate means of detecting, and which *being also annual in their period* (after rejecting whatever is merely casual and momentary) mix themselves intimately with the matter of our inquiry. Refraction too, besides its casual variations from night to night, which a long series of observations would eliminate, depends for its theoretical expression on the constitution of the strata of our atmosphere, and the law of the distribution of heat and moisture at different elevations, which cannot be unaffected by difference of season. No wonder then that mere meridional observations should, almost up to the present time, have proved insufficient, except in one very remarkable instance, to afford unquestionable evidence, and satisfactory quantitative measurement of the parallax of any fixed star.

(807.) The instance referred to is that of α Centauri, one of the brightest and for many other reasons, one of the most remarkable of the southern stars. From a series of observations of this star, made at the Royal Observatory of the Cape of Good Hope in the years 1832 and 1833, by Professor Henderson, with the mural circle of that establishment, a parallax to the amount of an entire second was concluded on his reduction of the observations in question after his return to England. Subsequent observations by Mr. Maclear,

N N

partly with the same, and partly with a new and far more
efficiently constructed instrument of the same description
made in the years 1839 and 1840, have fully confirmed the
reality of the parallax indicated by Professor Henderson's ob
servations, though with a slight diminution in its concluded
amount, which comes out equal to $0''\cdot9128$ at about $\frac{10}{11}$ths of
a second ; *bright stars in its immediate neighbourhood being
unaffected by a similar periodical displacement, and thus
affording satisfactory proof that the displacement indicated in
the case of the star in question is not merely a result of annual
variations of temperature.* As it is impossible at present to
answer for so minute a quantity as that by which this result
differs from an exact second, we may consider the distance of
this star as approximately expressed by the parallactic unit
of distance referred to in art. 804.

(808.) A short time previous to the publication * of this
important result, the detection of a sensible and measurable
amount of parallax in the star N° 61 Cygni of Flamsteed's
catalogue of stars was announced by the celebrated astronomer of Königsberg, the late M. Bessel.† This is a
small and inconspicuous star, hardly exceeding the sixth
magnitude, but which had been pointed out for especial observation by the remarkable circumstance of its being affected
by a *proper motion* (see art. 852.), *i. e.* a regular and continually progressive annual displacement among the surrounding
stars to the extent of more than $5''$ per annum, a quantity so
very much exceeding the average of similar minute annual
displacements which many other stars exhibit, as to lead to a
suspicion of its being actually nearer to our system. It is
not a little remarkable that a similar presumption of proximity exists also in the case of α Centauri, whose unusually
large proper motion of nearly $4''$ per annum is stated by
Professor Henderson to have been the motive which induced
him to subject his observations of that star to that severe discussion which led to the detection of its parallax. M.

---

* Prof. Henderson's paper was read before the Astronomical Society of
London, Jan. 3. 1839.   It bears date Dec. 24. 1838.
† Astronomische Nachrichten, Nos. 365, 366. Dec. 13. 1838.

Bessel's observations of 61 Cygni were commenced in August 1837, immediately on the establishment at the Königsberg observatory of a magnificent heliometer, the workmanship of the celebrated optician Fraunhofer, of Munich, an instrument especially fitted for the system of observation adopted ; which being totally different from that of direct meridional observation, more refined in its conception, and susceptible of far greater accuracy in its practical application, we must now explain.

(809.) Parallax, proper motion, and specific aberration (denoting by the latter phrase that part of the aberration of a star's light which may be supposed to arise from its individual peculiarities, and which we have every reason to believe at all events an exceedingly minute fraction of the whole,) are the only uranographical corrections which do not necessarily affect alike the apparent places of two stars situated in, or *very nearly in*, the same visual line.   Supposing then two stars at an immense distance, the one behind the other, but otherwise so situated as to appear very nearly along the same visual line, they will constitute what is called a star *optically double*, to distinguish it from a star *physically double*, of which more hereafter.   Aberration (that which is common to all stars), precession, nutation, *nay, even refraction, and instrumental causes of apparent displacement, will affect them alike,* or so very nearly alike (if the minute difference of their apparent places be taken into account) as to admit of the difference being neglected, or very accurately allowed for, by an easy calculation.   If then, instead of attempting to determine by observation the place of the nearer of two *very unequal* stars (which will probably be the larger) by direct observation of its right ascension and polar distance, we content ourselves with referring its place to that of its remoter and smaller companion by *differential observation, i. e.* by measuring only its *difference* of situation from the *latter*, we are at once relieved of the necessity of making these corrections, and from all uncertainty as to their influence on the result.   And for the very same reason, errors of adjustment (art. 136.), of graduation, and a host of instrumental errors,

which would for this delicate purpose fatally affect the
absolute determination of either star's place, are harmless
when only the difference of their places, each equally affected
by such causes, is required to be known.

(810.) Throwing aside therefore the consideration of all
these errors and corrections, and disregarding for the present

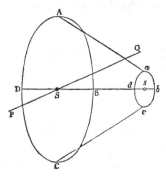

the minute effect of specific aberration and the uniformly progressive effect of proper motion, let us trace the effect of the
differences of the parallaxes of two stars thus juxtaposed, or
their apparent relative distance and position at various seasons
of the year.   Now the parallax being inversely as the distance,
the dimensions of the small ellipses apparently described
(art. 805.) by each star on the concave surface of the heavens
by parallactic displacement will differ, — the nearer star
describing the larger ellipse.   But both stars lying very
nearly in the same direction from the sun, these ellipses will
be similar and similarly situated.   Suppose S and s to be the
positions of the two stars as seen from the sun, and let
A B C D, a b c d, be their parallactic ellipses; then, since
they will be at all times similarly situated in these ellipses,
when the one star is seen at A, the other will be seen at a.
When the earth has made a quarter of a revolution in its
orbit, their apparent places will be B b; when another
quarter, C c ; and when another, D d.   If, then, we measure
carefully, with micrometers adapted for the purpose, their
apparent situation with respect to each other, at different
times of the year, we should perceive a periodical change,

both in the *direction* of the line joining them, and in the *distance* between their centers. For the lines A *a* and C *c* cannot be parallel, nor the lines B *b* and D *d* equal, unless the ellipses be of equal dimensions, *i. e.* unless the two stars have the same parallax, or are equidistant from the earth.

(811.) Now, micrometers, properly mounted, enable us to measure very exactly both the distance between two objects which can be seen together in the same field of a telescope, and the position of the line joining them with respect to the horizon, or the meridian, or any other determinate direction in the heavens. The double image micrometer, and especially the heliometer (art. 200, 201.) is peculiarly adapted for this purpose. The images of the two stars formed side by side, or in the same line prolonged, however momentarily displaced by temporary refraction or instrumental tremor, *move together*, preserving their relative situation, the judgment of which is no way disturbed by such irregular movements. The heliometer also, taking in a greater range than ordinary micrometers, enables us to compare one large star with more than one adjacent small one, and to select such of the latter among many near it, as shall be most favourably situated for the detection of any motion in the large one, not participated in by its neighbours.

(812.) The star examined by Bessel has two such neighbours, both very minute, and therefore probably very distant, most favourably situated, the one (*s*) at a distance of 7′ 42″, the other (*s′*) at 11′ 46″ from the large star, and so situated, that their directions from that star make nearly a right angle with each other. The effect of parallax therefore would necessarily cause the two distances S *s* and S *s′* to vary so as to attain their maximum and minimum values alternately at three-monthly intervals, and this is what was actually observed to take place, the one distance being always most rapidly on the increase or decrease when the other was stationary (the uniform effect of proper motion being understood of course to be always duly accounted for). This alternation, though so small in amount as to indicate, as a final result, a parallax, or rather a difference of parallaxes between the large and small stars of hardly more than one third of a second, was maintained with such regularity as to leave no room for reasonable doubt as to its cause, and having been confirmed by the further continuance of these observations, and quite recently by the exact coincidence between the result thus obtained, and that deduced by M. Peters from observations of the same star at the observatory of Pulkova*, is considered on all hands as fully established. The parallax of this star finally resulting from Bessel's observation is 0″·348 so that its distance from our system is very nearly three parallactic units. (Art. 804.)

(813.) The bright star α Lyræ has also near it, at only 43″ distance (and therefore within the reach of the parallel wire or ordinary double image micrometer) a very minute star, which has been subjected since 1835 to a severe and assiduous scrutiny by M. Struve, on the same principle of differential observation. He has thus established the existence of a measurable amount of parallax in the large star, less indeed than that of 61 Cygni (being only about ¼ of a second), but yet sufficient (such was the delicacy of his measurements) to justify this excellent observer in announcing the result as at least highly probable, on the strength of only five nights' observation, in 1835 and 1836. This probability, the continuation of the measures to the end of 1838 and the corroborative, though not in this case precisely coincident, result of Mr. Peters's investigations have converted into a certainty. M. Struve has the merit of being the first to bring into practical application this method of observation, which, though proposed for the purpose, and its great advantages pointed out by Sir William Herschel so early as 1781 †, remained long unproductive of any result, owing partly to the imperfection of micrometers for the measurement of

* With the great vertical circle by Ertel.
† It has been referred even to Galileo. But the general explanation of Parallax in the Systema Cosmicum, Dial. iii. p. 271 (Leyden edit. 1699) to which the reference applies, does not touch any of the peculiar features of the case, or meet any of its difficulties.

distance, and partly to a reason which we shall presently have occasion to refer to.

(814.) If the component individuals S, *s* (*fig.* art. 810.) be (as is often the case) very close to each other, the parallactic variation of their *angle of position,* or the extreme angle included between the lines A *a,* C *c,* may be very considerable, even for a small amount of difference of parallaxes between the large and small stars. For instance in the case of two adjacent stars 15″ asunder, and otherwise favourably situated for observation, an annual fluctuation to and fro in the apparent direction of their line of junction to the extent of half a degree (a quantity which could not escape notice in the means of numerous and careful measurements) would correspond to a difference of parallax of only $\frac{1}{8}$ of a second. A difference of 1″ between two stars apparently situated at 5″ distance might cause an oscillation in that line to the extent of no less than 11°, and if nearer one proportionally still greater. This mode of observation has not yet been put in practice, but seems to offer great advantages.*

(815.) The following is a list of stars to which parallax has been up to the present time more or less probably assigned:

				″
α Centauri	-	-	-	- 0·913 (Henderson.)
61 Cygni	-	-	-	- 0·348 (Bessel.)
α Lyræ	•	-	-	- 0·261 (Struve.)
Sirius	-	-	-	- 0·230 (Henderson.)
1830 Groombridge †	-	-	-	0·226 (Peters.)
Ursæ Majoris	-	-	-	- 0·133 ditto.
Arcturus	-	-	-	- 0·127 ditto.
Polaris	-	-	-	- 0·067 ditto.
Capella	-	-	-	- 0·046 ditto.

Although the extreme minuteness of the last four of these results deprives them of much numerical reliance, it is at least certain that the parallaxes by no means follow the order of magnitudes, and this is farther shown by the fact that α Cygni, one of M. Peters's stars, shows absolutely no indications of any measurable parallax whatever.

* See Phil. Trans. 1826, p. 266. *et seq.* and 1827, for a list of stars well adapted for such observation, with the times of the year most favourable. — The list in Phil. Trans. 1826, is incorrect.

† Groombridge's catalogue of circumpolar stars.

(816.) From the distance of the stars we are naturally led to the consideration of their real magnitudes. But here a difficulty arises, which, so far as we can judge of what optical instruments are capable of effecting, must always remain insuperable. Telescopes afford us only negative information as to the apparent angular diameter of any star. The round, well-defined, planetary discs which good telescopes show when turned upon any of the brighter stars are phænomena of diffraction, dependent, though at present somewhat enigmatically, on the mutual interference of the rays of light. They are consequently, so far as this inquiry is concerned, mere optical illusions, and have therefore been termed *spurious* discs. The proof of this is that telescopes of different apertures and magnifying powers, when applied for the purpose of measuring their angular diameters, give different results, the greater aperture (even with the same magnifying power) giving the smaller disc. That the true disc of even a large and bright star can have but a very minute angular measure, appears from the fact that in the occultation of such a star by the moon, its extinction is *absolutely instantaneous,* not the smallest trace of gradual diminution of light being perceptible. The apparent or spurious disc also remains *perfectly round* and *of its full size* up to the instant of disappearance, which could not be the case were it a real object. If our sun were removed to the distance expressed by our parallactic unit (art. 804.), its apparent diameter of 32′ 3″ would be reduced to only 0″·0093, or less than the hundredth of a second, a quantity which we have not the smallest reason to hope any practical improvement in telescopes will ever show as an object having distinguishable *form*.

(817.) There remains therefore only the indication which the quantity of light they send to us may afford. But here again another difficulty besets us. The light of the sun is so immensely superior in intensity to that of any star, that it is impracticable to obtain any direct comparison between them. But by using the moon as an intermediate term of comparison it may be done, not indeed with much precision, but sufficiently well to satisfy in some degree our curiosity on the subject.

Now α Centauri has been directly compared with the moon by the method explained in Art. 783.   By a mean of eleven such comparisons made in various states of the moon, duly reduced and making the proper allowance on photometric principles for the moon's light lost by transmission through the lens and prism, it appears that the mean quantity of light sent to the earth by a full moon exceeds that sent by α Centauri in the proportion of 27408 to 1.   Now Wollaston, by a method apparently unobjectionable, found * the proportion of the sun's light to that of the full moon to be that of 801072 to 1. Combining these results, we find the light sent us by the sun to be to that sent by α Centauri as 21,955,000,000, or about twenty-two thousand millions to 1.   Hence from the parallax assigned above to that star, it is easy to conclude that its intrinsic splendour, as compared with that of our sun at equal distances, is 2·3247, that of the sun being unity.†

(818.)   The light of Sirius is four times that of α Centauri and its parallax only 0″·230. (Art. 230.)   This in effect ascribes to it an intrinsic splendour equal to 63·02 times that of our sun. ‡

* Wollaston, Phil. Trans. 1829. p. 27.
† *Results of Astronomical Observations at the Cape of Good Hope, &c.* Art. 278. p. 363.   If only the results obtained near the quadratures of the moon (which is the situation most favourable to exactness) be used, the resulting value of the intrinsic light of the star (the sun being unity) is 4·1586.   On the other hand, if only those procured near the full moon (the worst time for observation) be employed, the result is 1·4017.   Discordances of this kind will startle no one conversant with Photometry.   That α Centauri really emits more light than our sun must, we conceive, be regarded as an established fact.   To those who may refer to the work cited it is necessary to mention that the quantity there designated by M, expresses, on the scale there adopted, 500 times the actual illuminating power of the moon at the time of observation, that of the mean full moon being unity.
‡ See the work above cited, p. 367.—Wollaston makes the light of Sirius one 20,000-millionth of the sun's.   Steinheil by a very uncertain method found ⊙= (3286500)² × Arcturus.

CHAPTER XVI.

(819.) Now, for what purpose are we to suppose such
magnificent bodies scattered through the abyss of space?
Surely not to illuminate *our* nights, which an additional moon
of the thousandth part of the size of our own would do much
better, nor to sparkle as a pageant void of meaning and reality, and bewilder us among vain conjectures. Useful, it is
true, they are to man as points of exact and permanent
reference; but he must have studied astronomy to little
purpose, who can suppose man to be the only object of his
Creator's care, or who does not see in the vast and wonderful
apparatus around us provision for other races of animated
beings. The planets, as we have seen, derive their light from
the sun; but that cannot be the case with the stars. These
doubtless, then, are themselves suns, and may, perhaps, each

in its sphere, be the presiding center round which other planets, or bodies of which we can form no conception from any analogy offered by our own system, may be circulating.

(820.) Analogies, however, more than conjectural, are not wanting to indicate a correspondence between the dynamical laws which prevail in the remote regions of the stars and those which govern the motions of our own system. Wherever we can trace the law of periodicity—the regular recurrence of the same phænomena in the same times—we are strongly impressed with the idea of rotatory or orbitual motion. Among the stars are several which, though no way distinguishable from others by any apparent change of place, nor by any difference of appearance in telescopes, yet undergo a more or less regular periodical increase and diminution of lustre, involving in one or two cases a complete extinction and revival. These are called periodical stars. The longest known and one of the most remarkable is the star *Omicron*, in the constellation Cetus (sometimes called Mira Ceti), which was first noticed as variable by Fabricius in 1596. It appears about twelve times in eleven years, or more exactly in a period of 331^d 15^h 7^m; remains at its greatest brightness about a fortnight, being then on some occasions equal to a large star of the second magnitude; decreases during about three months, till it becomes completely invisible to the naked eye, in which state it remains about five months: and continues increasing during the remainder of its period. Such is the general course of its phases. It does not always however return to the same degree of brightness, nor increase and diminish by the same gradations, neither are the successive intervals of its maxima equal. From the recent observations and inquiries into its history by M. Argelander, the mean period above assigned would appear to be subject to a cyclical fluctuation embracing eighty-eight such periods, and having the effect of gradually lengthening and shortening alternately those intervals to the extent of twenty-five days one way and the other.* The irregularities in the degree of brightness attained at the maximum are probably also periodical.

* Astronom. Nachr. No. 624.

Hevelius relates* that during the four years between October 1672 and December 1676 it did not appear at all. It was unusually bright on October 5. 1839 (the epoch of its maximum for that year according to M. Argelander's observations) when it exceeded α Ceti and equalled β Aurigæ in lustre.

(821.) Another very remarkable periodical star is that called Algol, or β Persei. It is usually visible as a star of the second magnitude, and such it continues for the space of $2^d 13\frac{1}{2}^h$, when it suddenly begins to diminish in splendour, and in about $3\frac{1}{2}$ hours is reduced to the fourth magnitude, at which it continues about 15^m. It then begins again to increase, and in $3\frac{1}{2}$ hours more is restored to its usual brightness, going through all its changes in $2^d 20^h 48^m 58^s\cdot5$. This remarkable law of variation certainly appears strongly to suggest the revolution round it of some opaque body, which when interposed between us and Algol, cuts off a large portion of its light; and this is accordingly the view taken of the matter by Goodricke, to whom we owe the discovery of this remarkable fact †, in the year 1782; since which time the same phænomena have continued to be observed, but with this remarkable additional point of interest; viz. that the more recent observations as compared with the earlier ones indicate a diminution in the periodic time. The latest observations of Argelander, Heis, and Schmidt, even go to prove that this diminution is not uniformly progressive, but is actually proceeding with accelerated rapidity, which however will probably not continue, but, like other cyclical combinations in astronomy, will by degrees relax, and then be changed into an increase, according to laws of periodicity which, as well as their causes, remain to be discovered. The

* Lalande's Astronomy, Art. 794.

† The same discovery appears to have been made nearly about the same time by Palitzch, a farmer of Prolitz, near Dresden, — a peasant by station, an astronomer by nature, — who, from his familiar acquaintance with the aspect of the heavens, had been led to notice among so many thousand stars this one as distinguished from the rest by its variation, and had ascertained its period. The same Palitzch was also the first to re-discover the predicted comet of Halley in 1759, which he saw nearly a month before any of the astronomers, who, armed with their telescopes, were anxiously watching its return. These anecdotes carry us back to the era of the Chaldean shepherds.

first minimum of this star in the year 1844 occurred on Jan. 3. at 4^h 14^m Greenwich mean time.*

(822). The star δ in the constellation Cepheus is also subject to periodical variations, which, from the epoch of its first observation by Goodricke in 1784 to the present time, have been continued with perfect regularity. Its period from minimum to minimum is 5^d 8^h 47^m $39^s\cdot5$, the first or epochal minimum for 1849 falling on Jan. 2. 3^h 13^m 37^s M. T. at Greenwich. The extent of its variation is from the fifth to between the third and fourth magnitudes. Its increase is more rapid than its diminution, the interval between the minimum and maximum of its light being only 1^d 14^h, while that from the maximum to the minimum is 3^d 19^h.

(823.) The periodical star β Lyræ, discovered by Goodricke also in 1784, has a period which has been usually stated at from 6^d 9^h to 6^d 11^h, and there is no doubt that in about this interval of time its light undergoes a remarkable diminution and recovery. The more accurate observations of M. Argelander however have led him to conclude † the true period to be 12^d 21^h 53^m 10^s, and that in this period a double maximum and minimum takes place, the two maxima being nearly equal and both about the 3·4 magnitude, but the minima considerably unequal, viz. 4·3 and 4·5m. In addition to this curious subdivision of the whole interval of change into two semi-periods, we are presented in the case of this star with another instance of slow alteration of period, which has all the appearance of being itself periodical. From the epoch of its discovery in 1784 to the year 1840 the period was continually lengthening, but more and more slowly, till at the last-mentioned epoch it ceased to increase, and has since been slowly on the decrease. As an epoch for the least or absolute minimum of this star, M. Argelander's calculations enable us to assign 1846 January 3^d 0^h 9^m 53^s G. M. T.

(824.) Another periodical star whose changes have been

---

* Ast. Nach. No. 472.

† Astron. Nachr. No. 624. See also the valuable papers by this excellent astronomer in A. N. Nos. 417, 455, &c.

carefully observed is η Aquilæ or Antinoi, first pointed out by Pigott in 1784 (a year fertile in such discoveries) as belonging to that class. Its period is 7^d 4^h 13^m 53^s, the first minimum for 1849 occurring on Jan. 2. at 19^h 22^m 55^s G. M.T. It occupies fifty-seven hours in its increase from 5m to 4·3m, and 115 hours in its decrease.

(825.) These are all the variable stars which have been observed with sufficient care and for a sufficient length of time to enable us to speak with precision as to their periods, epochs, and phases of brightness. But the number of those whose period is approximately or roughly known is considerable, and of those whose change is certain, though its period and limits are as yet unknown, still more so. The following table includes the principal among them, though each year adds to their number : —

Star.	Period.	Change of Mag.		Discovered by
	d. dec.	from	to	
β Persei (Algol) - -	2·8673	2	4	Goodricke, 1782.
λ Tauri - » - -	4±	4	5·4	Baxendell, 1848.
Cephei - - - -	5·3664	3·4	5	Goodricke, 1784.
η Aquilæ - - - -	7·1763	3·4	4·5	Pigott, 1784.
* Cancri R. A. (1800)= $8^h 32^m ·5$ N. P. D. 70° 15′	9·015	7·8	10	Hind, 1848.
ζ Geminorum - - -	10·2	4·3	4·5	Schmidt, 1847.
β Lyræ - - - -	12·9119	3·4	4·5	Goodricke, 1784.
α Herculis - - - -	63±	3	4	Herschel, 1796.
59 B. Scuti R. A. 1801 = $18^h 37^m$; N. P. D =95° 57′	71·200	5	0	Pigott, 1795.
ε Aurigæ - - - -	250±	3	4	Heis, 1846.
ο Ceti (Mira) - - -	331·63	2	0	Fabricius, 1596.
* Serpentis R. A. 1828 = $15^h 46^m 45^s$; P. D. 74° 20′ 30″	335±	7?	0	Harding, 1826.
χ Cygni - - - -	396·875	6	11	Kirch, 1687.
υ Hydræ (B. A. C. 4501.) -	494±	4	10	Maraldi, 1704.
* Cephei (B. A. C. 7582.) -	5 or 6 years	3	6	Herschel, 1782.
34 Cygni (B. A. C. 6990.) -	18 years ±	6	0	Janson, 1600.
* Leonis (B. A. C. 3345.) -	Many years	6	0	Koch, 1782.
κ Sagittarii - - -	Ditto	3	6	Halley, 1676.
ψ Leonis - - - -	Ditto	6	0	Montanari, 1667.
η Cygni - - - -	Ditto	4·5	5·6	Herschel Junr., 1842 ?
* Virginis R. A. (1840)= $12^h 3^m$; N. P. D. 82° 8′	145 days	6·7	0	Harding, 1814.
* Coronæ Bor. (B. A. C. 5236)	10½ months	6	0	Pigott, 1795.
7 Arietis (B. A. C. 581.) -	5 years ?	6	8	Piazzi, 1798.
η Argûs - - - -	Irregular	1	4	Burchell, 1827.

Star.	Period.	Change of Mag.		Discovered by
		from	to	
α Orionis - - - -	Irregular	1	1 2	Herschel Junr., 1836.
α Ursæ Majoris - - -	Some years	1·2	2	Ditto, 1846.
η Ursæ Majoris - - -	Ditto	1·2	2	Ditto, 1846.
β Ursæ Minoris - - -	2 or 3 years?	2	2·3	Struve, 1838.
α Cassiopeiæ - - -	225 days?	2	2·3	Herschel Junr., 1838.
α Hydræ - - - -	29 or 30 days?	2·3	3	Ditto, 1837.
* R. A. (1847.) = 22ʰ 58ᵐ 57ˢ·9 N. P. D. = 80° 17′ 30″	Unknown	8?	0	Hind, 1848.
* R. A. (1848.) = 7ʰ 33ᵐ 55ˢ·2 N. P. D. = 66° 11′ 56″	Ditto	9	0	Ditto, 1848.
* R. A. (1848.) = 7ʰ 40ᵐ 10ˢ·3 N. P. D. = 65° 53′ 29″	Ditto	9	0	Ditto, 1848.
Near * R. A. 22ʰ 21ᵐ 0ˢ·4 (1848.) N. P. D. 100° 42′ 40″	Ditto	7·8	0	Rümker.
* R. A. (1848.) 14ʰ 44ᵐ 39ˢ·6 N. P. D. 101° 45′ 25″	Ditto	8	9·10	Schumacher.
δ Ursæ Majoris - - -	Many years	2?	2·3	Matter of general remark.

N. B. In the above list the letters B. A. C. indicate the catalogue of the British Association, B. the catalogue of Bode. Numbers before the name of the constellation (as 34 Cygni) denote Flamsteed's stars. Since this table was drawn up, four additional stars, variable from the 8th or 9th magnitude to 0, have been communicated to us by Mr. Hind, whose places are as follows: (1.) R. A. 1ʰ 38ᵐ 24ˢ; N. P. D. 81° 9′ 39″; (2.) 4ʰ 50ᵐ 42ˢ, 82° 6′ 36″ (1846); (3.) 8ʰ 43ᵐ 8ˢ, 86° 11′ (1800); (4.) 22ʰ 12ᵐ 9ˢ, 82° 59′ 24″ (1800). Mr. Hind remarks that about several variable stars some degree of haziness is perceptible at their minimum. Have they clouds revolving round them as planetary or cometary attendants? He also draws attention to the fact that the red colour predominates among variable stars generally. The double star, No 2718 of Struve's Catalogue, R. A. 20ʰ 34ᵐ, P. D. 77° 54′, is stated by the author to be variable. Captain Smyth (Celestial Cycle, i. 274.) mentions also 3 Leonis and 18 Leonis as variable, the former from 6ᵐ to 0, P=78 days, the latter from 5ᵐ to 10ᵐ, P = 311ᵈ 23ʰ, but without citing any authority. Piazzi sets down 96 and 97 Virginis and 38 Herculis as variable stars.

(826.) Irregularities similar to those which have been noticed in the case of *o* Ceti, in respect of the maxima and minima of brightness attained in successive periods, have been also observed in several others of the stars in the foregoing list. χ Cygni, for example, is stated by Cassini to have been scarcely visible throughout the years 1699, 1700, 1701, at those times when it was expected to be most conspicuous. No. 59 Scuti is sometimes visible to the naked eye at its minimum, and sometimes not so, and its maximum is also very irregular. Pigott's variable star in Corona is stated by M.

Argelander to vary for the most part so little that the unaided
eye can hardly decide on its maxima and minima, while yet
after the lapse of whole years of these slight fluctuations, they
suddenly become so great that the star completely vanishes.
The variations of α Orionis, which were most striking and unequivocal in the years 1836—1840, within the years since
elapsed became much less conspicuous.   They seem now
(Jan. 1849) to have recommenced.

(827.) These irregularities prepare us for other phænomena
of stellar variation, which have hitherto been reduced to no
law of periodicity, and must be looked upon, in relation to
our ignorance and inexperience, as altogether casual; or, if
periodic, of periods too long to have occurred more than once
within the limits of recorded observation.   The phænomena
we allude to are those of *Temporary Stars*, which have appeared, from time to time, in different parts of the heavens,
blazing forth with extraordinary lustre; and after remaining
awhile apparently immovable, have died away, and left no
trace.   Such is the star which, suddenly appearing some time
about the year 125 B. C., and which was visible in the day
time, is said to have attracted the attention of Hipparchus,
and led him to draw up a catalogue of stars, the earliest on
record.   Such, too, was the star which appeared, A. D.
389, near α Aquilæ, remaining for three weeks as bright as
Venus, and disappearing entirely.   In the years 945, 1264,
and 1572, brilliant stars appeared in the region of the heavens between Cepheus and Cassiopeia; and, from the imperfect account we have of the places of the two earlier, as compared with that of the last, which was well determined, as
well as from the tolerably near coincidence of the intervals
of their appearance, we may suspect them, with Goodricke,
to be one and the same star, with a period of 312 or perhaps
of 156 years.   The appearance of the star of 1572 was so
sudden, that Tycho Brahe, a celebrated Danish astronomer,
returning one evening (the 11th of November) from his la-
boratory to his dwelling-house, was surprised to find a group
of country people gazing at a star, which he was sure did not
exist half an hour before.   This was the star in question.   It

was then as bright as Sirius, and continued to increase till it surpassed Jupiter when brightest, and was visible at midday. It began to diminish in December of the same year, and in March, 1574, had entirely disappeared. So, also, on the 10th of October, 1604, a star of this kind, and not less brilliant, burst forth in the constellation of Serpentarius, which continued visible till October, 1605.

(828.) Similar phænomena, though of a less splendid character, have taken place more recently, as in the case of the star of the third magnitude discovered in 1670, by Anthelm, in the head of the Swan; which, after becoming completely invisible, re-appeared, and, after undergoing one or two singular fluctuations of light, during two years, at last died away entirely, and has not since been seen.

(829.) On the night of the 28th of April, 1848, Mr. Hind observed a star of the fifth magnitude or 5·4 (very conspicuous to the naked eye) in a part of the constellation Ophiu-chus (R.A. 16^h 51^m 1^s·5. N.P.D. 102° 39′ 14″), where, from perfect familiarity with that region, he was certain that up to the 5th of that month no star so bright as 9·10 m. previously existed. Neither has any record been discovered of a star being there observed at any previous time. From the time of its discovery it continued to diminish, without any alteration of place, and before the advance of the season rendered further observation impracticable, was nearly extinct. Its colour was ruddy, and was thought by many observers to undergo remarkable changes, an effect probably of its low situation.

(830.) The alterations of brightness in the southern star η Argûs, which have been recorded, are very singular and surprising. In the time of Halley (1677) it appeared as a star of the fourth magnitude. Lacaille, in 1751, observed it of the second. In the interval from 1811 to 1815, it was again of the fourth; and again from 1822 to 1826 of the second. On the 1st of February, 1827, it was noticed by Mr. Burchell to have increased to the first magnitude, and to equal α Crucis. Thence again it receded to the second; and so continued until the end of 1837. All at once in the beginning of 1838 it suddenly increased in lustre so as to surpass all the stars of the first magnitude except Sirius, Canopus, and α Centauri, which last star it nearly equalled. Thence it again diminished, but this time not below the first magnitude until April, 1843, when it had again increased so as to surpass Canopus, and nearly equal Sirius in splendour. "A strange field of speculation," it has been remarked, "is opened by this phænomenon. The temporary stars heretofore recorded have all become totally extinct. Variable stars, so far as they have been carefully attended to, have exhibited periodical alternations, in some degree at least regular, of splendour and comparative obscurity. But here we have a star fitfully variable to an astonishing extent, and whose fluctuations are spread over centuries, apparently in no settled period, and with no regularity of progression. What origin can we ascribe to these sudden flashes and relapses? What conclusions are we to draw as to the comfort or habitability of a system depending for its supply of light and heat on so uncertain a source?" Speculations of this kind can hardly be termed visionary, when we consider that, from what has before been said, we are compelled to admit a community of nature between the fixed stars and our own sun; and when we reflect that geology testifies to the fact of extensive changes having taken place at epochs of the most remote antiquity in the climate and temperature of our globe; changes difficult to reconcile with the operation of secondary causes, such as a different distribution of sea and land, but which would find an easy and natural explanation in a slow variation of the supply of light and heat afforded primarily by the sun itself.

(831.) The Chinese annals of Ma-touan-lin *, in which stand *officially* recorded, though rudely, remarkable astronomical phænomena, supply a long list of "strange stars," among which, though the greater part are evidently comets, some may be recognized as belonging in all probability to the class of *Temporary Stars* as above characterized. Such is that which is recorded to have appeared in A. D. 173, between α

* Translated by M. Edward Biot, Connoissance des Temps, 1846.

and β *Centauri,* which (no doubt, scintillating from its low situation) exhibited " the five colours," and remained visible from December in that year till July in the next. And another which these annals assign to A. D. 1011, and which would seem to be identical with a star elsewhere referred to A. D. 1012, " which was of extraordinary brilliancy, and remained visible in the southern part of the heavens during three months," * a situation agreeing with the Chinese record, which places it low in Sagittarius. Among several less unequivocal is one referred to B. C. 134, in Scorpio, which may possibly have been Hipparchus's star. None of the stars of A. D. 389, 945, 1264, and 1572, however, are noticed in these records. It is worthy of especial notice, that all the stars of this kind on record, of which the places are distinctly indicated, have occurred, *without exception,* in or close upon the borders of the Milky Way, and that only within the following semicircle, the preceding having offered no example of the kind.

(832.) On a careful re-examination of the heavens, and a comparison of catalogues, many stars are now found to be missing; and although there is no doubt that these losses have arisen in the great majority of instances from mistaken entries, and in some from planets having been mistaken for stars, yet in some it is equally certain that there is no mistake in the observation or entry, and that the star has really been observed, and as really has disappeared from the heavens. The whole subject of variable stars is a branch of practical astronomy which has been too little followed up, and it is precisely that in which amateurs of the science, and especially voyagers at sea, provided with only good eyes, or moderate instruments, might employ their time to excellent advantage. It holds out a sure promise of rich discovery, and is one in which astronomers in established observatories are almost of necessity precluded from taking a part by the nature of the observations required. Catalogues of the com-

---

* Hind, Notices of the Astronomical Society, viii. 156., citing Hepidannus. He places the Chinese star of 173 B.C. between α and β *Canis Minoris,* but M. Biot distinctly says α, β *pied oriental du Centaure.*

parative brightness of the stars in each constellation have been constructed by Sir Wm. Herschel, with the express object of facilitating these researches, and the reader will find them, and a full account of his method of comparison, in the Phil. Trans. 1796, and subsequent years.

(833.) We come now to a class of phænomena of quite a different character, and which give us a real and positive insight into the nature of at least some among the stars, and enable us unhesitatingly to declare them subject to the same dynamical laws, and obedient to the same power of gravitation, which governs our own system. Many of the stars, when examined with telescopes, are found to be double, *i.e.* to consist of two (in some cases three or more) individuals placed near together. This might be attributed to accidental proximity, did it occur only in a few instances; but the frequency of this companionship, the extreme closeness, and, in many cases, the near equality of the stars so conjoined, would alone lead to a strong suspicion of a more near and intimate relation than mere casual juxtaposition. The bright star Castor, for example, when much magnified, is found to consist of two stars of nearly the third magnitude, within 5″ of each other. Stars of this magnitude, however, are not so common in the heavens as to render it otherwise than excessively improbable that, if scattered at random, they would fall so near. But this improbability becomes immensely increased by a consideration of the fact, that this is only one out of a great many similar instances. Mitchell, in 1767, applying the rules for the calculation of probabilities to the case of the six brightest stars in the group called the Pleiades, found the odds to be 500000 to 1 against their proximity being the mere result of a random scattering of 1500 stars (which he supposed to be the total number of stars of that magnitude in the celestial sphere *) over the heavens. Speculating further on this, as an indication of physical connexion rather than fortuitous assemblage, he was led to surmise the possibility (since converted into a certainty, but at that time,

* This number is considerably too small, and in consequence, Mitchell's odds in this case materially overrated. But enough will remain, if this be rectified, fully to bear out his argument. Phil. Trans. vol. 57.

antecedent to any observation) of the existence of compound stars revolving about one another, or rather about their common center of gravity. M. Struve, pursuing the same train of thought as applied specially to the cases of double and triple combinations of stars, and grounding his computations on a more perfect enumeration of the stars visible down to the 7th magnitude, in the part of the heavens visible at Dorpat, calculates that the odds are 9570 to 1 against any two stars, from the 1st to the 7th magnitude inclusive, out of the whole possible number of binary combinations then visible, falling (if fortuitously scattered) within 4″ of each other. Now the number of instances of such binary combinations actually observed at the date of this calculation was already 91, and many more have since been added to the list. Again, he calculates that the odds against any such stars fortuitously scattered, falling within 32″ of a third, so as to constitute a triple star, is not less than 173524 to 1. Now, four such combinations occur in the heavens; viz. θ Orionis, σ Orionis, 11 Monocerotis, and ζ Cancri. The conclusion of a physical connexion of some kind or other is therefore unavoidable.

(834.) Presumptive evidence of another kind is furnished by the following consideration. Both α Centauri and 61 Cygni are " Double Stars." Both consist of two individuals, nearly equal, and separated from each other by an interval of about a quarter of a minute. In the case of 61 Cygni, the stars exceeding the 7th magnitude, there is already a primâ facie probability of 9578 to 1 against their apparent proximity. The two stars of α Centauri are both at least of the 2nd magnitude, of which altogether not more than about 50 or 60 exist in the whole heavens. But, waving this consideration, both these stars, as we have already seen, have a proper motion so considerable that, supposing the constituent individuals unconnected, one would speedily leave the other behind. Yet at the earliest dates at which they were respectively observed these stars were not perceived to be double, and it is only to the employment of telescopes magnifying at least 8 or 10 times, that we owe the knowledge we now possess of their being so. With such a telescope Lacaille, in

1751, was barely able to perceive the separation of the two
constituents of α Centauri, whereas, had one of them only
been affected with the observed proper motion, they should
then have been 6′ asunder.   In these cases then some physical connexion may be regarded as proved by this fact alone.

(835.) Sir William Herschel has enumerated upwards of
500 double stars, of which the individuals are less than 32″
asunder.   M. Struve, prosecuting the inquiry with instruments more conveniently mounted for the purpose, and
wrought to an astonishing pitch of optical perfection, has
added more than five times that number.   And other observers have extended still further the catalogue of " Double
Stars," without exhausting the fertility of the heavens.
Among these are a great many in which the distance between
the component individuals does not exceed a single second.
They are divided into classes by M. Struve (the first living
authority in this department of Astronomy) according to the
proximity of their component individuals.   The first class
comprises those only in which the distance does not exceed
1″; the 2nd those in which it exceeds 1″ but falls short of 2″;
the 3rd class extends from 2″ to 4″ distance; the 4th from
4″ to 8″; the 5th from 8″ to 12″; the 6th from 12″ to 16″;
the 7th from 16″ to 24″, and the 8th from 24″ to 32″. Each
class he again subdivides into two sub-classes of which the
one under the appellation of *conspicuous* double stars (duplices
lucidæ) comprehends those in which both individuals exceed
the $8\frac{1}{4}$ magnitude, that is to say, are *separately* bright enough
to be easily seen in any moderately good telescope.   All
others, in which one or both the constituents are below this
limit of easy visibility, are collected into another sub-class,
which he terms *residuary (Duplices reliquæ)*.   This arrangement is so far convenient, that after a little practice in the
use of telescopes as applied to such objects, it is easy to judge
what optical power will probably suffice to resolve a star of
any proposed class and either sub-class, or would at least be
so if the second or residuary sub-class were further subdivided
by placing in a third sub-class " delicate " double stars, or
those in which the companion star is so very minute as to

require a high degree of optical power to perceive it, of which instances will presently be given.

(836.) The following may be taken as specimens of each class. They are all taken from among the lucid, or conspicuous stars, and to such of our readers as may be in possession of telescopes, and may be disposed to try them on such objects, will afford him a ready test of their degree of efficiency.

Class I., 0″ to 1″.

γ Coronæ Bor.	η Coronæ.	ꓶ Ophiuchi.	Atlas Pleiadum.
γ Centauri.	η Herculis.	φ Draconis.	4 Aquarii.
γ Lupi.	λ Cassiopeiæ.	φ Ursæ Majoris.	42 Comæ.
ε Arietis.	λ Ophiuchi.	χ Aquilæ.	52 Arietis.
ζ Herculis.	π Lupi.	ω Leonis.	66 Piscium.

Class II., 1″ to 2″.

γ Circini.	ζ Bootis.	ξ Ursæ Majoris.	2 Camelopardi.
δ Cygni.	ι Cassiopeiæ.	π Aquilæ.	32 Orionis.
ε Chamæleontis.	ι 2 Cancri.	σ Coronæ Bor.	52 Orionis.

Class III., 2″ to 4″.

α Piscium.	γ Virginis.	ζ Aquarii.	μ Draconis.
β Hydræ.	δ Serpentis.	ζ Orionis.	μ Canis.
γ Ceti.	ε Bootis.	ι Leonis.	ρ Herculis.
γ Leonis.	ε Draconis.	ι Trianguli.	σ Cassiopeiæ.
γ Coronæ Aus.	ε Hydræ.	κ Leporis.	44 Bootis.

Class IV., 4″ to 8″.

α Crucis.	θ Phœnicis.	ξ Cephei.	μ Eridani.
α Herculis.	κ Cephei.	π Bootis.	70 Ophiuchi.
α Geminorum.	λ Orionis.	ρ Capricorni.	12 Eridani.
δ Geminorum.	μ Cygni.	υ Argûs.	32 Eridani.
ζ Coronæ Bor.	ξ Bootis.	ω Aurigæ.	95 Herculis.

Class V., 8″ to 12″.

β Orionis.	ζ Antliæ.	ι Orionis.
γ Arietis.	η Cassiopeiæ.	f Eridani.
γ Delphini.	θ Eridani.	2 Canum Ven.

Class VI., 12″ to 16″.

α Centauri.	γ Volantis.	κ Bootis.
β Cephei.	η Lupi.	8 Monocerotis.
β Scorpii.	ζ Ursæ Major.	61 Cygni.

Class VII., 16″ to 24″.

α Canum Ven.	θ Serpentis.	24 Comæ.
ε Normæ.	κ Coronæ Aus.	41 Draconis.
ζ Piscium.	χ Tauri.	61 Ophiuchi.

Class VIII., 24″ to 32″.

δ Herculis.	κ Herculis.	χ Cygni.
η Lyræ.	κ Cephei.	23 Orionis.
ι Cancri.	ψ Draconis.	

(837.) Among the most remarkable triple, quadruple, or multiple stars (for such also occur), may be enumerated,

α Andromedæ.	θ Orionis.	ξ Scorpii.
ε Lyræ.	μ Lupi.	11 Monocerotis.
ζ Cancri.	μ Bootis.	12 Lyncis.

Of these, α Andromedæ, μ Bootis, and μ Lupi appear in telescopes, even of considerable optical power, only as ordinary double stars; and it is only when excellent instruments are used that their smaller companions are subdivided and found to be, in fact, extremely close double stars. ε Lyræ offers the remarkable combination of a double-double star. Viewed with a telescope of low power it appears as a coarse and easily divided double star, but on increasing the magnifying power, each individual is perceived to be beautifully and closely double, the one pair being about $2\frac{1}{2}''$, the other about $3''$ asunder. Each of the stars ζ Cancri, ξ Scorpii, 11 Monocerotis, and 12 Lyncis consists of a principal star, closely double, and a smaller and more distant attendant, while θ Orionis presents the phænomenon of four brilliant principal stars, of the respective 4th, 6th, 7th, and 8th magnitudes, forming a trapezium, the longest diagonal of which is $21''\cdot4$, and ac-

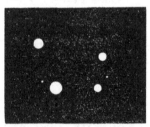

companied by two excessively minute and very close companions (as in the annexed figure), to perceive *both* which is one of the severest tests which can be applied to a telescope.

(838.) Of the "delicate" sub-class of double stars, or those consisting of very large and conspicuous principal stars, accompanied by very minute companions, the following specimens may suffice:

α 2 Cancri.	α Polaris.	κ Circini.	φ Virginis.
α 2 Capricorni.	β Aquarii.	κ Geminorum.	χ Eridani.
α Indi.	γ Hydræ.	μ Persei.	16 Aurigæ.
α Lyræ.	ι Ursæ Majoris.	η Bootis.	94 Ceti.

(839.) To the amateur of Astronomy the double stars offer a subject of very pleasing interest, as tests of the performance of his telescopes, and by reason of the finely contrasted colours which many of them exhibit, of which more hereafter. But it is the high degree of physical interest which attaches to them, which assigns them a conspicuous place in modern Astronomy, and justifies the minute attention and unwearied diligence bestowed on the measurement of their angles of position and distances, and the continual enlargement of our catalogues of them by the discovery of new ones. It was, as we have seen, under an impression that such combinations, if diligently observed, might afford a measure of parallax through the periodical variations it might be expected to produce in the relative situation of the small attendant star, that Sir W. Herschel was induced (between the years 1779 and 1784) to form the first extensive catalogues of them, under the scrutiny of higher magnifying powers than had ever previously been applied to such purposes. In the pursuit of this object, the end to which it was instituted as a means was necessarily laid aside for a time, until the accumulation of more abundant materials should have afforded a choice of stars favourably circumstanced for systematic observation. Epochal measures however, of each star, were secured, and, on resuming the subject, his attention was altogether diverted from the original object of the inquiry by phænomena of a very unexpected character, which at once engrossed his whole attention. Instead of finding, as he expected, that annual fluctuation to and fro of one star of a double star with respect to the other, — that alternate annual increase and decrease of their distance and angle of position, which the parallax of the earth's annual motion would produce, — he observed, in many instances, a regular progressive change; in some cases bearing chiefly on their distance, — in others on their position, and advancing steadily in one direction, so as clearly to indicate either a real motion of the stars themselves, or a general rectilinear motion of the sun and whole solar system, producing a parallax of a higher order than would arise from the earth's orbital motion, and which might be called systematic parallax.

(840.) Supposing the two stars, and also the sun, in motion independently of each other, it is clear that for the interval of several years, these motions must be regarded as rectilinear and uniform. Hence, a very slight acquaintance with geometry will suffice to show that the *apparent motion* of one star of a double star, referred to the other as a center, and mapped down, as it were, on a plane in which that other shall be taken for a fixed or zero point, can be no other than a right line. This, at least, must be the case if the stars be independent of each other; but it will be otherwise if they have a physical connexion, such as, for instance, real proximity and mutual gravitation would establish. In that case, they would describe orbits round each other, and round their common center of gravity; and therefore the apparent path of either, referred to the other as fixed, instead of being a portion of a straight line, would be bent into a curve concave towards that other. The observed motions, however, were so slow, that many years' observation was required to ascertain this point; and it was not, therefore, until the year 1803, twenty-five years from the commencement of the inquiry, that any thing like a positive conclusion could be come to respecting the rectilinear or orbitual character of the observed changes of position.

(841.) In that, and the subsequent year, it was distinctly announced by him, in two papers, which will be found in the Transactions of the Royal Society for those years*, that there exist sidereal systems, composed of two stars revolving about each other in regular orbits, and constituting what may be termed *binary stars*, to distinguish them from double stars generally so called, in which these physically connected stars are confounded, perhaps, with others only *optically* double, or casually juxtaposed in the heavens at different distances from the eye; whereas the individuals of a binary star are, of course, equidistant from the eye, or, at least, cannot differ more in distance than the semi-diameter of the orbit they describe about each other, which is quite insignificant compared with the immense distance between them and the earth.

---

* The announcement was in fact made in 1802, but unaccompanied by the observations establishing the fact.

Between fifty and sixty instances of changes, to a greater or less amount, in the angles of position of double stars, are adduced in the memoirs above mentioned; many of which are too decided, and too regularly progressive, to allow of their nature being misconceived. In particular, among the more conspicuous stars, — Castor, γ Virginis, ξ Ursæ, 70 Ophiuchi, σ and η Coronæ, ξ Bootis, η Cassiopeiæ, γ Leonis, ζ Herculis, δ Cygni, μ Bootis, ϵ 4 and ϵ 5 Lyræ, λ Ophiuchi, μ Draconis, and ζ Aquarii, are enumerated as among the most remarkable instances of the observed motion; and to some of them even periodic times of revolution are assigned; approximative only, of course, and rather to be regarded as rough guesses than as results of any exact calculation, for which the data were at the time quite inadequate. For instance, the revolution of Castor is set down at 334 years, that of γ Virginis at 708, and that of γ Leonis at 1200 years.

(842.) Subsequent observation has fully confirmed these results. Of all the stars above named, there is not one which is not found to be fully entitled to be regarded as binary; and, in fact, this list comprises nearly all the most considerable objects of that description which have yet been detected, though (as attention has been closely drawn to the subject, and observations have multiplied) it has, of late, received large accessions. Upwards of a hundred double stars, certainly known to possess this character, were enumerated by M. Mädler in 1841 *, and more are emerging into notice with every fresh mass of observations which come before the public. They require excellent telescopes for their effective observation, being for the most part so close as to necessitate the use of very high magnifiers (such as would be considered extremely powerful microscopes if employed to examine objects within our reach), to perceive an interval between the individuals which compose them.

(843.) It may easily be supposed, that phænomena of this kind would not pass without attempts to connect them with dynamical theories. From their first discovery, they were naturally referred to the agency of some power, like that of

* Dorpat Observations, vol. ix. 1840 and 1841.

gravitation, connecting the stars thus demonstrated to be in a state of circulation about each other; and the extension of the Newtonian law of gravitation to these remote systems was a step so obvious, and so well warranted by our experience of its all-sufficient agency in our own, as to have been expressly or tacitly made by every one who has given the subject any share of his attention. We owe, however, the first distinct system of calculation, by which the elliptic elements of the orbit of a binary star could be deduced from observations of its angle of position and distance at different epochs, to M. Savary, who showed[*], that the motions of one of the most remarkable among them (ξ Ursæ) were explicable, within the limits allowable for error of observation, on the supposition of an elliptic orbit described in the short period of $58\frac{1}{4}$ years. A different process of computation conducted Professor Encke[†] to an elliptic orbit for 70 Ophiuchi, described in a period of seventy-four years. M. Mädler has especially signalized himself in this line of inquiry (see note). Several orbits have also been calculated by Mr. Hind and Captain Smyth, and the author of these pages has himself attempted to contribute his mite to these interesting investigations. The following may be stated as the chief results which have been hitherto obtained in this branch of astronomy : —

[*] Connoiss. des Temps, 1830.   [†] Berlin Ephem. 1832.

The elements Nos. 1, 2, 3, 4 c, 5, 6 c, 7, 11 b, 12 a, are extracted from M. Mädler's synoptic view of the history of double stars in vol. ix. of the Dorpat Observations: 4 a, from the Connoiss. des Temps, 1830 : 4 b, 6 b, and 11 a, from vol. v. Trans. Astron. Soc. Lond.: 6 a, from Berlin Ephemeris, 1832 : No. 8. from Trans. Astron. Soc. vol. vi. : No. 9, 11 c, 12 b, and 13 from Notices of the Astronomical Society, vol. vii. p. 22., and viii. p. 159., and No. 10 from the author's " Results of Astronomical Observations, &c. at the Cape of Good Hope," p. 297. The Σ prefixed to No. 7. denotes the number of the star in M. Struve's Dorpat Catalogue (Catalogus Novus Stellarum Duplicium, &c. Dorpat. 1827), which contains the places for 1826 of 3112 of these objects.

The "position of the node " in col. 4. expresses the angle of position (see Art. 204.) of the line of intersection of the plane of the orbit, with the plane of the heavens on which it is seen projected. The "inclination " in col. 6. is the inclination of these two planes to one another. Col. 5. shows the angle actually included *in the plane of the orbit*, between the line of nodes (defined as above) and the line of apsides. The elements assigned in this table to ω Leonis, ξ Bootis, and Castor must be considered as very doubtful, and the same may perhaps be said of those ascribed to μ 2 Bootis, which rest on too small an arc of the orbit, and that too imperfectly observed, to afford a secure basis of calculation.

Star's Name.	Apparent semi-axis.	Excentricity.	Position of Node.	Perihelion from Node or Orbit.	Inclination.	Period in Years.	Perihelion Passage.	By whom computed.
1. Herculis	1"·189	0·44454	39° 26'	262° 4'	50° 53'	31·468	1829·50	Mädler.
2. η Coronæ B.	1·088	0·33760	24 18	261 21	71 8	43·246	1815·23	Ditto.
3. ζ Cancri	1·292	0·23486	1 28	266 0	63 17	58·910	1853·37	Ditto.
4. a. ξ Ursæ Majoris	3·857	0·41640	95 22	131 38	50 40	58·262	1817·25	Savary.
4. b. Ditto	3·278	0·37770	97 47	134 22	56 6	60·720	1816·73	Herschel, junior.
4. c. Ditto	2·417	0·41350	98 52	130 48	54 56	61·464	1816·44	Mädler.
5. ω Leonis	0·857	0·64338	135 11	185 27	46 33	82·533	1849·76	Ditto.
6. a. p Ophiuchi	4·328	0·43007	147 12	125 22	46 25	73·862	1806·88	Encke.
6. b. Ditto	4·392	0·46670	137 2	145 46	48 5	80·340	1807·06	Herschel, junior.
6. c. Ditto	4·192	0·44380	126 55	142 53	64 51	92·870	1812·73	Mädler.
7. Σ 3062	1·255	0·44958	15 3	137 27	35 31	94·765	1837·41	Ditto.
8. ξ Bootis	12·560	0·59374	359 59	100 59	80 5	117·140	1779·88	Herschel, junior.
9. δ Cygni	1·811	0·60667	24 54	243 24	46 23	178·700	1862·87	Hind.
10. γ Virginis	3·580	0·87952	5 33	313 45	23 36	182·120	1836·43	Herschel, junior.
11. a. Castor	8·086	0·75820	58 6	97 29	70 3	252·660	1855·83	Ditto.
11. b. Ditto	7·008	0·79725	23 5	87 37	70 58	232·124	1913·90	Mädler.
11. c. Ditto	6·300	0·24050	11 24	356 22	43 14	632·270	1699·26	Hind.
12. a. σ Coronæ B.	3·918	0·69978	25 7	64 38	29 29	608·450	1826·60	Mädler.
12. b. Ditto	5·194	0·72560	21 3	69 24	25 39	736·880	1826·48	Hind.
13. μ 2 Bootis	3·218	0·84010	117 21	103 17	46 57	649·720	1852·50	Ditto.
14. α Centauri	15·500	0·95000	86 7	291 22	47 56	77·000	1851·50	Jacob.

(844.) Of the stars in the above list, that which has been most assiduously watched, and has offered phænomena of the greatest interest, is γ Virginis. It is a star of the vulgar 3rd magnitude (3·08 = Photom. 3·494), and its component individuals are very nearly equal, and as it would seem in some slight degree variable, since, according to the observations of M. Struve, the one is alternately a little greater, and a little less than the other, and occasionally exactly equal to it. It has been known to consist of two stars since the beginning of the eighteenth century; the distance being then between six and seven seconds, so that any tolerably good telescope would resolve it. When observed by Herschel in 1780, it was 5″·66, and continued to decrease gradually and regularly till at length, in 1836, the two stars had approached so closely as to appear perfectly round and single under the highest magnifying power which could be applied to most excellent instruments — the great refractor at Pulkowa alone, with a magnifying power of 1000, continuing to indicate by the wedge-shaped form of the disc of the star its composite nature. By estimating the ratio of its length to its breadth and measuring the former, M. Struve concludes that, at this epoch (1836·41), the distance of the two stars, center from center, might be stated at 0″·22. From that time the star again opened, and at present (1849) the individuals are more than 2″ asunder. This very remarkable diminution and subsequent increase of distance has been accompanied by a corresponding and equally remarkable increase and subsequent diminution of relative angular motion. Thus, in the year 1783 the apparent angular motion hardly amounted to half a degree per annum, while in 1830 it had increased to 5°, in 1834 to 20°, in 1835 to 40°, and about the middle of 1836 to upwards of 70° per annum, or at the rate of a degree in five days. This is in entire conformity with the principles of Dynamics, which establish a necessary connexion between the angular velocity and the distance, as well in the apparent as in the real orbit of one body revolving about another under the influence of mutual attraction; the former varying inversely as the square of the latter, whatever be the curve described and whatever the law of the attractive force. It fortunately happens that Bradley, in 1718, had noticed and recorded in the margin of one of his observation books, the apparent direction of the line of junction of the two stars, as seen on the meridian in his transit telescope, viz., parallel to the line joining two conspicuous stars α and δ of the same constellation, as seen by the naked eye. This note, rescued from oblivion by the late Professor Rigaud, has proved of singular service in the verification of the elements above assigned to the orbit, which represent the whole series of recorded observations that date up to the end of 1846 (comprising an angular movement of nearly nine-tenths of a complete circuit), both in angle and distance, with a degree of exactness *fully equal to that of observation itself.* No doubt can, therefore, remain as to the prevalence in this remote system of the Newtonian law of gravitation.

(845.) The observations of ξ Ursæ Majoris are equally well represented by M. Mädler's elements (4 c of our table), thus fully justifying the assumption of the Newtonian law as that which regulates the motions of their binary systems. And even should it be the case, as M. Mädler appears to consider, that in one instance at least (that of p Ophiuchi), deviations from elliptic motion, too considerable to arise from mere error of observation, exist (a position we are by no means prepared to grant*), we should rather be disposed to look for the cause of such deviations in perturbations arising (as Bessel has suggested) from the large or central star itself being actually a close and hitherto unrecognized double star than in any defect of generality in the Newtonian law.

(846.) If the great length of the periods of some of these bodies be remarkable, the shortness of those of others is hardly less so. ζ Herculis has already completed two revo-

---

* p Ophiuchi belongs to the class of very unequal double stars, the magnitudes of the individuals being 4 and 7. Such stars present difficulties in the exact measurement of their angles of position which even yet continue to embarrass the observer, though, owing to later improvements in the art of executing such measurements, their influence is confined within much narrower limits than in the earlier history of the subject. In simply placing a fine single wire parallel to the line of junction of two such stars it is easily possible to commit an error of 3° or 4°. By placing them between two parallel thick wires such errors are in great measure obviated.

lutions since the epoch of its first discovery, exhibiting in
its course the extraordinary spectacle of a sidereal occultation,
the small star having twice been completely hidden behind
the large one. η Coronæ, ζ Cancri, and ξ Ursæ have each
performed more than one entire circuit, and 70 Ophiuchi and
γ Virginis have accomplished by far the larger portion of
one in angular motion. If any doubt, therefore, could remain
as to the reality of their orbitual motions, or any idea of explaining them by mere parallactic changes, or by any other
hypothesis than the agency of centripetal force, these facts
must suffice for their complete dissipation. We have the
same evidence, indeed, of their rotations about each other,
that we have of those of Uranus and Neptune about the
sun; and the correspondence between their calculated and
observed places in such very elongated ellipses, must be
admitted to carry with it proof of the prevalence of the
Newtonian law of gravity in their systems, of the very same
nature and cogency as that of the calculated and observed
places of comets round the central body of our own.

(847.) But it is not with the revolutions of bodies of a
planetary or cometary nature round a solar center that we
are now concerned; it is with that of sun round sun — each,
perhaps, at least in some binary systems where the individuals
are very remote and their period of revolution very long,
accompanied with its train of planets and *their* satellites,
closely shrouded from our view by the splendour of their
respective suns, and crowded into a space bearing hardly
a greater proportion to the enormous interval which separates
them, than the distances of the satellites of our planets from
their primaries bear to their distances from the sun itself.
A less distinctly characterized subordination would be incompatible with the stability of their systems, and with
the planetary nature of their orbits. Unless closely nestled
under the protecting wing of their immediate superior, the
sweep of their other sun in its perihelion passage round their
own might carry them off, or whirl them into orbits utterly
incompatible with the conditions necessary for the existence
of their inhabitants. It must be confessed, that we have

here a strangely wide and novel field for speculative excursions, and one which it is not easy to avoid luxuriating in.

(848.) The discovery of the parallaxes of α Centauri and 61 Cygni, both which are above enumerated among the "conspicuous" double stars of the 6th class (a distinction fully merited in the case of the former by the brilliancy of both its constituents), enables us to speak with an approach to certainty as to the absolute dimensions of both their orbits, and thence to form a probable opinion as to the general scale on which these astonishing systems are constructed.   The distance of the two stars of 61 Cygni subtends at the earth an angle which, since the earliest micrometrical measures in 1781, has varied hardly half a second from a mean value 15″·5.   On the other hand, the angle of position has altered since the same epoch by nearly 50°, so that it would appear probable that the true form of the orbit is not far from circular, its situation at right angles to the visual line, and its periodic time probably not short of 500 years.   Now, as the ascertained parallax of this star is 0″·348, which is, therefore, the angle the radius of the earth's orbit would subtend if equally remote, it follows that the mean distance between the stars is to that radius, as 15″·5 : 0″·348, or as 44·54 : 1. The orbit described by these two stars about each other undoubtedly, therefore, greatly exceeds in dimensions that described by Neptune about the sun.   Moreover, supposing the period to be five centuries (and the distance being actually on the increase, it can hardly be less) the general propositions laid down by Newton[*], taken in conjunction with Kepler's third law, enable us to calculate the sum of the masses of the two stars, which, on these data, we find to be 0·353, the mass of our sun being 1.   The sun, therefore, is neither vastly greater nor vastly less than the stars composing 61 Cygni.

(849.) The data in the case of α Centauri are more uncertain.   Since the year 1822, the distance has been steadily and pretty rapidly decreasing at the rate of about half a second per annum, and that with very little change in the

[*] Principia, l. i.   Prop. 57, 58, 59.

angle of position. Hence, it follows evidently that the plane of its orbit passes nearly through the earth, and (the distance about the middle of 1834 having been $17\frac{1}{2}''$) it is very probable that either an occultation, like that observed in ζ Herculis, on a close appulse of the two stars, will take place about the year 1867. As the observations we possess afford no sufficient grounds for a satisfactory calculation of elliptic elements, we must be content to assume what, at all events, they fully justify, viz., that the major semiaxis must exceed $12''$, and is very probably considerably greater. Now this with a parallax of $0''\cdot 913$ would give for the real value of the semiaxis $13\cdot 15$ radii of the earth's orbit, as a minimum. The real dimensions of their ellipse, therefore, cannot be so small as the orbit of Saturn; in all probability exceeds that of Uranus; and may possibly be much greater than either.

(850.) The parallel between these two double stars is a remarkable one. Owing no doubt to their *comparative* proximity to our system, their apparent proper motions are both unusually great, and for the same reason probably rather than owing to unusually large dimensions, their orbits appear to us under what, for binary double stars, we must call unusually large angles. Each consists, moreover, of stars, not very unequal in brightness, and in each both the stars are of a high yellow approaching to orange colour, the smaller individual, in each case, being also of a deeper tint. Whatever the diversity, therefore, which may obtain among other sidereal objects, these would appear to belong to the same family or genus. *

(851.) Many of the double stars exhibit the curious and beautiful phænomenon of contrasted or complementary colours. † In such instances, the larger star is usually of a

---

* Similar combinations are very numerous. Many remarkable instances occur among the double stars catalogued by the author in the 2nd, 3rd, 4th, 6th and 9th volumes of Trans. Roy. Ast. Soc. and in the volume of Southern observations already cited. See Nos. 121, 375, 1066, 1907, 2030, 2146, 2244, 2772, 3853, 3395, 3998, 4000, 4055, 4196, 4210, 4615, 4649, 4765, 5003, 5012, of these catalogues. The fine binary star, B. A. C. No. 4923, has its constituents $15''$ apart the one $6m$. yellow, the other $7m$. orange.

† " ———— other suns, perhaps,
　　 With their attendant moons thou wilt descry,

ruddy or orange hue, while the smaller one appears blue or green, probably in virtue of that general law of optics, which provides, that when the retina is under the influence of excitement by any bright, coloured light; feebler lights, which seen alone would produce no sensation but of whiteness, shall for the time appear coloured with the tint complementary to that of the brighter. Thus a yellow colour predominating in the light of the brighter star, that of the less bright one in the same field of view will appear blue; while, if the tint of the brighter star verge to crimson, that of the other will exhibit a tendency to green — or even appear as a vivid green, under favourable circumstances. The former contrast is beautifully exhibited by ι Cancri — the latter by γ Andromedæ*, both fine double stars. If, however, the coloured star be much the less bright of the two, it will not materially affect the other. Thus, for instance, η Cassiopeiæ exhibits the beautiful combination of a large white star, and a small one of a rich ruddy purple. It is by no means, however, intended to say, that in all such cases one of the colours is a mere effect of contrast, and it may be easier suggested in words, than conceived in imagination, what variety of illumination *two suns* — a red and a green, or a yellow and a blue one — must afford a planet circulating about either; and what charming contrasts and " grateful vicissitudes," — a red and a green day, for instance, alternating with a white one and with darkness, — might arise from the presence or absence of one or other, or both, above the horizon. Insulated stars of a red colour, almost as deep as that of blood†, occur in many parts of the heavens,

> Communicating male and female light,
> (Which two great sexes animate the world,)
> Stored in each orb, perhaps, with some that live."
> *Paradise Lost*, viii. 148.

* The small star of γ Andromedæ is close double. Both its individuals are green: a similar combination, with even more decided colours, is presented by the double star, h. 881.

† The following are the R. ascensions and N. P. distances for 1830, of some of the most remarkable of these sanguine or ruby stars : —

but no green or blue *star* (of any decided hue) has, we believe, ever been noticed unassociated with a companion brighter than itself.

(852.) Another very interesting subject of inquiry, in the physical history of the stars, is their proper motion. It was first noticed by Halley, that three principal stars, Sirius, Arcturus, and Aldebaran, are placed by Ptolomy, on the strength of observations made by Hipparchus, 130 years B.C., in latitudes respectively 20′, 22′, and 33′ more *northerly* than he actually found them in 1717.* Making due allowance for the diminution of obliquity of the ecliptic in the interval (1847 years) they ought to have stood, if really fixed, respectively 10′, 14′, and 0′ more *southerly*. As the circumstances of the statement exclude the supposition of error of transcription in the MSS., we are necessitated to admit a southward motion in latitude in these stars to the very considerable extent, respectively, of 37′, 42′, and 33′, and this is corroborated by an observation of Aldebaran at Athens, in the year A. D. 509, which star, on the 11th of March in that year, was seen immediately after its emergence from occultation by the moon, in such a position as it could not have had if the occultation were not nearly central. Now, from the knowledge we have of the lunar motions, this could not have been the case had Aldebaran at that time so much southern latitude as at present. *A priori*, it might be expected that apparent motions of some kind or other should be detected among so great a multitude of individuals scattered through space, and with nothing to keep them fixed. Their mutual attractions even, however inconceivably enfeebled by distance, and counteracted by opposing attractions from opposite quarters, must in the lapse of countless ages produce

R. A.	N. P. D.	R. A.	N. P. D.	R. A.	N. P. D.
h. m. s.	° ′ ″	h. m. s.	° ′ ″	h. m. s.	° ′ ″
4 40 53	61 46 21	10 52 10	107 24 40	20 7 8	111 50 11
5 38 29	136 32 15	12 37 31	148 45 47	21 37 18	31 59 47
9 27 56	152 2 48	16 29 44	122 2 0	21 37 20	52 54 47
9 48 31	130 47 12				

Of these No. 5. (in order of right ascension) is in the same field of view with α Hydræ, and No. 9. with β Crucis.

* Phil. Trans. 1717, vol. xxx. fo. 736.

some movements — some change of internal arrangement — resulting from the difference of the opposing actions. And it is a fact, that such apparent motions are really proved to exist by the exact observations of modern astronomy. Thus, as we have seen, the two stars of 61 Cygni have remained constantly at the same, or very nearly the same, distance, of 15″, for at least fifty years past, although they have shifted their local situation in the heavens, in this interval of time, through no less than 4′ 23″, the annual proper motion of each star being 5″·3; by which quantity (exceeding a third of their interval) this system is every year carried bodily along in some unknown path, by a motion which, for many centuries, must be regarded as uniform and rectilinear. Among stars not double, and no way differing from the rest in any other obvious particular, ε Indi* and μ Cassiopeiæ are to be remarked as having the greatest proper motions of any yet ascertained, amounting respectively to 7″·74 and 3″·74 of annual displacement. And a great many others have been observed to be thus constantly carried away from their places by smaller, but not less unequivocal motions. †

(853.) Motions which require whole centuries to accumulate before they produce changes of arrangement, such as the naked eye can detect, though quite sufficient to destroy that idea of mathematical fixity which precludes speculation, are yet too trifling, as far as practical applications go, to induce a change of language, and lead us to speak of the stars in common parlance as otherwise than fixed. Small as they are, however, astronomers, once assured of their reality, have not been wanting in attempts to explain and reduce them to general laws. No one, who reflects with due attention on the subject, will be inclined to deny the high probability, nay certainty, that the *sun* as well as the stars must have a proper motion in *some* direction; and the inevitable consequence of such a motion, if unparticipated by the rest, must

---

* D'Arrest. Astr. Nachr., No. 618.

† The reader may consult " a list of 314 stars having, or supposed to have, a proper motion of not less than about 0″·5 of a great circle " (*per annum*) by the late F. Baily, Esq. *Trans. Ast. Soc.* v. p. 158.

be a slow *average* apparent tendency of all the stars to the vanishing point of lines parallel to that direction, and to the region which he is leaving, however greatly individual stars might differ from such average by reason of their own peculiar proper motion. This is the necessary effect of perspective; and it is certain that it must be detected by observation, if we knew accurately the apparent proper motions of all the stars, and if we were sure that they were independent, *i.e.* that the whole firmament, or at least all that part which we see in our own neighbourhood, were not drifting along together, by a general *set* as it were, in one direction, the result of unknown processes and slow internal changes going on in the sidereal stratum to which our system belongs, as we see motes sailing in a current of air, and keeping nearly the same relative situation with respect to one another.

(854.) It was on this assumption, tacitly made indeed, but necessarily implied in every step of his reasoning, that Sir William Herschel, in 1783, on a consideration of the apparent proper motions of such stars as could at that period be considered as tolerably (though still imperfectly) ascertained, arrived at the conclusion that a relative motion of the sun, among the fixed stars in the direction of a point or parallactic apex, situated near λ Herculis, that is to say, in R. A. 17^h $22^m = 260°$ $34'$, N.P.D. $63°$ $43'$ (1790), would account for the chief observed apparent motions, leaving, however, some still outstanding and not explicable by this cause; and in the same year Prevost, taking nearly the same view of the subject, arrived at a conclusion as to the solar apex (or point of the sphere towards which the sun relatively advances), agreeing nearly in polar distance with the foregoing, but differing from it about 27° in right ascension. Since that time methods of calculation have been improved and concinnated, our knowledge of the proper motions of the stars has been rendered more precise, and a greater number of cases of such motions have been recorded. The subject has been resumed by several eminent astronomers and mathematicians: viz. 1st, by M. Argelander, who, from the consideration of the proper motions of 21 stars exceeding

1″ per annum in arc, has placed the solar apex in R. A. 256° 25′, N. P. D. 51° 23′; from those of 50 stars between 0″·5 and 1″·0, in 255° 10′, 51° 26′; and from those of 319 stars having motions between 0″·1 and 0″·5 per annum, in 261° 11′, 59° 2′: 2ndly, by M. Luhndahl, whose calculations, founded on the proper motions of 147 stars, give 252° 53′, 75° 34′: and 3rdly, by M. Otto Struve, whose result 261° 22′, 62° 24′, emerges from a very elaborate discussion of the proper motions of 392 stars. All these places are for A. D. 1790.

(855.) The most probable mean of the results obtained by these three astronomers, is (for the same epoch) R. A. = 259° 9′, N. P. D. 55° 23′. Their researches, however, extending only to stars visible in European observatories, it became a point of high interest to ascertain how far the stars of the southern hemisphere not so visible, treated independently on the same system of procedure, would corroborate or controvert their conclusion. The observations of Lacaille, at the Cape of Good Hope, in 1751 and 1752, compared with those of Mr. Johnson at St. Helena, in 1829–33, and of Henderson at the Cape in 1830 and 1831, have afforded the means of deciding this question. The task has very recently been executed in a masterly manner by Mr. Galloway, in a paper published in the Philosophical Transactions for 1847 (to which we may also refer the reader for a more particular account of the history of the subject than our limits allow us to give.) On comparing the records, Mr. Galloway finds eighty-one southern stars not employed in the previous investigations above referred to, whose proper motions in the intervals elapsed appear considerable enough to assure us that they have not originated in error of the earlier observations. Subjecting these to the same process of computation he concludes for the place of the solar apex, for 1790, as follows: viz. R. A. 260° 1′, N. P. D. 55° 37′, a result so nearly identical with that afforded by the northern hemisphere, as to afford a full conviction of its near approach to truth, and what may fairly be considered a demonstration of the physical cause assigned.

(856.) Of the mathematical conduct of this inquiry the nature of this work precludes our giving any account; but as the philosophical principle on which it is based has been misconceived, it is necessary to say a few words in explanation of it.   Almost all the greatest discoveries in astronomy have resulted from the consideration of what we have elsewhere termed RESIDUAL PHÆNOMENA*, of a quantitative or numerical kind, that is to say, of such portions of the numerical or quantitative results of observation as remain outstanding and unaccounted for after subducting and allowing for all that would result from the strict application of known principles.   It was thus that the grand discovery of the precession of the equinoxes resulted as a residual phænomenon, from the imperfect explanation of the return of the seasons by the return of the sun to the same apparent place among the fixed stars.   Thus, also, aberration and nutation resulted as residual phænomena from that portion of the changes of the apparent places of the fixed stars which was left unaccounted for by precession.   And thus again the *apparent* proper motions of the stars are the observed *residues* of their apparent movements outstanding and unaccounted for by strict calculation of the effects of precession, nutation, and aberration.   The nearest approach which human theories can make to perfection is to diminish this residue, this *caput mortuum* of observation, as it may be considered, as much as practicable, and, if possible, to reduce it to nothing, either by showing that something has been neglected in our estimation of known causes, or by reasoning upon it as a new fact, and on the principle of the inductive philosophy ascending from the effect to its cause or causes.   On the suggestion of any new cause hitherto unresorted to for its explanation, our first object must of course be to decide whether such a cause would produce *such* a result *in kind:* the next, to assign to it such an intensity as shall account for the greatest possible amount of the residual matter in hand.   The proper motion of the sun being suggested as such a cause, we have two

---

* Discourse on the Study of Natural Philosophy.   *Cab. Cyclopædia*, No. 14.

things disposable — its direction and velocity, both which it is evident, if they ever became known to us at all, can only be so by the consideration of the very phænomenon in question.   Our object, of course, is to account, if possible, for the *whole* of the observed proper motions by the proper assumption of these elements.   If this be impracticable, what remains unaccounted for is a residue of a more recondite kind, but which, so long as it *is* unaccounted for, we must regard as purely casual, seeing that, for anything we can perceive to the contrary, it might with equal probability be one way as the other.   The theory of chances, therefore, necessitates (as it does in all such cases) the application of a general mathematical process, known as " the method of least squares," which leads, as a matter of strict geometrical conclusion, to the values of the elements sought, *which, under all the circumstances, are the most probable.*

(857.) This is the process resorted to by all the geometers we have enumerated in the foregoing articles (art. 854,855). It gives not only the direction in space, but also the *velocity* of the solar motion, estimated on a scale conformable to that in which the velocity of the sidereal motions to be explained are given ; *i.e.* in seconds of arc as subtended at the average distance of the stars concerned, by its annual motion in space. But here a consideration occurs which tends materially to complicate the problem, and to introduce into its solution an element depending on suppositions more or less arbitrary. The distance of the stars being, except in two or three instances, unknown, we are compelled either to restrict our inquiry to *these*, which are too few to ground any result on, or to make some supposition as to the relative distances of the several stars employed.   In this we have nothing but general probability to guide us, and two courses only present themselves, either, 1st, To class the distances of the stars according to their magnitudes, or apparent brightnesses, and to institute separate and independent calculations for each class, including stars assumed to be equidistant, or nearly so : or, 2dly, To class them according to the observed amount of their apparent proper motions, on the presumption that those which appear to move fastest are really nearest to us. The former is the course pursued by M. Otto Struve, the latter by M. Argelander. With regard to this latter principle of classification, however, two considerations interfere with its applicability, viz. 1st, that we see the *real* motion of the stars foreshortened by the effect of perspective; and 2dly, that that portion of the total *apparent* proper motion which arises from the real motion of the sun depends, not simply on the absolute distance of the star from the sun, but also on its angular apparent distance from the solar apex, being, *cæteris paribus,* as the sine of that angle. To execute such a classification correctly, therefore, we ought to know both these particulars for each star. The first is evidently out of our reach. We are therefore, for that very reason, compelled to regard it as casual, and to assume that on the average of a great number of stars it would be uninfluential on the result. But the second cannot be so summarily disposed of. By the aid of an approximate knowledge of the solar apex, it is true, approximate values may be found of the simply apparent portions of the proper motions, supposing all the stars equidistant, and these being subducted from the total observed motions, the residues might afford ground for the classification in question.* This, however, would be a long, and to a certain extent precarious system of procedure. On the other hand, the classification by apparent brightness is open to no such difficulties, since we are fully justified in assuming that, on a general average, the brighter stars are the nearer, and that the exceptions to this rule are casual in that sense of the word which it always bears in such inquiries, expressing solely our ignorance of any ground for assuming a bias one way or other on either side of a determinate numerical rule. In Mr. Galloway's discussion of the southern stars the consideration of distance is waived altogether, which is equivalent to an admission of complete ignorance on this point, as well

---

* M. Argelander's classes, however, are constructed without reference to this consideration, on the sole basis of the total apparent amount of proper motion, and are, therefore, *pro tanto*, questionable. It is the more satisfactory then to find so considerable an agreement among his partial results as actually obtains.

as respecting the real directions and velocities of the individual motions.

(858.) The velocity of the solar motion which results from M. Otto Struve's calculations is such as would carry it over an angular subtense of 0″·3392 if seen at right angles from the average distance of a star of the first magnitude. If we take, with M. Struve, senior, the parallax of such a star as probably equal to 0″·209*, we shall at once be enabled to compare this annual motion with the radius of the earth's orbit, the result being 1·623 of such units. The sun then advances through space (relatively, at least, among the stars), carrying with it the whole planetary and cometary system with a velocity of 1·623 radii of the earth's orbit, or 154,185,000 miles *per annum*, or 422,000 miles (that is to say, nearly its own semi-diameter) *per diem*: in other words, with a velocity a very little greater than one-fourth of the earth's annual motion in its orbit.

(859.) Another generation of astronomers, perhaps many, must pass away before we are in a condition to decide from a more precise and extensive knowledge of the proper motions of the stars than we at present possess, how far the direction and velocity above assigned to the solar motion deviates from exactness, whether it continue uniform, and whether it show any sign of deflection from rectilinearity; so as to hold out a prospect of one day being enabled to trace out an arc of the solar orbit, and to indicate the direction in which the preponderant gravitation of the sidereal firmament is urging the central body of our system. An analogy for such deviation from uniformity would seem to present itself in the alleged existence of a similar deviation in the proper motions of Sirius and Procyon, both which stars were considered, up to a very recent period, and on very high astronomical authority, to have varied sensibly in this respect within the limits of authentic and dependable observation. Such, indeed, appeared to be the amount of evidence for this as a matter of *fact* as to give rise to a speculation on the probable circulation of these stars round opaque (and therefore invisible) bodies

* Etudes d'Astronomie Stellaire, p. 107.

at no great distances from them respectively, in the manner
of binary stars.   M. Struve, however (in his work already so
frequently cited), has destroyed this conclusion by instituting
a most searching and rigorous inquiry into all the circumstances of either case, and has succeeded in demonstrating
that the supposed anomalies have arisen solely from the
effects of instrumental error and imperfect determination of
the coefficients of the uranographical corrections.

(860.) The whole of the reasoning upon which the determination of the solar motion in space rests, is based upon the
entire exclusion of any *law* either derived from observation
or assumed in theory, affecting the amount and direction of
the real motions both of the sun and stars.   It supposes an
absolute non-recognition, in those motions, of any general
directive cause, such as, for example, a common circulation
of all about a common center.   Any such limitation introduced into the conditions of the problem of the solar motion
would alter *in toto* both its nature and the form of its solution.
Suppose for instance that, conformably to the speculations of
several astronomers, the whole system of the Milky Way,
including our sun, and the stars, our more immediate neighbours, which constitute our sidereal firmament, should have a
general movement of rotation in the plane of the galactic circle
(any other would be exceedingly improbable, indeed hardly
reconcilable with dynamical principles), being held together in
opposition to the centrifugal force thus generated by the mutual
gravitation of its constituent stars.   Except we at the same
time admitted that the scale on which this movement proceeds is so enormous that all the stars whose proper motions
we include in our calculations go together in a body, so far
as that movement is concerned (as forming too small an in-
tegrant portion of the whole to differ sensibly in their relation to its central point); we stand precluded from drawing
any conclusion whatever, not only respecting the absolute
motion of the sun, but respecting even its relative movement
among those stars, until we have established some law, or at
all events framed some hypothesis having the provisional force
of a law, connecting the whole, or a part of the motion of
each individual with its situation in space.

(861.) Speculations of this kind have not been wanting in astronomy, and recently an attempt has been made by M. Mädler to assign the local center in space, round which the sun and stars revolve, which he places in the group of the Pleiades, a situation in itself improbable, lying as it does no less than 26° out of the plane of the galactic circle, out of which plane it is almost inconceivable that any *general* circulation can take place. In the present defective state of our knowledge respecting the proper motion of the smaller stars, especially in right ascension, (an element for the most part far less exactly ascertainable than the polar distance, or at least which has been hitherto far less accurately ascertained,) we cannot but regard all attempts of the kind as to a certain extent premature, though by no means to be discouraged as forerunners of something more decisive. The question, as a matter of fact, whether a rotation of the galaxy in its own plane exist or not might be at once resolved by the assiduous observation both in R. A. and polar distance of a considerable number of stars of the Milky Way, judiciously selected for the purpose, and *including all magnitudes*, down to the smallest distinctly identifiable, and capable of being observed with normal accuracy: and we would recommend the inquiry to the special attention of directors of permanent observatories, provided with adequate instrumental means, in both hemispheres. Thirty or forty years of observation perseveringly directed to the object in view, could not fail to settle the question.*

(862.) The solar motion through space, if real and not simply relative, must give rise to uranographical corrections analogous to parallax and aberration. The solar or systematic parallax is no other than that part of the proper motion of each star which is simply apparent, arising from the sun's motion, and until the distances of the stars be known, must

---

* An examination of the proper motions of the stars of the B. Assoc. Catal. in the portion of the Milky Way nearest either pole (where the motion should be almost wholly in R A) indicates no distinct symptom of such a rotation. If the question be taken up fundamentally, it will involve a redetermination from the recorded proper motions, both of the precession of the equinoxes and the change of obliquity of the ecliptic.

remain inextricably mixed up with the other or real portion. The systematic aberration, amounting at its maximum (for stars 90° from the solar apex to about 5″) displaces all the stars in great circles diverging from that apex through angles proportional to the sines of their respective distances from it. This displacement, however, is permanent, and therefore uncognizable by any phænomenon, so long as the solar motion remains invariable; but should it, in the course of ages, alter its direction and velocity, both the direction and amount of the displacement in question would alter with it. The change, however, would become mixed up with other changes in the apparent proper motions of the stars, and it would seem hopeless to attempt disentangling them.

(863.) A singular, and at first sight paradoxical effect of the progressive movement of light, combined with the proper motion of the stars, is, that it alters the apparent periodic time in which the individuals of a binary star circulate about each other.* To make this apparent, suppose them to circulate round each other in a plane perpendicular to the visual ray in a period of 10,000 days. Then if both the sun and the center of gravity of the binary system remained fixed in space, the relative apparent situation of the stars would be exactly restored to its former state after the lapse of this interval, and if the angle of position were 0° at first, after 10,000 days it would again be so. But now suppose that the center of gravity of the star were in the act of receding in a direct line from the sun with a velocity of one-tenth part of the radius of the earth's orbit *per diem.* Then at the expiration of 10,000 days it would be more remote from us by 1000 such radii, a space which light would require 57 days to traverse. Although really, therefore, the stars would have arrived at the position 0° at the exact expiration of 10,000 days, it would require 57 days more for the notice of that fact to reach our system. In other words, the period would appear to us to be 10,057 days, since we could only conclude the period to be completed when to us as observers the original angle of position was again restored. A contrary motion would produce a contrary effect.

* Astronomische Nachrichten, No. 520.

CHAPTER XVII.

OF CLUSTERS OF STARS AND NEBULÆ.

OF CLUSTERING GROUPS OF STARS. — GLOBULAR CLUSTERS. — THEIR
STABILITY DYNAMICALLY POSSIBLE. — LIST OF THE MOST REMARK-
ABLE. — CLASSIFICATION OF NEBULÆ AND CLUSTERS. — THEIR
DISTRIBUTION OVER THE HEAVENS. — IRREGULAR CLUSTERS. —
RESOLVABILITY OF NEBULÆ. — THEORY OF THE FORMATION OF
CLUSTERS BY NEBULOUS SUBSIDENCE. — OF ELLIPTIC NEBULÆ.
— THAT OF ANDROMEDA. — ANNULAR AND PLANETARY NEBULÆ.
—DOUBLE NEBULÆ. — NEBULOUS STARS.—CONNEXION OF NEBULÆ
WITH DOUBLE STARS. — INSULATED NEBULÆ OF FORMS NOT
WHOLLY IRREGULAR. — OF AMORPHOUS NEBULÆ. — THEIR LAW
OF DISTRIBUTION MARKS THEM AS OUTLIERS OF THE GALAXY.
— NEBULÆ, AND NEBULOUS GROUP OF ORION — OF ARGO — OF
SAGITTARIUS — OF CYGNUS. — THE MAGELLANIC CLOUDS. — SIN-
GULAR NEBULA IN THE GREATER OF THEM. — THE ZODIACAL
LIGHT. — SHOOTING STARS.

(864.) WHEN we cast our eyes over the concave of the
heavens in a clear night, we do not fail to observe that here
and there are groups of stars which seem to be compressed
together in a more condensed manner than in the neighbour-
ing parts, forming bright patches and clusters, which attract
attention, as if they were there brought together by some
general cause other than casual distribution. There is a
group, called the Pleiades, in which six or seven stars may be
noticed, if the eye be directed full upon it; and many more
if the *eye be turned carelessly aside,* while *the attention* is kept
directed* upon the group. Telescopes show fifty or sixty

---

* It is a very remarkable fact, that the center of the visual area is far less
sensible to feeble impressions of light, than the exterior portions of the retina.
Few persons are aware of the extent to which this comparative insensibility
extends, previous to trial. To estimate it, let the reader look alternately
full at a star of the fifth magnitude, and beside it; or choose two, equally
bright, and about 3° or 4° apart, and look full at one of them, the probability
is, he will see *only the other.* The fact accounts for the multitude of stars with
which we are impressed by a general view of the heavens; their paucity when we
come to count them.

large stars thus crowded together in a very moderate space, comparatively insulated from the rest of the heavens. The constellation called Coma Berenices is another such group, more diffused, and consisting on the whole of larger stars.

(865.) In the constellation Cancer, there is a somewhat similar, but less definite, luminous spot, called Præsepe, or the bee-hive, which a very moderate telescope, — an ordinary night-glass for instance, — resolves entirely into stars. In the sword-handle of Perseus, also, is another such spot, crowded with stars, which requires rather a better telescope to resolve into individuals separated from each other. These are called clusters of stars; and, whatever be their nature, it is certain that other laws of aggregation subsist in these spots, than those which have determined the scattering of stars over the general surface of the sky. This conclusion is still more strongly pressed upon us, when we come to bring very powerful telescopes to bear on these and similar spots. There are a great number of objects which have been mistaken for comets, and, in fact, have very much the appearance of comets without tails: small round, or oval nebulous specks, which telescopes of moderate power only show as such. Messier has given, in the *Connois. des Temps* for 1784, a list of the places of 103 objects of this sort; which all those who search for comets ought to be familiar with, to avoid being misled by their similarity of appearance. That they are not, however, comets, their fixity sufficiently proves; and when we come to examine them with instruments of great power, — such as reflectors of eighteen inches, two feet, or more in aperture, — any such idea is completely destroyed. They are then, for the most part, perceived to consist entirely of stars crowded together so as to occupy almost a definite outline, and to run up to a blaze of light in the centre, where their condensation is usually the greatest. (See *fig.* 1. pl. II., which represents (somewhat rudely) the thirteenth nebula of Messier's list (described by him as *nébuleuse sans étoiles*), as seen in a reflector of 18 inches aperture and 20 feet focal length.) Many of them, indeed, are of an exactly round figure, and convey the complete idea of a globular space filled full of stars, insulated in the heavens, and constituting in itself a family or society apart from the rest, and subject only to its own internal laws. It would be a vain task to attempt to count the stars in one of these *globular clusters.* They are not to be reckoned by hundreds; and on a rough calculation, grounded on the apparent intervals between them at the borders, and the angular diameter of the whole group, it would appear that many clusters of this description must contain, at least, five thousand stars, compacted and wedged together in a round space, whose angular diameter does not exceed eight or ten minutes; that is to say, in an area not more than a tenth part of that covered by the moon.

(866.) Perhaps it may be thought to savour of the gigantesque to look upon the individuals of such a group as suns like our own, and their mutual distances as equal to those which separate our sun from the nearest fixed star: yet, when we consider that their united lustre affects the eye with a less impression of light than a star of the fourth magnitude, (for the largest of these clusters is barely visible to the naked eye,) the idea we are thus compelled to form of their distance from us may prepare us for almost any estimate of their dimensions. At all events, we can hardly look upon a group thus insulated, thus *in seipso totus, teres, atque rotundus,* as not forming a system of a peculiar and definite character. Their round figure clearly indicates the existence of some general bond of union in the nature of an attractive force; and, in many of them, there is an evident acceleration in the rate of condensation as we approach the center, which is not referable to a merely uniform distribution of equidistant stars through a globular space, but marks an intrinsic *density* in their state of aggregation, greater in the center than at the surface of the mass. It is difficult to form any conception of the dynamical state of such a system. On the one hand, without a rotatory motion and a centrifugal force, it is hardly possible not to regard them as in a state of progressive collapse. On the other, granting such a motion and such a force, we find it no less difficult to reconcile the apparent sphericity of their form with a rotation of the whole system

Q Q

round any single axis, without which internal collisions might
at first sight appear to be inevitable. If we suppose a
globular space filled with equal stars, uniformly dispersed
through it, and very numerous, each of them attracting
every other with a force inversely as the square of the
distance, the resultant force by which any one of them (those
at the surface alone excepted) will be urged, in virtue of
their joint attractions, will be directed towards the common
center of the sphere, and will be directly as the distance
therefrom. This follows from what Newton has proved of
the *internal* attraction of a homogeneous sphere. (See also
note on Art. 735.) Now, under such a law of force, each
particular star would describe a perfect ellipse about the
common center of gravity as its center, and *that*, in whatever
plane and whatever direction it might revolve. The condition, therefore, of a rotation of the cluster, as a mass, about
a single axis would be unnecessary. Each ellipse, whatever
might be the proportion of its axis, or the inclination of its
plane to the others, would be invariable *in every particular*,
and all would be described in one common period, so that at
the end of every such period, or *annus magnus* of the system,
every star of the cluster (except the superficial ones) would
be exactly re-established in its original position, thence to set
out afresh, and run the same unvarying round for an indefinite succession of ages. Supposing their motions, therefore, to be so adjusted at any one moment as that the orbits
should not intersect each other, and so that the magnitude of
each star, and the sphere of its more intense attraction, should
bear but a small proportion to the distance separating the
individuals, such a system, it is obvious, might subsist, and
realize, in great measure, that abstract and ideal harmony,
which Newton, in the 89th Proposition of the First Book
of the *Principia*, has shown to characterize a law of force
directly as the distance.*

(867.) The following are the places, for 1830, of the
principal of these remarkable objects, as specimens of their
class: —

* See also *Quarterly Review*, No. 94. p. 540.

R. A.			N. P. D.		R. A.			N. P. D.		R. A.			N. P. D.	
h.	m.	s.	°	'	h.	m.	s.	°	'	h.	m.	s.	°	'
0	16	25	163	2	15	9	56	87	16	17	26	51	143	34
9	8	33	154	10	15	34	56	127	13	17	28	42	93	8
12	47	41	159	57	16	6	55	112	33	11	26	4	114	2
13	4	30	70	55	16	23	2	102	40	18	55	49	150	14
13	16	38	136	35	16	35	37	53	13	21	21	43	78	34
13	34	10	60	46	16	50	24	119	51	21	24	40	91	34

Of these, by far the most conspicuous and remarkable is ω *Centauri* the fifth of the list in order of Right Ascension. It is visible to the naked eye as a dim round cometic object about equal to a star 4·5 m., though probably if concentered in a single point, the impression on the eye would be much greater. Viewed in a powerful telescope it appears as a globe of fully 20' in diameter, very gradually increasing in brightness to the center, and composed of innumerable stars of the 13th and 15th magnitudes (the former probably being two or more of the latter closely juxtaposed). The 11th in order of the list (R. A. 16^h 35^m) is also visible to the naked eye in *very* fine nights, between η and ζ Herculis, and is a superb object in a large telescope. Both were discovered by Halley, the former in 1677, and the latter in 1714.

(868.) It is to Sir William Herschel that we owe the most complete analysis of the great variety of those objects which are generally classed under the common head of Nebulæ, but which have been separated by him into—1st. Clusters of stars, in which the stars are clearly distinguishable; and these, again, into globular and irregular clusters; 2d. Resolvable nebulæ, or such as excite a suspicion that they consist of stars, and which any increase of the optical power of the telescope may be expected to resolve into distinct stars; 3d. Nebulæ, properly so called, in which there is no appearance whatever of stars; which, again, have been subdivided into subordinate uses, according to their brightness and size; 4th. Planetary nebulæ; 5th. Stellar nebulæ; and, 6th. Nebulous stars. The great power of his telescopes disclosed the existence of an immense number of these objects before unknown, and showed them to be distributed over the heavens, not by any means uniformly, but with a marked preference to a certain district, extending over the northern pole of the galactic circle, and occupying the constellations Leo, Leo Minor, the body, tail, and hind legs of Ursa Major, Canes Venatici, Coma Berenices, the preceding leg of Bootes, and the head, wings, and shoulder of Virgo. In this region, occupying about one-eighth of the whole surface of the sphere, one-third of the entire nebulous contents of the heavens are congregated. On the other hand, they are very sparingly scattered over the constellations Aries, Taurus, the head and shoulders of Orion, Auriga, Perseus, Camelopardalus, Draco, Hercules, the northern part of Serpentarius, the tail of Serpens, that of Aquila, and the whole of Lyra. The hours 3, 4, 5, and 16, 17, 18, of right ascension in the northern hemisphere are singularly poor, and, on the other hand, the hours 10, 11, and 12 (but especially 12), extraordinarily rich in these objects. In the southern hemisphere a much greater uniformity of distribution prevails, and with exception of two very remarkable centers of accumulation, called the Magellanic clouds (of which more presently), there is no very decided tendency to their assemblage in any particular region.

(869.) Clusters of stars are either globular, such as we have already described, or of irregular figure. These latter are, generally speaking, less rich in stars, and especially less condensed towards the center. They are also less definite in outline; so that it is often not easy to say where they terminate, or whether they are to be regarded otherwise than as merely richer parts of the heavens than those around them. Many, indeed the greater proportion of them, are situated in or close on the borders of the Milky Way. In some of them the stars are nearly all of a size, in others extremely different; and it is no uncommon thing to find a very red star much brighter than the rest, occupying a conspicuous situation in them. Sir William Herschel regards these as globular clusters in a less advanced state of condensation, conceiving all such groups as approaching, by their mutual attraction, to the globular figure, and assembling themselves together from all the surrounding region, under laws of which we have, it is true, no other proof than the observance of a gradation by which their characters shade into one another, so that it is impossible to say where one species ends and the other begins. Among the most beautiful objects of this class is that which surrounds the star κ Crucis, set down as a nebula by Lacaille. It occupies an area of about one 48th part of a square degree, and consists of about 110 stars from the 7th magnitude downwards, eight of the more conspicuous of which are coloured with various shades of red, green, and blue, so as to give to the whole the appearance of a rich piece of jewellery.

(870.) Resolvable nebulæ can, of course, only be considered as clusters either too remote, or consisting of stars intrinsically too faint to affect us by their individual light, unless where two or three happen to be close enough to make a joint impression, and give the idea of a point brighter than the rest. They are almost universally round or oval—their loose appendages, and irregularities of form, being as it were extinguished by the distance, and the only general figure of the more condensed parts being discernible. It is under the appearance of objects of this character that all the greater globular clusters exhibit themselves in telescopes of insufficient optical power to show them well; and the conclusion is obvious, that those which the most powerful can barely render resolvable, and even those which, with such powers as are usually applied, show no sign of being composed of stars, would be completely resolved by a further increase of optical power. In fact, this probability has almost been converted into a certainty by the magnificent reflecting telescope constructed by Lord Rosse, of six feet in aperture, which has resolved or rendered resolvable multitudes of nebulæ which had resisted all inferior powers. The sublimity of the spectacle afforded by that instrument of some of the larger globular and other clusters enumerated in the list given in Art. 867. is declared by all who have witnessed it to be such as no words can express.

(871.) Although, therefore, nebulæ do exist, which even in

this powerful telescope appear *as* nebulæ, without any sign of resolution, it may very reasonably be doubted whether there be really any essential physical distinction between nebulæ and clusters of stars, at least in the nature of the matter of which they consist, and whether the distinction between such nebulæ as are easily resolved, barely resolvable with excellent telescopes, and altogether irresolvable with the best, be any thing else than one of degree, arising merely from the excessive minuteness and multitude of the stars, of which the latter, as compared with the former, consist. The first impression which Halley, and other early discoverers of nebulous objects received from their peculiar aspect, so different from the keen, concentrated light of mere stars, was that of a phosphorescent vapour (like the matter of a comet's tail) or a gaseous and (so to speak) elementary form of luminous sidereal matter.* Admitting the existence of such a medium, dispersed in some cases irregularly through vast regions in space, in others confined to narrower and more definite limits, Sir W. Herschel was led to speculate on its gradual subsidence and condensation by the effect of its own gravity, into more or less regular spherical or spheroidal forms, denser (as they must in that case be) towards the center. Assuming that in the progress of this subsidence, local centers of condensation, subordinate to the general tendency, would not be wanting, he conceived that in this way solid nuclei might arise, whose local gravitation still further condensing, and so absorbing the nebulous matter, each in its immediate neighbourhood, might ultimately become stars, and the whole nebulæ finally take on the state of a cluster of stars. Among the multitude of nebulæ revealed by his telescopes, every stage of this process might be considered as displayed to our eyes, and in every modification of form to which the general principle might be conceived to apply. The more or less advanced state of a nebula towards its segregation into discrete stars, and of these stars themselves towards a denser state of aggregation round a central nucleus, would thus be in some sort an indication of age.

* Halley, Phil. Trans., xxix. p. 390.

Neither is there any variety of aspect which nebulæ offer, which stands at all in contradiction to this view. Even though we should feel ourselves compelled to reject the idea of a gaseous or vaporous "nebulous matter," it loses little or none of its force. Subsidence, and the central aggregation consequent on subsidence, may go on quite as well among a multitude of discrete bodies under the influence of mutual attraction, and feeble or partially opposing projectile motions, as among the particles of a gaseous fluid.

(872.) The "*nebular hypothesis*," as it has been termed, and the *theory of sidereal aggregation* stand, in fact, quite independent of each other, the one as a physical conception of processes which may yet, for aught we know, have formed part of that mysterious chain of causes and effects antecedent to the existence of separate self-luminous solid bodies; the other, as an application of dynamical principles to cases of a very complicated nature no doubt, but in which the possibility or impossibility, at least, of certain general results may be determined on perfectly legitimate principles. Among a crowd of solid bodies of whatever size, animated by independent and partially opposing impulses, motions opposite to each other *must* produce collision, destruction of velocity, and subsidence or near approach towards the center of preponderant attraction; while those which conspire, or which remain outstanding after such conflicts, *must* ultimately give rise to circulation of a permanent character. Whatever we may think of such collisions as events, there is nothing in this conception contrary to sound mechanical principles. It will be recollected that the appearance of central condensation among a multitude of separate bodies in motion, by no means implies permanent proximity to the center in each; any more than the habitually crowded state of a market place, to which a large proportion of the inhabitants of a town must frequently or occasionally resort, implies the permanent residence of each individual within its area. It is a fact that clusters thus centrally crowded do exist, and therefore the conditions of their existence must be dynamically possible, and in what has been said we may at least perceive

Q Q 4

some glimpses of the manner in which they are so. The actual intervals between the stars, even in the most crowded parts of a resolved nebula, to be seen at all by us, must be enormous. Ages, which to us may well appear indefinite, may easily be conceived to pass without a single instance of collision, in the nature of a catastrophe. Such may have gradually become rarer as the system has emerged from what must be considered its chaotic state, till at length, in the fulness of time, and under the pre-arranging guidance of that DESIGN which pervades universal nature, each individual may have taken up such a course as to annul the possibility of further destructive interference.

(873.) But to return from the regions of speculation to the description of facts. Next in regularity of form to the globular clusters, whose consideration has led us into this digression, are elliptic nebulæ, more or less elongated. And of these it may be generally remarked, as a fact undoubtedly connected in some very intimate manner with the dynamical conditions of their subsistence, that such nebulæ are, for the most part, beyond comparison more difficult of resolution than those of globular form. They are of all degrees of excentricity, from moderately oval forms to ellipses so elongated as to be almost linear, which are, no doubt, edge-views of very flat ellipsoids. In all of them the density increases towards the centre, and as a general law it may be remarked that, so far as we can judge from their telescopic appearance, their internal strata approach more nearly to the spherical form than their external. Their resolvability, too, is greater in the central parts, whether owing to a real superiority of size in the central stars or to the greater frequency of cases of close juxta-position of individuals, so that two or three united appear as one. In some the condensation is slight and gradual, in others great and sudden: so sudden, indeed, as to offer the appearance of a dull and blotted star, standing in the midst of a faint, nearly equable elliptic nebulosity, of which two remarkable specimens occur in R. A. 12^h 10^m 33^s, N. P. D. 41° $46'$, and in 13^h 27^m 28^s, 119° $0'$ (1830).

(874.) The largest and finest specimens of elliptic nebulæ which the heavens afford are that in the girdle of Andromeda (near the star ν of that constellation) and that discovered in 1783, by Miss Carolina Herschel, in R. A. 0^h 39^m 12^s, N.P.D. 116° 13′. The nebula in Andromeda (Plate II. fig. 3.) is visible to the naked eye, and is continually mistaken for a comet by those unacquainted with the heavens. Simon Marius, who noticed it in 1612 (though it appears also to have been seen and described as oval, in 995), describes its appearance as that of a candle shining through horn, and the resemblance is not inapt. Its form, as seen through ordinary telescopes, is a pretty long oval, increasing by insensible gradations of brightness, at first very gradually, but at last more rapidly, up to a central point, which, though very much brighter than the rest, is decidedly not a star, but nebula of the same general character with the rest in a state of extreme condensation. Casual stars are scattered over it, but with a reflector of 18 inches in diameter, there is nothing to excite any suspicion of its consisting of stars. Examined with instruments of superior defining power, however, the evidence of its resolvability into stars, may be regarded as decisive. Mr. G. P. Bond, assistant at the observatory of Cambridge, U.S., describes and figures it as extending nearly $2\frac{1}{2}°$ in length, and upwards of a degree in breadth (so as to include two other smaller adjacent nebulæ), of a form, generally speaking, oval, but with a considerably protuberant irregularity at its north following extremity, very suddenly condensed at the nucleus almost to the semblance of a star, and though not itself clearly resolved, yet thickly sown over with visible minute stars, so numerous as to allow of 200 being counted within a field of 20′ diameter in the richest parts. But the most remarkable feature in his description is that of two perfectly straight, narrow, and comparatively or totally obscure streaks which run nearly the whole length of one side of the nebula, and (though slightly divergent from each other) nearly parallel to its longer axis. These streaks (which obviously indicate a stratified structure in the nebula, if, indeed, they do not originate in the interposition of imperfectly transparent matter between us and it) are not seen on a general and cursory view of the nebula; they require attention to distinguish them*, and this circumstance must be borne in mind when inspecting the very extraordinary engraving which illustrates Mr. Bond's account. The figure in given our Plate II. fig. 3., is from a rather hasty sketch, and makes no pretensions to exactness. A similar, but much more strongly marked case of parallel arrangement than that noticed by Mr. Bond in this, is one in which the two semi-ovals of an elliptically formed nebula appear cut asunder and separated by a broad obscure band parallel to the larger axis of the nebula, in the midst of which a faint streak of light parallel to the sides of the cut appears, is seen in the southern hemisphere in R. A. $13^h 15^m 31^s$, N.P.D. $132° 8'$ (1830). The nebulæ in $12^h 27^m 3^s$, $63° 5'$, and $12^h 31^m 11^s$, $100° 40'$ present analagous features.

(875.) Annular nebulæ also exist, but are among the rarest objects in the heavens. The most conspicuous of this class is to be found almost exactly half way between β and γ Lyræ, and may be seen with a telescope of moderate power. It is small and particularly well defined, so as to have more the appearance of a flat oval solid ring than of a nebula. The axes of the ellipse are to each other in the proportion of about 4 to 5, and the opening occupies about half or rather more than half the diameter. The central vacuity is not quite dark, but is filled in with faint nebula, like a gauze stretched over a hoop. The powerful telescopes of Lord Rosse resolve this object into excessively minute stars, and show filaments of stars adhering to its edges.†

(876.) PLANETARY NEBULÆ are very extraordinary objects. They have, as their name imports, a near, in some instances, a perfect resemblance to planets, presenting discs round, or slightly oval, in some quite sharply terminated,

---

* Account of the nebula in Andromeda, by G. P. Bond, Assistant at the Cambridge Observatory, U. S.   Trans. American Acad., vol. iii. p. 80.

† The places of the annular nebulæ, at present known (for 1830) are,

	R. A.			N.P.D.			R. A.			N.P.D.	
1.	17^h	10^m	39^s	$128°$	$18'$	3.	18^h	47^m	13^s	$57°$	$11'$
2.	17	19	2	113	37	4.	20	9	33	59	57

in others a little hazy or softened at the borders. Their light is in some perfectly equable, in others mottled and of a very peculiar *texture*, as if curdled. They are comparatively rare objects, not above four or five and twenty having been hitherto observed, and of these nearly three fourths are situated in the southern hemisphere. Being very interesting objects we subjoin a list of the most remarkable.* Among these may be more particularly specified the sixth in order, situated in the Cross. Its light is about equal to that of a star of the 6·7 magnitude, its diameter about 12″, its disc circular or very slightly elliptic, and with a clear, sharp, well-defined outline, having exactly the appearance of a planet with the exception only of its colour, which is a fine and full blue verging somewhat upon green. And it is not a little remarkable that this phænomenon of a blue colour, which is so rare among *stars* (except when in the immediate proximity of yellow stars) occurs, though less strikingly, in three other objects of this class, viz. in No. 4, whose colour is sky-blue, and in Nos. 11 and 12, where the tint, though paler, is still evident. Nos. 2, 7, 9, and 12, are also exceedingly characteristic objects of this class. Nos. 3, 4, and 11 (the latter in the parallel of ν Aquarii, and about 5^m preceding that star), are considerably elliptic, and (respectively) about 38″, 30″ and 15″ in diameter. On the disc of No. 3, and very nearly in the center of the ellipse, is a star 9m, and the texture of its light, being velvety, or as if formed of fine dust, clearly indicates its resolvability into stars. The largest of these objects is No. 5, situated somewhat south of the parallel of β Ursæ Majoris and about 12^m following that star. Its apparent diameter is 2′ 40″, which, supposing it

* Places for 1830 of twelve of the most remarkable planetary nebulæ.

R.A.	N.P.D.	R.A.	N.P.D.	R.A.	N.P.D.
h. m. s.	° ′	h. m. s.	° ′	h. m. s.	° ′
1. 7 34 2	104 20	5. 11 4 49	34 4	9. 19 34 21	104 33
2. 9 16 39	147 35	6. 11 41 56	146 14	10. 19 40 19	39 54
3. 9 59 52	129 36	7. 15 5 18	135 1	11. 20 54 53	102 2
4. 10 16 36	107 47	8. 19 10 9	83 46	12. 23 17 44	48 24

placed at a distance from us not more than that of 61
Cygni, would imply a linear one seven times greater than
that of the orbit of Neptune. The light of this stupendous
globe is perfectly equable (except just at the edge, where it is
slightly softened), and of considerable brightness. Such an
appearance would not be presented by a globular space
uniformly filled with stars or luminous matter, which structure would necessarily give rise to an apparent increase of
brightness towards the center in proportion to the thickness
traversed by the visual ray. We might, therefore, be induced to conclude its real constitution to be either that of a
hollow spherical shell or of a flat disc, presented to us (by a
highly improbable coincidence) in a plane precisely perpendicular to the visual ray.

(877.) Whatever idea we may form of the real nature
of such a body, or of the planetary nebulæ in general,
which all agree in the absence of central condensation, it
is evident that the intrinsic splendour of their surfaces, *if
continuous*, must be almost infinitely less than that of the
sun. A circular portion of the sun's disc, subtending an
angle of 1′, would give a light equal to that of 780 full
moons; while among all the objects in question there is not
one which can be seen with the naked eye. M. Arago has
surmised that they may possibly be envelopes shining by
reflected light, from a solar body placed in their center, invisible to us by the effect of its excessive distance; removing,
or attempting to remove the apparent paradox of such an
explanation, by the optical principle that an illuminated
surface is equally *bright* at all distances, and, therefore, if
large enough to subtend a measurable angle, can be equally
well seen, whereas the central body, subtending no such
angle, has its effect on our sight diminished in the inverse
ratio of the square of its distance.* The assiduous applica-

* With due deference to so high an authority we must demur to the conclusion. Even supposing the envelope to reflect and scatter (equally in all directions) *all* the light of the central sun, the portion of the light so scattered which
would fall to our share, could not exceed that which that sun itself would send
to us by direct radiation. But this, *ex hypothesi*, is too small to affect the eye
with any luminous perception, much less then could it do so if spread over a

tion of the immense optical powers recently brought to bear on the heavens, will probably remove some portion of the mystery which at present hangs about these enigmatical objects.

(878.) Double nebulæ occasionally occur—and when such is the case, the constituents most commonly belong to the class of spherical nebulæ, and are in some instances undoubtedly globular clusters. All the varieties of double stars, in fact, as to distance, position, and relative brightness, have their counterparts in double nebulæ; besides which the varieties of form and gradation of light in the latter afford room for combinations peculiar to this class of objects. Though the conclusive evidence of observed relative motion be yet wanting, and though from the vast scale on which such systems are constructed, and the probable extreme slowness of the angular motion, it may continue for ages to be so, yet it is impossible, when we cast our eyes upon such objects, or on the figures which have been given of them *, to doubt their physical connexion. The argument drawn from the comparative rarity of the objects in proportion to the whole extent of the heavens, so cogent in the case of the double stars, is infinitely more so in that of the double nebulæ. Nothing more magnificent can be presented to our consideration, than such combinations. Their stupendous scale, the multitude of individuals they involve, the perfect symmetry and regularity which many of them present, the utter disregard of complication in thus heaping together system upon system, and construction upon construction, leave us lost in wonder and admiration at the evidence they afford of infinite power and unfathomable design.

(879.) Nebulæ of regular forms often stand in marked and symmetrical relation to stars, both single and double. Thus we are occasionally presented with the beautiful and striking phænomenon of a sharp and brilliant star concentrically surrounded by a perfectly circular disc or atmosphere of faint

---

surface many million times exceeding in angular area the apparent disc of the central sun itself. (See Annuaire du Bureau des Longitudes, 1842, p. 409, 410, 411.) M. Arago *is expressly contending for reflected* light. If the envelope be self-luminous, his reasoning is perfectly sound.

* Phil. Trans., 1833. Plate vii.

light, in some cases dying away insensibly on all sides, in others almost suddenly terminated. These are *Nebulous Stars.* Fine examples of this kind are the 45th and 69th nebulæ of Sir Wm. Herschel's fourth class * (R. A. 7^h 19^m 8^s, N. P. D. 68° 45', and 3^h 58^m 36^s, 59° 40'), in which stars of the 8th magnitude are surrounded by photospheres of the kind above described respectively of 12" and 25" in diameter. Among stars of larger magnitudes, 55 Andromedæ and 8 Canum Venaticorum may be named as exhibiting the same phænomenon with more brilliancy, but perhaps with less perfect regularity.

(880.) The connexion of nebulæ with double stars is in many instances extremely remarkable. Thus in R. A. 18^h 7^m 1^s, N. P. D. 109° 56', occurs an elliptic nebula having its longer axis about 50", in which, symmetrically placed, and rather nearer the vertices than the foci of the ellipse, are the equal individuals of a double star, each of the 10th magnitude. In a similar combination noticed by M. Struve (in R. A. 18^h 25^m, N. P. D. 25° 7'), the stars are unequal and situated precisely at the two extremities of the major axis. In R. A. 13^h 47^m 33^s, N. P. D. 129° 9', an oval nebula of 2' in diameter has very near its center a close double star, the individuals of which, slightly unequal, and about the 9·10 magnitude, are not more than 2" asunder. The nucleus of Messier's 64th nebula is " strongly suspected " to be a close double star — and several other instances might be cited.

(881.) Among the nebulæ which, though deviating more from symmetry of form, are yet not wanting in a certain regularity of figure, and which seem clearly entitled to be regarded as systems of a definite nature, however mysterious their structure and destination, by far the most remarkable are the 27th and 51st of Messier's Catalogue.† This consists

---

* The classes here referred to are not the species described in Art. 868., but lists of nebulæ, eight in number arranged according to brightness, size, density of clustering, &c., in one or other of which all nebulæ were originally classed by him.   Class I. contains " Bright nebulæ ; " II. " Faint do. ; " III. " Very faint do. ; " IV. " Planetary nebulæ, stars with bars, milky chevelures, short rays, remarkable shapes, &c. ; " V " Very large nebulæ ; " VI. " Very compressed rich clusters ; " VII. " Pretty much compressed do. ; " VIII. " Coarsely scattered clusters."

† Place for 1830 : R. A. 19^h 52^m 12^s, N. P. D. 67° 44', and R. A. 13^h 22^m 39^s, N.P.D. 41° 56'.

of two round or somewhat oval nebulous masses united by a short neck of nearly the same density. Both this and the masses graduate off however into a fainter nebulous envelope which completes the figure into an elliptic form, of which the interior masses with their connexion occupy the lesser axis. Seen in a reflector of 18 inches in aperture, the form has considerable regularity; and though a few stars are here and there scattered over it, it is unresolved. Lord Rosse, viewing it with a reflector of double that aperture, describes and figures it as resolved into numerous stars with much intermixed nebula; while the symmetry of form by rendering visible features too faint to be seen with inferior power, is rendered considerably less striking, though by no means obliterated.

(882.) The 51st nebula of Messier, viewed through an 18-inch reflector, presents the appearance of a large and bright globular nebula, surrounded by a ring at a considerable distance from the globe, very unequal in brightness in its different parts, and subdivided through about two-fifths of its circumference as if into two laminæ, one of which appears as if turned up towards the eye out of the plane of the rest. Near it (at about a radius of the ring distant) is a small bright round nebula. Viewed through the 6-feet reflector of Lord Rosse the aspect is much altered. The interior, or what appeared the upturned portion of the ring, assumes the aspect of a nebulous coil or convolution tending in a spiral form towards the center, and a general tendency to a spiroid arrangement of the streaks of nebula connecting the ring and central mass which this power brings into view, becomes apparent, and forms a very striking feature. The outlying nebula is also perceived to be connected by a narrow, curved band of nebulous light with the ring, and the whole, if not clearly resolved into stars, has a "resolvable" character which evidently indicates its composition.*

(883.) We come now to a class of nebulæ of totally differ-

---

* This description is from the recollection of a sketch exhibited by his Lordship at the British Association. Every astronomer must long for the publication of his own account of the wonders disclosed by this magnificent instrument.

ent character. They are of very great extent, utterly devoid
of all symmetry of form, — on the contrary, irregular and
capricious in their shapes and convolutions to a most extraordinary degree, and no less so in the distribution of their
light. No two of them can be said to present any similarity
of figure or aspect, but they have one important character in
common. They are all situated in, or very near, the borders
of the Milky Way. The most remote from it is that in the
sword handle of Orion, which being 20° from the galactic
circle, and 15° from the visible border of the Via Lactea, might
seem to form an exception, though not a striking one. But
this very situation may be adduced as a corroboration of the
general view which this principle of localization suggests.
For the place in question is situated in the prolongation of
that faint offset of the Milky Way which we traced (Art. 787.)
from α and ε Persei towards Aldebaran and the Hyades, and
in the zone of Great Stars noticed in Art. 785. as an appendage of, and probably bearing relation to that stratum.

(884.) From this it would appear to follow, almost as a matter
of course, that they must be regarded as outlying, very distant,
and as it were detached fragments of the great stratum of the
Galaxy, and this view of the subject is strengthened when
we find on mapping down their places that they may all be
grouped in four great masses or nebulous regions, — that of
Orion, of Argo, of Sagittarius, and of Cygnus. And thus,
inductively, we may gather some information respecting the
structure and form of the Galaxy itself, which, could we view
it as a whole, from a distance such as that which separates us
from these objects, would very probably present itself under
an aspect quite as complicated and irregular.

(885.) The great nebula surrounding the stars marked θ 1
in the sword handle of Orion was discovered by Huyghens
in 1656, and has been repeatedly figured and described by
astronomers since that time. Its appearance varies greatly
(as that of all nebulous objects does) with the instrumental
power applied, so that it is difficult to recognize in representations made with inferior telescopes, even principal features, to say nothing of subordinate details.   Until this

became well understood, it was supposed to have changed very materially, both in form and extent, during the interval elapsed since its first discovery. No doubt, however, now remains that these supposed changes have originated partly from the cause above-mentioned, partly from the difficulty of correctly drawing, and, still more, engraving such objects, and partly from a want of sufficient care in the earlier delineators themselves in faithfully copying that which they really did see. Our figure (Plate IV., *fig.* 1.) is reduced from a larger one made under very favourable circumstances, from drawings taken with an 18-inch reflector at the Cape of Good Hope, where its meridian altitude greatly exceeds what it has at European stations. The area occupied by this figure is about one 25th part of a square degree, extending in R. A. (or horizontally) 2^m of time, equivalent almost exactly to 30′ in arc, the object being very near the equator, and 24′ vertically, or in polar distance. The figure shows it reversed in both directions, the northern side being lowermost, and the preceding towards the left hand. In form, the brightest portion offers a resemblance to the head and yawning jaws of some monstrous animal, with a sort of proboscis running out from the snout. Many stars are scattered over it, which for the most part appear to have no connexion with it, and the remarkable sextuple star θ 1 Orionis, of which mention has already been made (Art. 837.), occupies a most conspicuous situation close to the brightest portion, at almost the edge of the opening of the jaws. It is remarkable, however, that within the area of the trapezium no *nebula* exists. The general aspect of the less luminous and cirrous portion is simply nebulous and irresolvable, but the brighter portion immediately adjacent to the trapezium, forming the square front of the head, is shown with the 18-inch reflector broken up into masses (very imperfectly represented in the figure), whose mottled and curdling light evidently indicates by a sort of granular texture its consisting of stars, and when examined under the great light of Lord Rosse's reflector, or the exquisite defining power of the great achromatic at Cambridge, U. S., is evidently perceived to con-

R R

sist of clustering stars. There can therefore be little doubt as to the whole consisting of stars, too minute to be discerned individually even with these powerful aids, but which become visible as points of light when closely adjacent in the more crowded parts in the mode already more than once suggested.

(886.) The nebula is not confined to the limits of our figure. Northward of θ about 33', and nearly on the same meridian are two stars marked C 1 and C 2 Orionis, involved in a bright and branching nebula of very singular form, and south of it is the star ι Orionis, which is also involved in strong nebula. Careful examination with powerful telescopes has traced out a continuity of nebulous light between the great nebula and both these objects, and there can be little doubt that the nebulous region extends northwards, as far as ε in the belt of Orion, which is involved in strong nebulosity, as well as several smaller stars in the immediate neighbourhood. Professor Bond has given a beautiful figure of the great nebula in Trans. American Acad. of Arts and Sciences, new series, vol. iii.

(887.) The remarkable variation in lustre of the bright star η in Argo, has been already mentioned. This star is situated in the most condensed region of a very extensive nebula or congeries of nebular masses, streaks and branches, a portion of which is represented in fig. 2. Plate IV. The whole nebula is spread over an area of fully a square degree in extent, of which that included in the figure occupies about one-fourth, that is to say, 28' in polar distance, and 32' of arc in R. A., the portion not included being, though fainter, even more capriciously contorted than that here depicted, in which it should be observed that the preceding side is towards the right hand, and the southern uppermost. Viewed with an 18-inch reflector, no part of this strange object shows any sign of resolution into stars, nor in the brightest and most condensed portion adjacent to the singular oval vacancy in the middle of the figure is there any of that curdled appearance, or that tendency to break up into bright knots with intervening darker portions which characterize the nebula of Orion, and indicate its resolvability. The whole is situated in a very rich and brilliant part of the Milky Way, so thickly strewed with stars (omitted in the figure), that in the area occupied by the nebula, not less than 1200 have been actually counted, and their places in R. A. and P. D. determined. Yet it is obvious that these have no connexion whatever with the nebula, being, in fact, only a simple continuation over it of the general ground of the galaxy, which on an average of two hours in Right Ascension in this period of its course contains no less than 3138 stars to the square degree, all, however, distinct, and (except where the object in question is situated) seen projected on a perfectly dark heaven, without any appearance of intermixed nebulosity. The conclusion can hardly be avoided, that in looking at it we see through, and beyond the Milky Way, far out into space, through a starless region, disconnecting it altogether from our system. " It is not easy for language to convey a full impression of the beauty and sublimity of the spectacle which this nebula offers, as it enters the field of view of a telescope fixed in Right Ascension, by the diurnal motion, ushered in as it is by so glorious and innumerable a procession of stars, to which it forms a sort of climax," and in a part of the heavens otherwise full of interest. One other bright and very remarkably formed nebula of considerable magnitude precedes it nearly on the same parallel, but without any traceable connexion between them.

(888.) The nebulous group of Sagittarius consists of several conspicuous nebulæ * of very extraordinary forms by no means easy to give an idea of by mere description. One of them (_h_, 1991 †) is singularly trifid, consisting of three bright and irregularly formed nebulous masses, graduating away insensibly externally, but coming up to a great intensity of

* About R. A. 17^h 52^m, N.P D. 113° 1′, four nebulæ, No. 41 of Sir Wm. Herschel's 4th class, and Nos. 1, 2, 3, of his 5th all connected into one great complex nebula. — In R. A. 17^h 53^m 27^s, N.P.D. 114° 21′, the 8th, and in 18^h 11^m, 106° 15′, the 17th of Messier's Catalogue.

† This number refers to the catalogue of nebulæ in Phil. Trans., 1833. The reader will find figures of the several nebulæ of this group in that volume, pl. iv. fig. 35 , in the Author's " Results of Observations, &c., at the Cape of Good Hope," Plates i. fig. 1., and ii. figs. 1 and 2, and in Mason's Memoir in the collection of the American Phil. Soc., vol. vii. art. xiii.

light at their interior edges, where they enclose and surround a sort of three-forked rift, or vacant area, abruptly and uncouthly crooked, and quite void of nebulous light. A beautiful triple star is situated precisely on the edge of one of these nebulous masses just where the interior vacancy forks out into two channels. A fourth nebulous mass spreads like a fan or downy plume from a star at a little distance from the triple nebula.

(889.) Nearly adjacent to the last described nebula, and no doubt connected with it, though the connexion has not yet been traced, is situated the 8th nebula of Messier's Catalogue. It is a collection of nebulous folds and masses, surrounding and including a number of oval dark vacancies, and in one place coming up to so great a degree of brightness, as to offer the appearance of an elongated nucleus. Superposed upon this nebula, and extending in one direction beyond its area, is a fine and rich cluster of scattered stars, which seem to have no connexion with it, as the nebula does not, as in the region of Orion, show any tendency to congregate about the stars.

(890.) The 19th nebula of Messier's Catalogue, though some degrees remote from the others, evidently belongs to this group. Its form is very remarkable, consisting of two loops like capital Greek Omegas, the one bright, the other exceedingly faint, connected at their bases by a broad and very bright band of nebula, insulated within which by a narrow comparatively obscure border, stands a bright, resolvable knot, or what is probably a cluster of exceedingly minute stars. A very faint round nebula stands in connexion with the upper or convex portion of the brighter loop.

(891.) The nebulous group of Cygnus consists of several large and irregular nebulæ, one of which passes through the double star k Cygni, as a long, crooked, narrow streak, forking out in two or three places. The others *, observed in the first instance by Sir W. Herschel and by the author of this work as separate nebulæ, have been traced into connexion by Mr. Mason, and shown to form part of a curious and intricate nebulous system, consisting, 1st, of a long, narrow, curved,

* R. A. 20ʰ 49ᵐ 20ˢ, N. P. D. 58° 27′.

and forked streak, and 2dly, of a cellular effusion of great extent, in which the nebula occurs intermixed with, and adhering to stars around the borders of the cells, while their interior is free from nebula, and almost so from stars.

(892.) The Magellanic clouds, or the nubeculæ (major and minor), as they are called in the celestial maps and charts, are, as their name imports, two nebulous or cloudy masses of light, conspicuously visible to the naked eye, in the southern hemisphere, in the appearance and brightness of their light not unlike portions of the Milky Way of the same apparent size. They are, generally speaking, round, or somewhat oval, and the larger, which deviates most from the circular form, exhibits the appearance of an axis of light, very ill defined, and by no means strongly distinguished from the general mass, which seems to open out at its extremities into somewhat oval sweeps, constituting the preceding and following portions of its circumference. A small patch, visibly brighter than the general light around, in its following part, indicates to the naked eye the situation of a very remarkable nebula (No. 30. Doradûs of Bode's catologue), of which more hereafter. The greater nubecula is situated between the meridians of 4^h 40^m and 6^h 0^m and the parallels of 156° and 162° of N.P.D., and occupies an area of about 42 square degrees. The lesser, between the meridians * 0^h 28^m and 1^h 15^m and the parallels of 162° and 165° N.P.D., covers about ten square degrees. Their degree of brightness may be judged of from the effect of strong moonlight, which totally obliterates the lesser, but not quite the greater.

(893.) When examined through powerful telescopes, the constitution of the nubeculæ, and especially of the nubecula major, is found to be of astonishing complexity. The general ground of both consists of large tracts and patches of nebulosity in every stage of resolution, from light, irresolvable with 18 inches of reflecting aperture, up to perfectly separated stars like the Milky Way, and clustering groups sufficiently insulated and condensed to come under the designation of irregular, and in some cases pretty rich clusters. But be-

* It is laid down nearly an hour wrong in all the celestial charts and globes.

sides those, there are also nebulæ in abundance, both regular
and irregular; globular clusters in every state of condensation;
and objects of a nebulous character quite peculiar, and which
have no analogue in any other region of the heavens.   Such
is the concentration of these objects, that in the area occupied
by the nubecula major, not fewer than 278 nebulæ and
clusters have been enumerated, besides 50 or 60 outliers,
which (considering the general barrenness in such objects of
the immediate neighbourhood) ought certainly to be reckoned
as its appendages, being about 6½ per square degree, which very
far exceeds the average of any other, even the most crowded
parts of the nebulous heavens.   In the nubecula minor, the
concentration of such objects is less, though still very striking,
37 having been observed within its area, and 6 adjacent, but
outlying.   The nubeculæ, then, combine, each within its own
area, characters which in the rest of the heavens are no less
strikingly separated,—viz., those of the galactic and the nebular system.   Globular clusters (except in one region of small
extent) and nebulæ of regular elliptic forms are comparatively rare in the Milky Way, and are found congregated in
the greatest abundance in a part of the heavens, the most
remote possible from that circle; whereas, in the nubeculæ,
they are indiscriminately mixed with the general starry
ground, and with irregular though small nebulæ.

(894.) This combination of characters, rightly considered,
is in a high degree instructive, affording an insight into the
probable comparative distance of *stars* and *nebulæ*, and the
real brightness of individual stars as compared one with
another.   Taking the apparent semidiameter of the nubecula
major at 3°, and regarding its solid form as, roughly speaking,
spherical, its nearest and most remote parts differ in their
distance from us by a little more than a tenth part of our
distance from its center.   The brightness of objects situated
in its nearer portions, therefore, cannot be *much* exaggerated,
nor that of its remoter *much* enfeebled, by their difference of
distance; yet within this globular space, we have collected
upwards of 600 stars of the 7th, 8th, 9th, and 10th magnitudes, nearly 300 nebulæ, and globular and other clusters,

of all degrees of resolubility, and smaller scattered stars innumerable of every inferior magnitude, from the 10th to such as by their multitude and minuteness constitute irresolvable nebulosity, extending over tracts of many square degrees. Were there but one such object, it might be maintained without utter improbability that its apparent sphericity is only an effect of foreshortening, and that in reality a much greater proportional difference of distance between its nearer and more remote parts exists. But such an adjustment, improbable enough in one case, must be rejected as too much so for fair argument in two. It must, therefore, be taken as a demonstrated fact, that stars of the 7th or 8th magnitude and irresolvable nebula may co-exist within limits of distance not differing in proportion more than as 9 to 10, a conclusion which must inspire some degree of caution in admitting, *as certain*, many of the consequences which have been rather strongly dwelt upon in the foregoing pages.

(895.) Immediately preceding the center of the nubecula minor, and undoubtedly belonging to the same group, occurs the superb globular cluster, No. 47. Toucani of Bode, very visible to the naked eye, and one of the finest objects of this kind in the heavens. It consists of a very condensed spherical mass of stars, *of a pale rose colour*, concentrically enclosed in a much less condensed globe of white ones, 15′ or 20′ in diameter. This is the first in order of the list of such clusters in Art. 867.

(896.) Within the nubecula major, as already mentioned, and faintly visible to the naked eye, is the singular nebula (marked as the star 30 Doradûs in Bode's Catalogue) noticed by Lacaille as resembling the nucleus of a small comet. It occupies about one-500th part of the whole area of the nubecula, and is so satisfactorily represented in plate V., fig. 1., as to render further description superfluous.

(897.) We shall conclude this chapter by the mention of two phænomena, which seem to indicate the existence of some slight degree of nebulosity about the sun itself, and even to place it in the list of nebulous stars. The first is that called the zodiacal light, which may be seen any very clear

evening soon after sunset, about the months of March, April,
and May, or at the opposite seasons before sunrise, as a cone
or lenticularly-shaped light, extending from the horizon obliquely upwards, and following generally the course of the
ecliptic, or rather that of the sun's equator. The apparent
angular distance of its vertex from the sun varies, according
to circumstances, from 40° to 90°, and the breadth of its base
perpendicular to its axis from 8° to 30°. It is extremely
faint and ill defined, at least in this climate, though better
seen in tropical regions, but cannot be mistaken for any
atmospheric meteor or aurora borealis. It is manifestly in
the nature of a lenticularly-formed envelope, surrounding the
sun, and extending beyond the orbits of Mercury and Venus,
and nearly, perhaps quite, attaining that of the earth, since its
vertex has been seen fully 90° from the sun's place in a great
circle. It may be conjectured to be no other than the denser
part of that medium, which, we have some reason to believe,
resists the motion of comets; loaded, perhaps, with the actual
materials of the tails of millions of those bodies, of which
they have been stripped in their successive perihelion passages (Art. 566.). An *atmosphere* of the sun, in any proper
sense of the word, it cannot be, since the existence of a
gaseous envelope propagating pressure from part to part;
subject to mutual friction in its strata, and therefore rotating
in the same or nearly the same time with the central body;
and of such dimensions and ellipticity, is utterly incompatible
with dynamical laws. If its particles have inertia, they must
necessarily stand with respect to the sun in the relation of separate and independent minute planets, each having its own
orbit, plane of motion, and periodic time. The total mass being
almost nothing compared to that of the sun, mutual *perturbation* is out of the question, though *collisions* among such as
may cross each other's paths may operate in the course of
indefinite ages to effect a subsidence of at least some portion
of it into the body of the sun or those of the planets.

(898.) Nothing prevents that these particles, or some
among them, may have some tangible size, and be at very
great distances from each other. Compared with planets

visible in our most powerful telescopes, rocks and stony masses of great size and weight would be but as the impalpable dust which a sunbeam renders visible as a sheet of light when streaming through a narrow chink into a dark chamber. It is a fact, established by the most indisputable evidence, that stony masses and lumps of iron do occasionally, and indeed by no means unfrequently, fall upon the earth from the higher regions of our atmosphere (where it is obviously impossible they can have been generated), and that they have done so from the earliest times of history. Four instances are recorded of persons being killed by their fall. A block of stone fell at Ægos Potamos, B.C. 465, as large as two mill-stones ; another at Narni, in 921, projected, like a rock, four feet above the surface of the river, into which it was seen to fall. The emperor Jehangire had a sword forged from a mass of meteoric iron which fell, in 1620, at Jahlinder, in the Punjab.* Sixteen instances of the fall of stones in the British Isles are well authenticated to have occurred since 1620, one of them in London. In 1803, on the 26th of April, thousands of stones were scattered by the explosion into fragments of a large fiery globe over a region of twenty or thirty square miles around the town of L'Aigle, in Normandy. The fact occurred at mid-day, and the circumstances were officially verified by a commission of the French government.† These, and innumerable other instances‡, fully establish the general fact ; and after vain attempts to account for it by volcanic projection, either from the earth or the moon, the planetary nature of these bodies seems at length to be almost generally admitted. The heat which they possess when fallen, the igneous phænomena which accompany them, their explosion on arriving within the denser regions of our atmosphere, &c., are all sufficiently accounted for on physical principles, by the condensation of the air before

* See the emperor's own very remarkable account of the occurrence, translated in Phil. Trans. 1793, p. 202.
† See M. Biot's report in Mém. de l'Institut. 1806.
‡ See a list of upwards of 200, published by Chladni, Annales du Bureau des Longitudes de France, 1825.

them in consequence of their enormous velocity, and by the relations of air in a highly attenuated state to heat.*

(899.) Besides stony and metallic masses, however, it is probable that bodies of very different natures, or at least states of aggregation, are thus circulating round the sun. Shooting stars, often followed by long trains of light, and those great fiery globes, of more rare, but not *very* uncommon occurrence, which are seen traversing the upper regions of our atmosphere, sometimes leaving trains behind them remaining for many minutes, sometimes bursting with a loud explosion, sometimes becoming quietly extinct, may not unreasonably be presumed to be bodies extraneous to our planet, which only become visible when in the act of grazing the confines of our atmosphere.   Among the last mentioned meteors are some which can hardly be supposed solid masses. The remarkable meteor of Aug. 18. 1783, traversed the whole of Europe, from Shetland to Rome, with a velocity of about 30 miles per second, at a height of 50 miles from the surface of the earth, with a light greatly surpassing that of the full moon, and a real diameter of fully half a mile.   Yet with these vast dimensions, it changed its form visibly, and at length quietly separated into several distinct bodies, accompanying each other in parallel courses, and each followed by a tail or train.

(900.) There are circumstances in the history of shooting stars, which very strongly corroborate the idea of their extraneous or *cosmical* origin, and their circulation round the sun in definite orbits.   On several occasions they have been observed to appear in unusual, and, indeed, astonishing numbers, so as to convey the idea of a shower of rockets, or of snow-flakes falling, and brilliantly illuminating the whole heavens for hours together, and that not in one locality, but over whole continents and oceans, and even (in one instance) in both hemispheres.   Now it is extremely remarkable that, whenever this great display has been exhibited (at least in

* Edinburgh Review, Jan. 1848, p. 195.   It is very remarkable that no new chemical element has been detected in any of the numerous meteorolites which have been subjected to analysis.

modern times), it has uniformly happened on the night between the 12th and 13th, or on that between the 13th and
14th of November. Such cases occurred in 1799, 1823,
1832, 1833, and 1834. On tracing back the records of
similar phænomena, it has been ascertained, moreover, that
more often those identical nights, but sometimes those immediately adjacent, have been, time out of mind, habitually
signalized by such exhibitions. Another annually recurring
epoch, in which, though far less brilliant, the display of
meteors is more *certain* (for that of November is often interrupted for a great many years), is that of the 10th of August,
on which night, and on the 9th and 11th, numerous, large,
and bright shooting stars, with trains, are almost sure to be
seen. Other epochs of periodic recurrence, less marked than
the above, have also been to a certain extent established.

(901.) It is impossible to attribute such a recurrence of
identical dates of very remarkable phænomena to accident.
Annual periodicity, irrespective of geographical position,
refers us at once to the place occupied by the earth in its
annual orbit, and leads direct to the conclusion that at that
place the earth incurs a liability to *frequent* encounters or
concurrences with a stream of meteors in their progress of
circulation round the sun. Let us test this idea by pursuing
it into some of its consequences. In the first place then,
supposing the earth to plunge, in its yearly circuit, into a
uniform *ring* of innumerable small meteor-planets, of such
breadth as would be traversed by it in one or two days;
since during this small time the motions, whether of the
earth or of each individual meteor, may be taken as uniform
and rectilinear, and those of all the latter (at the place and
time) parallel, or very nearly so, it will follow that the relative
motion of the meteors referred to the earth as at rest, will be
also uniform, rectilinear, and *parallel.* Viewed, therefore,
from the center of the earth (or from any point in its circumference, if we neglect the diurnal velocity as very small
compared with the annual) they will all appear to diverge
from a common point, *fixed in relation to the celestial sphere,*
as if emanating from a sidereal apex (Art. 115.).

(902.) Now this is precisely what actually happens. The meteors of the 12th—14th of November, or at least the vast majority of them, describe apparently arcs of great circles, passing through or near γ Leonis. No matter what the situation of that star with respect to the horizon or to its east and west points may be at the time of observation, the paths of the meteors all appear to diverge from that star. On the 9th—11th of August the geometrical fact is the same, the apex only differing; B Camelopardali being for that epoch the point of divergence. As we need not suppose the meteoric ring coincident in its plane with the ecliptic, and as for a *ring* of meteors we may substitute an elliptic annulus of any reasonable excentricity, so that both the velocity and direction of each meteor may differ to any extent from the earth's, there is nothing in the great and obvious difference *in latitude* of these apices at all militating against the conclusion.

(903.) If the meteors be uniformly distributed in such a ring or elliptic annulus, the earth's encounter with them in every revolution will be certain, if it occur once. But if the ring be broken, if it be a succession of groupes revolving in an ellipse in a period *not* identical with that of the earth, years may pass without a rencontre; and when such happen, they may differ to any extent in their intensity of character, according as richer or poorer groupes have been encountered.

(904.) No other plausible explanation of these highly characteristic features (the annual periodicity, and divergence from a common apex, *always alike for each respective epoch*) has been even attempted, and accordingly the opinion is generally gaining ground among astronomers that shooting stars belong to their department of science, and great interest is excited in their observation and the further development of their laws. The most connected and systematic series of observations of them, having for their object to trace out their *relative* paths with respect to the earth, are those of Benzenberg and Brandes, who, by noting the instants and apparent places of appearance and extinction, as well as the precise

apparent paths among the stars, of individual meteors, from
the extremities of a measured base line nearly 50,000 feet
in length, were led to conclude that their heights at the
instant of their appearance and disappearance vary from
16 miles to 140, and their relative velocities from 18 to
36 miles per second, velocities so great as clearly to indicate
an independent planetary circulation round the sun.

(905.) It is by no means, however, inconceivable that the
earth approaching to such as differ but little from it in direction and velocity, may have attached many of them to
it as permanent satellites, and of these there *may* be some so
large, and of such texture and solidity, as to shine by reflected
light, and become visible (such, at least, as are very near the
earth) for a brief moment, suffering extinction by plunging
into the earth's shadow; in other words, undergoing total
eclipse.   Sir John Lubbock is of opinion that such is the
case, and has given geometrical formulæ for calculating their
distances from observations of this nature.*   The observations
of M. Petit, director of the observatory of Toulouse, would
lead us to believe in the existence of at least one such body,
revolving round the earth, as a satellite, in about 3 hours
20 minutes, and therefore at a distance equal to 2·513 radii of
the earth from its center, or 5000 miles above its surface.†

* Phil. Mag., Lond. Ed. Dub. 1848, p. 80.
† Comptes Rendus, Oct. 12. 1846, and Aug. 9. 1847.

PART IV.

OF THE ACCOUNT OF TIME.

CHAPTER XVIII.

NATURAL UNITS OF TIME. — RELATION OF THE SIDEREAL TO THE
SOLAR DAY AFFECTED BY PRECESSION. — INCOMMENSURABILITY
OF THE DAY AND YEAR. — ITS INCONVENIENCE. — HOW OB-
VIATED. — THE JULIAN CALENDAR. — IRREGULARITIES AT ITS
FIRST INTRODUCTION. — REFORMED BY AUGUSTUS. — GREGORIAN
REFORMATION. — SOLAR AND LUNAR CYCLES. — INDICTION. —
JULIAN PERIOD. — TABLE OF CHRONOLOGICAL ERAS. — RULES
FOR CALCULATING THE DAYS ELAPSED BETWEEN GIVEN DATES.
— EQUINOCTIAL TIME.

(906). Time, like distance, may be measured by comparison
with standards of any length, and all that is requisite for
ascertaining correctly the length of any interval, is to be
able to apply the standard to the interval throughout its
whole extent, without overlapping on the one hand, or
leaving unmeasured vacancies on the other; to determine,
without the possible error of a unit, the number of integer
standards which the interval admits of being interposed
between its beginning and end; and to estimate precisely
the fraction, over and above an integer, which remains when
all the possible integers are subtracted.

(907). But though all standard units of time are equally
possible, theoretically speaking, yet all are not, practically,
equally convenient. The solar day is a natural interval
which the wants and occupations of man in every state of
society force upon him, and compel him to adopt as his
fundamental unit of time. Its length as estimated from the
departure of the sun from a given meridian, and its next
return to the same, is subject, it is true, to an annual fluctuation in excess and defect of its mean value, amounting at its

maximum to full half a minute. But except for astronomical purposes, this is too small a change to interfere in the slightest degree with its use, or to attract any attention, and the tacit substitution of its mean for its true (or variable) value may be considered as having been made from the earliest ages, by the ignorance of mankind that any such fluctuation existed.

(908). The time occupied by one complete rotation of the earth on its axis, or the mean * sidereal day, may be shewn, on dynamical principles, to be subject to no variation from any external cause, and although its duration would be shortened by contraction in the dimensions of the globe itself, such as might arise from the gradual escape of its internal heat, and consequent refrigeration and shrinking of the whole mass, yet theory, on the one hand, has rendered it almost certain that this cause cannot have effected any perceptible amount of change during the history of the human race; and, on the other, the comparison of ancient and modern observations affords every corroboration to this conclusion. From such comparisons, Laplace has concluded that the sidereal day has not changed by so much as one hundredth of a second since the time of Hipparchus. The mean sidereal day therefore possesses in perfection the essential quality of a standard unit, that of complete *invariability*. The same is true of the mean sidereal year, if estimated upon an average sufficiently large to compensate the minute fluctuations arising from the periodical variations of the major axis of the earth's orbit due to planetary perturbation (Art. 668.).

(909.) The mean solar day is an immediate derivative of the sidereal day and year, being connected with them by the same relation which determines the synodic from the sidereal revolutions of any two planets or other revolving bodies (Art. 418.). The *exact* determination of the ratio of the sidereal to the solar day, which is a point of the utmost importance in astronomy, is however, in some degree, complicated by the effect of precession, which renders it necessary

---

* The true sidereal day is variable by the effect of nutation ; but the variation (an excessively minute fraction of the whole) compensates itself in a revolution of the moon's nodes.

to distinguish between the absolute time of the earth's
rotation on its axis, (the real natural and invariable standard
of comparison,) and the *mean* interval between two successive
returns of a given star to the same meridian, or rather of a
given meridian to the same star, which not only differs by a
minute quantity from the sidereal day, but is actually not
the same for all stars.   As this is a point to which a
little difficulty of conception is apt to attach, it will be
necessary to explain it in some detail.   Suppose then π the
pole of the ecliptic, and P that of the equinoctial, A B C D
the solstitial and equinoctial colures at any given epoch, and
P $p\,q\,r$ the small circle described by P about π in one
revolution of the equinoxes, *i. e.* in 25870 years, or 9448300
solar days, all projected on the plane of the ecliptic A B C D.
Let S be a star anywhere situated on the ecliptic, or *between*
it and the small circle P $q\,r$.   Then if the pole P were at

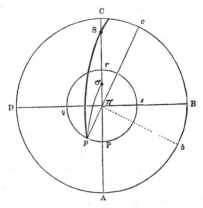

rest, a meridian of the earth setting out from P S C, and
revolving in the direction C D, will come again to the star
after the exact lapse of one sidereal day, or one rotation of
the earth on its axis.   But P is not at rest.   After the lapse
of one such day it will have come into the situation (suppose)
p, the vernal equinox B having retreated to b, and the
colure P C having taken up the new position $p\,c$.   Now a
conical movement impressed on the axis of rotation of a
globe already rotating is equivalent to a rotation impressed

on the whole globe round the axis of the cone, in addition to that which the globe has and retains round its own independent axis of revolution. Such a new rotation, in transferring P to p, being performed round an axis passing through π, will not alter the situation of that point of the globe which has π in its zenith. Hence it follows that $p\,\pi\,c$ passing through π will be the position taken up by the meridian P π C after the lapse of an exact sidereal day. But this does not pass through S, but falls short of it by the hour-angle $\pi\,p$ S, which is yet to be described before the meridian comes up to the star. The meridian, then, has lost so much on, or lagged so much behind, the star in the lapse of that interval. The same is true whatever be the arc P p. After the lapse of any number of days, the pole being transferred to p, the spherical angle $\pi\,p$ S will measure the total hour angle which the meridian has lost on the star. Now where S lies any where between C and r, this angle continually increases (though not uniformly), attaining 180° when p comes to r, and still (as will appear by following out the movement beyond r) increasing thence till it attains 360° when p has completed its circuit. Thus in a whole revolution of the equinoxes, the meridian will have lost one exact revolution upon the star, or in 9448300 sidereal days, will have re-attained the star only 9448299 times: in other words, the length of the day measured by the mean of the successive arrivals of any star outside of the circle P $p\,q\,r$ on one and the same meridian is to the absolute time of rotation of the earth on its axis as 9448300 : 9448299, or as 1·00000011 to 1.

(910.) It is otherwise of a star situated *within* this circle, as at σ. For such a star the angle $\pi\,p\,\sigma$, expressing the lagging of the meridian, increases to a maximum for some situation of p between q and r, and decreases again to o at r; after which it takes an opposite direction, and the meridian begins to get in advance of the star, and continues to get more and more so, till p has attained some point between s and P, where the advance is a maximum, and thence decreases again to o when p has completed its circuit. For

any star so situated, then, the mean of all the days so
estimated through a whole period of the equinoxes is an
absolute sidereal day, as if precession had no existence.

(911.) If we compare the sun with a star situated in the
ecliptic, the sidereal year is the mean of all the intervals of
its arrival at that star throughout indefinite ages, or (without
fear of sensible error) throughout recorded history.   Now, if
we would calculate the synodic sidereal revolution of the sun
and of a meridian of the earth *by reference to a star so situated,*
according to the principles of Art. 418., we must proceed as
follows : Let D be the length of the mean solar day (or
synodic day in question) d the mean sidereal revolution of
the meridian *with reference to the same star*, and y the sidereal
year.   Then the arcs described by the sun and the meridian
in the interval D will be respectively $360° \dfrac{D}{y}$ and $360° \dfrac{D}{d}$.
And since the latter of these exceeds the former by precisely
360°, we have

$$360° \frac{D}{d} = 360° \frac{D}{y} + 360°;$$

whence it follows that

$$\frac{D}{d} = 1 + \frac{D}{y} = 1·00273780,$$

taking the value of the sidereal year y as given in Art. 383,
viz. $365^d\ 6^h\ 9^m\ 9·6^s$.   But, as we have seen, d is not the absolute *sidereal day*, but exceeds it in the ratio 1·00000011 : 1.
Hence to get the value of the mean solar as expressed in
absolute sidereal days, the number above set down must be
increased in the same ratio, which brings it to 1·00273791,
which is the ratio of the solar to the sidereal day actually
in use among astronomers.

(912). It would be well for chronology if mankind would,
or could have contented themselves with this one invariable,
natural, and convenient standard in their reckoning of time.
The ancient Egyptians did so, and by their adoption of an
historical and official year of 365 days have afforded the only
example of a practical chronology, free from all obscurity or
complication.   But the return of the seasons, on which

depend all the more important arrangements and business of cultivated life is not conformable to such a multiple of the diurnal unit. Their return is regulated by the *tropical year*, or the interval between two successive arrivals of the sun at the vernal equinox, which, as we have seen (Art. 383.), differs from the sidereal year by reason of the motion of the equinoctial points. Now this motion is not absolutely uniform, because the ecliptic, upon which it is estimated, is gradually, though very slowly, changing its situation in space under the disturbing influence of the planets (Art. 640.). And thus arises a variation in the *tropical* year, which is dependent on the place of the equinox (Art. 383.). The *tropical* year is actually about $4 \cdot 21^s$ shorter than it was in the time of Hipparchus. This absence of the most essential requisite for a standard, viz. invariability, renders it necessary, since we cannot help employing the tropical year in our reckoning of time, to adopt an arbitrary or artificial value for it, so near the truth, as not to admit of the accumulation of its error for several centuries producing any practical mischief, and thus satisfying the ordinary wants of civil life ; while, for scientific purposes, the tropical year, so adopted, is considered only as the representative of a certain number of integer days and a fraction — the day being, in effect, the only standard employed. The case is nearly analagous to the reckoning of value by guineas and shillings, an artificial relation of the two coins being fixed by law, near to, but scarcely ever exactly coincident with, the natural one, determined by the relative market price of gold and silver, of which either the one or the other — whichever is really the most invariable, or the most in use with other nations, — may be assumed as the true theoretical standard of value.

(913). The other inconvenience of the tropical year as a greater unit is its incommensurability with the lesser. In our measure of space all our subdivisions are into aliquot parts: a yard is three feet, a mile eight furlongs, &c. But a year is no *exact* number of days, nor an integer number with any exact fraction, as one third or one fourth, over and above; but the surplus is an *incommensurable* fraction, composed of hours, minutes, seconds, &c., which produces the same kind of inconvenience in the reckoning of time that it would do in that of money, if we had gold coins of the value of twenty-one shillings, with odd pence and farthings, and a fraction of a farthing over. For this, however, there is no remedy but to keep a strict register of the surplus fractions; and, when they amount to a whole day, cast them over into the integer account.

(914). To do this in the simplest and most convenient manner is the object of a well-adjusted calendar. In the Gregorian calendar, which we follow, it is accomplished with as much simplicity and neatness as the case admits, by carrying a little farther than is done above, the principle of an assumed or artificial year, and adopting *two* such years, both consisting of an exact integer number of days, viz. one of 365 and the other of 366, and laying down a simple and easily remembered rule for the order in which these years shall succeed each other in the civil reckoning of time, so that during the lapse of at least some thousands of years the sum of the integer artificial, or Gregorian, years elapsed shall not differ from the same number of real tropical years by a whole day. By this contrivance, the equinoxes and solstices will always fall on days similarly situated, and bearing the same name in each Gregorian year; and the seasons will for ever correspond to the same months, instead of running the round of the whole year, as they must do upon any other system of reckoning, and used, in fact, to do before this was adopted as a matter of ignorant haphazard in the Greek and Roman chronology, and of strictly defined and superstitiously rigorous observance in the Egyptian.

(915.) The Gregorian rule is as follows :— The years are denominated *as years current* (*not as years elapsed*) from the midnight between the 31st of December and the 1st of January immediately subsequent to the birth of Christ, according to the chronological determination of that event by Dionysius Exiguus. Every year whose number is not divisible by 4 without remainder, consists of 365 days; every year which *is* so divisible, but is not divisible by 100, of 366; every year divisible by 100, but not by 400, again of 365; and every year divisible by 400, again of 366. For example, the year 1833, not being divisible by 4, consists of 365 days; 1836 of 366; 1800 and 1900 of 365 each; but 2000 of 366. In order to see how near this rule will bring us to the truth, let us see what number of days 10000 Gregorian years will contain, beginning with the year A.D. 1. Now, in 10000, the numbers not divisible by 4 will be ¾ of 10000 or 7500; those divisible by 100, but not by 400, will in like manner be ¾ of 100, or 75; so that, in the 10000 years in question, 7575 consist of 366, and the remaining 2425 of 365, producing in all 3652425 days, which would give for an average of each year, one with another, 365^{d}·2425. The actual value of the tropical year, (art. 383.) reduced into a decimal fraction, is 365·24224, so the error in the Gregorian rule on 10000 of the present tropical years, is 2·6, or 2^{d} 14^{h} 24^{m}; that is to say, less than a day in 3000 years; which is more than sufficient for all human purposes, those of the astronomer excepted, who is in no danger of being led into error from this cause. Even this error is avoided by extending the wording of the Gregorian rule one step farther than its contrivers probably thought it worth while to go, and declaring that years divisible by 4000 should consist of 365 days. This would take off two integer days from the above calculated number, and 2·5 from a larger average; making the sum of days in 100000 Gregorian years, 36524225, which differs only by a single day from 100000 real tropical years, such as they exist at present.

(916.) In the historical dating of events there is no year A.D. 0. The year immediately previous to A.D. 1, is always called B.C. 1. This must always be borne in mind in reckoning chronological and astronomical intervals. The sum of the nominal years B.C. and A.D. must be diminished by 1. Thus, from Jan. 1. B.C. 4713, to Jan 1. 1582, the years elapsed are not 6295, but 6294.

(917.) As any distance along a high road might, though in a rather inconvenient and roundabout way, be expressed without introducing error by setting up a series of milestones,

at intervals of unequal lengths, so that every fourth mile, for
instance, should be a yard longer than the rest, or according
to any other fixed rule; taking care only to mark the stones
so as to leave room for no mistake, and to advertise all
travellers of the difference of lengths and their order of succession; so may any interval of time be expressed correctly by
stating in what Gregorian years it begins and ends, *and whereabouts in each.* For this statement coupled with the decla-
ratory rule, enables us to say how many integer years are to
be reckoned at 365, and how many at 366 days. The latter
years are called bissextiles, or leap-years, and the surplus days
thus thrown into the reckoning are called *intercalary or leap-
days.*

(918.) If the Gregorian rule, as above stated, had always
and in all countries been known and followed, nothing would
be easier than to reckon the number of days elapsed between
the present time, and any historical recorded event. But this
is not the case; and the history of the calendar, with reference to chronology, or to the calculation of ancient observations, may be compared to that of a clock, going regularly
when left to itself, but sometimes forgotten to be wound up;
and when wound, sometimes set forward, sometimes backward, either to serve particular purposes and private interests,
or to rectify blunders in setting. Such, at least, appears to
have been the case with the Roman calendar, in which our own
originates, from the time of Numa to that of Julius Cæsar,
when the lunar year of 13 months, or 355 days, was augmented
at pleasure to correspond to the solar, by which the seasons are
determined, by the arbitrary intercalations of the priests, and
the usurpations of the decemvirs and other magistrates, till
the confusion became inextricable. To Julius Cæsar, assisted
by Sosigenes, an eminent Alexandrian astronomer and
mathematician, we owe the neat contrivance of the two years
of 365 and 366 days, and the insertion of one bissextile after
three common years. This important change took place in
the 45th year before Christ, which he ordered to commence
on the 1st of January, *being the day of the new moon immediately following the winter solstice of the year before.* We may

judge of the state into which the reckoning of time had fallen, by the fact, that to introduce the new system it was necessary to enact that the previous year, 46 B.C., should consist of 445 days, a circumstance which obtained for it the epithet of " the year of confusion."

(919.) Had Cæsar lived to carry out into practical effect, as Chief Pontiff, his own reformation, an inconvenience would have been avoided, which at the very outset threw the whole matter into confusion. The words of his edict, establishing the Julian system have not been handed down to us, but it is probable that they contained some expression equivalent to " every fourth year," which the priests misinterpreting after his death to mean (according to the sacerdotal system of numeration) as *counting the leap year newly elapsed as* No. 1. *of the four,* intercalated *every third* instead of every 4th year. This erroneous practice continued during 36 years, in which therefore 12 instead of 9 days were intercalated, and an error of three days produced; to rectify which, Augustus ordered the suspension of all intercalation during three complete *quadriennia,* — thus restoring, as may be presumed his intention to have been, the Julian dates for the future, and re-establishing the Julian system, which was never afterwards vitiated by any error, till the epoch when its own inherent defects gave occasion to the Gregorian reformation. According to the Augustan reform the years A.U.C. 761, 765, 769, &c., which we now call A.D. 8, 12, 16, &c., are leap years. And starting from *this* as a certain fact, (for the statements of the transaction by classical authors are not so precise as to leave *absolutely no doubt* as to the previous intermediate years,) astronomers and chronologists have agreed to reckon backwards in unbroken succession on this principle, and thus to carry the Julian chronology into past time, *as if* it had never suffered such interruption, and *as if* it were certain (which it is not, though we conceive the balance of probabilities to incline that way *) that Cæsar,

---

* With Scaliger, Ideler, and all the best authorities. Yet it has been argued that Cæsar would naturally begin his first *quadriennium* with three ordinary years, deferring the rectification of their accumulated error to the fourth, by

by way of securing the intercalation as a matter of precedent, made his initial year 45 B. C. a leap year. Whenever, therefore, in the relation of any event, either in ancient history, or in modern, previous to the change of style, the time is specified in our modern nomenclature, it is always to be understood as having been identified with the assigned date by threading the mazes (often very tangled and obscure ones) of special and national chronology, and referring the day of its occurrence to its place in the Julian system *so interpreted.*

(920.) Different nations in different ages of the world have of course reckoned their time in different ways, and from different epochs, and it is therefore a matter of great convenience that astronomers and chronologists (as they have agreed on the uniform adoption of the Julian system of years and months) should also agree on an epoch antecedent to them all, to which, as to a fixed point in time, the whole list of chronological eras can be differentially referred. Such an epoch is the noon of the 1st of January, B. C. 4713, which is called the epoch of the Julian period, a cycle of 7980 Julian years, to understand the origin of which, we must explain that of three subordinate cycles, from whose combination it takes its rise, by the multiplication together of the numbers of years severally contained in them, viz : — the Solar and Lunar cycles, and that of the indictions.

(921.) The Solar cycle consists of 28 Julian years, after the lapse of which the same days of the week on the Julian system would always return to the same days of each month throughout the year. For four such years consisting of 1461 days, which is not a multiple of 7, it is evident that the least number of years which will fulfil this condition must be seven times that interval, or 28 years. The place in this cycle for any year A. D., as 1849, is found by adding 9 to the year, and dividing by 28. The remainder is the number sought, 0 being counted as 28.

---

inserting *there* the intercalary day. For the correction of Roman dates during the fifty-two years between the Julian and Augustan reformations, see Ideler, " Handbuch der Mathematischen und Technischen Chronologie," which we take for our guide throughout this chapter.

(922.) The Lunar cycle consists of 19 years or 235 lunations, which differ from 19 Julian years of $365\frac{1}{4}$ days only by about an hour and a half, so that, supposing the new moon to happen on the first of January, in the first year of the cycle, it will happen on that day (or within a very short time of its beginning or ending) again after a lapse of 19 years, and almost certainly on that day, and within an hour and a half of the same hour of the day, after the lapse of four such cycles, or 76 years; and all the new moons in the interval will run on the same days of the month as in the preceding cycle. This period of 19 years is sometimes called the Metonic cycle, from its discoverer Meton, an Athenian mathematician, a discovery duly appretiated by his countrymen, as ensuring the correspondence between the lunar and solar years, the former of which was followed by the Greeks. Public honours were decreed to him for this discovery, a circumstance very expressive of the annoyance which a lunar year of necessity inflicts on a civilized people, to whom a regular and simple calendar is one of the first necessities of life. The cycle of 76 years, a great improvement on the Metonic cycle, was first proposed by Callippus, and is therefore called the Callippic cycle. To find the place of a given year in the lunar cycle, (or as it is called the Golden Number,) add 1 to the number of the year A.D., and divide by 19, the remainder (or 19 if exactly divisible,) is the Golden Number.

(923.) The cycle of the indictions is a period of 15 years used in the courts of law and in the fiscal organization of the Roman empire, under Constantine and his successors, and thence introduced into legal dates, as the Golden Number, serving to determine Easter, was into ecclesiastical ones. To find the place of a year in the indiction cycle, add 3 and divide by 15. The remainder (or 15 if 0 remain) is the number of the indictional year.

(924.) If we multiply together the numbers 28, 19, and 15, we get 7980, and, therefore, a period or cycle of 7980 years will bring round the years of the three cycles again in the same order, so that each year shall hold the same place in all the three cycles as the corresponding year in the foregoing period. As none of the three numbers in question have any common factor, it is evident that no two years in the same compound period can agree in all the three particulars: so that to specify the numbers of a year in each of these cycles is, in fact, to specify the year, if within that long period; which embraces the entire of authentic chronology. The period thus arising of 7980 Julian years, is called the Julian period, and it has been found so useful, that the most competent authorities have not hesitated to declare that, through its employment, light and order were first introduced into chronology.* We owe its invention or revival to Joseph Scaliger, who is said to have received it from the Greeks of Constantinople. The first year of the current Julian period, or that of which the number in each of the three subordinate cycles is 1, was the year 4713 B. C., and the noon of the 1st of January of that year, for the meridian of Alexandria, is the chronological epoch, to which, by all historical eras, are most readily and intelligibly referred, by computing the number of integer days intervening between that epoch and the noon (for Alexandria) of the day, which is reckoned to be the first of the particular era in question. The meridian of Alexandria is chosen as that to which Ptolemy refers the commencement of the era of Nabonassar, the basis of all his calculations.

(925.) Given the year of the Julian period, those of the subordinate cycles are easily determined as above. Conversely, given the years of the solar and lunar cycles, and of the indiction, to determine the year of the Julian period proceed as follows : — Multiply the number of the year in the solar cycle by 4845, in the lunar by 4200, and in the Cycle of the Indictions by 6916, divide the sum of the products by 7980, and the remainder is the year of the Julian period sought.

(926.) The following table contains these intervals for some of the more important historical eras : —

* Ideler, Handbuch, &c. vol. i. p. 77.

Intervals in Days between the Commencement of the Julian Period, and that of some other remarkable chronological and astronomical Eras.

Names by which the Era is usually cited.	First Day current of the Era.	Chronological Designation of the Year.	Current Year of the Julian Period.	Interval Days.
Julian Epochs.	*Julian Dates.*			
Julian period - - - -	Jan. 1.	B C. 4713	1	0
Creation of the world (Usher) -	(Jan. 1.)	4004	710	258,963
Era of the Deluge (Aboulhassan Kuschiar)	Feb. 18.	3102	1612	588,466
Ditto Vulgar Computation - -	(Jan. 1.)	2348	2366	863,817
Era of Abraham (Sir H. Nicholas)	Oct. 1.	2015	2699	985,718
Destruction of Troy, (ditto) -	July 12.	1184	3530	1,289,160
Dedication of Solomon's Temple -	(May 1.)	1015	3699	1,350,815
Olympiads (mean epoch in general use)	July 1.	776	3938	1,438,171
Building of Rome (Varronian epoch, U. C.)	April 22.	753	3961	1,446,502
Era of Nabonassar - - -	Feb. 26.	747	3967	1,448,638
Metonic cycle (Astronomical epoch)	July 15.	432	4282	1,563,831
Callippic cycle Do. (Biot)	June 28.	330	4384	1,599,608
Philippic era, or era of Philip Aridæus	Nov. 12.	324	4390	1,603,398
Era of the Seleucidæ - - -	Oct. 1.	312	4402	1,607,739
Cæsarean era of Antioch - -	Sept. 1.	49	4665	1,703,770
Julian reformation of the Calendar -	Jan. 1.	45	4669	1,704,987
Spanish era - - - - -	Jan. 1.	38	4676	1,707,544
Actian era in Rome - - -	Jan. 1.	30	4684	1,710,466
Actian era of Alexandria - -	Aug. 29.	30	4684	1,710,706
Vulgar or Dionysian era - -	Jan. 1.	A D. 1	4714	1,721,424
Era of Diocletian - - -	Aug. 29.	284	4997	1,825,030
Hejira (astronomical epoch, new moon)	July 15.	622	5335	1,948,439
Era of Yezdegird - - - -	June 16.	632	5345	1,952,063
Gelalæan era (Sir H. Nicholas)	March 14.	1079	5792	2,115,285
Last day of Old Style (Catholic nations)	Oct. 4.	1582	6295	2,299,160
Last day of Old Style in England -	Sept. 2.	1752	6465	2,361,221
Gregorian Epochs.	*Gregorian Dates.*			
New Style in Catholic nations -	Oct. 15.	1582	6295	2,299,161
Ditto in England - - -	Sept. 14.	1752	6465	2,361,222
Commencement of the 19th century.	Jan. 1.	1801	6514	2,378,862
Epoch of Bode's catalogue of stars				
Epoch of the catalogue of stars of the R. Astronomical Society	Jan. 1.	1830	6543	2,389,454
Epoch of the catalogue of the British Association	Jan. 1.	1850	6563	2,396,759

N. B. The civil epochs of the Metonic cycle, and the Hejira, are each one day later than the astronomical, the latter being the epochs of the absolute *new moons*, the former those of the earliest possible visibility of the lunar crescent in a tropical sky. M. Biot has shown that the solstice and new moon not only coincided on the day here set down as the commencement of the Callippic cycle, but that, by a happy coincidence, a *bare* possibility existed of seeing the crescent moon at Athens *within that day, reckoned from midnight to midnight.*

(927.) The determination of the exact interval between any two given dates, is a matter of such importance, and, unless methodically performed, is so very liable to error, that the following rules will not be found out of place. In the first place it must be remarked, generally, that a date, whether of a day or year, always expresses the day or year *current* and not *elapsed*, and that the designation of a year by A. D. or B. C. is to be regarded as the *name* of that year, and *not as a mere number uninterruptedly designating the place of the year in the scale of time.* Thus, in the date, Jan. 5. B. C. 1, Jan. 5 does not mean that 5 days of January in the year in question have elapsed, but that 4 have elapsed, and the 5th is current. And the B. C. 1, indicates that *the first day* of the year so named, (the first year current before Christ,) preceded the first day of the vulgar era by one year. The scale of A. D. and B. C. is not continuous, the year 0 in both being wanting; so that (supposing the vulgar reckoning correct) our Saviour was born in the year B. C. 1.

(928.) *To find the year current of the Julian period,* (J. P.) *corresponding to any given year current* B. C. *or* A. D. If B. C., subtract the number of the year from 4714 : if A. D., add its number to 4713. For examples, see the foregoing table.

(929.) *To find the day current of the Julian period corresponding to any given date, Old Style.* Convert the year B. C. or A. D. into the corresponding year J. P. as above. Subtract 1 and divide the number so diminished by 4, and call Q the integer quotient, and R the remainder. Then will Q be the number of entire *quadriennia* of 1461 days each, and R the residual years, *the first of which is always a leap-year.* Convert Q into days by the help of the first of the annexed tables, and R by the second, and the sum will be the interval between the Julian epoch, and the commencement, Jan. 1. of the year. Then find the days intervening between the beginning of Jan. 1., and that of the date-day by the third table, using the column for a leap-year, where R=0, and that for a common year when R is 1, 2, or 3. Add the days so found to those in Q + R, and the sum will be the days elapsed of the Julian period, the number of which increased by 1 gives the day current.

TABLE 1. Multiples of 1461, the days in a Julian *Quadriennium*.					
1	1461	4	5844	7	10227
2	2922	5	7305	8	11688
3	4383	6	8766	9	13149

TABLE 2. Days in Residual years.	
0	0
1	366
2	731
3	1096

TABLE 3. — Days elapsed from Jan. 1. to the 1st of each Month.

	In a common Year.	In a leap Year.			In a common Year.	In a leap Year.
Jan. 1. -	0	0	July 1. -		181	182
Feb. 1. -	31	31	Aug. 1. -		212	213
March 1. -	59	60	Sept. 1. -		243	244
April 1. -	90	91	Oct. 1. -		273	274
May 1. -	120	121	Nov. 1. -		304	305
June 1. -	151	152	Dec. 1. -		334	335

EXAMPLE. — What is the current day of the Julian period corresponding to the last day of Old Style in England, on Sept. 2., A. D. 1752.

$$
\begin{array}{l}
1752 \\
\underline{4713} \\
6465 \text{ year current.} \\
\underline{1} \\
4)\overline{6464} \text{ years elapsed.} \\
Q=1616 \\
R=0
\end{array}
\qquad
\begin{array}{l}
1000 \\
600 \\
10 \\
6 \\
R=0 \\
\text{Jan. 1. to Sept. 1.} \\
\text{Sept. 1. to Sept. 2.}
\end{array}
\qquad
\begin{array}{l}
1,461,000 \\
876,600 \\
14,610 \\
8,766 \\
0 \\
244 \\
\underline{1} \\
2,361,221 \text{ days elapsed.}
\end{array}
$$

Current day the $2,361,222^d$.

(930.) To find the same for any given date, New Style. Proceed as above, considering the date as a Julian date, and disregarding the change of style. Then from the resulting days, subtract as follows: —

For any date of New Style, antecedent to March 1. A. D. 1700   - 10 days.
After Feb. 28. 1700 and before March 1. A. D. 1800 -   -   - 11 days.
    ,,   1800   ,,   ,,   1900 -   -   - 12 days.
    ,,   1900   ,,   ,,   2100 -   -   - 13 days, &c.

(931.) *To find the interval between any two dates, whether of Old or New Style, or one of one, and one of the other.* Find the day current of the Julian period corresponding to each date, and their difference is the interval required. If the dates contain hours, minutes, and seconds, they must be annexed to their respective days current, and the subtraction performed as usual.

(932.) The Julian rule made every fourth year, without exception, a bissextile. This is, in fact, an over-correction; it supposes the length of the tropical year to be $365\frac{1}{4}^d$, which is too great, and thereby induces an error of 7 days in 900 years, as will easily appear on trial. Accordingly, so early as the year 1414, it began to be perceived that the equinoxes were gradually creeping away from the 21st of March and September, where they ought to have always fallen had the Julian year been exact, and happening (as it appeared) too early. The necessity of a fresh and effectual reform in the calendar was from that time continually urged, and at length admitted. The change (which took place under the popedom of Gregory XIII.) consisted in the omission of ten nominal days after the 4th of October, 1582, (so that the next day was called the 15th, and not the 5th,) and the promulgation of the rule already explained for future regulation. The change was adopted immediately in all catholic countries; but more slowly in protestant. In England, " the change of style," as it was called, took place after the 2d of September, 1752, eleven nominal days being then struck out; so that, the last day of Old Style being the 2d, the first of New Style (the next day) was called the 14th, instead of the 3d. The same legislative enactment which established the Gregorian year in England in 1752, shortened the preceding year, 1751, by a full quarter. Previous to that time, the year was held to begin with the 25th March, and the year A.D. 1751 did so accordingly; but that year was not suffered to run out, but was supplanted on the 1st January by the year 1752, which it was enacted should commence on that day, as well as every subsequent year. Russia is now the only country in Europe in which the Old Style is still adhered to, and (another secular year having elapsed) the difference between the European and Russian dates amounts, at present, to 12 days.

(933.) It is fortunate for astronomy that the confusion of dates, and the irreconcilable contradictions which historical statements too often exhibit, when confronted with the best knowledge we possess of the ancient reckonings of time, affect recorded observations but little. An astronomical observation, of any striking and well-marked phænomenon, carries with it, in most cases, abundant means of recovering its exact date, when any tolerable approximation is afforded to it by chronological records; and, so far from being abjectly dependent on the obscure and often contradictory dates, which the comparison of ancient authorities indicates, is often itself the surest and most convincing evidence on which a chronological epoch can be brought to rest. Remarkable eclipses, for instance, now that the lunar theory is thoroughly understood, can be calculated back for several thousands of years, without the possibility of mistaking the day of their occurrence. And, whenever any such eclipse is so interwoven with the account given by an ancient author of some historical event, as to indicate precisely the interval of time between the eclipse and the event, and at the same time completely to identify the eclipse, that date is recovered and fixed for ever.*

(934.) The days thus parcelled out into years, the next step to a perfect knowledge of time is to secure the identification of each day, by imposing on it a name universally known and employed. Since, however, the days of a whole year are too numerous to admit of loading the memory with distinct names for each, all nations have felt the necessity of breaking them down into parcels of a more moderate extent; giving names to each of these parcels, and particularizing the days in each by numbers, or by some especial indication. The lunar month has been resorted to in many instances; and some nations have, in fact, preferred a lunar to a solar

* See the remarkable calculations of Mr. Baily relative to the celebrated solar eclipse which put an end to the battle between the kings of Media and Lydia, B. C. 610. Sept. 30. Phil. Trans. ci. 220.

chronology altogether, as the Turks and Jews continue to do to this day, making the year consist of 12 lunar months, or 354 days. Our own division into twelve unequal months is entirely arbitrary, and often productive of confusion, owing to the equivoque between the lunar and calendar month.* The intercalary day naturally attaches itself to February as the shortest.

(935.) Astronomical time, reckons from the noon of the current day; civil from the preceding midnight, so that the two dates coincide only during the earlier half of the astronomical, and the later of the civil day. This is an inconvenience which might be remedied by shifting the astronomical epoch to coincidence with the civil. There is, however, another inconvenience, and a very serious one, to which both are liable, inherent in the nature of the day itself, which is a local phænomenon, and commences at different instants of absolute time, under different meridians, whether we reckon from noon, midnight, sunrise, or sunset. In consequence, all astronomical observations require in addition to their date, to render them comparable with each other, the longitude of the place of observation from some meridian, commonly respected by all astronomers. For geographical longitudes, the Isle of Ferroe has been chosen by some as a common meridian, indifferent (and on that very account offensive) to all nations. Were astronomers to follow such an example, they would probably fix upon Alexandria, as that to which Ptolemy's observations and computations were reduced, and as claiming on that account the respect of all while offending the national egotism of none. But even this will not meet the whole difficulty. It will still remain doubtful, on a meridian 180° remote from that of Alexandria, what day is intended by any given date. Do what we will, when it is Monday the 1st of January, 1849, in one part of the world, it will be Sunday the 31st of December, 1848, in another, so long as time is reckoned by local hours. This equivoque, and the necessity

* " A month in law is a lunar month or twenty-eight days, (!!) unless otherwise expressed."— *Blackstone*, ii. chap. 9., " a lease for twelve months is only for forty-eight weeks." *Ibid.*

of specifying the geographical locality as an element of the date, can only be got over by a reckoning of time which refers itself to some event, real or imaginary, common to all the globe. Such an event is the passage of the sun through the vernal equinox, or rather the passage of an imaginary sun, supposed to move with perfect equality, through a vernal equinox supposed free from the inequalities of nutation, and receding upon the ecliptic with *perfect* uniformity. The actual equinox is variable, not only by the effect of nutation, but by that of the inequality of precession resulting from the change in the plane of the ecliptic due to planetary perturbation. Both variations are, however, periodical, the one, in the short period of 19 years, the other, in a period of enormous length, hitherto uncalculated, and whose maximum of fluctuation is also unknown. This would appear, at first sight, to render impracticable the attempt to obtain from the sun's motion any rigorously uniform measure of time. A little consideration, however, will satisfy us that such is not the case. The solar tables, by which the apparent place of the sun in the heavens is represented with almost absolute precision from the earliest ages to the present time, are constructed upon the supposition that a certain angle, which is called " the sun's mean longitude," (and which is in effect the sum of the mean sidereal motion of the sun, *plus* the mean sidereal motion of the equinox in the opposite direction, *as near as it can be obtained* from the accumulated observations of twenty-five centuries,) increases with rigorous uniformity as time advances. The conversion of this mean longitude into time at the rate of 360° to the mean tropical year, (such as the tables assume it,) will therefore give us both the unit of time, and the uniform measure of its lapse which we seek. It will also furnish us with an epoch, not indeed marked by any real event, but not on that account the less positively fixed, being connected, through the medium of the tables, with every single observation of the sun on which they have been constructed and with which compared.

(936.) Such is the simplest abstract conception of equinoctial time. It is the mean longitude of the sun of *some*

T T

one approved set of solar tables, converted into time at the
rate of 360° to the tropical year. Its unit is the mean
tropical year which those tables assume *and no other,* and its
epoch is the mean vernal equinox of these tables for the
current year, or the instant when the mean longitude of the
tables is rigorously 0, according to the assumed mean motion
of the sun and equinox, the assumed epoch of mean longitude,
and the assumed equinoctial point on which the tables have
been computed, *and no other.* To give complete effect to
this idea, it only remains to specify the particular tables fixed
upon for the purpose, which ought to be of great and
admitted excellence, since, once decided on, the very essence
of the conception is that *no subsequent alteration in any respect
should be made, even when the continual progress of astronomical
science shall have shown any one or all of the elements concerned
to be in some minute degree erroneous* (as necessarily they
must), *and shall have even ascertained the corrections they
require* (to be themselves again corrected, when another step
in refinement shall have been made).

(937.) Delambre's solar tables (in 1828) when this mode
of reckoning time was first introduced, appeared entitled to
this distinction.   According to these tables, the sun's mean
longitude was 0°, or the mean vernal equinox occurred, in the
year 1828, on the 22d of March at 1^h 2^m $59^s{\cdot}05$ mean time
at Greenwich, and therefore at 2^h 12^m $20^s{\cdot}55$ mean time at
Paris, or 2^h 56^m $34^s{\cdot}55$ mean time at Berlin, at which instant,
therefore, the equinoctial time was 0^d 0^h 0^m $0^s{\cdot}00$, being the
commencement of the 1828th year current of equinoctial
time, if we choose to date from the mean tabular equinox,
nearest to the vulgar era, or of the 6541st year of the Julian
period, if we prefer that of the first year of that period.

(938.) Equinoctial time then dates from the mean vernal
equinox of Delambre's solar tables, and its unit is the mean
tropical year of these tables ($365^d{\cdot}242264$).   Hence, having
the fractional part of a day expressing the difference between
the mean local time at any place (suppose Greenwich) on
any one day between two consecutive mean vernal equinoxes,
that difference will be the same for every other day in the

same interval. Thus, between the mean equinoxes of 1828 and 1829, the difference between equinoctial and Greenwich time is $0^d \cdot 956261$ or $0^d\ 22^h\ 57^m\ 0^s \cdot 95$, which expresses the equinoctial day, hour, minute, and second, corresponding to mean noon at Greenwich on March 23. 1828, and for the noons of the 24th, 25th, &c., we have only to substitute 1d, 2d, &c. for 0^d, retaining the same decimals of a day, or the same hours, minutes, &c., up to and including March 22. 1829. Between Greenwich noon of the 22d and 23d of March, 1829, the 1828th equinoctial year terminates, and the 1829th commences. This happens at $0^d \cdot 286003$, or at $6^h\ 51^m\ 50^s \cdot 66$ Greenwich mean time, after which hour, and until the next noon, the Greenwich hour added to equinoctial time $364^d \cdot 956261$ will amount to more than $365 \cdot 242264$, a complete year, which has therefore to be subtracted to get the equinoctial date in the next year, corresponding to the Greenwich time. For example, at $12^h\ 0^m\ 0^s$ Greenwich mean time, or $0^d \cdot 500000$, the equinoctial time will be $364 \cdot 956261 + 0 \cdot 500000 = 365 \cdot 456261$, which being greater than $365 \cdot 242264$, shows that the equinoctial year current has changed, and the latter number being subtracted, we get $0^d \cdot 213977$ for the equinoctial time of the 1829th year current corresponding to March 22., 12^h Greenwich mean time.

(939.) Having, therefore, the fractional part of a day for any one year expressing the equinoctial hour, &c., at the mean noon of any given place, that for succeeding years will be had by subtracting $0^d \cdot 242264$, and its multiples, from such fractional part (increased if necessary by unity), and for preceding years by adding them. Thus, having found $0 \cdot 198525$ for the fractional part for 1827, we find for the fractional parts for succeeding years up to 1853 as follows * : —

---

* These numbers differ from those in the Nautical Almanack, and would require to be substituted for them, to carry out the idea of equinoctial time as above laid down. In the years 1828—1833, the late eminent editor of that work used an equinox slightly differing from that of Delambre, which accounts for the difference in those years. In 1834, it would appear that a deviation both from the principle of the text and from the previous practice of that ephemeris took place, in deriving the fraction for 1834 from that for 1833, which has been ever since perpetuated. It consisted in rejecting the mean longitude of Delambre's tables, and adopting Bessel's correction of that element. The effect

1828	·956261	1835	·260413	1842	·564565	1848	·110981
1829	·713997	1836	·018149	1843	·322301	1849	·868717
1830	·471733	1837	·775885	1844	·080037	1850	·626453
1831	·229469	1838	·533621	1845	·837773	1851	·384189
1832	·987205	1839	·291357	1846	·595509	1852	·141925
1833	·744941	1840	·049093	1847	·353245	1853	·899661
1834	·502677	1841	·806829				

of this alteration was to insert $3^m 3^s·68$ *of purely imaginary time*, between the end of the equinoctial year 1833 and the beginning of 1834, or, in other words, to make the interval between the noons of March 22. and 23. 1834, $24^h 3^m 3^s·68$, when reckoned by equinoctial time. In 1835, and in all subsequent years, a further departure from the principle of the text took place by substituting Bessel's tropical year of 365·2422175, for Delambre's. Thus the whole subject has fallen into confusion, and we have to choose between reverting to the original design in its integrity, or to continue the present practice (eschewing all further change) for future years.

APPENDIX.

I. LISTS OF NORTHERN AND SOUTHERN STARS, WITH THEIR APPROXIMATE MAGNITUDES, ON THE VULGAR AND PHOTOMETRIC SCALE.

1. NORTHERN STARS.

Star	Vulg.	Phot.	Star	Vulg.	Phot.	Star	Vulg.	Phot.	Star	Vulg.	Phot.
Arcturus	0·77	1·18	γ Cassiopeiæ	2·52	2·93	η Draconis	3·02	3·43	ε Aurigæ (Var.)	3·37	3·78
Capella	1·0	1·4	α Andromedæ	2·54	2·95	β Draconis	3·06	3·47	γ Lyncis	3·39	3·80
Lyra	1·0	1·4	α Cassiopeiæ	2·57	2·98	β Arietis	3·09	3·50	ζ Draconis	3·40	3·81
Procyon	1·0	1·4	γ Geminorum	2·59	3·00	γ Pegasi	3·11	3·52	π Herculis	3·41	3·82
α Orionis (Var.)	1·0	1·43	Algol (Var.)	2·62	3·03	ε Virginis ?	3·14	3·55	β Canis Min. ?	3·41	3·82
Aldebaran	1·1	1·5	ε Pegasi	2·62	3·03	θ Aurigæ	3·17	3·58	ζ Tauri	3·42	3·83
α Aquilæ	1·28	1·69	γ Draconis	2·62	3·03	β Herculis	3·18	3·59	δ Draconis	3·42	3·83
Pollux	1·6	2·0	β Leonis	2·63	3·04	α Canum Ven.	3·22	3·63	μ Geminorum	3·42	3·83
Regulus	1·6	2·0	α Ophiuchi	2·63	3·04	β Ophiuchi	3·23	3·64	γ Boötis	3·43	3·84
α Cygni	1·90	2·31	β Cassiopeiæ	2·63	3·04	δ Cygni	3·24	3·65	ε Geminorum	3·43	3·84
Castor	1·94	2·35	γ Cygni	2·63	3·04	ε Persei	3·26	3·67	δ Herculis	3·44	3·85
ε Ursæ (Var.)	1·95	2·36	α Pegasi	2·65	3·06	η Tauri ?	3·26	3·67	δ Geminorum	3·44	3·85
α Ursæ (Var.)	1·96	2·37	β Pegasi	2·65	3·06	ζ Persei	3·27	3·68	η Orionis	3·45	3·86
α Persei	2·07	2·48	α Coronæ	2·69	3·10	ζ Herculis	3·28	3·69	β Cephei	3·45	3·86
γ Orionis (Var.)	2·18	2·59	β Ursæ	2·71	3·12	ι Aurigæ	3·29	3·70	θ Ursæ	3·45	3·86
β Tauri	2·28	2·69	ε Boötis	2·77	3·18	γ Ursæ Minor	3·30	3·71	ι Ursæ	3·46	3·87
Polaris	2·28	2·69	ε Cygni	2·80	3·21	β Pegasi	3·31	3·72	η Aurigæ	3·46	3·87
γ Leonis	2·34	2·75	α Cephei	2·88	3·29	ζ Aquilæ	3·32	3·73	γ Lyræ	3·47	3·88
α Arietis	2·40	2·81	α Serpentis	2·90	3·31	β Cygni	3·33	3·74	γ Geminorum	3·48	3·89
ζ Ursæ	2·43	2·84	δ Leonis	2·92	3·33	γ Persei	3·34	3·75	γ Cephei	3·48	3·89
β Andromedæ	2·45	2·86	γ Aquilæ	2·94	3·35	μ Ursæ	3·35	3·76	κ Ursæ	3·49	3·90
β Aurigæ	2·48	2·89	δ Cassiopeiæ	2·98	3·39	β Triang. Bor.	3·35	3·76	ε Cassiopeiæ	3·49	3·90
γ Andromedæ	2·50	2·91	η Boötis	3·01	3·42	δ Persei	3·36	3·77	δ Aquilæ	3·50	3·91
						ψ Ursæ	3·36	3·77			

2. SOUTHERN STARS.

Star	Mag. Vulg.	Mag. Phot.	Star	Mag. Vulg.	Mag. Phot.	Star	Mag. Vulg.	Mag. Phot.	Star	Mag. Vulg.	Mag. Phot.
Sirius	0·08	0·49	α Hydræ	2·30	2·71	η Centauri	2·91	3·32	β Eridani	3·26	3·67
η Argûs (Var.)	—	—	δ Canis	2·32	2·73	κ Argûs	2·94	3·35	θ Argûs	3·26	3·67
Canopus	0·29	0·70	α Pavonis	2·33	2·74	β Corvi	2·95	3·36	β Hydri	3·27	3·68
α Centauri	0·59	1·00	β Gruis	2·36	2·77	β Scorpii	2·96	3·37	ε Corvi	3·28	3·69
Rigel	0·82	1·23	σ Sagittarii	2·41	2·82	ζ Centauri	2·96	3·37	Aræ	3·31	3·72
α Eridani	1·09	1·50	δ Argûs	2·42	2·83	ζ Ophiuchi	2·97	3·38	α Toucani	3·32	3·73
β Centauri	1·17	1·58	β Ceti	2·46	2·87	α Aquarii	2·97	3·38	β Capricorni	3·32	3·73
α Crucis	1·2	1·6	λ Argûs	2·46	2·87	π Argûs	2·98	3·39	ρ Argûs	3·32	3·73
Antares	1·2	1·6	θ Centauri	2·54	2·95	δ Centauri	2·99	3·40	π Scorpii	3·35	3·76
Spica	1·38	1·79	β Canis	2·58	2·99	α Leporis	3·00	3·41	β Leporis	3·35	3·76
Fomalhaut	1·54	1·95	κ Orionis	2·59	3·00	δ Ophiuchi	3·00	3·41	γ Lupi	3·36	3·77
β Crucis	1·57	1·98	δ Orionis	2·61	3·02	ζ Sagittarii	3·01	3·42	υ Scorpii	3·37	3·78
α Gruis	1·66	2·07	γ Centauri	2·68	3·09	π Ophiuchi	3·05	3·46	ι Orionis	3·37	3·78
γ Crucis	1·73	2·14	ε Scorpii	2·71	3·12	β Libræ	3·07	3·48	α Aræ	3·40	3·81
ε Orionis	1·84	2·25	ζ Argûs	2·72	3·13	γ Virginis	3·08	3·49	π Sagittarii	3·40	3·81
ε Canis	1·86	2·27	α Phœnicis	2·78	3·19	μ Argûs	3·08	3·49	α Muscæ	3·43	3·84
λ Scorpii	1·87	2·28	ι Argûs	2·80	3·21	δ Sagittarii	3·11	3·52	α Hydri?	3·44	3·85
ζ Orionis	2·01	2·42	α Lupi	2·82	3·23	α Libræ	3·12	3·53	τ Scorpii	3·44	3·85
β Argûs	2·03	2·44	ε Centauri	2·82	3·23	λ Sagittarii	3·13	3·54	ζ Hydræ	3·45	3·86
γ Argûs	2·08	2·49	η Canis	2·85	3·26	β Lupi	3·14	3·55	γ Hydræ	3·46	3·87
ε Argûs	2·18	2·59	β Aquarii	2·86	3·27	α Columbæ	3·15	3·56	β Trianguli	3·46	3·87
α Trianguli A.	2·23	2·64	δ Scorpii	2·89	3·30	ι Centauri	3·20	3·61	σ Scorpii	3·50	3·91
ε Sagittarii	2·26	2·67	η Ophiuchi	2·90	3·31	δ Capricorni	3·20	3·61	τ Argûs	3·50	3·91
θ Scorpii	2·29	2·70	γ Corvi			δ Corvi	3·22	3·63			

II. Synoptic Table of the Elements of the Planetary System.

Name of Body.	Mean Distance from the Sun, or Semi-axis.	Mean Sidereal Period in Mean Solar Days.	Excentricity in parts of the Semi-axis.	Inclination of Orbit to the Ecliptic.			Longitude of the Ascending Node.		
				o	'	''	o	'	''
Sun									
Mercury	0·3870981	87·9692580	0·2055149	7	0	9·1	45	57	30·9
Venus	0·7233316	224·7007869	0·0068607	3	23	28·5	74	54	12·9
Earth	1·0000000	365·2563612	0·0167836						
Mars	1·5236923	686·9796458	0·0933070	1	51	6·2	48	0	3·5
Flora	2·2016870	1193·249	0·1565570	5	53	4·8	110	18	12·0
Vesta	2·3610810	1325·147	0·0895694	7	8	29·7	103	23	31·6
Iris	2·3806240	1341·636	0·2299424	5	28	15·9	259	48	10·2
Metis	2·3856070	1345·850	0·1202532	5	34	27·8	68	32	17·4
Hebe	2·4257866	1379·994	0·2001805	14	47	5·6	138	29	42·6
Astraea	2·5770470	1511·095	0·1880586	5	19	22·7	141	25	14·6
Juno	2·6708370	1594·296	0·2548847	13	3	22·1	170	54	45·6
Ceres	2·7680510	1682·125	0·0766523	10	37	4·4	80	48	46·6
Pallas	2·7728580	1686·510	0·2398150	34	37	33·1	172	43	59·7
Jupiter	5·2027760	4332·5848212	0·0481621	1	18	51·3	98	26	18·9
Saturn	9·5387861	10759·2198174	0·0561505	2	29	35·7	111	56	37·4
Uranus	19·1823900	30686·8208296	0·0466794	0	46	28·4	72	59	35·3
Neptune	30·0368000	60126·7100000	0·0087195	1	46	59·0	130	5	11·0

ELEMENTS OF THE PLANETARY SYSTEM.

Name of Body.	Longitude of the Perihelion. (° , ")	Mean {Longitude (L)} {Anomaly (A)} at Epoch. (° , ")	Epoch of the Elements in M.T. at {Greenwich (G)} {Berlin (B)}	Mass (denominator of fraction, the Sun being 1.).	Diameter in Miles.	Density.	Time of Rotation on Axis. (h. m.)
Sun	-	-	1801. Jan. 1. 0h G.	1	882000	0·25	607 48
Mercury	74 21 46·9	L=166 0 48·6	Do.	4865751	3140	1·12	24 5
Venus	128 43 53·1	11 33 3·0	Do.	401839	7800	0·92	23 21
Earth	99 30 5·0	100 39 10·2	Do.	389551	7926	1·00	24 0
Mars	332 23 56·6	64 22 55·5		2680337	4100	0·95	24 37
Flora	33 0 40·8	A=35 48 7·0	1848. Jan. 1. 0h B.				
Vesta	250 46 32·2	225 44 18·8	1850. Jan. 9. 0h B.	-	250 ?		
Iris	41 41 13·5	330 41 54·0	1848. Jan. 1. 0h B.				
Metis	70 33 42·8	146 30 18·5	1847. May 5. 12h B.				
Hebe	14 50 50·3	275 8 51·3	1847. Jan. 1. 0h B.				
Astraea	135 20 47·0	318 45 3·3	1846. Jan. 1. 0h B.		79 ?	- -	27 0 ?
Juno	54 24 12·8	124 31 10·8	1850. Apr. 8. 0h B.	-	163 ?		
Ceres	147 46 12·4	219 6 29·5	1850. Sept. 25. 0h B.	-			
Pallas	121 21 48·5	217 31 10·6	1850. Aug. 23. 0h B.				
Jupiter	11 8 34·6	L=112 15 23·0	1801. Jan. 1. 0h G.	1047·871	87000	0·24	9 56
Saturn	89 9 29·8	135 20 6·5	Do.	3501·600	79160	0·14	10 29
Uranus	167 31 16·1	177 48 23·0	Do.	24905	34500	0·24	9 30 ?
Neptune	47 12 56·7	330 44 41·8	1848. Jan. 1. 0h G.	18780	41500	0·14	

Note.—The elements of the orbits of Mercury, Venus, the Earth, Mars, Jupiter, Saturn, and Uranus, are those given by the late F. Baily, Esq., in his " Astronomical Tables and Formulæ," and are the same with those which form the basis of Delambre's tables, embodying the formulæ of Laplace. The elements of Uranus and Neptune can only be regarded as provisional; those of the former requiring considerable corrections, necessitated by the discovery of Neptune, but which, not being yet finally ascertained, by reason of the uncertainty still attending on the mass and elements of the latter planet, it was thought better to leave the old elements untouched than to give an imperfect rectification of them. The masses of the planets are those most recently adopted by Encke, (Ast. Nachr., No. 443.) on mature consideration of all the authorities, that of Neptune excepted, which is Prof. Peirce's determination from Bond's and Lassell's observation of the satellite discovered by the latter. The densities are Hansen's (Ast. Nachr. No. 443.).

The Elements of Vesta, Juno, Ceres, and Pallas, are the osculating elements for 1850, computed by Encke (Ast. Nachr. No. 636.). Those of Flora, Iris, Metis, Hebe, and Astræa, are from the respective computations of Brunnow, (Ast. Nachr. No. 645.), Galle (Ast. Nachr., No. 643.), Sontag (Ast. Nachr., No. 644.), Lehman (Ast. Nachr., No. 636.), and D'Arrest (Ast. Nachr., No. 626.). The five last-named planets being so recently discovered, these elements may undergo material rectification from future observation.

III.

SYNOPTIC TABLE OF THE ELEMENTS OF THE ORBITS OF THE SATELLITES, SO FAR AS THEY ARE KNOWN.

N. B. — The distances are expressed in equatorial radii of the primaries. The epoch is Jan. 1. 1801, unless otherwise expressed. The periods, &c. are expressed in mean solar days.

1. THE MOON.

Mean distance from earth	59^r·96435000
Mean sidereal revolution	27^d·321661418
Mean synodical ditto	29^d·530588715
Excentricity of orbit	0·054844200
Mean revolution of nodes	6793^d·391080
Mean revolution of apogee	3232^d·575343
Mean longitude of node at epoch	$13°$ $53'$ $17''$·7
Mean longitude of perigee at do.	266 10 7 ·5
Mean inclination of orbit	5 8 47 ·9
Mean longitude of moon at epoch	118 17 8 ·3
Mass, that of earth being 1,	0·011399
Diameter in miles	2153
Density, that of the earth being 1,	0·5657

2. SATELLITES OF JUPITER.

Sat.	Sidereal Revolution.				Mean Distance.	Inclination of Orbit to a fixed Plane proper to each.	Inclination of the fixed Plane to Jupiter's Equator.	Retrograde Revolution of Nodes on the fixed Plane.	Mass, that of Jupiter being 1,000,000,000.
	d	h	m	s		° ′ ″	° ′ ″	Years.	
1	1	18	27	33·506	6·04853	0 0 0	0 0 6	-	17328
2	3	13	14	36·393	9·62347	0 27 50	0 1 5	29·9142	23235
3	7	3	42	33·362	15·35024	0 12 20	0 5 2	141·7390	88497
4	16	16	31	49·702	26·99835	0 14 58	0 24 4	531·0000	42659

The excentricities of the 1st and 2d Satellites are insensible, those of the 3d and 4th small, but variable, in consequence of their mutual perturbation.

3. SATELLITES OF SATURN.

Name and order of Satellite.	Sidereal Revolution.				Mean distance.	Epoch of Elements.	Mean Longitude at the Epoch.			Excentricity.	Perisaturnium.		
	d	h	m	s			°	′	″		°	′	″
1. Mimas	0	22	37	22·9	3·3607	1790·0	256	58	48				
2. Enceladus	1	8	53	6·7	4·3125	1836·0	67	41	36				
3. Tethys	1	21	18	25·7	5·3396	Ditto	313	43	48	0·04 ?	54 ?		
4. Dione	2	17	41	8·9	6·8398	Ditto	327	40	48	0·02 ?	42 ?		
5. Rhea -	4	12	25	10·8	9·5528	Ditto	353	44	0	0·02 ?	95 ?		
6. Titan -	15	22	41	25·2	22·1450	1830·0	137	21	24	0·029314	256	38	11
7. Hyperion -	22	12	?		28±	-	-						
8. Iapetus -	79	7	53	40·4	64·3590	1790·0	269	37	48				

The longitudes are reckoned in the plane of the ring from its *descending* node with the ecliptic. The first seven satellites move in, or very nearly in, its plane; that of the 8th is inclined to it at an angle about half way intermediate between the planes of the ring and of the planet's orbit. The apsides of Titan have a direct motion of 30′ 28″ per annum in longitude (on the ecliptic).

The discovery of Hyperion is quite recent, having been made on the same night (Sept. 19. 1848), by Mr. Lassell, of Liverpool, and Prof. Bond, of Cambridge, U. S. Its distance and period are as yet hardly more than conjecture. Messrs. Kater, Encke, and Lassell agree in representing the ring of Saturn as subdivided by several narrow dark lines, besides the broad black divisions which ordinary telescopes show.

4. Satellites of Uranus.

Sat.	Sidereal Revolution.				Mean distance.	Epochs of Passage and most ascending Node of Orbits. G. T.	Nodes and Inclinations.
	d	h	m	s			
1	4	?			17·0		The orbits are inclined at an angle of about 78° 58′ to the ecliptic in a plane whose ascending node is in long. 165° 30′ (Equinox of 1798). Their motion is *retrograde*. The orbits are nearly circular.
2	8	16	56	31·3	19·8 ?	1787. Feb. 16th. 0h 10m	
3	10	23	?		22·8 ?		
4	13	11	7	12·6	45·5 ?	1787. Jan. 7th 0h 28m	
5	38	2	?		91·0 ?		
6	107	12	?				

5. Satellites of Neptune.

One only has been *certainly* observed,—its approximate period being 5d 20h 50m 45s,—distance about 12 radii of the planet.

IV.
Elements of Periodical Comets at their Last Appearance.

	Halley's.	Encke's.	Biela's.	Faye's.	De Vico's.	Brorsen's.
τ	1835. Nov. 15. 22^h 41^m 22^s	1845. Aug. 9. 15^h 11^m 11^s	1846. Feb. 11. 0^h 2^m 50^s	1843. Oct. 17. 9^h 42^m 16^s	1844. Sept. 2. 11^h 36^m 53^s	1846. Feb. 25. 9^h 13^m 35^s
ϖ	304° 31′ 32″	157° 44′ 21″	109° 5′ 47″	49° 34′ 19″	342° 31′ 15″	116° 28′ 34″
☊	55 9 59	334 19 33	245 56 58	209 29 19	63 49 31	102 39 36
ι	17 45 5	13 7 34	12 34 14	11 22 31	2 54 45	30 55 7
a	17·98796	2·21640	3·50182	3·81179	3·09946	3·15021
ε	0·967391	0·847436	0·755471	0·555962	0·617256	0·793629
P	27865^d·74	1205^d·23	2393^d·52	2718^d·26	1993^d·09	2042·24
	Retrograde.	Direct.	Direct.	Direct.	Direct.	Direct.

τ is the time of perihelion passage; ϖ the longitude of the perihelion; and ☊ that of the ascending node for the epoch of the perihelion; ι, the inclination to the ecliptic; a, the semi-axis; ε, the excentricity; P, the period in days.

N. B. The reader will find a complete list of elements of all known comets up to June, 1847, by all their several computors, in Prof. Encke's edition of Olbers's "Abhandlung über die leichteste und bequemste Methode die Bahn eines Cometen zu berechnen." The list is compiled by Dr. Galle. It contains orbits of 178 distinct comets. From an examination of these orbits we collect the following, as a more correct statement of cometary statistics than that in art. 601. viz.:— Retrograde comets under 10° inclination, 3 out of 15; under 20°, 9 out of 29. Retrograde comets, moving in orbits sensibly elliptic, under 17° inclination, 0 out of 9. In such orbits, of all inclinations from 0 to 90°, 11 out of 37. Thus we see that the induction of that article is materially strengthened by the enlarged field of comparison.

INDEX.

B.

Barometer, nature of its indication, 33.
Use in calculating refraction, 43. In
determining heights, 287.
Belts of Jupiter, 512. Of Saturn, 514.
Benzenberg's principle of collimation,
179.
Bessel, his results respecting the figure
of the earth, 220. Discovers parallax
of 61 Cygni, 812.
Biela's comet, 579...
Biot, his aëronautic ascent, 32.
Bode, his (so called) law of planetary
distances, 505. Violated in the case
of Neptune, 507.
Borda, his principle of repetition, 198.
Bouvard, his suspicion of extraneous
influence on Uranus, 760.

C.

Cæsar, his reform of the Roman calen-
dar, 917.
Calendar, Julian, 917.   Gregorian'
914...
Cause and effect, 439. and note.
Center of the earth, 80. Of the sun, 462.
Of gravity, 360. Revolution about,
452.
Centrifugal force. Elliptic form of earth
produced by, 224. Illustrated, 225.
Compared with gravity, 229. Of a
body revolving on the earth's sur-
face, 452.
Ceres, discovery of, 505.
Challis, Prof., 506, note.
Charts, celestial, 111.   Construction
of, 291... Bremiker's, 506, and note.
Chinese records of comets, 574   Of
irregular stars, 831.
Chronometers, how used for determining
differences of longitude, 255.
Circle, arctic and antarctic, 94. Verti-
cal, 100. Hour, 106. Divided, 163.
Meridian, 174. Reflecting, 197. Re-
peating, 198.   Galactic, 793.
Clepsydra, 150.
Clock, 151.   Error and rate of, how
found, 253.
Clouds, greatest height of, 34.   Magel-
lanic, 892...
Clusters of stars, 864... Globular, 867.
Irregular, 869.
Collimation, line of, 155.
Collimator, 178...
Coloured stars, 851...
Colures, 307.
Comets, 554.   Seen in day-time, 555.
590.   Tails of, 556...566. 599.   Ex-

treme tenuity of, 558. General description of, 560. Motions of, and
described, 561...   Parabolic, 564.
Elliptic, 567... Hyperbolic, 564. Dimensions of, 565.   Of Halley, 567...
Of Cæsar, 573.   Of Encke, 576.   Of
Biela, 579.   Of Faye, 584.   Of
Lexell, 585.   Of De Vico, 586.   Of
Brorsen, 587.   Of Peters, 588.   Sy-
nopsis of elements (Appendix).   Increase of visible dimensions in receding from the sun, 571.580.   Great,
of 1843, 589... Its supposed identity
with many others, 594... Interest attached to subject, 597.   Cometary
statistics, and conclusions therefrom,
601.
Commensurability (near) of mean motions; of Saturn's satellites, 550. Of
Uranus and Neptune, 669. and note.
Of Jupiter and Saturn, 720.   Earth
and Venus, 726.   Effects of, 719.
Compensation of disturbances, how effected, 719. 725.
Compression of terrestrial spheroid, 221.
Configurations, inequalities depending
on, 655...
Conjunctions, superior and inferior, 473.
Perturbations chiefly produced at, 713.
Consciousness of effect when force is
exerted, 439.
Constellations, 60. 301.   How brought
into view by change of latitude, 52.
Rising and setting of, 58.
Copernican explanation of diurnal mo-
tion, 76.   Of apparent motions of
sun and planets, 77.
Correction of astronomical observations,
324... s.   Uranographical summary,
view of, 342...
Culminations, 125.   Upper and lower,
126.
Cycle, of conjunctions of disturbing
and disturbed planets, 719.   Metonic, 926.   Callippic, *ib.*   Solar, 921.
Lunar, 922.   Of indictions, 923.

D.

Day, solar, lunar, and sidereal, 143.
Ratio of sidereal to solar, 305. 909.
911.   Solar unequal, 146.   Mean
ditto, invariable, 908.   Civil and
astronomical, 147.   Intercalary, 916.
Days elapsed between principal chronological eras, 926.   Rules for reckoning between given dates, 927.
Declination, 105.   How obtained, 295.
Definitions, 82...

THE END.

Plate II.

Fig. 1

Fig. 2.

Fig. 3

H. Adlard, sc.

London; Longman, Brown, Green, & Longmans.

Plate III.

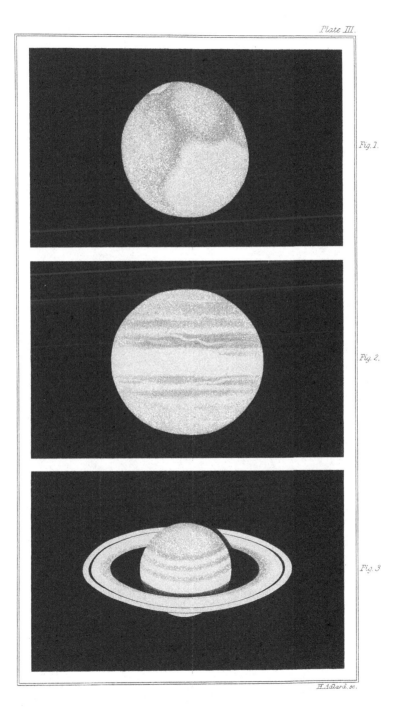

Fig. 1.

Fig. 2.

Fig. 3

H. Adlard, sc.

London; Longman, Brown, Green, & Longmans.

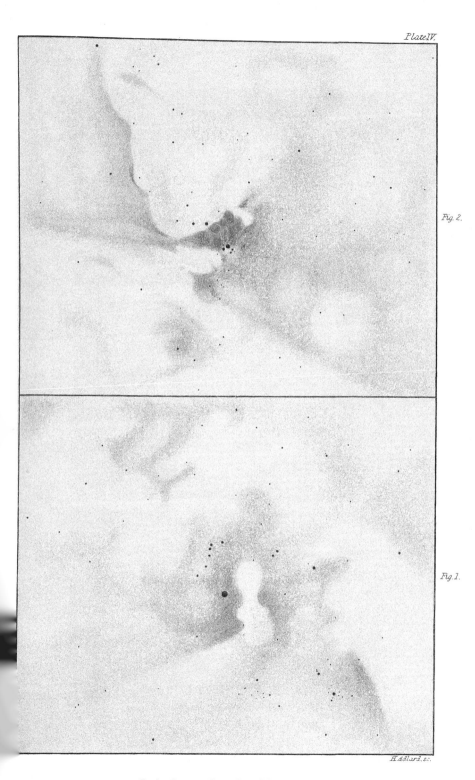

London: Longman, Brown, Green, & Longmans.

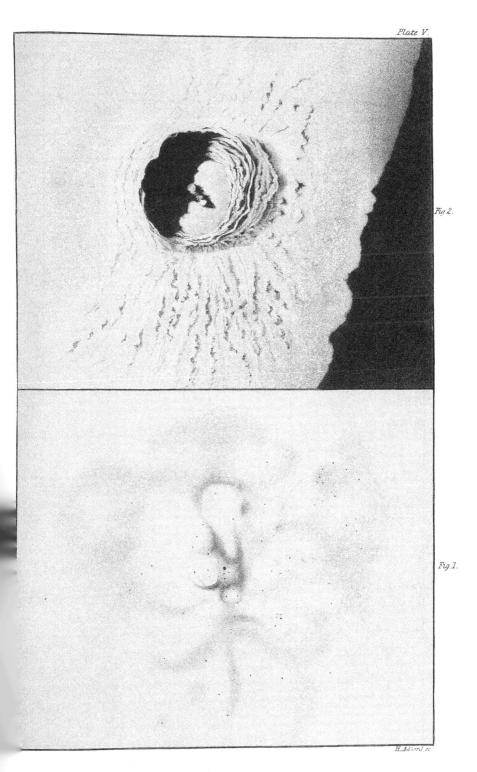

Plate V.

Fig 2.

Fig. 1.

H. Adlard. sc.

London; Longman, Brown, Green & Longmans.

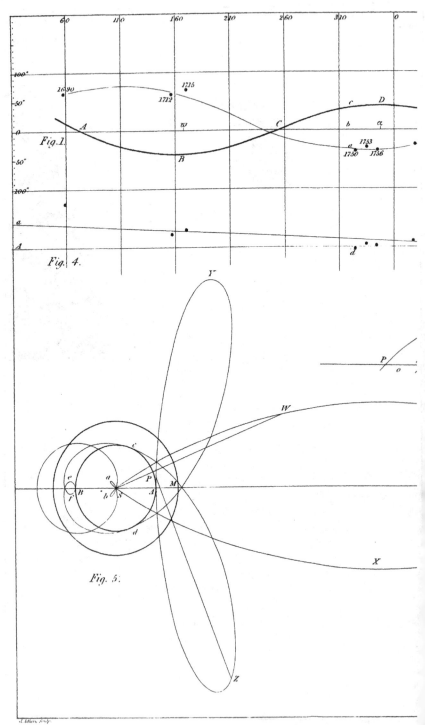

Fig. 1.

Fig. 4.

Fig. 5.

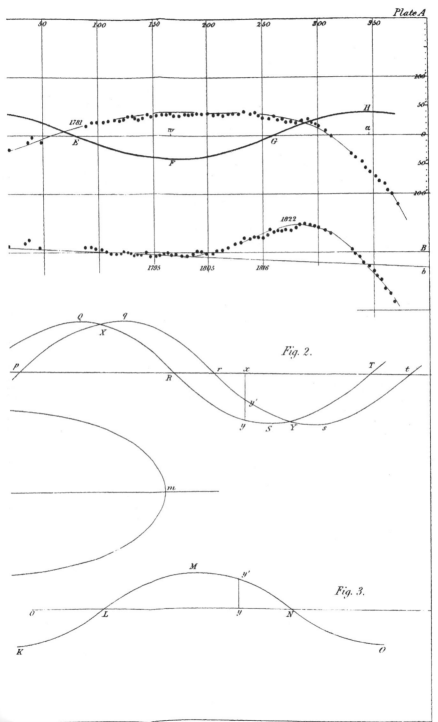

Fig. 2.

Fig. 3.

wn, Green & Longmans.

To fold out opposite page 514.

Printed in the United States
By Bookmasters